Springer Wien New York

Rotterdam
March 2013

Edition Angewandte
Book Series of the
University of Applied Arts Vienna
Edited by Gerald Bast, Rector

edition: 'ʌngewʌndtə

Georg Glaeser

Geometry and its Applications in Arts, Nature and Technology

Springer Wien New York

Georg Glaeser, Institute for Art and Technology,
University of Applied Arts Vienna, Austria

Translation from German language edition (with 60 additional pages):
Geometrie und ihre Anwendungen in Kunst, Natur und Technik
by Georg Glaeser

Printed in Austria

Springer Wien New York is part of
Springer Science+Business Media
springer.at

Graphic Design: Georg Glaeser
Translation: Peter Calvache, George Campbell
Copy editing: Georg Glaeser
Cover Design: Peter Calvache
Druck / Printed by: Holzhausen Druck GmbH, 1140 Wien, Austria

Printed on acid-free and chlorine-free bleached paper
SPIN: 86165199

With approx. 900 figures

Library of Congress Control Number: 2012950025

ISSN 1866-248X
ISBN 978-3-7091-1450-6 Springer Wien New York

Preface

Self-sufficient geometry?

Geometry is usually considered as a branch of mathematics. Indeed, the kind of mental procedure that is required for geometry is closely connected to mathematical logic, and due to its enormous breadth, it is often thought of as a separate science. In any case, geometry and mathematics complement each other quite nicely. Nevertheless, no special mathematical knowledge is required in order to read this book. Whatever mathematical equipment we need will arise by itself.
Geometry can be conducted in several ways, but the kind of geometry that appears in this book *seeks to utilize the reader's three-dimensional imagination as often as possible.* Such imagination is present to different extents in different people, but research has shown that a preoccupation with geometry can improve one's spatial imagination considerably. You will sometimes find fine-printed notes about analytical geometry or calculus, but these instances will only serve to reinforce ideas that should already have been established by your three-dimensional imagination.

A sixth sense

Since it not only utilizes but improves this "sixth sense" of spatial imagination, younger readers often find it easier to progress in geometry compared to other branches of mathematics. The latter requires very solid foundations: Errors may lurk in every line of calculation and are more often overlooked by inexperienced eyes. The kind of geometry with which we are concerned, however, always yields results which can, to some extent, be verified by our imagination. It is for this reason that geometric ("synthetic") proofs are usually preferred to the strictly mathematical ("analytic") proofs. It is by no means rare that these proofs contain such deep levels of understanding that new or additional results inevitably come to the attentive mind.

Descriptive geometry and more

My concept is based on the ideas of my academic teacher Walter Wunderlich and my predecessor Erich Frisch, who both have, in traditional ways, constructed a simple yet complete educational edifice concerning "descriptive geometry". But since time never stands still, and the triumph of computing led to a revolution of geometry in particular, their concepts were strongly modified and expanded upon. I wish to maintain a strong connection to the related fields of *computer graphics* and *computer geometry*, which both owe their roots to classical geometry. Furthermore, the geometry that is described in this book shall not limit itself to mere depiction, especially since the methodology for the creation of images has so profoundly changed in the digital age.

Acknowledgments – the network

I would like to thank my staff Boris Odehnal, Peter Calvache, and Christian Perrelli (for several free-hand drawings) for the many ideas and fruitful discussions that have contributed to this book, as well as for their help with the translation and corrections of the manuscripts themselves. The final proofreading of the English version was done by George Campbell and Eugenie Maria Theuer.

My former co-workers Franz Gruber, Hans-Peter Schröcker, Gerhard Karlhuber, and Thomas Backmeister have contributed images to this book. Further help came from Wilhelm Fuhs, Herbert Löffler, and Michael Schrott. Several illustrations were created by Harald Andreas Korvas, Zorica Nicolic, Otmar Öhlinger, Markus Roskar, Paulo Tosold, and Stefan Wirnsperger.

I was assisted in this endeavor by several colleagues at the Technical University of Vienna (Andreas Asperl, Fritz Manhart, and Walter Jank). Walter Hofmann made a valuable contribution to the topic of sundials. Andreas Rüdinger, who represented the German publisher on this project, has been very constructive throughout this book's development.

I am privileged to teach at the University of Applied Arts in Vienna, where I can count renowned artists like Zaha Hadid, Hans Hollein, Bernhard Leitner, Greg Lynn, Paolo Piva, Wolf D. Prix, Hani Rashid, and Boris Sipek as my (partly former) colleagues. Several examples in this book are inspired by their creative work. I am also grateful to have enjoyed further assistance by my colleagues Klaus Bollinger, Ernst Maczek-Mateovits, Roland Burgard, and Marcus Bruckmann. When questions of physics were on the table, I could always count on the competence of Georg Fuchs and my brother Othmar Glaeser.

Furthermore, I would like to thank the students at the University of Applied Arts Vienna, who, in a generous interplay of give and take, were always keen to contribute ideas, several of which eventually came to fruition. Some of their work is presented in this book.

The "acrobatic interludes" on some photos depict my athletically-minded nephews (Figures 1.43, 10.1 and 10.1).

Finally, my at that time 12-year-old daughter Sophie has contributed several drawings (e.g., Fig. 3.54), several models (Fig. A.4), and even a tiling pattern (Fig. 12.14)! It is to her and to my wife Romana's patience in putting up with my unceasing enthusiasm for photography that I am extremely grateful. Many ideas for photos are wholly to their credit, and it is due to their creativity that this has become a sort-of "picture book of geometry". Good photos rarely emerge "by themselves", or require the unwieldy power of expensive photographic equipment. What is needed is a lot of patience, a good eye, and – as I will show in the appendix – an understanding of geometry.

Vienna, in the November of 2012

Table of Contents

Introduction

Where geometry matters

Naturally, there exist many cross-connections between geometry and technical disciplines such as *mechanical engineering*, *architecture*, and *civil engineering*. The fields of *fine arts* and *design* are likewise influenced by or dependent on geometry, to say nothing of *physics*, *astronomy*, *geography*, *chemistry*, *biology*, and *molecular biology*. All of these fields of creativity utilize geometry to derive new insights or to visualize existing ones. Even the harmonic relationships of *music* can be depicted in purely geometrical terms.

- When should geometric questions be pursued at a lower dimension? The rotation about a straight line can be imagined much more effectively if the line appears as a point. If the underlying procedure is properly understood, many problems can be simplified.
- When, on the other hand, should we look at a problem from a "higher level"? The planar conic sections – when spatially considered – are barely different. Beings who inhabit higher dimensions would, thus, consider many of our differentiations to be superfluous.

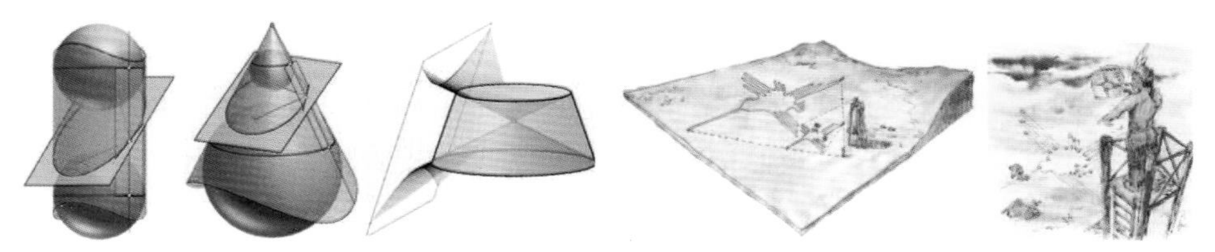

Conic sections

Creation of a geoglyphs

- How could the ancient Peruvians have created their enormous geoglyphs without the need to invent the equivalent of a hot-air-balloon? A combination of central projection and central dilation makes it possible.

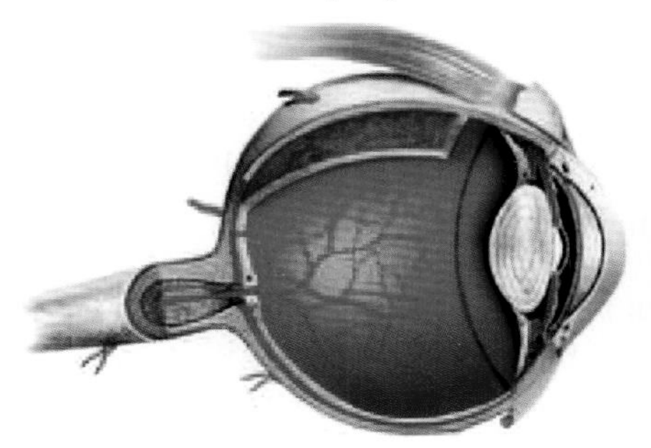

Light refraction on the eye

How many photos are required?

- How does vision actually work, and how is it reproduced by a camera? What conditions need to be met so that the image in a photograph seems "natural" and not distorted or flattened? Five hundred years ago, Leonardo da Vinci and Leone Battista *Alberti*, who worked with exact spatial projections to planes, already asked these questions.

- When can a distorted photo be "rectified"? How many photos are required in order to reconstruct the true shape of an object? Such questions are not only interesting to architects, but also to the designers of 3D scanners.

- How does a map look like where the shortest distances between pairs of points on the globe appear as straight lines or as circular arcs? From what maps can the course angle be directly deduced? Such problems arise from the fact that spheres cannot be unfolded into planes without certain distortions. All of cartography depends on such geometrical insights.

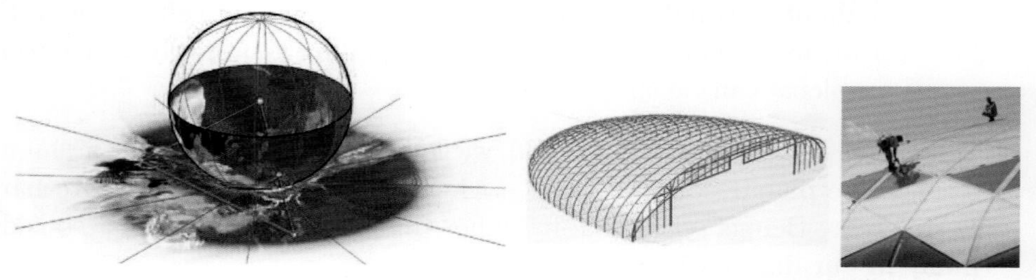

Maps and shortest distances   Good triangulation

- Modern architects are often motivated to design curved buildings. Can geometry be used to minimize the potentially horrendous construction costs?

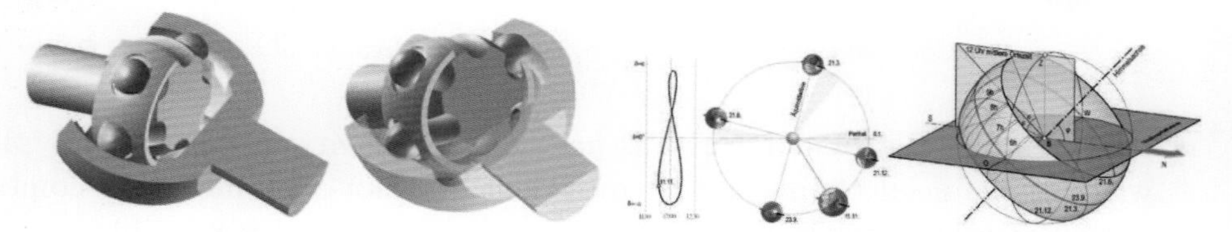

Constant velocity joint   Motion of the earth about the sun

- How can the rotation about an axis be transferred to another axis, and how can this transition be accomplished in a smooth manner? This question arises frequently in mechanical engineering.

- Why does it not always take the equal amount of time for the sun to reach its point of culmination? The question is partially physical in nature, but involves geometry to a much greater degree.

- "Through each point there exists exactly one parallel to a predefined line." What happens if one removes this unprovable geometrical axiom? It results in the outlandish field of "non-Euclidean geometry", whose insights are even relevant to modern physics!

Hyperbolic geometry

Snails and horns

- What can be deduced about their growth from the general shape of snail shells? How can the double helix be conceived? This particular structure of DNA has become the symbol of modern biotechnology.

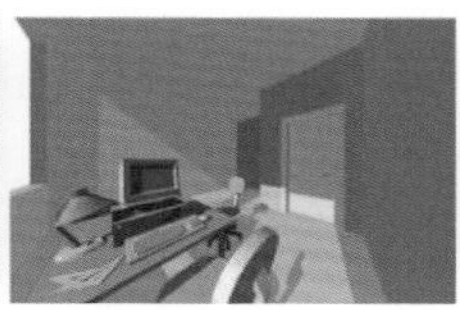

Extreme perspectives

The trick with tilt-shift lenses

- Using simple geometrical tricks, how can you create pictures which would otherwise require an expensive tilt-shift lens? Why is the correct selection of focal distance not only a matter of "visibility"? When are ultra-perspectives not only allowed but necessary? Why does sharpness depend on the aperture?

These and many other questions will be broached and answered in this book. Contemporary students have the advantage of computer programs which allow them to visualize graphics that they would otherwise have to draw painstakingly by hand. Therefore, the balance of geometrical education should always tilt in favor of geometrical comprehension and the analysis of its many processes. Such a change in curriculum allows teachers to address topics that, while relevant to the later working practice of students, would not have otherwise been addressed due to time restrictions.

Level of detail

I have made a great effort to describe laws and rules *as simply as possible*, before delving into additional details, exceptions, and sophistications.
For example, I will first state that (p. 56): "If two lines are parallel in two different parallel projections, then they are, in general, also parallel in three-dimensional space." The exceptions are addressed subsequently – in this important case, with much greater detail.
Further examples: If a curve is "not bent", it means that it has a mathematical curvature of zero. Or: "Each planar section of a sphere is a circle." Of course: If the plane is merely touching the sphere, then the circle shrinks to a so-called "zero-circle", which can also be considered a point. But what if the plane

does not intersect the sphere? Mathematicians can prove that this yields an imaginary circle. But how can such an abstract idea be expressed without overstraining the layman? If such exceptions are left unsaid, the interested reader will still understand the idea, and the fundamental assertion remains *concise and to the point*!

The website `www.uni-ak.ac.at/geometrie`

This books is accompanied by a website. There you will find updates, weblinks to the many specialized topics, and further examples of geometry in action. Finally, a plethora of interactive programs is provided that let you visualize many complex sequences of motion and physical simulations.
The advantage of such additional material is obvious: First of all, the website is constantly enriched, and the prior material is always kept updated without having to change the book to which it refers. I hope that you will find much useful material on the website and provide me with feedback from which I can further improve this service!

The geometrical network ...

1 An idealized world made of simple elements

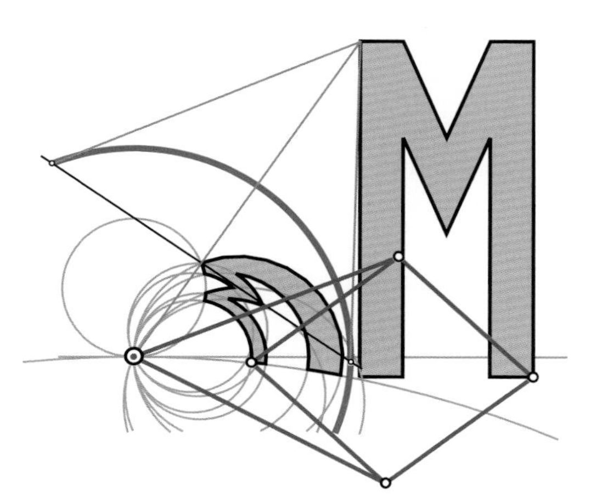

In this chapter, we will learn about the most important building blocks of the geometric world: Points, straight lines, and circles in the two-dimensional drawing plane as well as planes, circles, spheres, cylinders, and cones of revolution in three-dimensional space.

Points in geometry are "infinitely small" and have, therefore, no expansion. In the same sense, straight lines are always "infinitely thin" – as well as "infinitely long". Analogies can be made for planes, cylinders and so on: Mathematically idealized versions simplify how we deal with these objects in our imagination.

In geometry, we are not afraid to utter the word *infinity*. Likewise, the abstract concept of *dimension* is very frequently invoked: A set of planes through space forms a three-dimensional *plane space*. The theoretical foundation for this statement is given by *projective geometry*, with which we shall deal in the last section.

First, we will define terms in the drawing plane and further deepen our understanding by applying remarkable examples of the geometry of triangles. We should then have enough knowledge to switch smoothly to spatial geometry. During this process, we shall gain numerous insights by generalizing our results of planar geometry.

Survey

1.1 Points, straight lines, and circles in the drawing plane

Dimensionless spheres

Points in geometry (and anywhere in mathematics) are "infinitely small" and therefore have no expansion. As such, they can be seen as *zero-dimensional* objects. When we label points, we will make use of European convention and utilize Latin capital letters such as A, B, P, Q, etc.

Rays without beginning or end

By stringing together an infinite number of points – in a distinct direction – we get an infinitely thin and infinitely long *one-dimensional* straight line. It can be said that a straight line is "generated" by translating a point. According to European convention, straight lines are labelled using Latin lower-case letters such as a, b, g, h, etc.
Straight lines are not bounded segments: Every *line segment* $\overline{PQ}$ lies on an infinitely long *carrier straight line* $g = PQ$.
In order to get accustomed to geometrical language, we will now repeat some terms that we have already used so far.

Circles

Circles are so-called *lines of constant distance.* All points on the circle k have the same constant distance from the midpoint M, namely the radius r. In this book, we will usually notate circles using the letter k. When we speak of more than one circle, they are notated as $k_1, k_2, \ldots$ or $k_A, k_B, k_M, \ldots$ (given the midpoints A, B or M). Circles possess the attribute of being uniformly curved and, like all curves, they are also infinitely thin.

The most natural curve

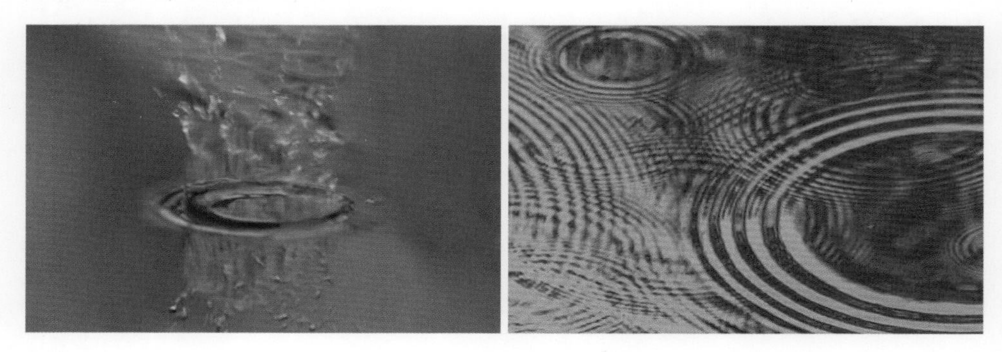

Fig. 1.1 Water waves

Circles are very popular elements of construction because it is comparatively easy to work accurately with them. Straight lines are rarely found in nature – circles in the plane or spheres in space are frequent by comparison. One only needs to throw stone pebbles into a lake to be amazed by the concentric

rings that are then formed by the wave peaks. All types of waves, including light waves and sonic waves, expand in a circular manner in the plane and spherically in three-dimensional space.
It would be impossible to draw long straight lines on a desert island. On the other hand, an adequate crotch will let us draw circles rather easily. The supposed last words of *Archimedes* (about 287–212 B.C.) come to mind: "Do not disturb my circles!" (He was referring to his constructions in the sand, and directed his statement at the Roman soldier who, after the conquest of Syracuse, stabbed him to death despite an order to spare his life.)

Triangles

A *triangle* is a union of three points A, B, C. It can equally be described by the three line segments that are formed by the three points: $\overline{AB} = c$, $\overline{BC} = a$, and $\overline{CA} = b$. All other parameters, such as the angles $\alpha = \angle BAC$, $\beta = \angle ABC$, and $\gamma = \angle ACB$, can thus be calculated or constructed. This book prefers construction to calculation because, instead of yielding mere numbers, it produces a "residual" that may be useful for further contemplation.

Congruent and similar triangles

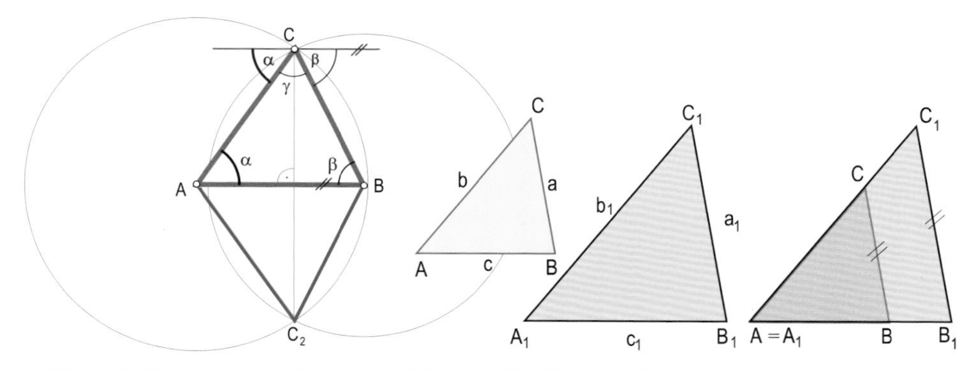

Fig. 1.2 Congruent solutions

Fig. 1.3 Similar triangles, theorem of intersecting lines

Let us construct a triangle ABC, given the lengths of its sides: We can orient the side $\overline{AB}$ in any way we like (such as horizontally) and draw two circles k_A and k_B with the radii $\overline{AC}$ and $\overline{BC}$. If the given lengths of the triangle's sides are mathematically possible (the sum of the length of two sides must exceed the third), we will get two solutions for the third vertex (C_1 and C_2, Fig. 1.2). These two solutions do not differ by size (hence, they are called *congruent*), but by "orientation": They are "reversely congruent". By drawing the parallel to AB through C, we can also notice that:

The sum of angles in a triangle equals 180°.

Fig. 1.4 shows both sides of a peacock's feather (see also Fig. 6.51). The patterns emerge not due to color but because of complex "twisting" of single hairs, and while they are not completely similar to each other, their outlines certainly are (the figures in the image are *reversely congruent*). When considered *spatially*, it, nevertheless, remains the same object which, interestingly enough, is *directly congruent.*

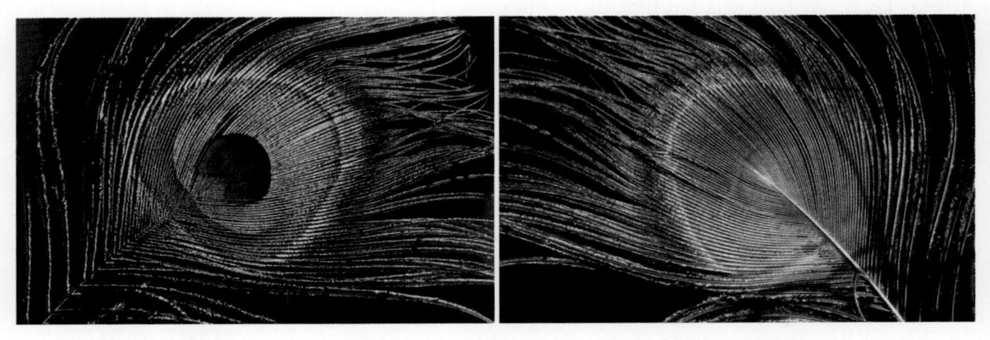

Fig. 1.4 Peacock's feather (Front and back side) – reversely congruent?

The term *similarity* generalizes the notion of congruence. Two triangles are *similar* if they possess identical angles. The side lengths may change, but a relationship can immediately be established using a *factor of similarity.* Here, the *theorems of intersecting lines* are applicable. These simple statements are surprisingly difficult to prove: For this, we need so-called *axioms*, which are unprovable, but commonsense statements that must be accepted as true.
Let us consider a triangle ABC, whose side lengths a, b, c are multiplied by a constant factor k (the factor of similarity), resulting in a similar triangle $A_1B_1C_1$ (Fig. 1.3): $a_1 = k\,a,\ b_1 = k\,b,\ c_1 = k\,c$. Triangles with a constant aspect ratio are considered similar.
Further we have

$$\overline{AB} : \overline{A_1B_1} = \overline{AC} : \overline{A_1C_1} = \overline{BC} : \overline{B_1C_1}$$

If we translate one triangle into the other in such a way that the legs of an angle of the first triangle overlap the respective angle of the second triangle, then the remaining two sides are parallel. Thus, we have the

Theorems of intersecting lines: If an angle is intersected by two parallel lines, then the ratio of the distances $\overline{AB}$ and $\overline{AB_1}$ on the first leg and the respective distances $\overline{AC}$ and $\overline{AC_1}$ on the second leg are equal. Furthermore, we have the same ratio on the parallel sections $\overline{BC}$ and $\overline{B_1C_1}$.

One of the oldest theorems of geometry

Contemporary geometry, with its strict requirements for proof, dates back to the ancient Greeks, whose science dominated the southern Mediterranean (from Sicily through Asia Minor right down to northern Egypt). Alexandria (in the Nile delta) was seen as their scientific capital. Even before Egypt became Hellenized, it was visited by *Thales* of Milet (625–546 B.C., Asia Minor), who gained much geometric knowledge from there.

The realization that a logical proof should be required for every geometric theorem was a great breakthrough for geometry and mathematics. Among other things, *Thales* proved the famous theorem that bears his name, but which was known long before:

Thales' theorem: All angles in a semicircle are right angles. Put in a different way: From each point on the circle, the diameter appears at a right angle.

Proof:
Let us consider an arbitrary rectangle $ABCD$, as in Fig. 1.5. Rectangles contain only right angles by definition ($AB \perp BC$). The diagonal $\overline{AC}$ has length d. It is obvious that this rectangle possesses a "circumcircle" with radius $d/2$, whose midpoint lies at the intersection of the diagonals. Thus, if seen from the point B (as well as D), $\overline{AC}$ appears at a right angle. If we now look at all rectangles with an equally long diagonal d, we get the same result every time. Of all these rectangles, we can always overlay both the diagonal and the circumcircle. B and D thus pass through the circumcircle. ◇

This theorem forms the basis for many further proofs.

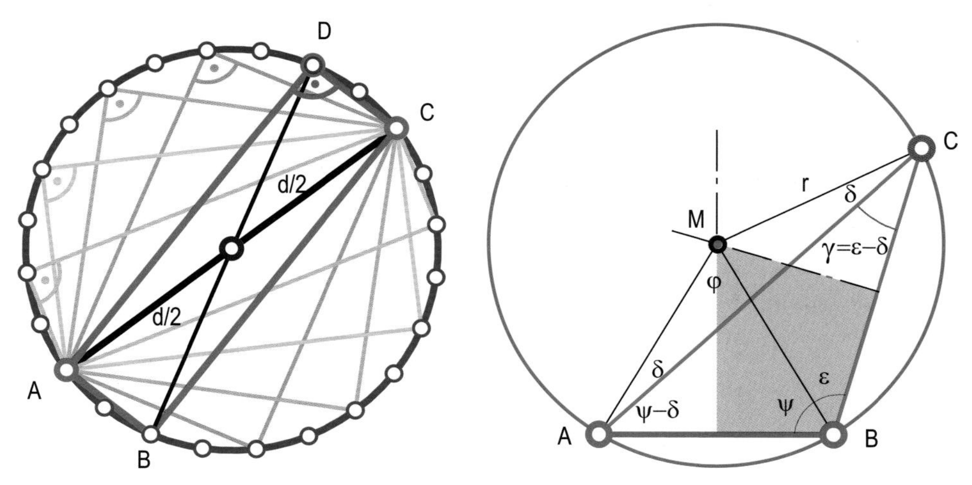

Fig. 1.5 *Thales'* theorem

Fig. 1.6 Theorem of the angle of circumference

Thales' theorem is only the special case of the following situation:

Theorem of the angle of circumference: From the vantage point of a circle, the chord of a circle is always seen at the same angle or from its supplementary angle.

Proof:
Let ABM be an isosceles triangle with $\angle AMB = \varphi$. Then $\angle BAM = \angle ABM = \psi = 1/2(180° - \varphi)$. Let us now draw a circle around M through A and B and pick an arbitrary point C on it. We now have two further isosceles triangles AMC and BMC. If δ and ε describe the same angles of these triangles, then $\gamma = \angle ACB = \varepsilon - \delta$.

The sum of all angles in a triangle ABC is $[\psi - \delta] + [\varepsilon + \psi] + [\varepsilon - \delta] = 180°$. Thus, we get $2(\varepsilon - \delta) = \varphi$ or $\gamma = \varphi/2$ (=constant). If C resides on the other side of AB, then the following is true for the angle sum of ABC: $[\delta - \psi] + [\delta + \varepsilon] + [\varepsilon - \psi] = 180°$ and thus $\gamma = \delta + \varepsilon = 180° - \varphi/2$. ◇

As a byproduct of this proof, we can derive the

Modification of the theorem of the angle of circumference: From the circumcenter of a triangle, each triangle side is visible under a central angle which is twice the inscribed angle in the opposite vertex.

It would not have been necessary to prove *Thales*' theorem: It is only a special case of the theorem of the angle of circumference. On the other hand, it is always useful to practice different techniques of proof. In [11] you can find a proof for this theorem that is "kinematic" in nature – much like the proof of Thales' theorem: If a parameter is changed, the point traces a curve, which in this case is a circle. In particular, we will benefit from such considerations in the chapter about kinematics.

The circle steps out of line

Circles play a fundamental role in the classical geometry of *Euclid*. They intersect a straight line at most two times (if the straight line merely touches the circle, the points are actually infinitely close to each other). Likewise, two distinct circles never share more than two points. This is as self-evident as it is unusual – at a later point, we will find out why this is so. In any case: Most classical constructions can be subdivided in such a way that they can be reassembled merely by intersecting straight lines and circles with other straight lines and circles. Such problems are then either totally *linear* (if no circles are required) or *quadratic* or even more. The construction of a triangle whose side lengths are given is quadratic because two (reversely congruent) solutions are obtained by intersecting circles.

Directly congruent figures, twisted into each other

We will need the following elementary theorem later:

Two directly congruent figures (same orientation) in the plane can always be transformed, one into the other, by an unambiguously defined translation or rotation.

Proof:
Let us pick two directly congruent triangles ABC and $A^*B^*C^*$ (Fig. 1.7 on the left side) on top of both congruent figures and try to overlap them via rotation. In order to determine the center of rotation, we will need to employ the *perpendicular bisector*. This is the straight line that runs through the midpoint of $\overline{PQ}$, at a right angle to PQ. It is the locus of all points which are equidistant to P and Q. If ABC and $A^*B^*C^*$ describe two congruent triangles, then the center Z of rotation has to be equidistant to A and A^*, i.e., on the perpendicular bisector of $\overline{AA^*}$. The same applies to B and B^*, C and C^*.

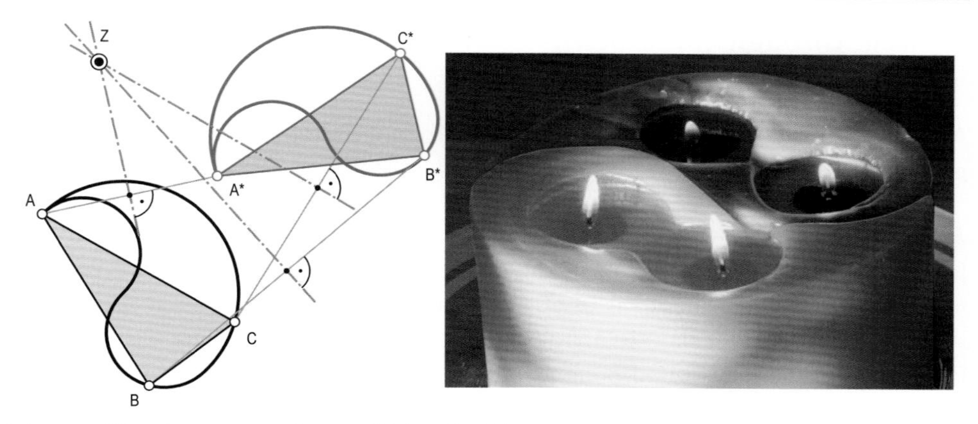

Fig. 1.7 Directly congruent figures. Right side: With two-sided and two-colored candles, it is not immediately evident that both parts were cast using the same mold.

The question is whether all perpendicular bisectors actually run through the center of rotation Z. Let Z be the intersection between the first two bisectors. Therefore, we can imagine Z as being connected to the straight line AB. The auxiliary triangle ABZ can be rotated into the triangle A^*B^*Z about the point Z. However, C is also fixed to AB, that is, the distance $\overline{CZ}$ does not change during the rotation. Thus, Z must also lie on the perpendicular bisector of C and C^*. This remains true for all respective point pairs on the congruent figures. ◇

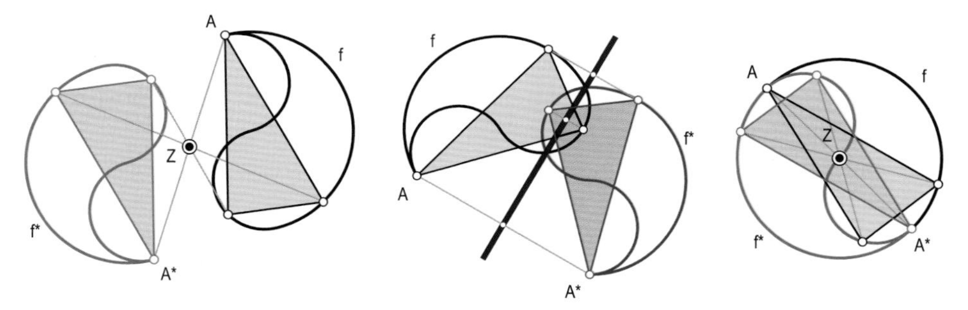

Fig. 1.8 Reflection in a point or in a straight line, rotation of 180°

If the figure f is rotated by 180° about the center Z, it becomes a directly congruent figure f^* and can equally be seen as the *reflection* of f. Point reflections in the plane are, therefore, directly oriented. Reflections at a straight line, on the other hand, are reversely oriented. (Fig. 1.8).

1.2 Special points inside the triangle

Every "real" triangle (A, B and C do not lie on a straight line) possesses an *orthocenter*, a *centroid*, and a *circumcenter*. This may be familiar to most people, but it is not at all trivial to prove. The *circumcenter* U is the easiest to see.

A circle through three points – the circumcenter

A circle can be defined by three points. To put it differently: Every triangle possesses a *circumcircle*.

Proof:
Let us first construct the center of the circumcircle as the intersection of two *perpendicular bisectors*: The locus of all points that are equidistant to A and B is a straight line perpendicular to AB at the midpoint M_{AB} of the line segment $\overline{AB}$. Furthermore, the locus of all points that are equidistant to B and C is a straight line that is perpendicular to BC at the midpoint M_{BC} of the line segment $\overline{BC}$. Both perpendicular bisectors intersect at U. Now we have $\overline{UA} = \overline{UB}$, and also $\overline{UB} = \overline{UC}$, from which we can conclude $\overline{UA} = \overline{UC}$. This means that the third perpendicular bisector must also pass through U. Consequently, U is the center of the circle through A, B, and C. ◇

The centroid: A point of physical significance

Centroids (or barycenters) play a significant role in physics, as they simplify many considerations. The existence of an unambiguous centroid S in a triangle is not trivial to prove. A vector-based proof for its existence can be found in [11]. Its construction, on the other hand, is very simple: One merely needs to connect the vertices of the triangle to the midpoints of the opposite sides and intersect these so-called *medians*.
The term *centroid* has its roots in physics. There, a distinction is made between vertex centroids, edge centroids, and volume centroids. For arbitrary figures or polygons, these centroids are not identical [11]. For triangles, edge centroids and volume centroids coincide, and for polygons with central symmetry, all centroids coincide. This is why it is so useful to work with triangles for technical applications. In three-dimensional space, these extraordinary properties are inherited by tetrahedra or polyhedra with central symmetry.

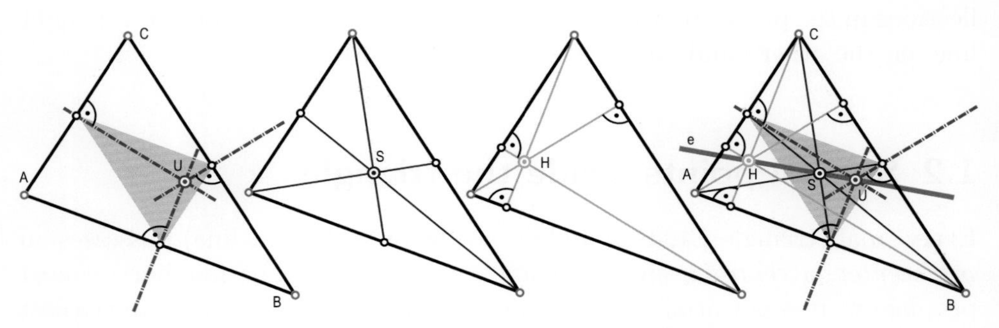

Fig. 1.9 Circumcenter, centroid, orthocenter, Euler line

The orthocenter: A point of geometric significance

The *altitude* of a triangle is the straight line that is orthogonal to a triangle side and passes through the opposite vertex. The following observation is important for many later considerations and – once again – not at all trivial:

All three altitudes of a triangle pass through a single point, the orthocenter.

Proof:
Let us consider Fig. 1.9 right: The midpoints of the triangle sides form the *medial triangle* that is similar to the original triangle, because all of its sides are parallel to the respective sides of the original triangle. The perpendicular bisectors of the original triangle intersect at a single point, as we have already proven, but they can also be interpreted as the *altitudes* of the newly created medial triangle, because they are (by definition of an "altitude") perpendicular to their triangle side and pass through the opposite vertex. ◇

This proof is not only elegant, but gives us a further advantage: It immediately implies the existence of the *Euler line*. (This is precisely what I mean when I say that synthetic proofs are more useful than the purely analytical ones.)

All three points are guaranteed to lie on a straight lines ...

Fig. 1.9 on the right shows that the side lengths of the original triangle ABC relate to the resulting pink medial triangle $M_{AB}M_{BC}M_{CA}$ at a ratio of $2:1$. What is more: As the respective vertices lie on rays through a fixed center – the centroid S –, they can be transferred into each other by means of *central similarity*. Given this central similarity, the orthocenter H of the big triangle (the *anti-complementary triangle*) transforms into the orthocenter of the medial triangle, which, in turn, is also the circumcenter U of the original triangle. The straight line through the center S that connects H and U is the Euler line. Furthermore, we have $\overline{HS} : \overline{SU} = 2 : 1$.

The incenter – or: a circle for three tangents

The question about a circle that touches all three triangle sides leads us to the incircle. Its center is determined by the intersection of at least two *angle bisectors*. Every circle that touches two straight lines has to be centered on such a bisector. If, e.g., we have intersected the angle bisectors through A and B, then we have found the center of a circle that touches both AC and AB, but also BC and BA, and thus, all three sides. Its midpoint must, thus, also belong to a third angle bisector – this means that all bisectors pass through a single point.
Thus, three straight lines (tangents) can define a circle, unless they do pass through a single point. However, this is not unambiguous: We have ignored that there always exists a *pair of angle bisectors* that forms a right angle. How many points do they define? Once again, it is possible to use just two

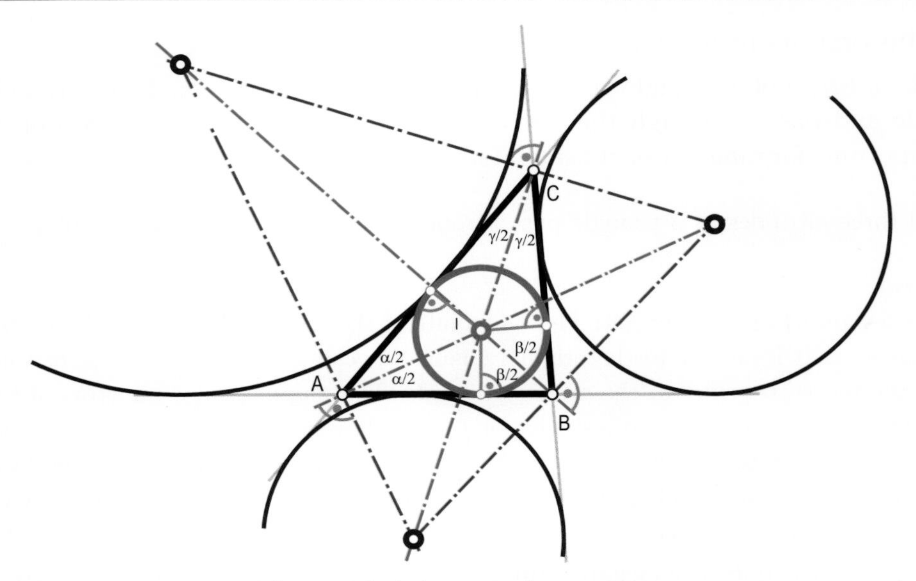

Fig. 1.10 Incircle center, three excenters

pairs for construction (the third pair provides no new solution). All in all, there exist four solutions to this construction, namely, the incircle and the three *excircles*. The question of finding a circle tangent to the side lines of a triangle is, thus, ambiguous. At least, there exists a "finite number" of solutions! By the way: In general, neither the incenter nor the centers of the excircles (the *excenters*) lie on the Euler line.

Two ways of defining a circle are still missing ...

Fig. 1.11 Secant and tangent of a circle ...

The following theorem is useful for some circle constructions:

For all secants of a circle through a fixed point P, the product of the distance of P to the intersection points remains constant. This is called the *power* of the point in relation to the circle.

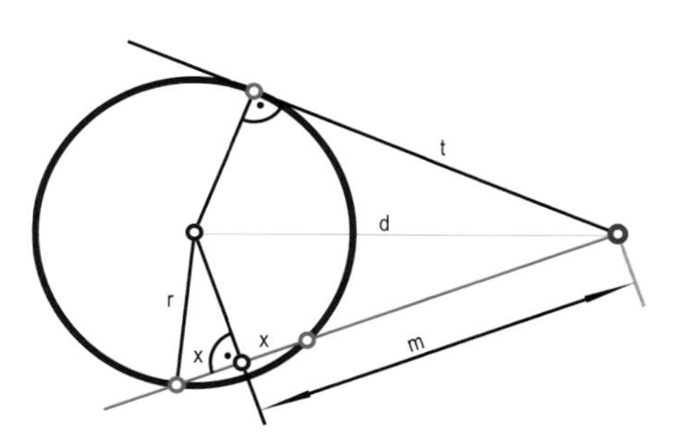

Fig. 1.12 The *power* of a point

Proof:
We use the notations of Fig. 1.12. According to *Pythagoras* the following is true: $m^2 + (r^2 - x^2) = d^2$, or $m^2 - x^2 = d^2 - r^2 =$ constant. However, $m^2 - x^2 = (m+x)\cdot(m-x)$ is exactly the product of the distances to the intersections (the power). If P lies outside the circle, there exist touching secants whose intersection points move together. In such cases, the product of the distances becomes the square of the tangent distance. ◇

• Two nontrivial circle constructions

Find a circle k that passes through two points P_1 and P_2 and furthermore touches a straight line t, or alternatively passes through a point P and touches two given straight lines t_1 and t_2.

Solution:
Let us start with the first task (Fig. 1.13): We intersect the line P_1P_2 of the given points with the given tangent t (Fig. 1.13), determine the power $\sqrt{\overline{ZP_1} \cdot \overline{ZP_2}}$ of the intersection Z that can easily be constructed using Euclid's theorem. We have now found the contact point T on t. The desired center of the circle lies on t's normal at the contact point T and the perpendicular bisector of P_1 and P_2. We shall not overlook the second solution: The tangent section can be determined on both sides.

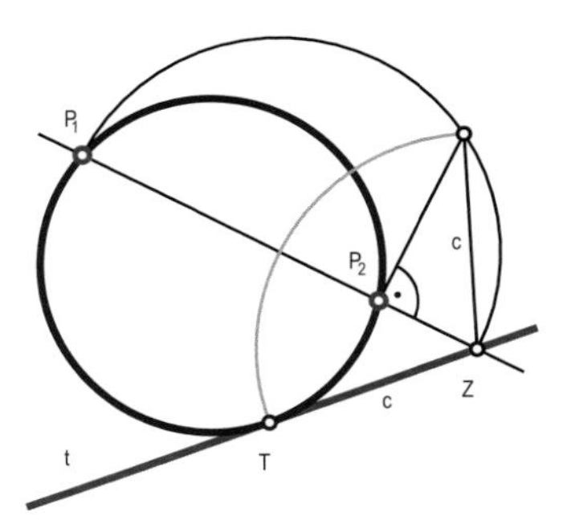

Fig. 1.13 2 points + 1 tangent

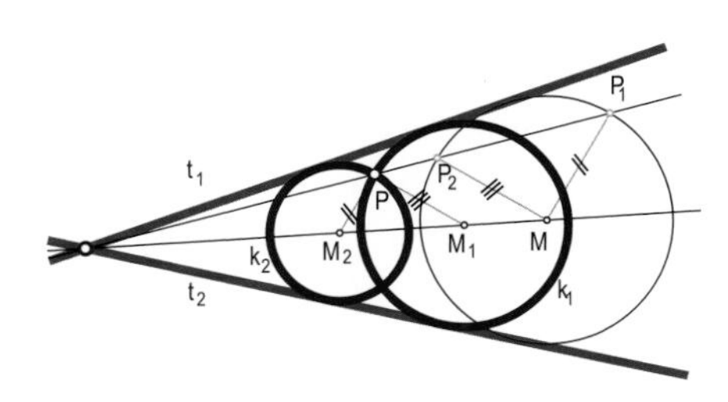

Fig. 1.14 1 point + 2 tangents

Now for the second problem (Fig. 1.14): The locus of the centers of all circles that touch both tangents t_1 and t_2 is the pair of angle bisectors (Fig. 1.14). The solution circles lie in the same sector as P that is formed by t_1 and t_2. Consequently, the centers lie on one particular "half-line". Let us draw an arbitrary circle k touching t_1 and t_2, with the center M located on this half-line. Relative to Z, it is similar to the touching circles. The point P is, thus, related, due to similarity, to two possible points P_1 or P_2 at the intersection of k with the ray ZP. The rays MP_1 and MP_2 are transformed into the parallel rays M_2P and M_1P. M_1 and M_2 are the desired centers of the two solutions k_1 and k_2. ♠

If, by now, you have gained an appetite for more oddities of triangles and circles, you are invited to ponder over three additional but no less intriguing theorems by *Feuerbach*, *Kiepert*, and *Morley*. You may also jump forward to the next section, where we will leave the drawing plane for a short excursion into Euclidean space in order to explain how these topics apply in three dimensions.

A circle through nine (!) points in the triangle

If we look at the triangle ABC, we may notice that the centroid S subdivides the medians at a ratio of $2:1$. According to our knowledge about the points on the *Euler line*, the ratio $\overline{SU} : \overline{SH} = 1 : 2$ holds for the orthocenter H, the centroid S, and the circumcenter U of a triangle ABC. If we reflect the triangle at S and shrink it at a ratio of $1:2$, we get a directly congruent triangle $A_2B_2C_2$ whose vertices are the midpoints of the triangle sides (Fig. 1.15).
If we also transform the circumcircle, we get a circle k that passes through the midpoints of the triangle edges. The circumcenter U corresponds to the midpoint $U_2 = F$ of the smaller circle. With $\overline{FS} = s$ we have $\overline{SU} = 2s$, $\overline{FU} = 3s$, and $\overline{HU} = 6s$. Thus, F is the midpoint of the line segment $\overline{HU}$. Given these circumstances, *Feuerbach* had the following brilliant idea: Shrinking the circumcircle from the orthocenter H at ratio $1:2$ (without reflection) results in the same circle k, because U travels towards F.
After this type of shrinking, the original triangle ABC corresponds to a triangle $A_3B_3C_3$ of half the size. The circle k contains these points as well. The triangles $A_3B_3C_3$ and $A_2B_2C_2$ are congruent due to the same scaling factors and have mutually parallel sides.
The six points are, thus, not *arbitrarily* distributed on the circle, but are mutually opposite to each other. Now we still want to show that the foot of each altitude A_1, B_1, and C_1 of the triangle ABC lies precisely on k: By definition, $AA_1 \perp BC$ and thus $A_1A_3 \perp A_1A_2$, which means that k is *Thales' circle* over A_3A_2 and thus A_1 also lies on k. The analogue is true for B_1 and C_1. Thus, we have found even nine points on the circle! For the purpose of illustration: Combinatorics tells us that if we pick any 9 points, 3 from this set can be combined into circles in $(9 \cdot 8 \cdot 7)/(3 \cdot 2 \cdot 1) = 84$ different ways.

The beer mat method

From the proof above, the following remarkable addition can be derived (as an “encore”, so to speak): If we scale the nine-point circle k from C by the

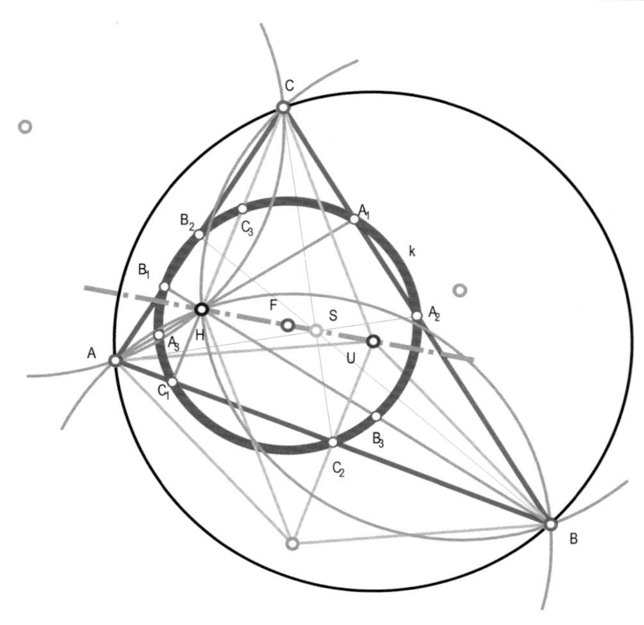

Fig. 1.15 Nine-point circle through the foot of each altitude, the side bisectors, and the vertices of the medial triangle

factor 2, the bisectors A_2 and B_2 pass over to A and B, as well as C_3 into H. Therefore, there exists a circle that is congruent to the circumcircle and contains two vertices of the triangle as well as the orthocenter. This statement is true for all triangle points. So, the next time you find yourself waiting for your meal at a restaurant, with three circular beer mats within reaching distance, mark three points ABC at the edges of a single beer mat (which then becomes the circumcircle) and position the other two mats so that they contain two of these points each. The remaining intersection of the two beer mats is the orthocenter ...

A point of high importance

Let us now imagine, as in Fig. 1.16, that we have drilled three tiny holes into a disk, through which we have pulled three strings of equal length that are tied together at the end G. (The triangle that is formed by the holes must not have an angle greater than 120°.) To the other ends of the strings, we attach three equal weights. The weights will immediately stretch the strings to form angles of 120°: The three equal forces are then in equilibrium.
In sum, the system tries to place the weights at the lowest possible location. With a given string length d and the distances x, y, and z of the point G from the triangle points A, B, and C – the weights hang at depths $d - x$, $d - y$, and $d - z$. The sum of these depths is, therefore, $3d - (x + y + z)$. If the weights have to hang as low as possible, then the expression $x + y + z$ must be a minimum. It is also possible to define G as the point where the *sum of distances to the triangle vertices is minimal.*

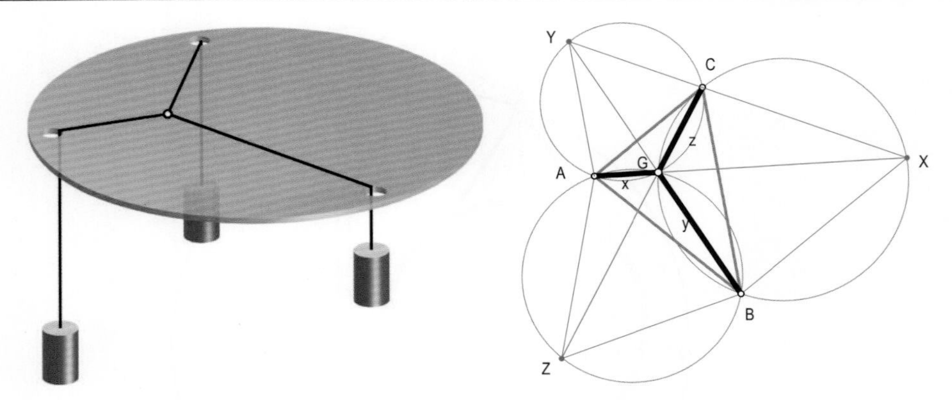

Fig. 1.16 Fermat point

The point G is ascribed to both *Fermat* and *Steiner*. It can be constructed by intersecting the two circles or by intersecting at least two circles of circumference or much more simply, by extending every triangle side outwards into equilateral triangles and connect the new points X, Y, and Z with the opposite triangle points A, B, and C: The circumcircle of the triangle ABZ is the circle of circumference of the segment $\overline{AB}$, from whose points the segment on the side of Z is visible at $60°$, and the segment on the side G at the supplementary angle of $180° - 60° = 120°$.
In this special case, it should be obvious that the three lines actually intersect at a point: If a point G is found as the intersection of two lines, then two triangle sides can be seen under $120°$ from G – and the third triangle side must automatically be $120°$, since the sum of these angles must add up to $360°$.
Let us now discuss a generalization of this method.

How can an arbitrary triangle be converted into an equilateral triangle?

Two triangles ABC and XYZ are *perspective* to each other, if the connecting straight lines AX, BY, and CZ pass through *one point.* A theorem that traces back to *Kiepert* now states that (Fig. 1.17 on the left, without proof):

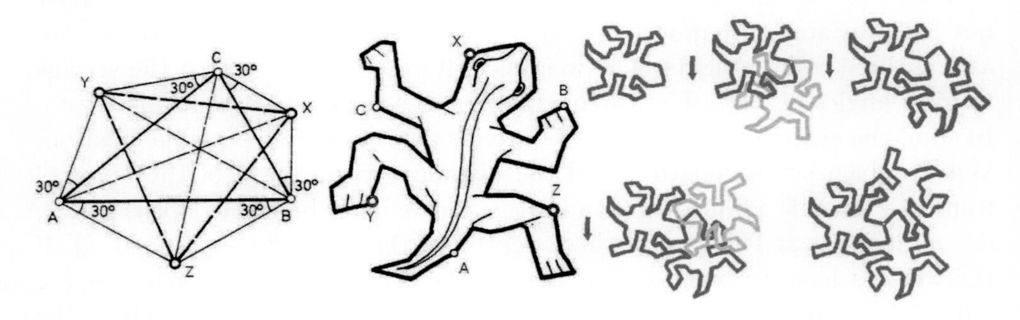

Fig. 1.17 *Kiepert*-triangle and *Escher*-lizard

Kiepert's Theorem: If isosceles triangles are constructed above the sides of an arbitrary triangle ABC and they each share a common base angle φ, then the new vertices XYZ form a triangle that is perspective to the original triangle. For $\varphi = 30°$, XYZ is an isosceles triangle.

- **An inventive tessellation**

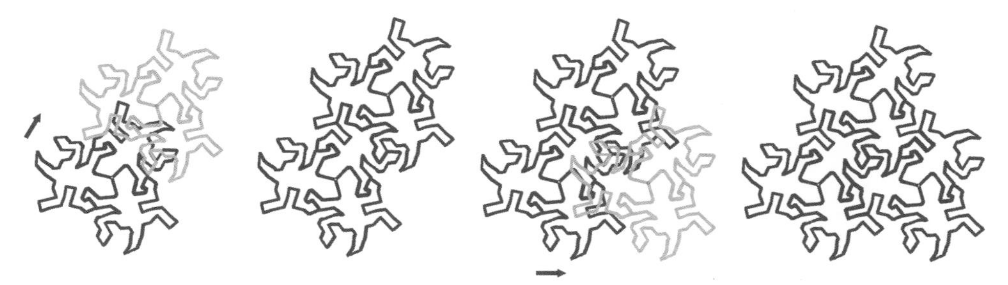

Fig. 1.18 Tessellation using the *Escher*-lizard

A plane can be "tessellated" using equilateral triangles. If one shapes the originating tiles in a particularly ingenious fashion – as M.C. *Escher* certainly did – and used them to fill the figure Fig. 1.17, one gets a remarkably non-trivial *tessellation*, as in Fig. 1.18. More information on tessellation will be provided in the Chapter 12. ♠

Do there exist other methods to convert an arbitrary triangle into an equilateral triangle?

Triangle geometry is an extremely "mature" discipline of science. To make a sensational discovery in this field in our day and age is close to a miracle. In the year 1904, the American mathematician Frank *Morley* discovered the following: A theorem whose aesthetics recalls the elegant proofs in Euclid's Elements – despite the fact that the theorem is not fundamentally geometrical, as we shall soon see.

Morley's theorem: If the interior angles of an arbitrary triangle are trisected, the intersections of the trisecting straight lines that are adjacent to the triangle sides form an equilateral triangle.

Proof:
The proof that follows is, in essence, due to D.J. Newman (1996). We put the cart before the horse, so to speak, and begin with the result – an equilateral triangle XYZ with side length 1. We then measure off, as in the lower figure, angles u, v, and w twice each. We require that $u+v+w = \dfrac{4\pi}{3}$. The resulting auxiliary triangles should not overlap. If we can show that $s = s^*$ and $t = t^*$, then we are almost done, because by "relabeling", we obtain equality for the remaining angles. First, we show that $s + t = s^* + t^*$.

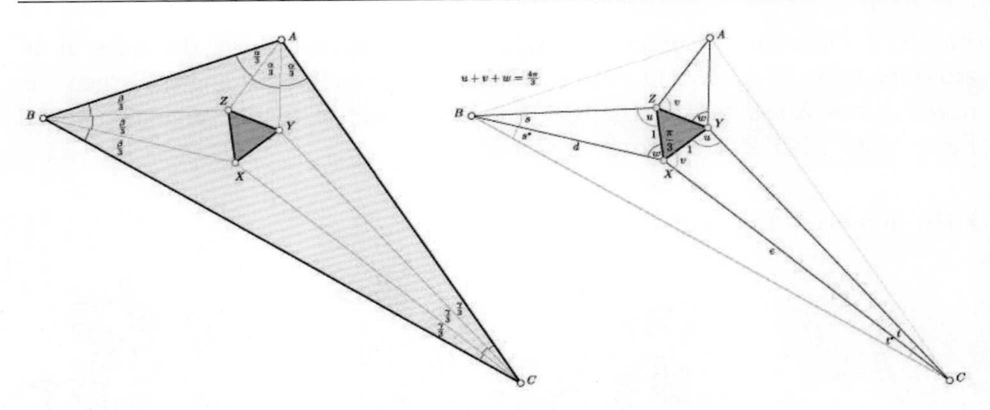

Fig. 1.19 *Morley*'s Theorem

This is a result of $s = \pi - u - w,\ t = \pi - u - v\ \Rightarrow\ s + t = \dfrac{2\pi}{3}$
and $s^* + t^* = \pi - \left(2\pi - \dfrac{\pi}{3} - (v + w)\right) = \dfrac{2\pi}{3} - u.$
Moreover, by applying the law of sines in the triangles XZB and XYC, we obtain

$$\frac{1}{\sin s} = \frac{d}{\sin u},\ \frac{1}{\sin t} = \frac{e}{\sin u} \Rightarrow \frac{\sin s}{\sin t} = \frac{e}{d}$$

and by applying the law of sines in BCX, we get

$$\frac{d}{\sin t^*} = \frac{e}{\sin s^*} \Rightarrow \frac{\sin s^*}{\sin t^*} = \frac{e}{d}.$$

We may now conclude (because the sine function is monotonic) that

$$s + t = s^* + t^*\ \wedge\ \frac{\sin s^*}{\sin t^*} = \frac{\sin s}{\sin t} \Rightarrow s = s^*\ \wedge\ t = t^*.$$

◇

The dilemma of angle trisection

Morley's theorem is not fundamentally geometrical – as we have already mentioned. This means that it can not be verified or "replicated" by compass and ruler alone. Why do such approaches fail?
The task involves the division of an arbitrary angle into three equal parts. This involves, from a mathematical standpoint, the solution of a third-order equation. Despite our best efforts, our compass and ruler are not enough to provide a solution in the general case.
And yet the problem seems so simple! How would you trisect an angle? With a set square and some quick mental arithmetic? *Archimedes* provided a solution that requires some constructional agility but allows us to find a mathematically correct solution:
Let α be an angle (Fig. 1.20) through an isosceles triangle AMB. Let us draw an auxiliary circle k (radius $r = \overline{AM}$) around M. Now we can mark two points P and Q on our ruler at a distance r and move the ruler such that P moves along the elongated triangle leg AM and Q moves on the circle k. If

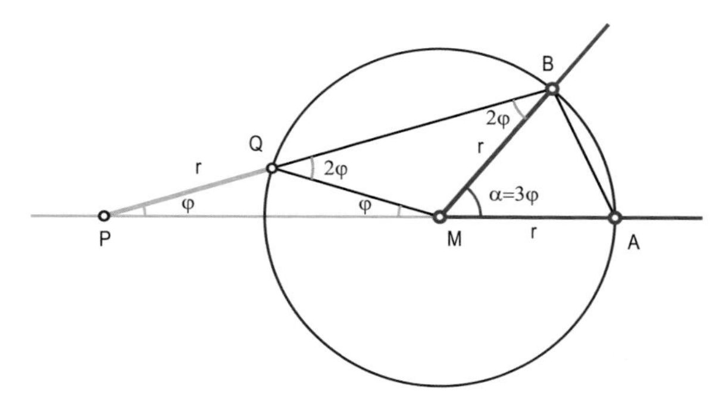

Fig. 1.20 Trisection according to *Archimedes*

the elongation of PQ passes through B, then the section MQ encompasses, along with the triangle leg AM, an angle of $\varphi = \alpha/3$.

Proof:

The proof of *Archimedes* is as ingenious as it is simple: We recognize two isosceles triangles MBQ and PQM (the equilateral legs have length r). From $\varphi = \angle PMQ$ it follows that $\angle MPQ = \varphi$. The angle $\angle MQB$ is twice as large (2φ) because the third angle in the triangle PMQ measures $180° - \varphi - \varphi$. This also means that $\angle MBQ = 2\varphi$, and furthermore that $\angle PMB = 180° - \varphi - 2\varphi$. The supplementary angle α is, therefore, 3φ. ◇

However, can this be considered a simple compass-and-ruler-based solution? Well, it is not as flawless as it may appear: Our meticulous fitting of the ruler is not a precise construction, but rather a matter of "trial and error". Let us try to be more precise and draw all positions of the ruler that pass through B. We now get a so-called pencil of rays. Each ray possesses an intersection Q with the circle k. From there we can determine a point P at distance r, resulting in a point P. If P lies on the elongation of AM, then everything went successfully. What, then, is the locus of all points P? If the locus were a circle or an arbitrary conic section, then everything would be simple – the intersection of a conic section and a straight line can be determined by elementary geometry. Alas: The curve is of a higher order – namely, it is a so-called conchoid, here a *limaçon* (Fig. 1.21 on the left).

A second try: Let us choose P on the elongated leg AM. We then determine a point P on k at distance r (this yields two symmetrical solutions) and we get Q. In the general case, the elongation PQ does not intersect k at B, but we can try this many times. This leads to PQ enveloping a curve. The right solution is the tangent from B to this. Yet, maybe this curve is "simple"? No such luck! This time, it is a so-called *astroid* (Fig. 1.21 on the right), which is also of a higher order.

Many a courageous mathematician has "bloodied his nose" on angle trisection – despite the fact that it has long been proven that there *can be no solution* by ruler and compass.

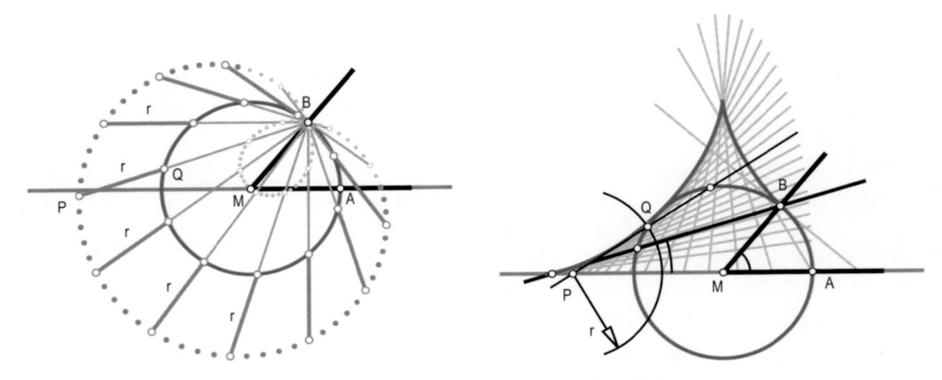

Fig. 1.21 A Limaçon and an Astroid

Points at infinity

Points can be infinitely far away. In such cases, mathematicians speak of *points at infinity*. (Should you find it difficult to imagine this concept, try imagining a distance of something like a light year, which, for all modest purposes, amounts to roughly the same thing.) Points at infinity are often labeled by a u in the subscript: Thus, P_u is, if not otherwise stated, a point at infinity. This may sound simpler than it is: We shall soon discover the mental stumbling blocks and geometrical advantages that it implies.
Every straight line has a point at infinity. But why only *one* and not two? Is it not true that straight lines proceed towards infinity in two directions? Are these two, by any chance, the same point? Can two points on a straight line one of which is located one light year in one direction and the other one light year in the opposite direction be identical?
Let us take some time and remember the following simple statement:

- Two different straight lines in the plane always possess an intersection.

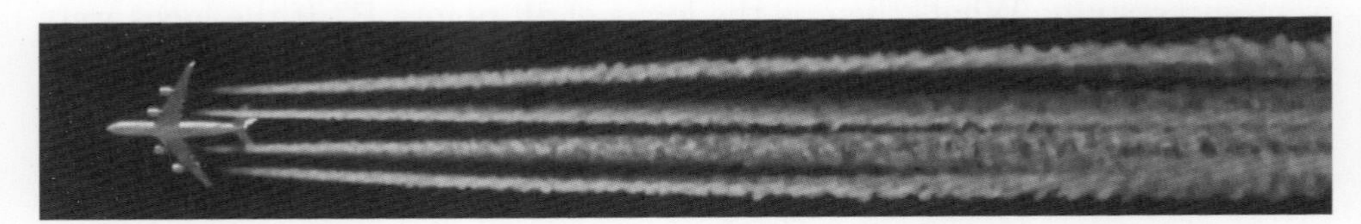

Fig. 1.22 Parallel lines intersect, by definition, exactly *once* – but only "at infinity"!

If we do not allow for points at infinity, we must immediately make an exception: Two straight lines in the plane always possess an intersection, *provided that they are not parallel*. If, on the other hand, we decide to be pragmatic and allow for points at infinity, we can state the following: Two parallel straight lines share one point at infinity (and not two, otherwise we would once again encounter an exception). We shall soon discover that there is a very reasonable explanation for this if we look at the problem "from a higher perspective". For now, let us remember:

- A point at infinity is determined by the direction of the straight line. Parallel straight lines determine the same point at infinity.

This whole issue appears much simpler when considered "from a higher perspective". Yet, in order to do this, we need to know more about planes, which are discussed in the next section.

1.3 Elementary building blocks in space

Let us now dare the jump from the drawing plane into three-dimensional space. This is not a dangerous step, because our bodies naturally occupy three-dimensional space. Our spatial imagination will permit us to comprehend the necessary trains of thought.
In space, points remain "infinitely small", having no extent whatsoever. But this is more of a philosophical statement. The stars of our universe could hardly be called infinitely small – on a photo from a great distance, however, they appear as points. Speaking of which: What appear to be stars in Fig. 1.23 are the relicts of fireworks on the sky of New Year's Eve ...

Fig. 1.23 "Points" and "straight lines" in space

Points may, once again, be located at infinity, and the familiar u in the subscript is used for their notation. We will soon explain the mental difficulties, as well as the advantages, of this type of thinking.
In the drawing plane, an infinite number of points results – if arranged in a certain way – in an infinitely thin and infinitely long *one-dimensional* straight line. We can also say that a straight line is "generated" by the translation of a point.
Straight lines should be thought of as unbounded: Every *line segment* $\overline{PQ}$ possesses an infinitely long straight *carrier line* $g = PQ$.
In space, however, we are still missing a third basic element:

Surfaces without curvature

In everyday speech, "plane" describes something horizontal: The High Plane of Tibet, the Pannonian Plane, etc. Sloped, uncurved surfaces are commonly

Fig. 1.24 A "plane" ...

referred to as "sloping sites", but not so in geometry: A plane is formed when we array an infinite number of parallel straight lines along another straight line. It can be said that a plane is "generated" by the translation of a straight line. The generating straight line and the directional straight line of the translation both "generate the plane".

Fig. 1.25 ... and a "sloped plane" in colloquial speech

In Europe, planes are denoted by lowercase Greek letters, such as α, β, ε, φ, etc. This sometimes leads to problems with the notation of angles, which are usually also named in this manner. For example: Image planes are often called π, while having nothing to do with the circle circumference-diameter ratio $\pi = 3{,}141592653589793\ldots$. It is, nevertheless, important to stick to this convention, since the alternative is even more confusing (many Greek capital letters, for instance, are identical to the Latin capital letters, leading to planes being possibly mistaken for points).

Let a be the generating straight line and b the straight line pointing in the direction of the translation. The straight lines a and b intersect at $S = a \cap b$ (read "a intersected with b"). The generated plane shall be called $\varepsilon = ab$ (read "a joined with b"). Every straight line g that intersects a and b lies entirely in ε, since it intersects every parallel of a in ε. One can therefore write $g \subset \varepsilon$ (read "g in ε").

Thus, an infinite number of straight lines lies in a plane. Each of these straight lines also implies a point at infinity. What, then, is the locus of all of these points? If we intersect all straight lines in the plane with another straight line u in the same plane that is "almost infinitely far away", then all intersections lie on this straight line. Therefore, it makes sense to conclude:

- All points at infinity of a plane lie on a straight line, the *line at infinity* of the plane.

We now have the following theorem without exception:

- Two different planes always have a straight line of intersection.

Every plane stance, thus, defines a line at infinity, and since infinitely many of such orientations are possible in space, the following question arises: What is the locus of all lines at infinity? Now we know how to proceed in an analogous way: We intersect all planes of space with an "almost infinitely distant plane" ω and get infinitely many of almost infinitely distant straight lines that, by definition, lie in ω which thus becomes the *plane at infinity.*
Mystical? Not at all! The next chapter deals with projections, and in it, we shall again consider planes at infinity, but from a higher position. What may seem confusing right now will by then seem completely normal.
The following theorems should already be perfectly comprehensible to us:

- A straight line g *always* shares an intersection point S with a plane ε. (If the straight line is located entirely in the plane, then every point on the straight line is an "intersection point".)

- Three different planes ε, φ, ψ *always* intersect in one point $S = \varepsilon \cap \varphi \cap \psi$. (If the point is a point at infinity, then the intersection lines of two planes are mutually parallel.)

- If two parallel planes $\varepsilon \parallel \varphi$ are intersected by a third arbitrary plane ψ, then both intersection lines of ψ are parallel.

  As a "proof", it could obviously be stated:

  The chain of evidence is as follows: The three planes ε, φ, ψ share a point at infinity. All intersecting lines of any two planes must pass through this point and must therefore be parallel. By the way: The third, not-so-evident intersection line $\varepsilon \cap \varphi$ is the line at infinity of the planes.

1.4 Euclidean space

We live in three-dimensional space and are, in general, more or less used to three-dimensional thinking. We call this space the "space of perception". Distances and angles are of great importance to us. A sheet of paper can be seen as the two-dimensional subspace of the space of perception. We are particularly familiar with geometry in two dimensions, as it is easy to "overview" its limited boundaries. However, if the sheet of paper takes on astronomical proportions, it becomes more complicated. After all, we are located on the globe, on a surface on which straight lines do not even exist.

Perfectly straight lines do not exist in nature either. From physics we know that not even light travels in perfectly straight directions: It is diverted by the mass of the stars and star systems in whose vicinity it passes. The term *gravitational lensing* is used to describe this phenomenon.
However, let us stay "on Earth", just like *Euclid*, one of the greatest thinkers in history. More than 2300 years ago, he operated a school of mathematics in Alexandria and condensed, in his 13-chapter "*Elements*" (which are treated as separate "books"), the entire mathematical and geometric knowledge of his time. The *Elements* survived the turmoils of history (including two fires at the Library of Alexandria) and were later translated – from Arabic into Latin – and published in Europe, where they have henceforth formed the basis for geometry right up until the 19th century.

Basic terms and axioms

First of all, *Euclid* defined two basic terms, namely, "points" and "straight lines", which can be put into geometrical relations by means of words like "lie", "between", "congruent", etc. He then defined the so-called *axioms of geometry*: Sensible, immediately obvious, and non-contradictory rules that can be assumed but cannot be proven, such as: "For two points there exists exactly one straight line on which they both lie", or "Of three points on a straight line, there is always one that lies between two others".
Now he defines new entities (circles, etc.), about which assertions can be made, all of which can be proven entirely by logical combinations of the aforementioned axioms. Thus, gradually, one brick is laid upon the other. It is essential that all definitions and proofs are based on secure knowledge that can be traced back to the axioms.
This method was groundbreaking for its day – moreover, it was capable of verifying the existing empirical rules and procedures of the Egyptians and the Babylonians. Some of these rules could actually be proven and were, thus, promoted to "theorems", while others were, nevertheless, determined to be useful approximations.

The standing leg that isn't

One of *Euclid*'s axioms says that through a point outside of a straight line, it is possible to draw a *unique parallel* to it. Mathematicians have been trying for many centuries to prove this axiom by means of others, thus promoting it to the status of a "theorem". By the mid-19th century, it was proven that while geometry is possible without this axiom, several well-known theorems would have to be discarded.
In the last section, we will learn more about *non-Euclidean geometry*. This particular axiom of parallels is problematic when considering the number of points at infinity of a straight line: If we do not want to accept that there exists only one point at infinity for each direction of a straight line, then much of space of perception geometry must inevitably collapse.

Fig. 1.26 Distance circle

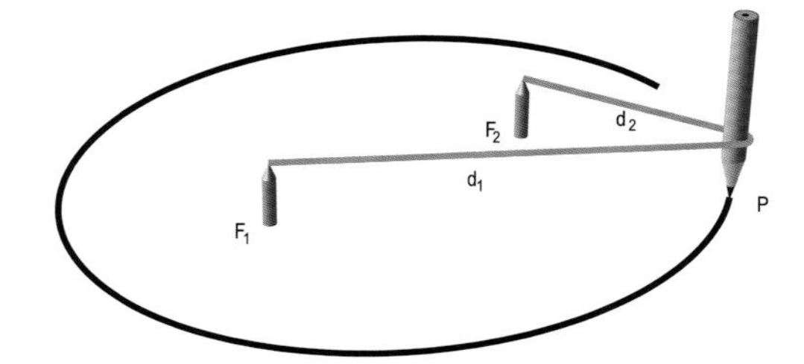

Fig. 1.27 String construction of an ellipse

Sum of distances

If we look for all points P in the plane that are at given distances d_1 and d_2 from points F_1 and F_2, then we must intersect the circles $k_1(F_1; d_1)$ and $k_2(F_2; d_2)$. As we know, this yields (in general) two solutions, or no (real) solution. If the circles merely touch, we speak of a *double solution.*
Thus, we can construct an ellipse as the locus of all points P, whose *sum of distances* from two fixed points F_1 and F_2 is constant ($= 2a$).
The principal apices A and B of the curve are found when the distance circles touch each other. The auxiliary apices C and D result when both radii equal a. F_1 and F_2 are called *focal points* of the ellipse. For the hyperbola, the *difference* of the distances from two fixed points is constant. You can find out more about conic sections from the chapter 5.

The Apollonian circles

The search for all points P in whose distances from two fixed points F_1 and F_2 form a constant ratio lead to the *Apollonian circles.* If this ratio is $1:1$, the circle "degenerates" into the *perpendicular bisector.* A straight line could, thus, be interpreted as a circle with an "infinitely large radius" and "a center at infinity".

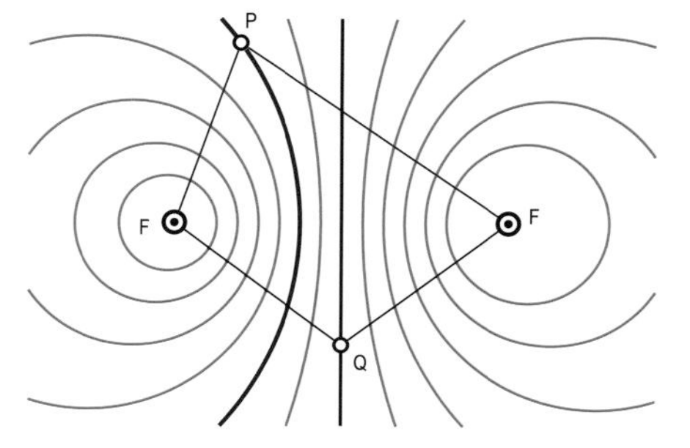

Fig. 1.28 Apollonian circles

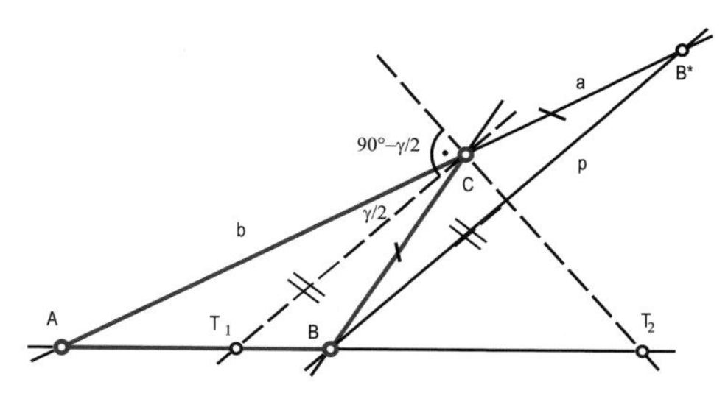

Fig. 1.29 Pair of angle bisectors

The perpendicular bisector is the set of points that is equidistant to two given points. While this may be obvious, Apollonius' more complex assertion needs to be proven. A "full-blooded mathematician" would, thus, draw a coordinate system that is defined by the following fixed points: Let $F_1(0/0)$ be the point of origin and $F_2(d/0)$ a position on the abscissa. The condition for a point of the plane $P(x/y)$ is then as follows:

$$\overline{PF_1} : \overline{PF_2} = \sqrt{x^2 + y^2} : \sqrt{(x-d)^2 + y^2} = c = \text{constant.}$$

By squaring the equation and proceeding with a slight rearrangement of the terms, one gets the equation of a circle with the midpoint on the x-axis. This may be elegant, but would have been entirely mysterious to Apollonius of Perga (born in Asia Minor 262 B.C., died in Alexandria 190 B.C.). Let us not forget that Apollonius not only proved this theorem, but also authored eight tomes of fundamental work on the "geometry of positions", which deal with conic sections (Conica) in particular.
He first considered the following theorem:

The angle bisectors of a triangle divide the opposite side in the ratio of the adjacent sides.

Proof:
The theorem is immediately obvious, given the appropriate sketch (Fig. 1.29). The two angle bisectors C form a right angle because they subdivide the angle γ and the supplementary angle $180° - \gamma$, and therefore $\gamma/2 + (180° - \gamma)/2 = 90°$. Let us now draw the parallel p through a triangle vertex B towards the inner bisector that is simultaneously orthogonal to the outer bisector. The straight line p intersects the extension of AC in a point B^*, and the triangle BCB^* is isosceles. On the side AC, $\overline{CA} = b$ and $\overline{CB^*} = a$. This *affine ratio* is transferred by means of the inner angle bisector and p onto AB according to the theorem of intersecting lines: $\overline{AT_1} : \overline{T_1B} = b : a$. The point B^* thus arrives at B while the point C at the division point T_1 on the inner angle bisector A remains fixed. According to the theorem of intersecting lines, the following is indeed true: $\overline{AT_1} : \overline{T_1B} = b : a$. ◇

Harmonic points

The exterior angle bisector divides the line segment by analogous reasoning, but with an inverted sign:

$$\overline{AT_2} : \overline{BT_2} = b : a \text{ or } \overline{AT_2} : \overline{T_2B} = b : (-a).$$

The pair (A, B) is said to separate the pair (T_1, T_2) *harmonically.* Harmonic pairs of points play a fundamental role in projective geometry, which is the subject of a later chapter.
However, let us return to Apollonius. The theorem that involved the angle bisectors might also have been proven by means of the sine law [11], of which he was probably aware – but this was not necessary, as he proceeded in his preferred way of "elementary geometry".
We are getting really close to the circle of Apollonius (Fig. 1.30): We have the triangle ABP and both angle bisectors that cut out the division points T_1 and T_2 from AB, whose position is exclusively dependent on the ratio

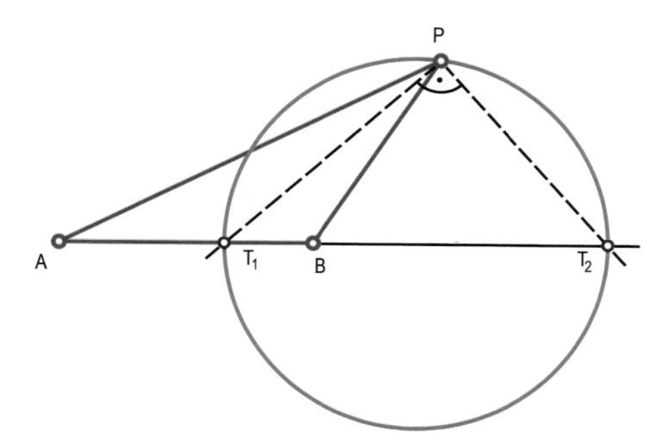

Fig. 1.30 Proof of Apollonius

$a : b$. The right angle that is formed by both angle bisectors literally "cries out" for *Thales'* theorem: C lies on the Thales circle above T_1T_2. Now we shall change the triangle in such a way that $a : b$ remains constant. Thus, T_1 and T_2 also remain constant, and the Thales circle, likewise, remains fixed. C moves on the Thales circle, which we may now call the *circle of Apollonius.*

All that's remaining is the product

Next to the ellipse (constant sum of distances to two fixed points) and the hyperbola (absolute of the difference of the distances is constant), Apollonius also managed to fit the circles (constant quotient) into this scheme. He has surely also tried to attain useful results from the product of the distances, but they simply do not yield such "beautiful" results (they are fourth-order curves, which would have caused problems for the ancient Greeks, as there were no computers to make sense of them).

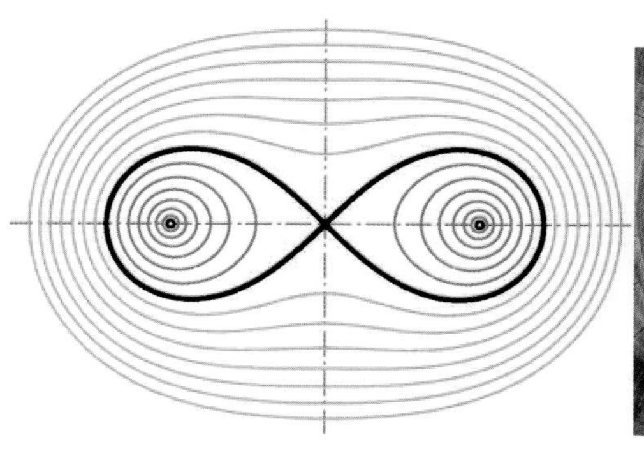

Fig. 1.31 Cassini's curves, defined as the locus of all points with a constant product of the distances from two fixed points ...

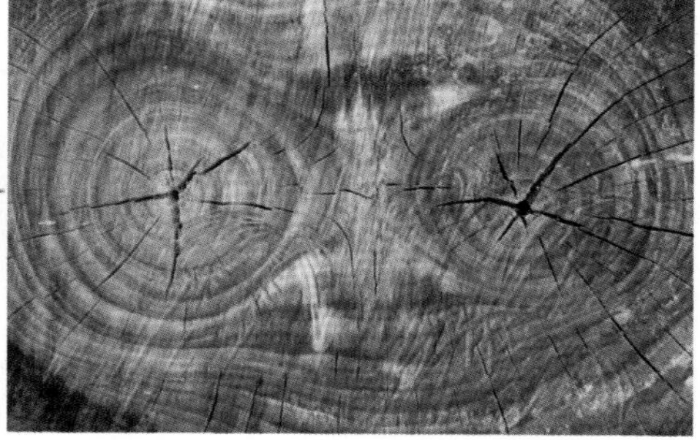

Fig. 1.32 ... and a stunning similarity to the cross-section of a pine tree trunk. Consider also Fig. 1.7 on the right.

If the product of the distances to two fixed points F_1 and F_2 should be constant for the points P, we are dealing with a so-called *Cassini oval*, which we will again encounter when discussing intersections of a torus (Fig. 6.65).

From the plane into three-dimensional space

All prior reflections about distances are equally applicable in space if we rotate the locus curves about the axis F_1F_2. Circles, whose centers lie on

the axis, turn into spheres. At their limiting cases (infinite radii) the planes appear orthogonal to the axis by rotation.
The intersection points of two such circles move, during rotation, in planes perpendicular to the axis. Ellipses trace ellipsoids, and arbitrary curves generate arbitrary surfaces of revolution. We will discuss these surfaces in great detail.

Analogies on the sphere

The equivalent solutions on the sphere are equally interesting. For instance, let us ask for the points on the Earth's surface that are equidistant from two fixed points: We then have to intersect two circles on the sphere. An analogous ellipse construction leads to a *spherical conic section*. For all points P on such a curve, the sum of minimal (spherical) distances from F_1 over P to F_2 is equal.

Other distance lines and distance surfaces

Apart from circles and spheres, there exist other important distance lines and distance surfaces.
What does the curve of points P in the plane look like if they have a fixed orthogonal distance d from a straight line f in this plane? This question yields a *parallel pair* of straight lines. By means of rotation of this situation about the axis f, we get a *cylinder of revolution* with radius d whereas the translation yields a *distance plane* to the given plane φ through f.

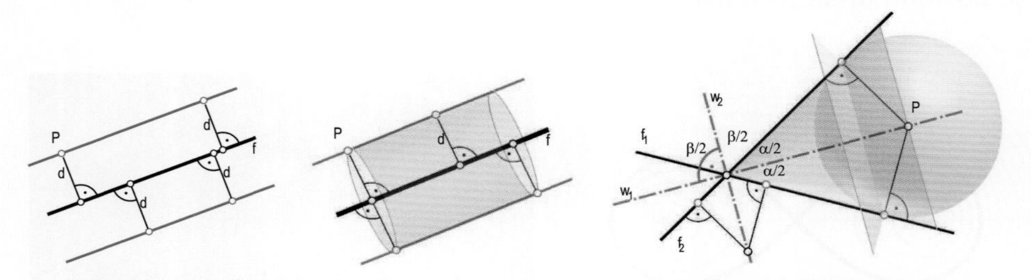

Fig. 1.33 Parallel pair and cylinder **Fig. 1.34** Angle bisector and cone

Given two fixed non-parallel straight lines f_1 and f_2 in the plane, we are looking for all points P that have the same orthogonal distance to both lines. A pair w_1, w_2 of angle bisectors can be found. Both angle bisectors w_1 and w_2 are always perpendicular: Let $\alpha = \angle f_1 f_2$ be the mutual angle of both given straight lines. Since these are not oriented, one could also have measured $\beta = 180° - \alpha$ (the supplementary angle). For the angle $\omega = \angle w_1 w_2$ the following is true: $\omega = \alpha/2 + \beta/2 = \alpha/2 + (180° - \alpha)/2 = 90°$.
In this case, a rotation about an axis does not result in an analogous theorem in spatial geometry. Nevertheless: If we rotate about one of the bisectors, such as w_1, the straight lines f_1 and f_2 will lie on the same *cone of revolution*. The circle P that touches both straight lines becomes an inscribed sphere that

touches the cone, and the rays through the contact points form an additional cone of revolution that intersects the first cone perpendicularly along a circle. The translation of the planar situation in the direction of the vertical to the carrier plane is much more interesting. It implies a pair of *angle bisector planes*, on which all points are located that are equidistant to two given planes.
We can conclude that it is often possible to derive spatial analogies from two-dimensional situations.

1.5 Polarity, duality and inversion

Infinity can mean many things

Until this point, we have uttered the term *dimension* many times. A point, for instance, is zero-dimensional – it has no extent. A straight line is one-dimensional – with extent in precisely one direction. A plane is two-dimensional. It can both be seen as a collection of points and a collection of straight lines. Let us attempt a comparison: Are there more points or straight lines in a plane?
The question may strike you as slightly insane. First of all: Why should we even want to find out? Secondly: It should be obvious that a plane contains an infinite number of points, as well as an infinite number of straight lines. Whatever do we want to differentiate with a term like *infinity*?
The answer to this question leads to unexpected relationships. Pondering such supposed absurdities is, indeed, worthwhile.

A close relationship

We define the following geometrical operations in the plane that we shall call *polarity with regard to the circle*:
Let us choose an arbitrary but fixed circle k with the midpoint M (Fig. 1.35). If a point P lies outside of this circle, then we have two tangents at k that touch k in two points Q and R. The connecting straight line $p = QR$ is assigned to the point P: $P \mapsto p$. It is perpendicular to the line MP: $p \perp MP$. The inverse mapping is also unambiguous: If an arbitrary straight line p intersects the circle at the points Q and R, then we can define tangents through these points that intersect at a point P: $p \mapsto P$. Once again, $MP \perp p$ applies. The assignment $p \leftrightarrow P$ is, thus, invertible, and we speak of *pole* P and *polar* p. Even points at infinity are captured: Their polars are the diameters of k.
The case of P lying on the circle k poses no additional difficulties. In such cases, p is the tangent of k in P.
If P lies inside k, we define its polar as follows: We draw an auxiliary straight line $p^* \perp MP$ through P, whose pole P^* is the foot of the normal of the

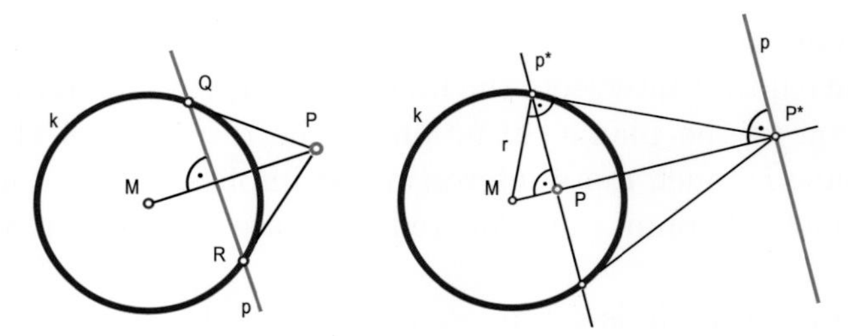

Fig. 1.35 Polarity in the plane

straight line p from M that is to be determined. In the reverse case, if p passes k, we determine the foot of the normal P^* of M on p, whose polar p^* intersects MP^* in the pole P. Even the line at infinity of the plane can be determined in this way: It is equal to the center M of the circle.
Therefore, the polarity of the circle consistently guarantees the unambiguous relationship between points and straight lines in the plane, from which it follows that the plane contains *precisely as many points as straight lines.*
It is, therefore, correct to speak of the plane as a two-dimensional *point space* as well as a two-dimensional *line space.* These two spaces are mutually *dual spaces.*
This circumstance can further be illuminated mathematically: The definition of a point in the plane requires two numbers (the x-coordinate and the y-coordinate). The same is true for the definition of a straight line (e.g., the y-value on the y-axis and the slope angle).

Let us, once again, attempt to derive a spatial analogy (Fig. 1.36). If we rotate our plane about the axis MP, then the circle k generates the sphere κ, the tangent on the circle becomes a cone of revolution, the contact points Q and R become contact circles of the cone and κ, and their carrier plane is the *polar plane* π of the point P. The assignment $\pi \mapsto P$ works in the same way. The polarity with respect to the sphere κ, once again, guarantees the consistent assignment between points and planes of space (as for the plane at infinity, it is the midpoint of the sphere). We can, thus, conclude that space contains *as many points as it contains planes.*

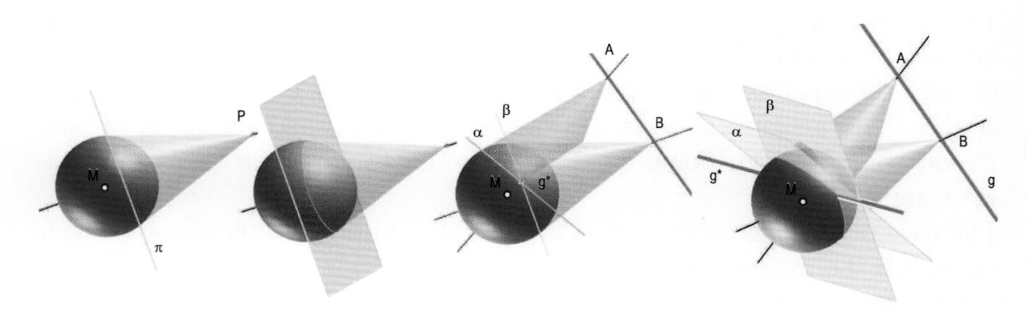

Fig. 1.36 Polarity in space

Thus, Euclidean space can be interpreted as both a three-dimensional *point space* as well as a three-dimensional *plane space.* Both spaces are mutually dual spaces.
Mathematically speaking, a point in space is defined by three numbers (e.g., through its Cartesian coordinates). The same is true for a plane (which is, e.g., determined by the normal distance from the origin and the inclination angles with respect to the xy-plane and the xz-plane).

Back to the two-dimensional case: If r is the radius of the circle, then we have a simple relation between the pole P and its polar $p \ni P^*$ according to Euclid's theorem (Fig. 1.35):

$$\overline{MP} \cdot \overline{Mp} = r^2.$$

With this we can proof the following theorem:

If a point A in the plane moves on a straight line b (Fig. 1.37), then its polar a passes through a pencil of straight lines that runs through the pole B of b. The assignment pole↔polar is, thus, invertible, and linear pencils correspond to linear pencils.

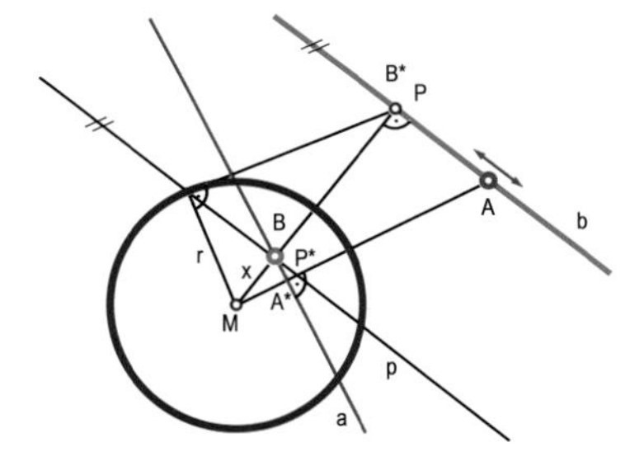

Fig. 1.37 The proof

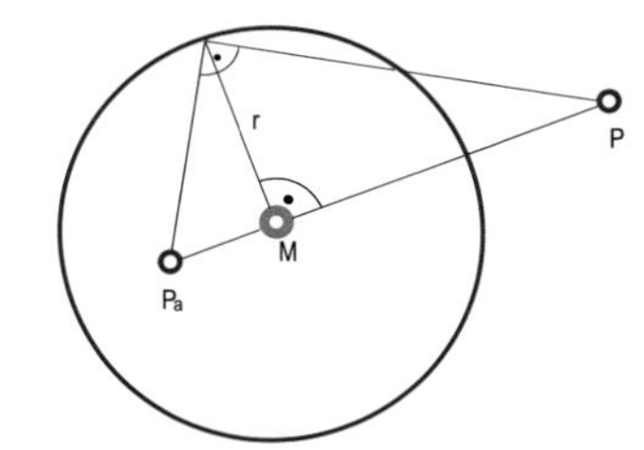

Fig. 1.38 Construction of the *anti-inverse* point

Proof:
Let B be a pole and $b \ni B^*$ its polar (Fig. 1.37); then it follows that $\overline{MB} = r^2/\overline{Mb}$. Further, let A be a point on b. For its polar $a \ni A^*$, $\overline{MA} = r^2/\overline{Ma}$. a intersects the ray MB at the distance x. The triangles MBA^* and MAB^* are similar; it is also true that

$$\overline{MA} : \overline{MB^*} = x : \overline{Ma} \Rightarrow x = r^2/\overline{MB^*} = \overline{MB}.$$

◇

The construction of polars of a point with respect to a circle so far involves tangents of circles. We have to differentiate between points inside and outside the circle. There exists a very simple substitute construction that always works: Simply construct a right-angle hook as in Fig. 1.38 that yields a point P_a that is "anti-inverse" to P. According to the altitude theorem, $\overline{MP} \cdot \overline{MP_a} = r^2$. The point, therefore, only needs to be reflected at the center of the circle in order to lie on the polar.

It appears that while making the leap from the plane into space, we have somehow lost track of straight lines; so let us catch up on that. A straight line g in space can be interpreted as the connection of two points A and B (Fig. 1.36 on the right). The polar planes α and β correspond to these points. Let us now consider the intersection line $g^* = \alpha \cap \beta$. If we are able to show that a polar plane γ through g^* ($\gamma \supset g^*$) can, indeed, be assigned to every point C on g, then it makes sense to define g and g^* as *reciprocal polars*. In that case, straight lines correspond to straight lines, and, therfore, line space is self-dual!

Proof:

Let us interpret the planar situation spatially. The circle can then be interpreted as a sphere contour. The line $g = AB$ lies in the image plane, the corresponding polar planes α and β are then "projecting" – this means that they appear as straight lines. Their intersection line g^* is perpendicular to the image plane and appears as a point at the distance $r^2/\overline{Mg}$. This distance is equal for all points of g. The linearity of planar polarity is, thus, transferred into three-dimensional space! ◇

The principle of duality

We now notice the following circumstance: By polarization with respect to a circle in the plane, the terms point and straight line are interchanged, as are the operations *connect* and *intersect*. Thus, if a theorem contains only such terms and operations, it can easily be converted into a correct dual theorem. Let us consider a simple example:

- The connection of two points is a straight line ↔ The intersection of two straight lines is a point.

The analogue is also possible in space: By polarization with respect to a sphere in space, the terms point and plane are interchanged, as are the operations *connect* and *intersect*. In this case, the term straight line remains fixed. Thus, if a theorem of spatial geometry contains only such terms and operations, it can easily be converted into a correct dual theorem.
Here are some simple examples that illustrate this:

- The connection of two points is a straight line. ↔ The intersection of two planes is a straight line.

- Three intersecting planes define a point. ↔ Three "connected" points define a plane.

- If two straight lines lie in a plane, then they intersect at a point. ↔ If two straight lines intersect at a point, then they lie in a plane.

The following theorem may not be as self-explanatory:

- Spatial curves can be generated by a moving point as well as by a moving plane.

We will often use this theorem: It holds the key to the so-called *developable surfaces*, which play an increasingly important role in modern architecture.

An infinite amount and infinitely more

Polarity has shown that space contains as many planes as points. From mathematics, we know that in a Cartesian coordinate system, every point can be described by three coordinates. This is why we speak of three-dimensional space (which should, more accurately, be referred to as a *point space*), and for the same reason a three-dimensional *plane space* exists as a dual counterpart of point space. But is the *line space* also three-dimensional?
Let us consider all straight lines determined by a fixed point P and an arbitrary point in a horizontal plane β ($P \notin \beta$) (Fig. 1.39). There exist as many lines as there are points in this plane, because any point in β can be joined with P. By comparison: In Fig. 1.40, there is also exactly one straight line through each point in the plane.
If we now move the *vertex of the pencil* P on an arbitrary straight line $g \parallel \beta$, we get for each of the infinite positions of P the same number of connecting lines. This number alone – of all straight lines that intersect g – is as large as the number of points in space! It would appear that there are *many more* straight lines in space than there are points.

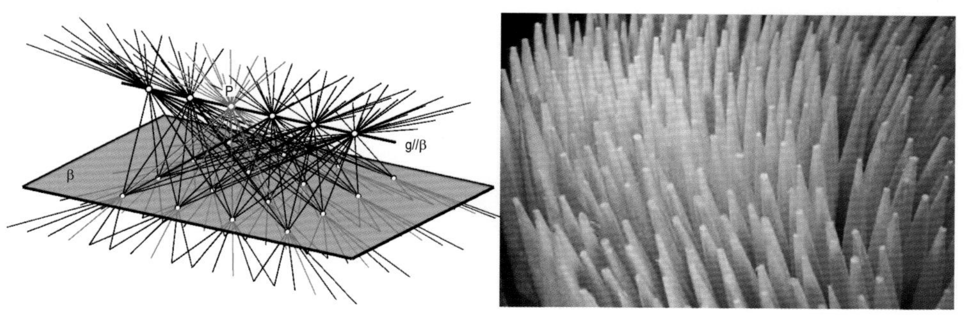

Fig. 1.39 Variation of the vertex of the pencil. Through each point P there exist as many straight lines as there are points in the plane.

Fig. 1.40 A small selection of all straight lines – in the particular case, it is known as the congruence of lines.

Now, the vertex of the pencil P should be able to vary all over the plane $\gamma \parallel \beta$. There, this point should have the two coordinates (u/v). The points in the base plane β also possess two-dimensional coordinates (x/y). Each choice of the four parameters $(u/v/x/y)$ yields a new straight line that is not identical to any other. These parameters could be interpreted as coordinates in a four-dimensional coordinate system (however it may look: we are not capable of imagining it visually). It is, therefore, said that: *The line space is four-dimensional.*
That may be easy to state, but it is quite something to consider that our space of perception should have *many more* straight lines than points and planes! Naturally, such an idea has consequences, and thus, we must not be surprised if, from

a geometrical point of view, many things that involve straight lines require more effort.
For instance, we will see that it is not at all easy to view an arbitrarily positioned straight line such that it appears as a single point (in other words that it is "projecting"). Moreover, if we want to imagine rotations about arbitrary axes, we have to be "geometrically trained", and pretty well at that. How simple it is, by comparison, to imagine a rotation about a vertical axis!
It is all the more important to visualize complicated situations through specific points of view, which is the task of descriptive geometry. I call this the "reduction of dimension", and the training of this skill is one of the highest goals of geometry education.
In algebraic geometry, dimensions are counted by "degrees of freedom": Three points in space define a plane. Each point in the plane has two degrees of freedom. Thus, the dimension of the plane space (considered as a space of points) can be calculated by $3 \cdot 3 - 3 \cdot 2 = 3$. By analogy, two spatial points define a straight line, where each point on the straight line possesses 1 degree of freedom. Thus, the dimension of the line space is $2 \cdot 3 - 2 \cdot 1 = 4$.

Inversion – not every transformation is linear

The polarity with respect to a circle $(M; r)$ in the plane defines a linear relation between points and straight lines on a plane. The pole P corresponds to the polar line $p \ni P^*$ where $\overline{MP} \cdot \overline{MP^*} = r^2$. We can also assign the point P^* to the point P. This relation is called *inversion* or *reflection in a circle.* While this correlation is invertible, it is, intriguingly enough, no longer linear: When P moves in a straight line, P^* moves on a circle.

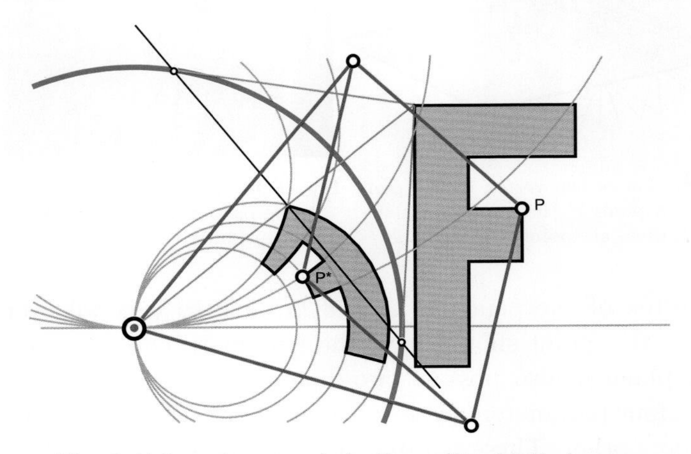

Fig. 1.41 Inversion on a circle, *Peaucellier-Lipkin* inversor

The proof for this is mathematically easy to derive and can be found in the [11]. Without going too deep into detail (we will talk about it later): The inverse of an algebraic curve of order n is (in general) a curve of order $2n$. The straight line is of first order, the circle of second order. A conic section corresponds (in general) to a curve of order four, etc. The inversion is, thus, called a *quadratic transformation.*

An allegedly inexplicable "anomaly"

An exception is particularly notable during inversion: A circle does not correspond, as we would expect, to a fourth order curve, but once again to a circle – and if the circle contains the circle midpoint M, it corresponds to a straight line (the proof for this will be given on P. 182)! Such things irritate at first: Are we not dealing with a quadratic transformation after all?

In order to explain this occurence, we would have to make an adventurous "excursion" into a non-real world – but this would exceed the boundaries of this book. Rest assured, the supposed anomaly can be brought under mathematical control.

A mechanism for the realization

The fact that circles correspond to circles is not at all self-evident, but it can be used to construct a mechanism that automatically executes the inversion (Fig. 1.41).

Here, one is justified to ask the question "why do I need this?". Let me explain: By using this inversion, a mechanism can be built that converts pure rotation into *exactly straightforward* motion. This problem is not to be underestimated in the field of mechanics: A rotational motion is something very practical: Every engine rotates a driveshaft in one form or the other. Using the mechanism, a point is to be moved in a straight line without having to be directly guided on something like a straight track. A certain E. *Hart* solved this problem through inversion ...

Spherical reflection

The inversion can be performed in three-dimensional space equally well. The inversion circle is replaced by an *inversion sphere* $(M; r)$. The points P and P^* are, once again, linked via the relation $\overline{MP} \cdot \overline{MP^*} = r^2$. P^* lies in the polar plane of P. The transformation $P \mapsto P^*$ is quadratic: Planes are, thus, mapped to spheres (in general). The proof can be made by rotating the planar situation straight line $\mapsto$ circle about the normal of the straight line through the point M.

Surfaces of n-th order (see P. 140) are transformed into surfaces of $2n$-th order. Planes are also considered as spheres with "infinitely large radius" and "center at infinity". Spheres and planes are transformed into spheres or planes, i.e., "inversions preserve spheres", see also Fig. 6.89).

By analogy to *circle reflection*, we may, at this point, speak of *reflection in a sphere.*

By the way: The physical reflection in a *spherical mirror* is, as a concept, very different to inversion: It transfers space onto the surface of a sphere. This means that three-dimensional space must be transfered onto a two-dimensional surface, and this loss of dimension implies a simultaneous loss of information: The situation in space cannot be reconstructed from the image on the sphere. The inversion in the sphere, on the other hand, is an invertible relation between points in space that does not cause the loss of information (Fig. 1.44).

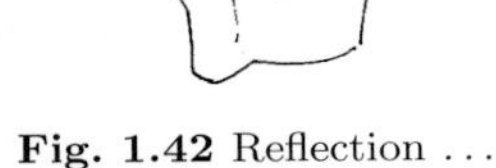

Fig. 1.42 Reflection ...

Fig. 1.43 ... in a sphere

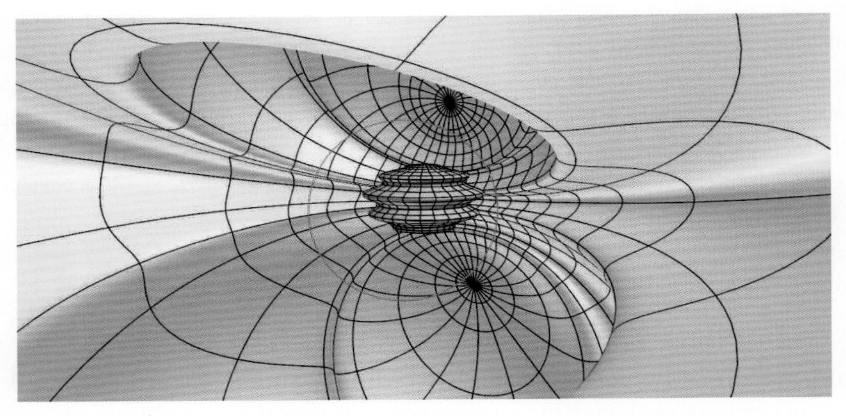

Fig. 1.44 Inversion of a rotational surface at a sphere

Purism

The ancient Greeks considered a problem to be "exactly solvable" if it could be constructed by compass and ruler. This is a justifiable demand for people without computers. For many centuries, however, it has not only been possible to exactly solve quadratic problems, but also those of third or fourth order – by a detour into complex algebra. All other methods are approximations at best, including the computer-assigned ones, even if they claim to be precise to 15 decimal places.
It is possible to be stricter still. Which problems are only solvable by compass? It can be shown that *every* problem that is solvable by compass and ruler is also solvable by *compass alone*. Of course, the separate steps become more elaborate. The following example is said to have been devised by a suggestion of *Napoleon* and is famous for this reason.
The allegedly simple problem reads as follows: How to find the center K of a given circle k without the assistance of a ruler?
By compass *and* ruler, it can be solved in only a few seconds: One simply has to choose two arbitrary points A and B on the circle k and use them as centers to draw two circles a and b of equal size. The line joining the intersections of a and b has to pass through the center. By repeating the construction with a third arbitrary point C instead of B, the final center can be found. It is, of course, the circumcenter of the triangle ABC.

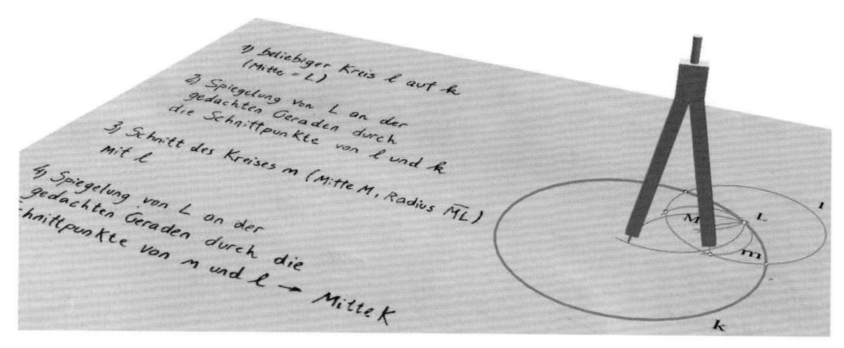

Fig. 1.45 The assignment of *Napoleon* including instructions

Without a ruler, it is not much more difficult. Six circles are needed in order to find the midpoint – easy if the instructions are followed (Fig. 1.45). However, the proof for this is pretty elaborate: Many properties are needed which we have come to know from inversion.

• **A proof that isn't: appearances can deceive ...**

In this chapter, we have proven several theorems in an elementary way, requiring, in addition, only our imagination and our common sense. Consider the following "proof" that "every angle is a right angle". Even experienced geometers often struggle with finding the error in the proof.

Assertion: *Every angle in the plane is a right angle*
Proof:
Let $ABCD$ be a quadrangle with the following properties (Fig. 1.46):
(1) $\angle ABC = 90°$, (2) $\angle BCD = \alpha \neq 90°$, (3) $\overline{AB} = \overline{CD}$.

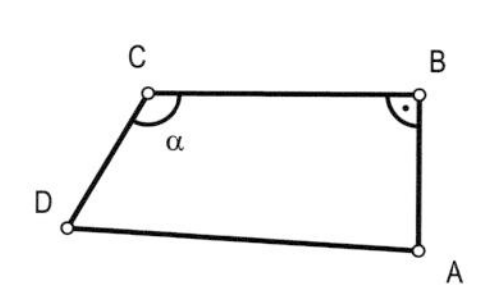

Fig. 1.46 Our quadrangle

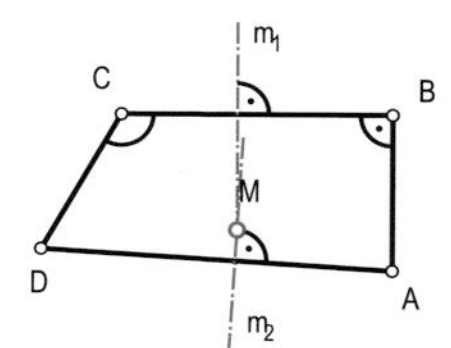

Fig. 1.47 The point M

The perpendicular bisectors m_1 and m_2 on BC and AD are not parallel due to (2) and (3) and thus intersect at a point M (Fig. 1.47).

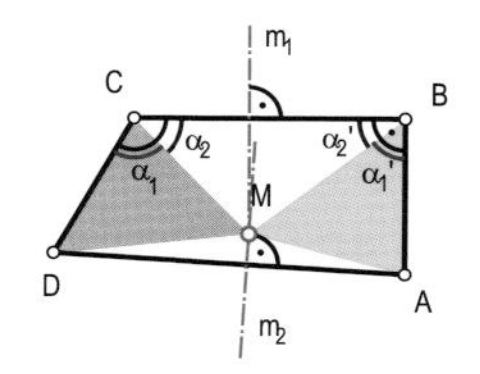

Fig. 1.48 Congruent triangles

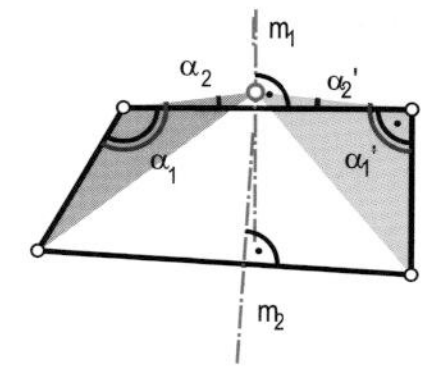

Fig. 1.49 M outside

Let us now compare the triangles ABM and DMC (Fig. 1.48). According to the precondition (3) $\overline{AB} = \overline{CD}$ and thus $\overline{BM} = \overline{CM}$ (M lies m_1) as well as $\overline{DM} = \overline{AM}$ (M lies on m_2). Both triangles are, thus, congruent and have, thus, the same angles:

$\alpha_1 = \angle MCD = \angle ABM = \alpha'_1$. Since MBC is isosceles, it follows that $\alpha_2 = \alpha'_2$ and finally $\alpha_1 + \alpha_2 = \alpha = \alpha'_1 + \alpha'_2 = 90°$.
Thus, the proof appears to be finished if M lies inside the quadrangle $ABCD$. This point may still lie outside of the quadrangle (Fig. 1.49). In this case, we get congruent triangles ABM and DCM as before and thus the equal angles $\angle DCM = \angle ABM$ as well as $\angle BCM = \angle CBM$ (the triangle MBC is isosceles). This time, if we calculate the difference of the angles, we also get $\angle BCD = \alpha = \angle ABC = 90°$. ◇

As we can see: Appearances can be deceiving – even in two dimensions!
You must surely be curious about the error in the proof. You will find it on the website to the book. ♠

1.6 Projective and non-Euclidean geometry

In this section, the geometrical circumstances will be presented from a different point of view. This way, projective geometry becomes the basis for many new insights – and the explanation for otherwise perplexing oddities of geometry.
Why, for example, is a circle defined by three points and a parabola by four, but it takes five points to define an ellipse and a hyperbola? Why do two circles possess a maximum of two intersections, but two arbitrary conic sections up to four?
The solution lies in two-fold arrangements: Firstly, the line at infinity should not be considered as unreachable, and secondly, non-real elements of the drawing plane should also be accepted, and complex conjugate points and straight lines in particular.
Working with elements at infinity can actually be made plausible through a "dimensional jump" from the drawing plane into space. The "complex extension" of the plane requires a jump into four-dimensional space which cannot be visually imagined.
Interestingly enough, geometries that deal with higher-dimensional spaces actually exist and achieve important results for "regular" geometry.

Let us discard lengths and angles

Projective geometry is a very important field of geometry – it limits regular geometry insofar as it completely discards all *congruence relations*. Terms such as the length of a line segment or the size of an angle become irrelevant. Hence, the concept of parallelism has to be changed. Pythagoras' theorem and the Law of Right Angles from descriptive geometry cannot be stated in projective geometry.
The *incidence relation* becomes all the more important: Do three points lie on a straight line? Do three planes intersect in a straight line? The conclusions can be directly carried over to Euclidean geometry, even if it permits points

at infinity, which we have defined as the intersections of two parallel straight lines.

Projective geometry makes use of terms such as point, straight line, and plane – as well as incidence relations such as "lies on" or "intersects". This immediately recalls the duality principle that we have already discussed.

For every theorem of projective geometry, the corresponding dual theorem is automatically correct and needs not be proven.

The proof of the dual statement can be omitted because the original statement needs to be dualized only, i.e., the terms points, lines, planes have to be changed to planes, lines, points, $\cup$ and $\cap$ have to be interchanged, while incidence relations are kept.

The dimensional jump

Let us consider the Euclidean plane π_e, this time "embedded" in space: It could, for example, be the locus of all points with height zero. Now let us project all elements of this plane from a center point Z that does not lie within it, such as a point on the z-axis other than the origin.

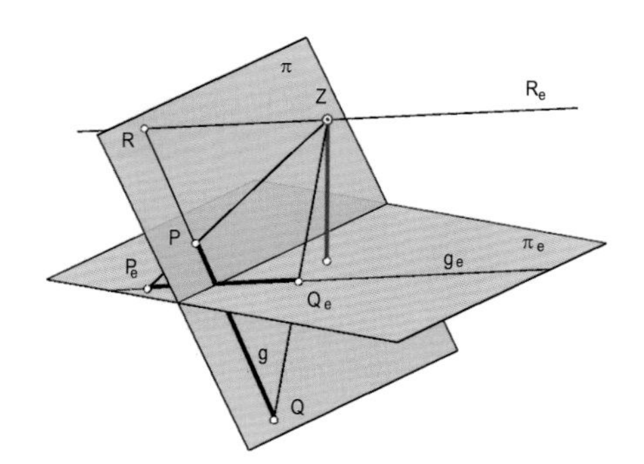

Fig. 1.50 We escape into space

Each point gives rise to a line (ray) that is the connection of a particular point with Z. Thus, lines become planes if projected from Z. The new objects are elements of a *star* with apex Z and planar geometry thus becomes the *geometry in a star* in space.

In order to understand the situation planimetrically, let us intersect the pencil with an *arbitrary* plane π that does not pass through Z. Now, any point in π_e corresponds to a point in π and any straight lines in π_e corresponds to a straight line in π.

Rows of points and pencils of straight lines

If a point P_e in π_e moves on a straight line g_e, then the corresponding pencil ray moves in a projection plane γ – thus, the corresponding point P in π moves on a straight line $g = \pi \cap \gamma$. This functions unambiguously in both

directions. The (dual) analogy may be drawn for pencils of straight lines in π_e and π.
With some help from Z, both "fields" π_e and π are thus related via a *perspective collineation.* The center Z is the *center of the collineation* and the intersection line $a = \pi_e \cap \pi$ is the *axis of collineation.*

Projectivity = likewise linearity

In a last step, let us, once again, rid ourselves from the idea that we have escaped to space. By arbitrarily positioning them next to each other, let us "merge" both planes. The alternating linear relationship (collineation) between both fields is preserved – but in general, we now neither have a center nor an axis of collineation. If we work in the Euclidean plane, we are performing Euclidean geometry, and if we work in the projective plane, we are performing projective geometry.

Conic section remains a conic section

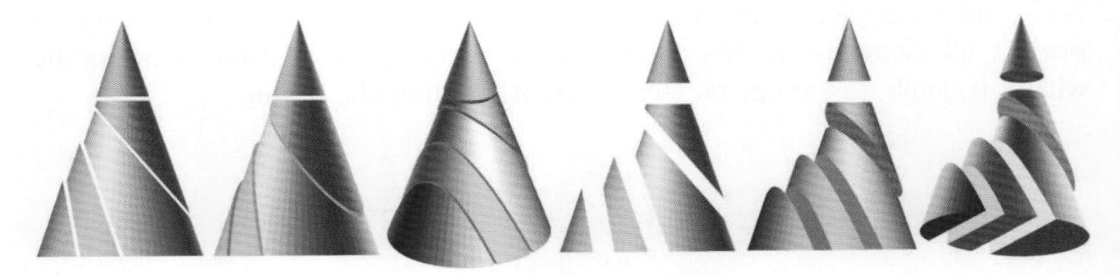

Fig. 1.51 Intersections of a cone of revolution

In the pencil through Z, a circle k_e in the Euclidean plane π_e corresponds to a circular cone that is (in general) not *right*, i.e., it is not a cone of revolution. This is what we call a *quadratic cone.* Common sense might suggest that the cone should be elliptic, but if the section plane π is "steep enough", it may yield a hyperbola as intersection curve or, in the special case, a parabola. We would then have to speak of a hyperbolic or a parabolic cone – though it is the same cone. Let us, therefore simply, state that the cone is quadratic, which relates to its algebraic property that it always possesses two intersection points with an arbitrary straight line (if mere touching counts as two intersections and non-real solutions are allowed).

In Euclidean space there exist no elliptic, hyperbolic or parabolic cones – only *quadratic cones.* It is, therefore, not necessary to distinguish between the various types of conic sections in projective geometry.

The line at infinity is suddenly "completely normal"

Is it really impossible to tell these types of conic sections apart? What is the difference, planimetrically speaking, between an ellipse, a hyperbola and

Fig. 1.52 Affinely distorted cones

a parabola? The number of points at infinity is obviously important: A hyperbola has exactly two, a parabola exactly one, and an ellipse none (in real space).

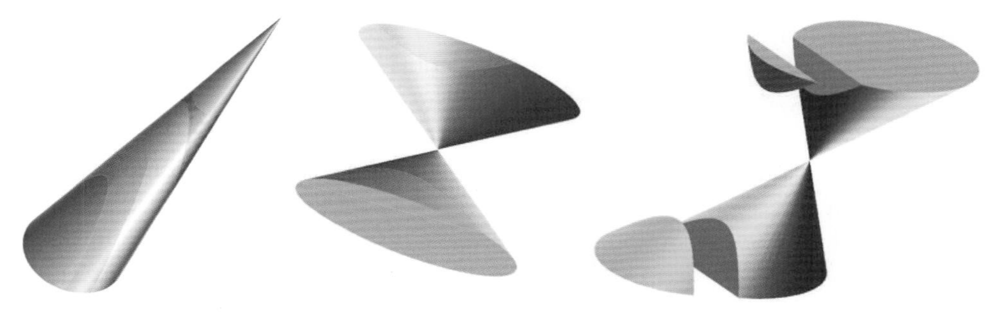

Fig. 1.53 Arbitrary quadratic cones

The Euclidean plane has an infinite amount of "points at infinity" – any point at infinity belongs to all straight lines parallel to a certain direction. In the first chapter, we have decided that each straight line only has *one* point at infinity even though we could have also assumed two. This will now prove to have been very reasonable.

If we now draw a straight line of each direction from Z, then they form a plane that is parallel to π_e. It intersects π in a "completely regular" line u (if π is not parallel to π_e), namely, the image of the *line at infinity* of π_e. The distance to an arbitrary point at infinity U_e of the Euclidean plane is, according to the construction, parallel to the pencil straight line ZU_e, which intersects the *projective plane* π in a point U on the image of the line at infinity u (Fig. 1.55).

Fig. 1.54 now shows symbolically how we can distinguish the three types of conic sections: By means of their relative position to a regular straight line u.

Non-Euclidean geometry

In the previous sections, we have discussed the geometry of *Euclid*, which is essentially equivalent to our everyday "familiar" geometry. It is based on

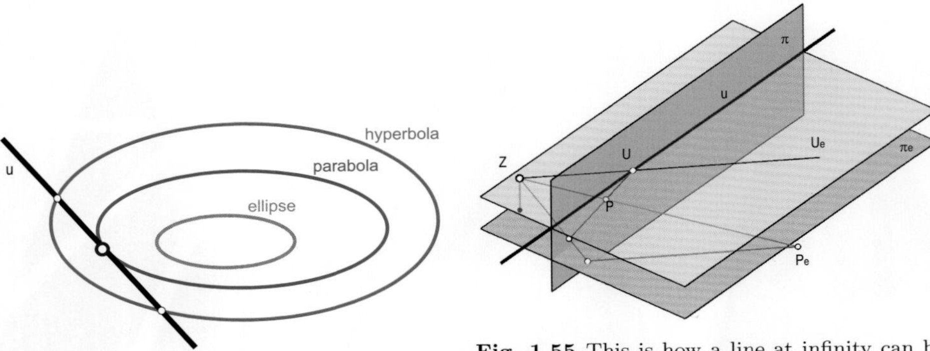

Fig. 1.54 Three types of conic sections

Fig. 1.55 This is how a line at infinity can be derived

a system of axioms, which are a collection of self-explanatory assumptions that are not further proven.
Theoretically, it is now possible to create new geometries that are based on other axioms. However, this entails the enormous restriction that the entire *system of axioms* must be completely non-contradictory. It should, therefore, be impossible to derive two theorems from the system that prove the mutually opposite.
Moreover, a system of axioms should, if possible, be *independent* and *complete* even though this is not essential.
In fact, such systems of axioms have already been found. The corresponding geometries can be seen – like the Euclidean – as some kind of playing around, with terms and axioms as rules of the game. As fascinating as this game may be: If the geometry in question stands in strong contradiction to our usual understanding, and if it does not produce any results that are worth mentioning, then it is damned to insignificance.
Euclid's system of axioms had a real "bone of contention", namely, the

Parallel postulate: In a plane there exists to each point exactly one straight line that is "parallel" to a given line which does not pass through the point.

For many centuries, nobody doubted this self-evident assertion. It has, however, been impossible to prove it by means of the other axioms, thus necessitating its status as an axiom of its own.
Gausz@Gauß, *Lobatschewskij*, and *Bolyai* realized almost at the same time (but independent of each other) that it could be replaced by contrary axioms without causing the entire system of axioms to become contradictory. The thus emerging geometries were called *non-Euclidean geometries.*
It is basically possible to differentiate between two types of non-Euclidean geometries: *hyperbolic geometry,* where there exists an infinite number of parallels to each point, and *elliptic geometry,* which was discovered somewhat later by *Riemann* and contains no parallels at all.

What's so fascinating about them?

What's fascinating about these geometries is that they play around with the concept of parallels – and thus, with the concept of infinity. "Locally speaking", "our" Euclidean geometries are barely different from their non-Euclidean cousins. Thus, any creature that inhabits its own small space and is not concerned by far-away objects would not be aware of the inherent difficulties. No wonder that modern physics, which concocts "wild theories" about the formation and the expansion of the universe, is very interested in these geometries.

We will further see that things appear "entirely normal" if we look at the issue "from a higher perspective": If we project the objects of the *elliptic plane* from a point off the plane, we get *Euclidean geometry in a star* that should be very familiar to us. If we now intersect the rays of the star with a sphere centered at the apex of the star, we are, once again, dealing with geometry in a sphere (Fig. 1.56).

If the elements of the *hyperbolic plane* are projected onto a surrounding sphere, then they can be thought of in much more traditional Euclidean terms. A hyperbolic creature can never leave the boundary of the sphere. A straight line of the hyperbolic plane corresponds to a projected circle on the sphere. It possesses two intersections with the world boundary, and thus, two "points at infinity". Through an arbitrary point on the sphere, there now exist *two different* projecting circles through these points at infinity. Every point on the straight line that lies outside the sphere is unreachable, and thus, a point at infinity. Therefore, it can be stated that all straight lines that occupy a whole sector are mutually parallel in the hyperbolic sense.

If we now dare a dimensional jump and take advantage of the conditions in three-dimensional non-Euclidean space, we have almost arrived at the ideas of *Steven Hawking*, who speaks of unimaginable cosmic phenomena that, if seen from a higher dimension, are not at all unusual.

Geometry on the sphere

Let us perform non-Euclidean geometry on the surface of a sphere. We define "line segments" as the shortest connections between points (or, great circles as geodesic lines). The equator and all circles of longitude are, thus, "straight lines"; arbitrary circles of latitude are not. By means of such straight lines, it is now possible to define objects such as triangles. It is thus possible to describe geometry on the sphere surface just as well as in the plane. It must be remembered that the sum of angles of such a spherical triangle always exceeds 180°. All Euclidean axioms are still applicable, except for the parallel postulate: On a sphere surface, two different straight lines (great circles) always share two common points and can, thus, never be "parallel".

The curvature of space

Going one dimension higher, we have reached outer space. Just as the "inhabitant of a sphere" cannot immediately detect that "his" supposedly straight

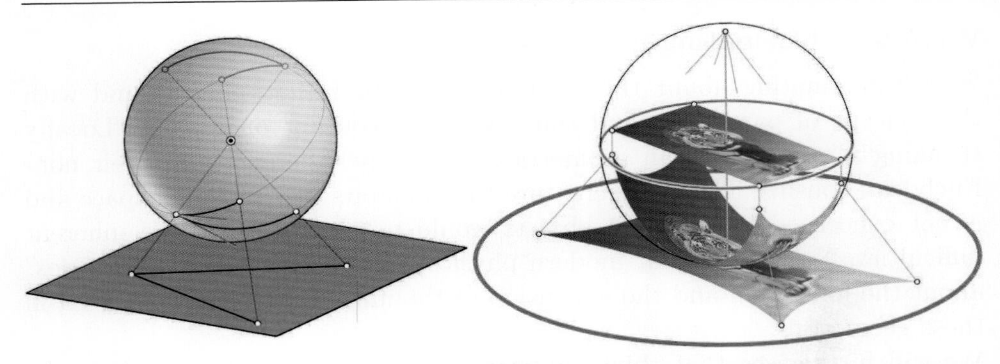

Fig. 1.56 On the left: sphere model of the elliptic plane (measurement of angles). On the right: sphere model and a planar, conformal model of the hyperbolic plane.

lines are actually curved (see also application p. 212), we do not really know the (eventual) trajectory of our straight lines (=light rays)! We know from modern physics that light is deflected by large mass (and even "swallowed" by black holes). *In the vicinity of large masses, space appears to be curved – and thus, non-Euclidean.* Since *Einstein*'s general theory of relativity, we have to look at things differently.

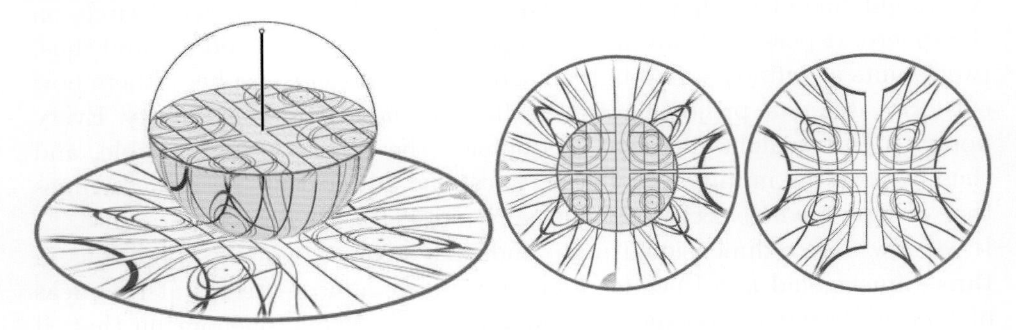

Fig. 1.57 Curved spaces

The question about the large-scale Euclidean nature of space is not yet conclusively answered, but as soon as astrophysicists surmount this mystery, they will also have answered a seemingly unrelated issue: Will our universe, which was created in a *Big Bang*, collapse in the far future in a *Big Crunch*?

Interestingly enough, the answer to these questions appears to lie in the determination of the mass of *Neutrinos*. Neutrinos are inconceivably small particles that travel through space at nearly the speed of light and effortlessly penetrate mass in the process. Thus, the question about the largest seems intimately linked to the question about the smallest ...

2 Projections and shadows: Reduction of the dimension

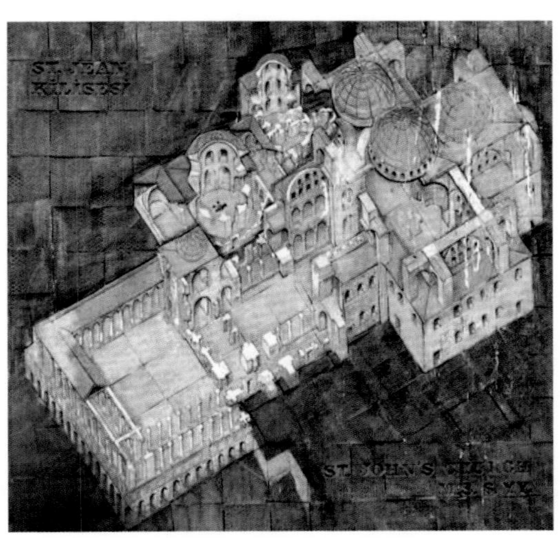

This chapter deals with projections from space to a plane.
Such a projection is the basic idea of perspective, which was first explored in the Renaissance. While this kind of projection makes things more vivid, it does not make them easier. Classical *Descriptive Geometry*, which emerged at the end of the 18th century, uses special kinds of projections.
In particular, these are the normal projections in the "principal directions".
This deliberate dimensional jump can often simplify the solution of problems. Shadows do not only make images more interesting, they also offer additional information: While a spatial situation can never be unambiguously reconstructed from a mere image, shadows make this task much easier.
One of the goals of Descriptive Geometry is to describe spatial objects by means of two "associated" normal projections. Therein it is possible to manipulate the images with well developed techniques in order to measure lengths and angles, and if necessary, to reconstruct the object in space.

Survey

2.1 The principle of the central projection

The points in space are pushed into a plane

Let us imagine an arbitrary fixed *projection plane* or *image plane* π in space and a fixed *center of projection* Z that does not lie in π ($Z \notin \pi$). A straight line $p = ZP$ through an arbitrary point P in space shall be called a *projection ray*. P must, therefore, not coincide with the center of projection ($P \neq Z$). The intersection of p with the image plane results in the projection P^c (the *image point*) of P.
It does not matter if the point is in front of the image plane, behind it, or if it is even behind the center of projection. All points that lie on the same projection ray have the same image point. How else could the entire space be squeezed into a single plane?

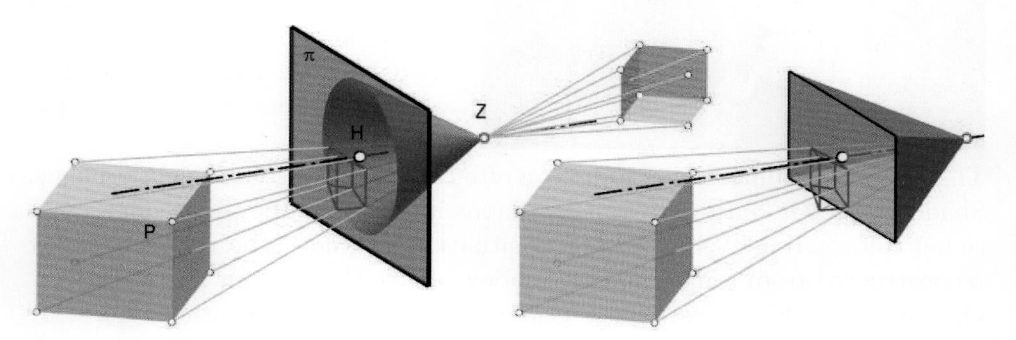

Fig. 2.1 Central projection of points with a viewing cone or a viewing pyramid

Photography is (essentially) a central projection. The eye (the optical center of the lens) lies between the object and the image plane (the light-sensitive surface).
Due to the limited viewing angle of the lens, only points within the "viewing cone" are actually pictured. Furthermore, due to the rectangular form of the negative (or the light-sensitive chip), this cone is further reduced to a frustum ("viewing pyramid").

Points at infinity are absolutely normal

If we are lucky, the projection ray p will intersect the image plane in a "proper point" $P^c = p \cap \pi$. This rule is equally valid for points at infinity. Points at infinity P_u are given by the direction of a straight line: The projection ray p is, then, simply parallel to this direction and intersects the image plane at a point P_u^c, which is no longer at infinity (Fig. 2.2 left). Points at infinity in the image plane are exceptions as they lie in π and thus do not have to be projected. Their image points are, thus, also points at infinity (Fig. 2.2 right).

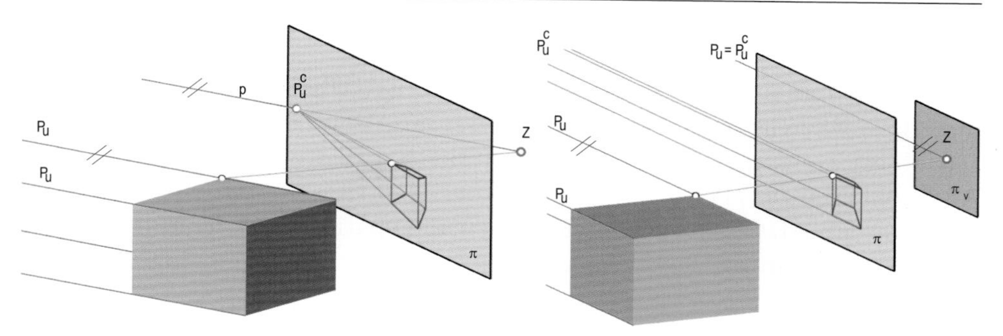

Fig. 2.2 After central projection, points at infinity become regular points (see left). An exception: The points at infinity in the image plane are identical to their projections, and thus remain fixed (see right).

Points can "disappear"

In unlucky cases, the projection ray p may lie parallel to the image plane. The image point thus resides at infinity even though the point itself does not need to be a point at infinity: It only needs to lie in the particular plane π_v that contains the center Z and lies parallel to π ($\pi_v \parallel \pi, \pi_v \ni Z$). Such a point is called the *vanishing point* and the plane π_v is called the *vanishing plane* (Fig. 2.2 right).
We thus see: All points in space (including the points at infinity) that do not lie in the vanishing plane have image points that are not points at infinity.

What happens to straight lines?

Let us now map a straight line in space, which is obviously done by mapping all of its points. You will now rightfully assume that it is only necessary to map *two* points A and B of a straight line g in order to get the connecting straight line of the image points as an image of the entire straight line: $g^c = A^cB^c$. This is easy to proof: The projection rays through all points of g form the plane Zg, which, in fact, intersects the image plane π in a straight line: $g^c = (Zg) \cap \pi$.

A linear mapping

Central projection is, thus, "line preserving". This should be no surprise as we know (from photography) that straight edges of buildings or floor ledges also appear as straight lines. Every day, we see hundreds of photographs in books and in magazines, and therefore, this observation appears "natural" to us. In a separate chapter concerning perspective, we will discuss what exactly is so "natural" about it.
In any case: Central projection is "simple" because it is *linear*, or line preserving. We have already dealt with a linear mapping, namely, the polarity with respect to a circle or a sphere, but while, in those cases, points were transformed to straight lines or planes, central projection maps points to points and straight lines to straight lines.

The mapping only works in one direction

There still remains a big difference to polarity: The latter works in both directions whereas central projection is not invertible. In other words: A spatial object cannot be "reconstructed" from *one* photograph (Fig. 2.3 left). Additional data is needed, such as whether or not the spatial object in question is a cuboid. If that is the case, then the spatial object can, indeed, be reconstructed, up to scale (Fig. 2.3 right).

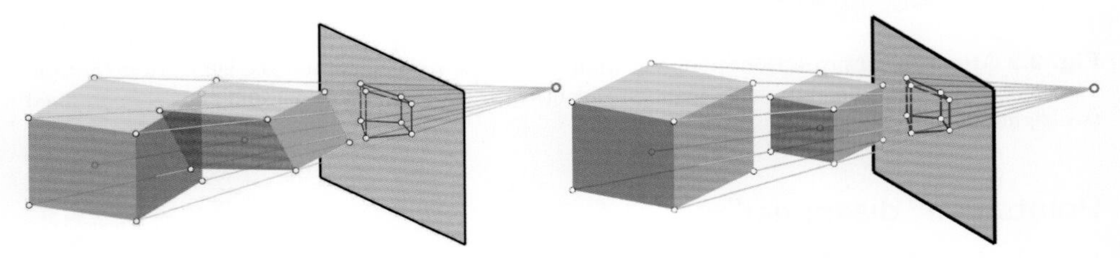

Fig. 2.3 Two pairs of objects with identical projections

"3D-scanning" has been a highly relevant problem recently, and one that has not yet been solved to complete satisfaction. Laser beams are often used to detect objects – a process that not only determines the geometric projection but also the distances from the projection center. However, this method is problematic for very large objects (such as reconstructions from the air) or microscopically small objects. For guaranteed accurate reconstruction of complicated objects, one usually needs a larger number of high-quality photographs.

From my perspective ...

Central projection (usually called *perspective*) will greatly bother us in Chapter 9. Perspective is of great importance in painting and architecture, and most people are attracted by "interesting perspectives". This may be due to the fact that perspective conveys a very subjective and personal impression. After all, it is often said that "from my perspective, things seem so and so."

Objects at infinity examined closer

Central projection can also serve as a linking tool between Euclidean and projective geometry, where different rules need to be established.
The mere fact that central projection allows us to pull points and lines at infinity into the image plane and thus within our reach is a fascinating fact, isn't it?. The usefulness of assigning only *one* point at infinity to a straight line once again becomes apparent (in central projection, it becomes a regular point on the straight line). The term line at infinity in a plane now becomes comprehensible: If one projects all points at infinity that lie in a plane ε that is not parallel to the image plane, then all projection rays lie in a plane $\overline{\varepsilon}$ through the center Z that is parallel to ε. It intersects the image plane in a "proper" straight line.

Centrally projected planes

The term *plane at infinity* is not as easily understood using central projection – although the plane at infinity can be mapped point by point, all these image points put together fill up the entire image plane. The planes that contain the center Z are exceptions: They contain an infinite number of projection rays that only have a single image point, and thus, the planes through Z appear as straight lines.

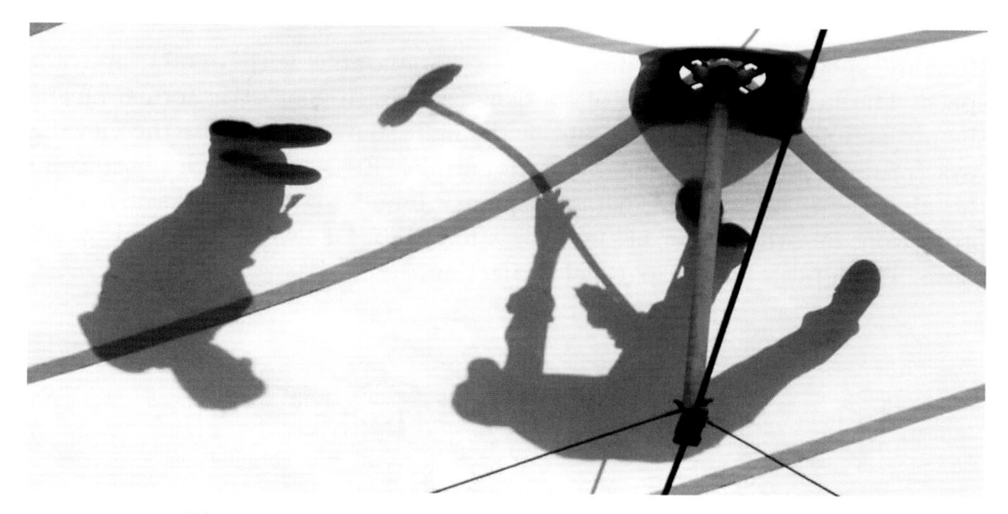

Fig. 2.4 Parallel lighting (view from below of the roof of a tent)

Central lighting

Tacitly, we have always considered an "eye point" or a center of sight instead of a general projection center. It should be noted that central projections often occur in relation to shadows. The light source needs not be infinitely far away (even though this is often assumed for sun and moon). Besides, many "finitely distant" light sources can only generously be simplified as points.

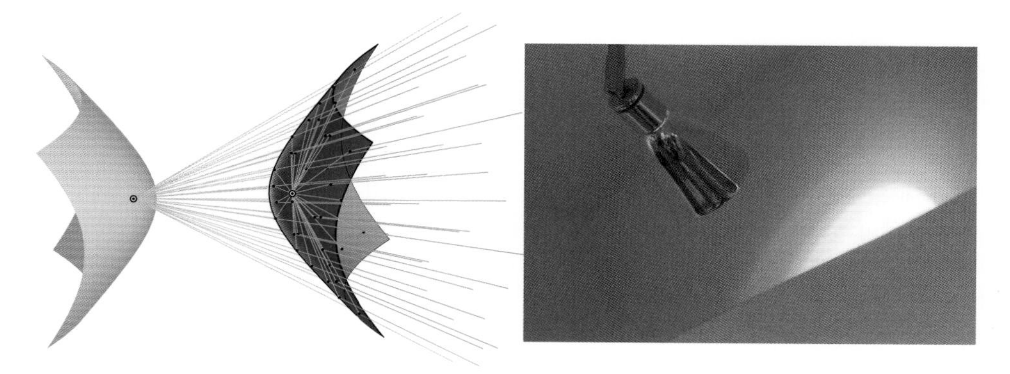

Fig. 2.5 Two-sheeted hyperboloid of revolution in theory and practice

Not even a conic spotlight provides light rays that perfectly originate in one point. Actually, here appears a so-called *two-sheeted hyperboloid of revolution* (Fig. 2.5), which we will discuss later. Still, it is true that every finite point-shaped light source produces a central projection from space to the shadow-receiving surface (which is often a shadow-receiving plane).

2.2 Through restrictions to parallel projection and normal projection

Roughly speaking, central projection is comparable with photography of space. The farther the center of the lens (which, in geometric terms, means the projection center, the eye point or the light source), the more the viewing rays (projection rays or light rays) are parallel. Fig. 2.8 shows a photo from the air where the relative distance is very large. Fig. 2.7 shows a normal projection, i.e., the center of the lens is infinitely far away, whereas Fig. 2.6 shows a general (oblique) parallel projection.

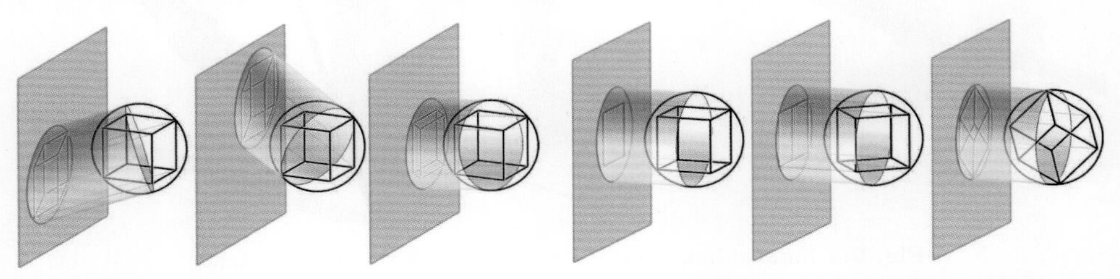

Fig. 2.6 Oblique parallel projection ... **Fig. 2.7** ... and normal projection

If the center of the projection (eye point or position of the light source) is a point at infinity, the rules of central projection still apply. However, the viewing rays or light rays are parallel, from which an important property emerges:

Parallel projection *preserves parallelity of lines*: If two straight lines in space are parallel, then their parallel projections are also parallel (unless they appear as points).

Proof:
Let $a \parallel b$ be two parallel straight lines in space. All projection rays through the points of a and b form two parallel planes α and β. These two planes are to be intersected with the projection plane π in order to find the images of a and b. The three planes α, β, and π share a point at infinity U, which remains fixed during projection as it already lies in π. The image lines $a^p = \alpha \cap \pi$ and $b^p = \beta \cap \pi$ pass through U and are, therefore, parallel. ◇

However, the converse of the theorem does *not* hold:
If two straight lines have parallel images, we may not conclude that the corresponding lines in space are parallel.

From the preservation of parallelity, we immediately infer *the preservation of affine ratios* under parallel projection:

Parallel projection *preserves affine ratios*, i.e., it preserves the ratio of lengths of segments on the same (or a parallel) straight line.

Proof:
Let g be a straight line with three points P, Q, and R on it. The *trace point* $G = g \cap \pi$ of the straight line remains fixed during (parallel) projection ($\Rightarrow \ g^p \ni G$). Both legs g and g^p lie in the plane and are now intersected by parallels (projection rays) through P, Q, and R. From the intercept theorem, we can conclude $\overline{PQ} : \overline{QR} = \overline{P^pQ^p} : \overline{Q^pR^p}$. ◇

Preservation of parallelity and preservation of affine ratios are invaluable when drawing an "oblique projection" or when reconstructing a spatial situation from an oblique projection. We consider a Cartesian coordinate system (and its origin U) to be attached to an object (such as a cuboid). The three coordinate axes x, y, and z are mutually perpendicular lines in space and concurrent in U. We fix the *unit points* on the axes.

- **A simple depiction by means of a horizontal projection**

In practice, one type of parallel projection has established itself in particular: the horizontal projection (also called horizontal axonometric projection). Therein, the image of the x-axis is chosen arbitrarily (for example, at an angle $\alpha \neq 0°$ to the horizontal, Fig. 2.9). The image of the y-axis is perpendicular to the chosen image of the x-axis, and the image of the z-axis "looks upwards". x- and y-values remain undistorted while the z-values may be arbitrarily shortened (although it is also possible to leave them undistorted).

Fig. 2.8 Normal projection ...

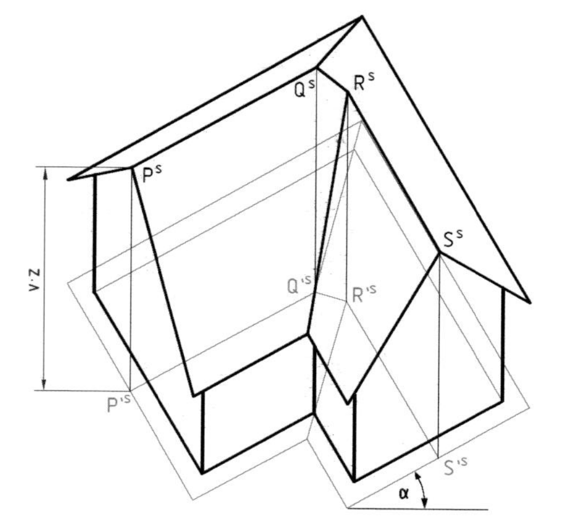

Fig. 2.9 ... and horizontal projection

Fig. 2.9 demonstrates how pictures of some descriptive quality can be "conjured up" from only a small number of lines. This procedure is especially useful

Fig. 2.10 Oblique or ...

Fig. 2.11 ... normal projection: informative

for objects that have a complicated floor plan but whose construction barely changes with height (such as prisms and cylinders of revolution). Obviously, this applies in particular to city maps:
All streets and floor plans can thus be depicted, and remain undistorted (except for the scale). It is, therefore, possible to measure lengths and angles in horizontal planes. The oblique view presents information in the same image about the relative height and appearance of the buildings. ♠

Fig. 2.12 Perspective views with horizontal axes come close to normal projections with a horizontal axis. On the left: Architectural model of Lukas Galehr. On the right: Since it can be assumed that the coral fishes are similar to each other, multiple views are combined in a single image.

Fig. 2.11 and Fig. 2.12 show (good approximations to) normal projections of a scene. These images allow conclusions that are much more difficult to deduce based on central projections. It can, for instance, immediately be said whether or not the three objects in Fig. 2.11 are positioned uniformly or if the layer of snow on all three chairs is equally thick.
Another example of the advantage of parallel projections:

- Reconstruction of a spatial situation based on shadows

How high above the ground is the cat in Fig. 2.13?

Solution:
The picture shows two near-parallel-projections (long telescopic lens) of a cat in "gallop" on a slightly inclined hill. Let us play *Sherlock Holmes* for a

Fig. 2.13 How high above the ground is the cat? The shadow provides the answer!

moment: The picture on the right almost seems like a composite photograph: Only on close inspection, we recognize a possible shadow on the right side. The left image (taken only moments later) confirms our assumption.
The sunlight, therefore, falls at a very flat angle. The scene must, therefore, have taken place only shortly after sunrise or shortly before sunset. The dew is still glittering on the grass – so it must have been morning. So then, is it possible to say how high the "flying" cat in the right image might be?
Let us be a little broad in our assumptions: The hill is obviously an inclined plane. The light comes pretty precisely from the side. One anticipates the shadow P^s of the left back paw P. We can now draw a triangle $PP'P^s$, whose side $P'P^s$ is parallel to the hill direction (PP' is perpendicular). The distance $\overline{PP'}$ indicates the distance of the jump: If the cat's pelvis height is assumed to be 30 cm (the animal appears to be relatively young), then the height of the jump would be almost half of that – perhaps about 12 cm.

Fig. 2.14 A scarab beetle is taking off from a car roof. Shadows and reflections provide useful information about the spatial situation.

Fig. 2.14 shows another situation where you can train your skills. Here, we have reflections in the base plane (a car roof) in addition to the cast shadow and therefore three projections in one image. ♠

In our considerations, we have used the preservation of parallelity of normal projection and also the preservation of affine ratios of parallel shadows. On an actual photograph that was taken with a physical lens, we would need to proceed with much greater precision. We will later return to this problem in greater detail.
The following theorem can sometimes be very useful:

Two straight lines in space are, in general, parallel if they appear parallel in two parallel projections. They are guaranteed to be parallel if they appear parallel in three "independent" parallel projections.

Proof:
If the images of two straight lines a and b are parallel under a parallel projection, then they lie on two parallel planes $\alpha_1 \parallel \beta_1$ that appear as straight lines. If the straight lines are parallel in a second parallel projection, then (in general) they also lie on two further parallel planes $\alpha_2 \parallel \beta_2$, and then, $a = \alpha_1 \cap \alpha_2 \parallel b = \beta_1 \cap \beta_2$.
However, it is possible that a and b are skew, and by complete chance, both projection directions are perpendicular to the common normal n of a and b. In this case, only a parallel projection that is not perpendicular to n is helpful. The three projection rays through a point in space must, therefore, not lie on a single plane.◇

Fig. 2.15 Dangling radar corner reflector on a ship

Fig. 2.16 "Cat's eyes": Different magnifications of a retroreflector on wheel spokes

• Radar corner reflectors and cat's eyes

A construction as in Fig. 2.15 is often found on smaller ships. It is to be shown that this reflector, composed of three mutually perpendicular squares, reflects all incoming rays (and radar rays in particular) "back to the sender". The retroreflectors on bicycles ("cat's eyes"), which reflect the headlights of cars, work in the same way.

Solution:
The two-dimensional analogue is known much better: A billiard ball is reflected by the "double cushion" and ideally returns parallel to the original

direction. The proof for this can be seen on the left of Fig. 2.17: If one takes into account the law of reflection ("angle of incidence = angle of reflection"), then the ray returns, with the respective denomination of angles, at an angle of $90° + (\varepsilon - 90°) = \varepsilon$.

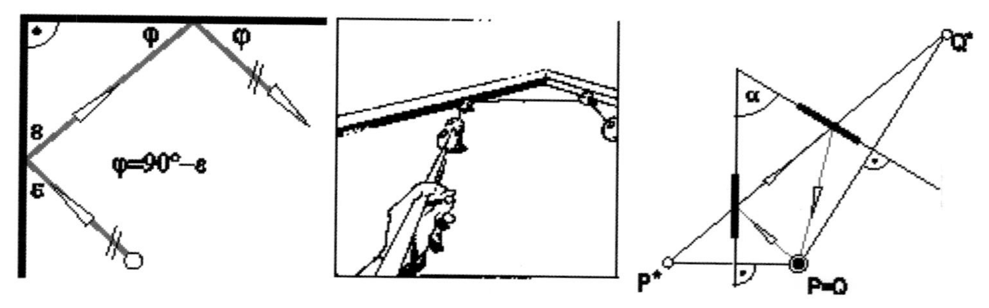

Fig. 2.17 Reflection of a billiard ball

And now the jump into space: Let us look at the situation such that two of the three reflecting planes appear as straight lines. Such a principal view produces precisely the aforementioned billiard-problem. The reflection in the projected planes corresponds to the planar double reflection, and the reflection on the third – frontal – plane is not apparent: The reflection normal appears as a point, and both the ray of incidence and the reflection ray lie on the same plane that appears as a straight line (Fig. 2.18 Center, left, and right)!

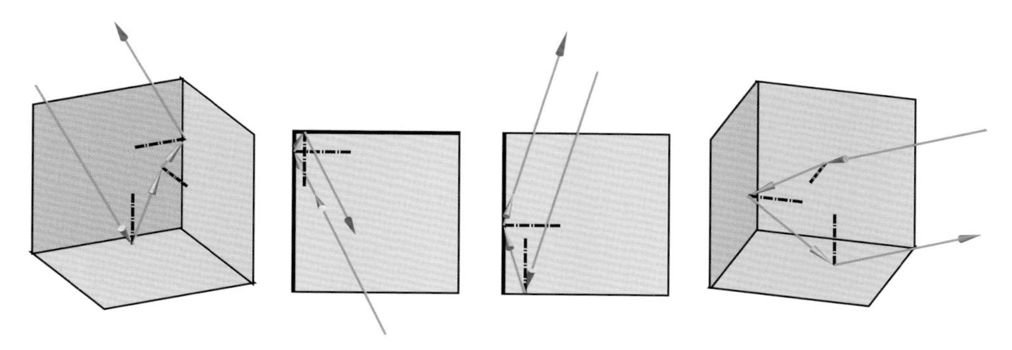

Fig. 2.18 Reflection in the corner of a cube (Center: plan view and front view)

Relative to the direction of incidence, we can, thus, see the ray as parallel in all principal views. However, if two straight lines appear parallel in three independent parallel projections, then they are also parallel in space!
Fig. 2.17 on the right resolves the following problem: How must a billiard ball be hit if it should return to the starting point P? The path of the ball (the angle of incidence) is first reflected by the first cushion. Whichever direction it is, it must pass through the mirrored starting point P^*. After having been reflected once, the ray must pass through the finishing point $Q = P$. It must also pass through the point Q^*, which was mirrored by the second cushion. This finally and unambiguously defines the once-reflected ray.

One may ask what happens if, instead of a corner reflector, we use a series of four-sided pyramids as reflectors. Fig. 2.19 on the right shows a simulation of the multiple reflection of light or sound waves. One can see that the incoming rays (marked in green) are split up into three main directions (marked in red). In fact, such "three-dimensional wallpapers" are used as sound absorbers, e.g., in exercise rooms for musicians. ♠

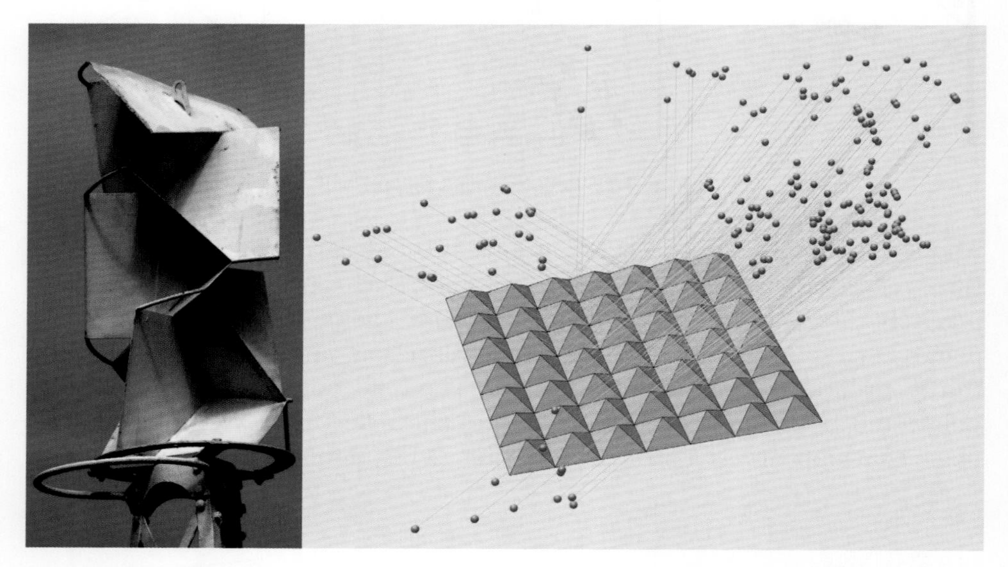

Fig. 2.19 Left: Series of corner reflector, used for navigation of ships along the shoreline. Right: A series of quadratic pyramids can be used as sound absorber.

2.3 Assigned normal projections

In order to capture an object unambiguously, we need, in general, at least two projections. Fig. 2.21 shows (almost) synchronous photographs (central projections) of a cat in mid-flight. Under ideal conditions (and given *completely* synchronous photos), it would be possible to reconstruct the positions of all visible body parts from several of such pictures.
Normal projections are much better suited for this than photographs. If, in addition, the projections are not picked arbitrarily but perpendicular to each other, it further simplifies many problems.
In most cases, we can easily attach a Cartesian coordinate system to an object (Fig. 2.20, Fig. 2.22). The normal projections to the coordinate planes thus become the *principal views*. The projection towards the z-axis becomes the

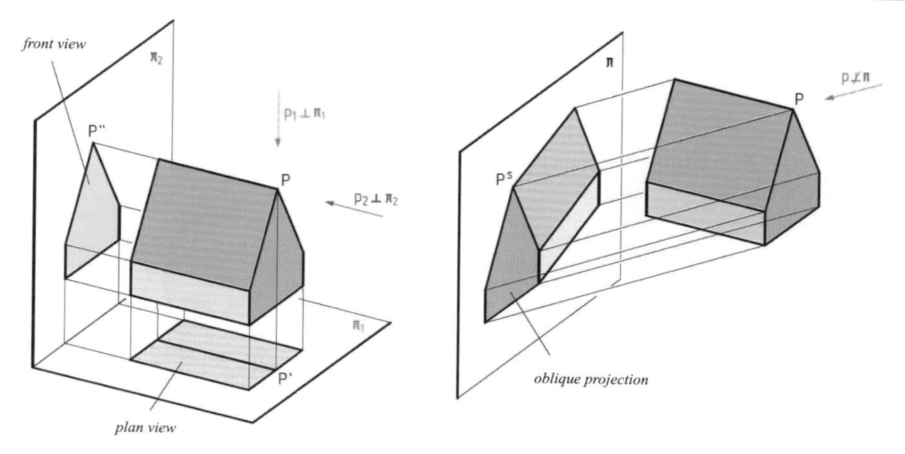

Fig. 2.20 Two projections convey much more information than just one.

Fig. 2.21 Two different (almost) simultaneous projections

plan view (or *top view*), towards the y-axis the *front view*, and towards the y-axis the *right side view* (or *profile view*).

If a point in space is defined by the coordinates $P(x/y/z)$, then its principal views have the coordinates $P'(x/y)$, $P''(y/z)$, and $P'''(x/z)$. Any two out of the three views share one coordinate, so that they can be *assigned* to each other. If the plan view is drawn underneath the front view, then there exists an *order line* from P' to P'' (Fig. 2.23). The coordinates of each point can be read from plan view and front view alone. The double y-coordinate defines the order line.

Likewise, there exists an order line between front view and right side view, and between the plan and a side view as in Fig. 2.27 left and middle. (For better understanding, a general normal projection is additionally depicted on the right side.) Both assigned normal projections of points in space always lie on order lines. The position of the coordinate origin – and thus, the position of the axis projections – is not essential for many applications. The relative

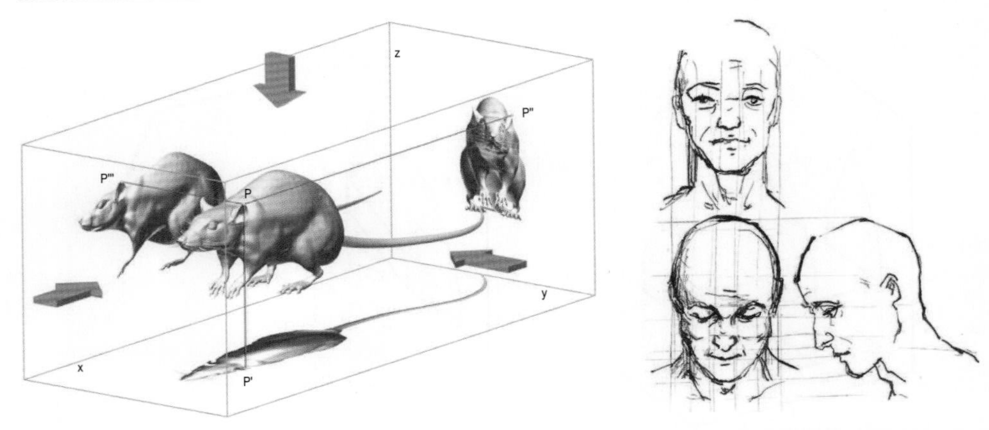

Fig. 2.22 On the left: The three principal views (normal projections on pairwise perpendicular planes). On the right: A famous study by Leonardo da Vinci, almost 300 years before Gaspard Monge (Fig. 2.23). The two views pictured underneath are assigned normal projections. The undistorted front view of the head above it is an additional assigned auxiliary view.

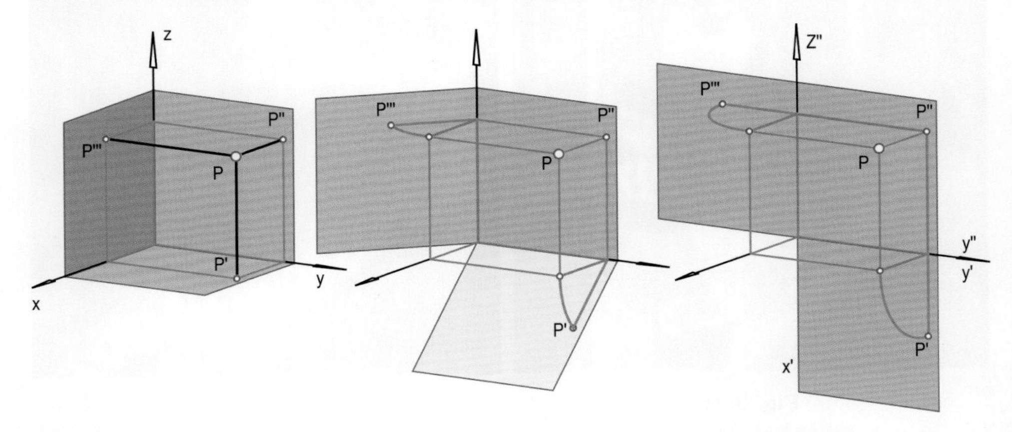

Fig. 2.23 The creation of plan view, front view, and right side view (principle of Gaspard *Monge*)

position to the *planes of reference* parallel to the coordinate planes are much more important.

- **Assignment problems?**

One often deals with two or more special views of an object that are put into arbitrary apposition. The object is much more easily imagined, however, if these views are brought into assigned position. By the way: This is an exquisite way to train one's spatial ability.

Fig. 2.24 illustrates this principle by means of two images of a snail shell, each in mutually perpendicular directions. Despite the fact that these are photographs – and not strict normal projections –, both pictures can easily be assigned (right). This is immediately helpful in the detection of assigned points.

The American norm speaks of the plan view, front view, and left side view and right side view. In contrast to Monge's model, the bottom views and back views are often

Fig. 2.24 Left, center: Two special views. Right: Assigned position.

invoked – a custom which can occasionally confuse. Computer rendering (with its capacity to provide arbitrary spatial views with "no additional effort") has, on the other hand, eroded these norms to some extent. ♠

• Which object is it?

What does a spatial object look like when its *contour* in the front view is a square, an isosceles triangle in a profile view, and a circle in the plan view? This may be quite a challenge for a beginner. Even when provided with the principal views of the wedge-shaped object in Fig. 2.25 – where, next to the contours, an additional edge and half an ellipse are visible –, it is not easy to deduce its actual appearance. The solution is a cylinder of revolution, which has been cut two times symmetrically. This procedure yields, as we will soon see (on p. 136), ellipses as edges. ♠

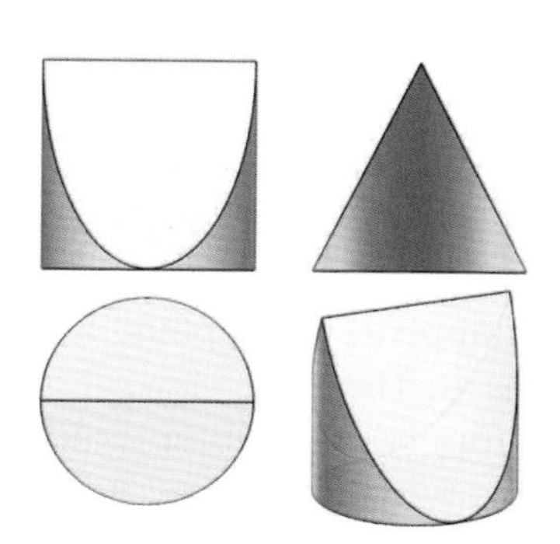

Fig. 2.25 Principal views and a general normal projection

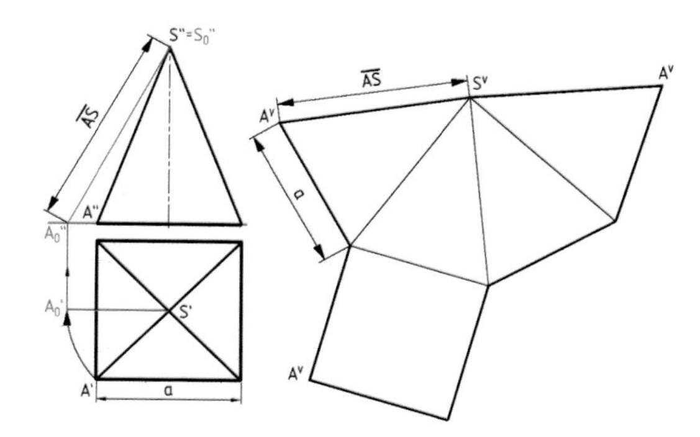

Fig. 2.26 Construction of the unfolding with the help of the principal views (Parallel rotation of AS)

• Simple construction of the unfolding (Fig. 2.26)

Given a quadratic pyramid, the *unfolding* (development) is to be constructed.

Solution:

Let us position the object in the most simple way. The base edges in the plan view appear undistorted. If we rotate an edge about the object axis into a plane parallel to the front plane (the rotation can be seen as such in the plan view), then the edge will appear undistorted in the front view. ♠

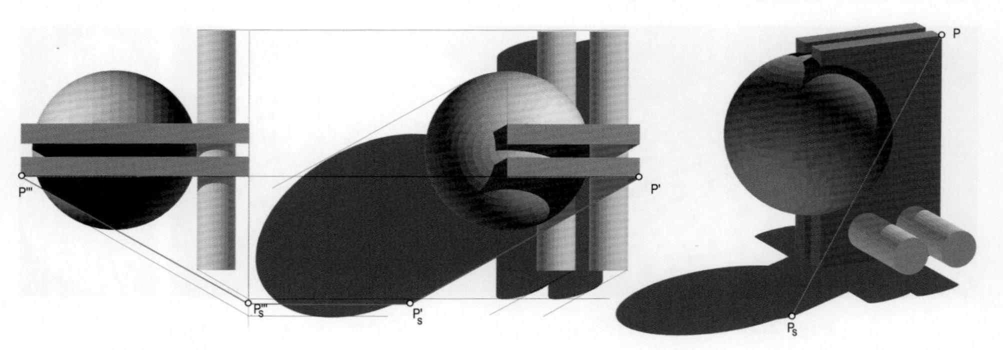

Fig. 2.27 Plan view (center) and the assigned normal projection (left) of a scene (right)

• Shadows on the coordinate-parallel screen planes

In the plan view of Fig. 2.27, the parallel shadow on the horizontal ground plane is to be determined with the help of an assigned *side view*. The direction of the light ray is given in the plan view and right side view. If this direction is offset through the assigned points P' and P''', then the image P_s''' of the intersection point P_s is already found in the right side. Its plan viewP_s' lies on the order line.

Without discussing the example in further detail (we have not yet discussed the shadow of spheres on planes or the shadow of straight lines of spheres), the bounding lines can be found immediately by looking at the light rays tangential to the sphere in the principal views. ♠

Side views are often utilized in Descriptive Geometry. The novel freehand sketch in Fig. 2.28 (found on a paper fragment) should demonstrate the principle: If one wishes to create a third view to two assigned normal views (such as the plan and front view), which should be assigned to one of the two given views (the front view, in this case), one simply picks a projecting base plane ε in the other (dropped) view, perpendicular to the direction of the order line (green). In the new view, it also appears as a point and perpendicular to the order line. On the order lines in the new view, it is possible to transcribe the distances of ε from the dropped view (blue). Side views will seldom be used to produce coherent images – rather, *special* views will need to be generated where (for instance) the planes appear as straight lines ([18]).

• Shadows provide additional information

Geometrically speaking, shadows do not differ from projections. If shadows can be detected in an image, then they provide the same kind of additional information as additional views. Fig. 2.29 shows a mathematically defined heart (the formulas were found at `www.mathematische_basteleien.de/herz.htm`). Due to the shadows on the coordinate planes (they are equivalent to the principal views), the overall shape of the object can be deduced with some accuracy. ♠

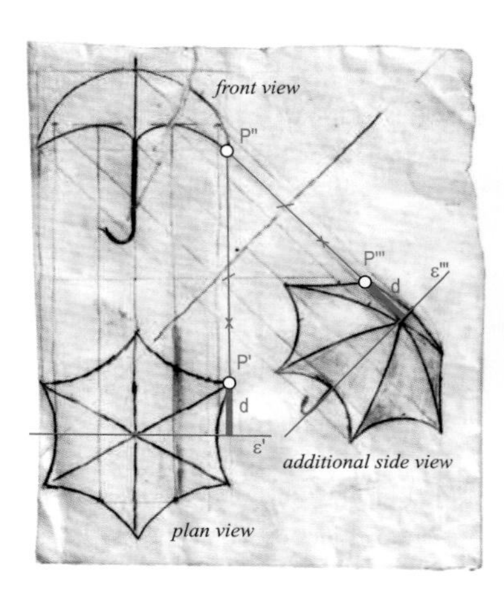

Fig. 2.28 Additional side view

Fig. 2.29 Principal projections as shadows

Fig. 2.30 Shadows provide additional information

A frequent problem is the so-called *determination of visibility*. Fig. 2.31 illustrates that the position of seeming collision points can sometimes not be judged from a photo. Yet, in some cases, the solution is provided by mere coincidence – in this case, through the shadow (Fig. 2.30). Let us ponder what is needed for such a determination in general.

- **Determination of visibility for skew straight lines**

Two straight lines have a point of intersection if they lie in a plane. If they do not lie in a plane, they are "skew".
If we imagine straight lines as materialized rods, then a straight line obscures a skew straight line at the apparent intersection point in every view, or it is itself obscured by the latter. In order to make the decision, we need *two* projections – ideally, assigned normal views.
Fig. 2.32 shows a plan view and a front view of two straight lines a and b. The intersection in the front view does not correspond to the intersection in the plan view, otherwise the points would have to lie on an order line. The lines a and b are, thus, skew. In the plan view, a point 1 on a and another point 2 on b seem to coincide. If we draw the order line through the apparent point of intersection $1' = 2'$ and look for the assigned points $1'' \in a''$ and

Fig. 2.31 Visibility is often difficult to determine. In the scene in the left image, the orca throws the "stuntman" into the air. Don't worry: No collision with the column actually took place. In the right image, it can be gathered from some details that the parachute does not become entangled.

$2'' \in b''$, we can decide which point "lies higher" – and is, thus, visible in the plan view. By analogy, the projections $3' \in a'$ and $4' \in b'$ can be determined for the overlapping points $3'' \in a'' = 4'' \in b''$. The point that lies further ahead is visible in the front view. We thus remember:

Visibility decisions in a normal projection are done in an assigned normal projection.

Fig. 2.32 Determination of visibility in theory ...

Fig. 2.33 ...and practice: The leg positions are confusing enough, but in what way do the ropes pass?

A draftsman suggests the overlapping of two straight lines by a slight interruption of the overlapped line – thus creating an "aura" around each straight line. Such subtle tricks contribute a lot towards helping spatial imagination. On the other hand, errors in such determinations of visibility have a negative effect on spatial imagination, and can even "destroy" the mental image of a three-dimensional object. ♠

The following important theorem occurs in many deliberations:

A straight line that is perpendicular to a plane is also perpendicular to each straight line in that plane (whether intersecting or not).

Fig. 2.34 Practical application of the constant right angle: The axis of rotation of the excavator scoop drawn in red is perpendicular to the desired ground plane. The excavator can now flatten the incline with the scoop by a rotation of the jib.

Proof:
Let us look at a right angle with the legs a and b (apex $S = a \cap b$). If we leave a fixed and rotate b about a, then b generates the normal plane $\nu \perp a$. Each arbitrary straight line in ν is parallel to one of the positions of b, and thus also forms a right angle with a. ◇

Fig. 2.34 illustrates the theorem: An excavator scoop, which rotates about a normal to the skew ground plane, can thus be used to flatten ground soil.

Principal lines and the Law of Right Angles

If a straight line is parallel to the image plane and thus perpendicular to the direction of the normal projection, it is called the *principal line.* Segments on the principal line are not distorted in a normal projection. For principal lines, the following theorem holds:

Law of Right Angles: A right angle appears as a right angle in a normal projection if, and only if, at least one of the legs is a principal line and no leg appears as a point.

Proof:
Let a and b be the legs once more, $S = a \cap b$ the apex of the right angle, and π the image plane. During rotation about a, b generates the normal plane $\nu \perp a$. If $a \parallel \pi$ is now parallel, then the plane ν appears as a straight line perpendicular to a in the image.
If $a \nparallel \pi$, then ν does not appear as a straight line. The normals $n \perp a$ then form different angles with π. It follows that the special normal $b \parallel \pi$ is perpendicular to every normal in the image, including a. ◇

A direct conclusion of the Law of Right Angles is the following fundamental rule of drawing that we shall discuss in the appendix about freehand drawing:

During normal projection, a circle appears as an ellipse, whose principal axis is perpendicular to the image of the circle's axis. The diameter of the circle equals the length of the principal axis.

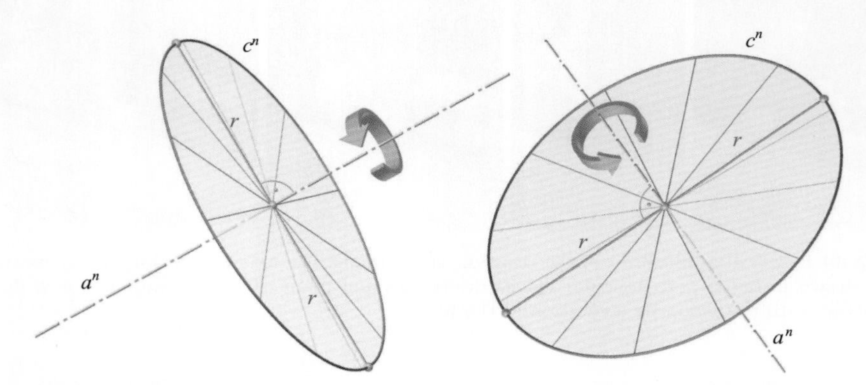

Fig. 2.35 In a normal projection the image of a circle c is an ellipse c^n. The minor axis of c^n is identical with the image a^n of the axis of the circle a. The semimajor axis equals the radius r of the circle.

Proof:
We assume beforehand that the image c^n of a circle is an ellipse (we will show this later). Among all diameters of c, only the one on the principal line appears undistorted (Fig. 2.35). Due to all other diameters appearing shortened, the main apices of the image ellipse lie on the principal line. The minor axis of the image ellipse is, therefore, the normal projection a^n of the circle axis. ◇

• **The crescent does not seem to face the sun (Fig. 2.36)**

At sunset, the sun disappears below the horizon. If we have waxing moon, it is still above the horizon. One would expect the crescent to be horizontal, as it "faces the sun". Yet it is tilted. Why is that?

Solution:
When we focus on the moon and zoom in, we have a near-orthogonal projection so that the Law of Right Angles applies. The contour of the waxing moon is composed of the contour c of the moon (a half circle) and the visible half of the "terminator" t, i.e., the circle that separates the illuminated and the dark side of the moon (see also Application 4.12).
In order to get the minor axis of the image ellipse of the terminator, the axis s of the circular terminator t has to be projected on the image plane π. The line s can also be interpreted as the sun ray through the center M of the moon. Fig. 2.36 illustrates how that can be done: Let Z be the spectator and S_∞ be the sun (almost) at infinity. The projection plane π is perpendicular to the principal projection ray MZ. The axis $s = MS_\infty$ of the terminator is intersected with π, which leads to an auxiliary point H with $H = H^n$. The projected ray $s^n = M^nH^n = M^nH$ obviously has a non-vanishing inclination

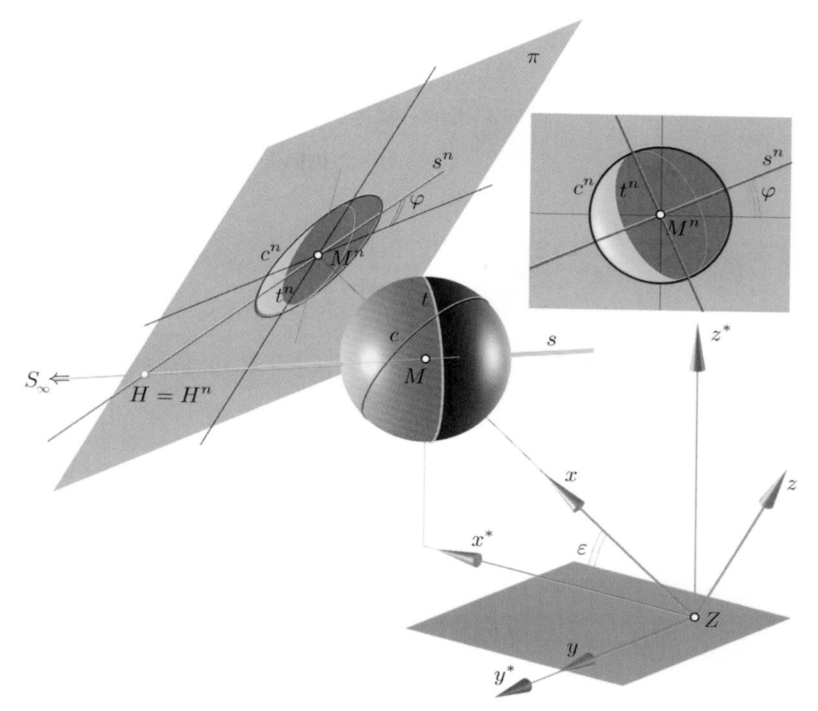

Fig. 2.36 The Law of Right Angles is responsible for the phenomenon that the crescent is tilted at sunset.

angle φ. The higher the moon stands, the larger is this alleged "error". More about the topic under `www1.uni-ak.ac.at/geom/files/moon-tilt.pdf`. ♠

• When exactly do spring and the other seasons begin (Fig. 2.37)?
The beginnings of spring and fall are characterized by the fact that day and night have the same duration (*equinox*). Summer and winter begin when the nights are longest or shortest respectively. How can we find the exact points of time?

Solution:
Due to Kepler's law, the orbit of the earth (*ecliptic*) is an ellipse with the sun in one focus. The axis direction of the earth stays constant during the year. If we consider a plan view onto the plane of the ecliptic, the sun rays that reach the earth are always principal lines.
The equinox happens when the terminator (see previous application) of the earth goes through the poles. In this moment, the earth axis and the sun rays are perpendicular. Thus, the point of time can be determined accurately by means of the Law of Right Angles: The corresponding earth positions lie on the normal to the plan view of the axis direction.
Perpendicular to this direction, we immediately get the positions where the angle σ between the sun rays and the axis direction is extreme – this leads to the longest days or nights.

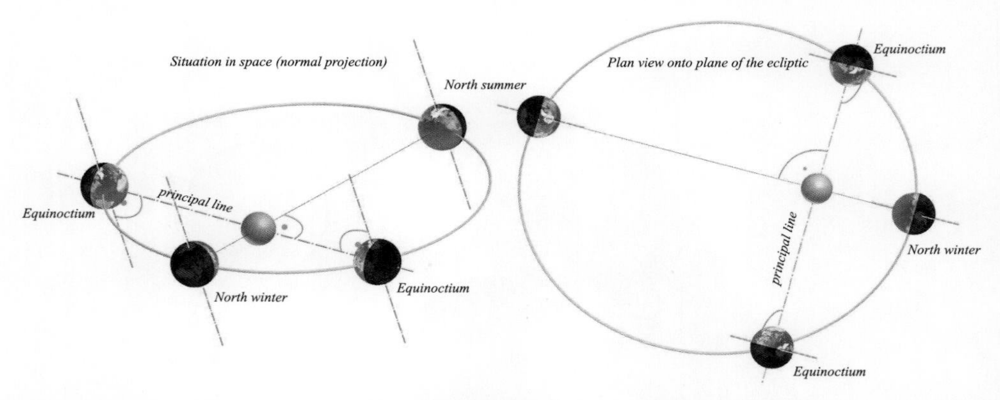

Fig. 2.37 The Law of Right Angles helps to find the beginnings of the seasons. Left: General normal projection, right: plan view. (The shape of the elliptic orbit is exaggerated to show the construction more clearly.)

To proof this, we need again the Law of Right Angles (Fig. 2.38): Let E be the position of the earth where the plan view of the axis direction runs through the sun S. σ is determined by a right-angle triangle EAC, where A is a fixed point on the axis ($\overline{EA}$ is constant) and C is its foot on s.

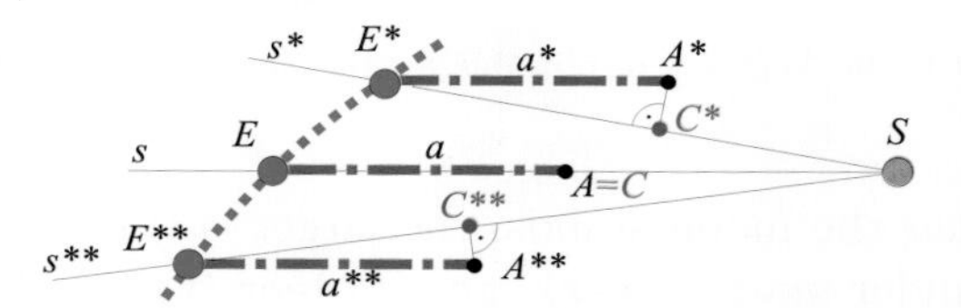

Fig. 2.38 The extreme angle between the sun rays and the earth axis

In the given position, the triangle appears as an edge. In any other neighboring position E^* or E^{**}, the foot C can be constructed as the planar foot of A, since s is a principal line. This means $\overline{AC}$ gets longer and thus σ smaller in both directions, so that σ has its maximum in E. ♠

• **Topographic projection**

In many areas of application, such as the building of roads (Fig. 5.42 on the right), a single normal projection – the plan view – is sufficient if one also works with "height markings". This procedure is called the *topographic projection*.

It allows for the solution of most common problems (including the involved complex surfaces) in disciplines such as street planning or cartography.

An "almost trivial" byproduct of topographic projection is the so-called *roof construction*, which allows for even complex buildings to be correctly roofed in a way that is geometrically correct and physically sensible. ♠

2.4 The difference about technical drawing

Classical descriptive geometry and technical drawing share many of their methods despite the fact that technical drawing utilizes many "drawing shortcuts" in addition to different "rules". This has enormous advantages during drawing, but requires certain foreknowledge from the reader, such as the awareness about the nature of a section (which we will discuss shortly). To imagine "actually" sectioned objects is to commit a fundamental mistake!
Technical drawing engineers tend to be very pragmatic. Rules are "suggestions" that should, if possible, be considered. Owing to the impossibility of adhering to all rules simultaneously, an engineer is, therefore, mindful of comments such as "if possible", "if there is enough space", etc.

Fig. 2.39 Left: A single, deliberate, additional view is optimal. Right: The cutlery (Doris Schamp) complicates the matter due to the presence of free-form surfaces.

• Special additional views

In technical drawing, one also speaks of plan views, front views, left and right views. Usually, as many views are drawn as are necessary.
"Devices" (the occasional name for such objects) can be viewed in any number of special and general views. Such images aid the spatial imagination of the beholder, but still, it is rare that these devices can actually be built from the images alone (Fig. 2.39 right).
Fig. 2.39 on the left shows a device pictured in a typical technical drawing: The front view and an additional assigned view are enough to determine the shape and describe all necessary lengths and angles undistorted.

♠

• Virtual sections

Technical drawing often confronts us with the problem of displaying objects that, even if given many projections, are still very difficult to picture and to dimension. Object *sections* were invented for this purpose – an *imaginary* decomposition along a section plane that is usually perpendicular to a body axis or contains the axis. Whatever is contained in the section plane is usually crosshatched (excepting longitudinal sections of braces, ribs, screws, nuts, flat washers, etc.). Whatever lies in front of the section plane is omitted – whatever lies behind is drawn.

Fig. 2.40 on the left shows a section through an elbow. In this case, a simple section (AB) is sufficient to depict the location of drilling holes and screws. The arrows refer to the viewing direction.

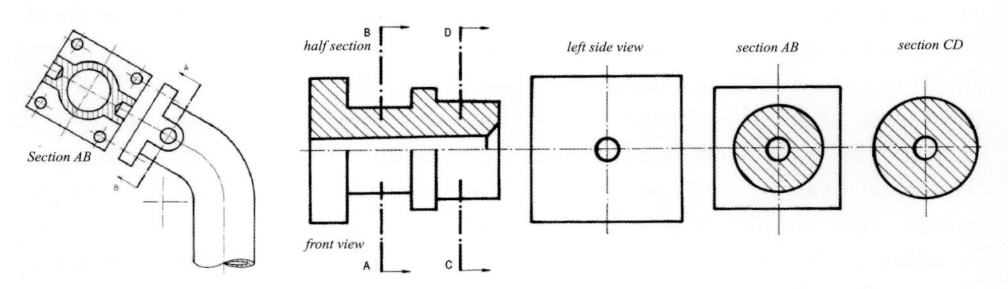

Fig. 2.40 Left: One simple section, right: several views and virtual sections

The drawing in Fig. 2.40 right shall illustrate the necessity of more than one section even when dealing with supposedly simple objects. The following rule holds: *With multiple sections, every section is done without reference to the other sections.* On the left is a front view of the object (a nozzle). Its symmetrical structure allows us to achieve two goals at the same time: One half of it – right up until the semi-colon axis of symmetry – is depicted as a view (whether it is the top or the bottom half depends on the norm involved). Since the other half is symmetric, it can be pictured as a so-called *half-section* in order to illustrate the – not all too complicated – interior of the nozzle. Here, the important rule must be noted that the section axis must not be drawn as a continuous line: The section is only virtual!
If all parts of the nozzle were cylindrical, we would already be finished. This is due to a common assumption in technical drawing that all objects are automatically cylindrical unless otherwise noted. A view from the left (usually drawn from the right side of the front view) shows, however, that at least the head of the nozzle is square. Whatever lies behind it may still be cylindrical or square. In the example, the part between A and B is cylindrical whereas the thickening to the right of it is square. This can be shown properly in a section from A to B ("full section"). Finally, a full section from C to D should indicate that the rest of the nozzle is, once again, cylindrical. Given enough space, overlapping sections are drawn "in sequence". ♠

• **Dimensioning**

Devices should primarily be dimensioned in places where it is expected (where the measurements are not shortened) and secondarily in a way that can be measured on the physical workpiece. For this reason, the diameters of circles are usually dimensioned instead of the radii – if necessary with a "measurement across space" (Fig. 2.41 on the left). The diameter sign Ø is added if the circle appears as an edge.
In order to dimension invisible drilling holes (and similar features), one works with full sections or half-sections. It is often enough to "rip" the object apart locally – in other words, to perform a *partial section.* If the object is too long to be displayed in full, it can be halted by a wavy line or an eight-loop.

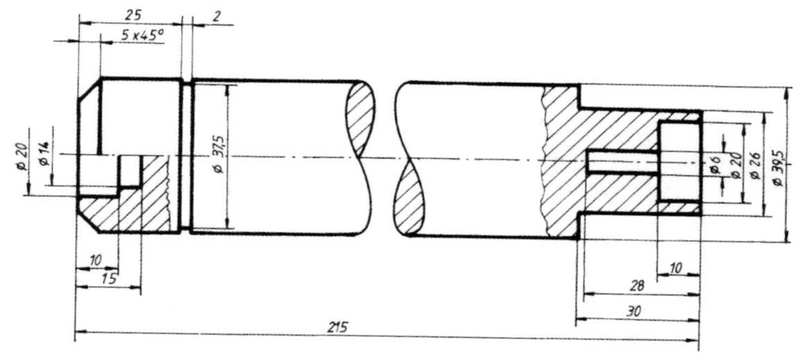

Fig. 2.41 Dimensioning of a workpiece

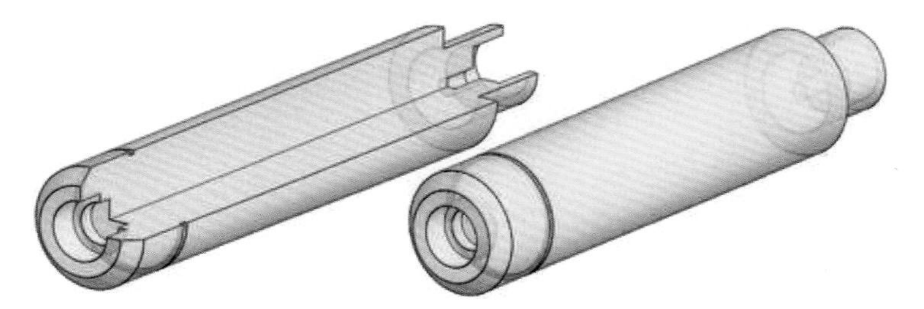

Fig. 2.42 The adjacently described workpiece, on the left with *real* sections (wrong!)

External dimensions are always important. Closed loops of dimensions are to be avoided. If possible, intersecting dimension lines are to be avoided. As simple as the workpiece in Fig. 2.41 may be, it is relatively difficult to create a comprehensible sketch which displays all details without ambiguity. The computer allows for the use of transparency in the drawing. If a quarter of the object is taken out, it might raise the clarity of the drawing, but can also be misunderstood! ♠

Sections in construction drawings

Fig. 2.43 Perspective view of the object in Fig. 2.44, placed within its planned future environment through compositing.

Construction drawings may differ considerably from technical drawings. Even the "norm" is differrent (1 cm instead of 1 mm). Furthermore, lines are used instead of dimensioning arrows. As with technical drawings, virtual sections are essential.

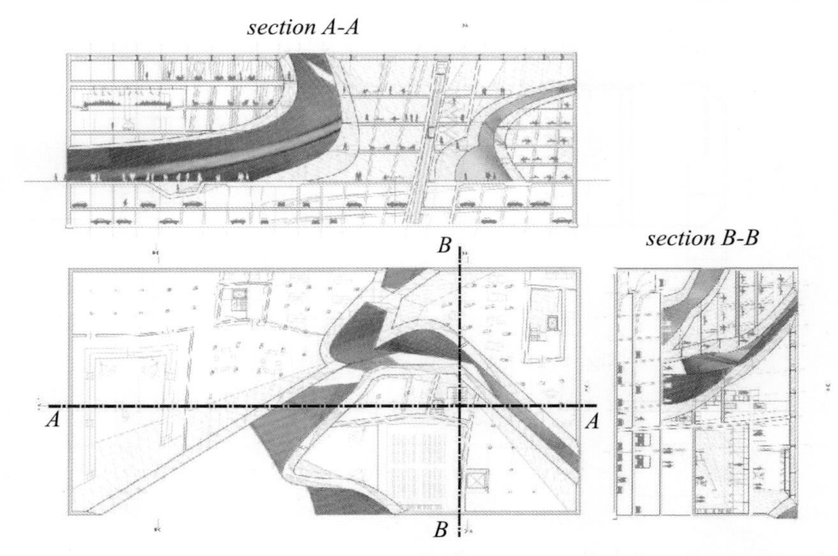

Fig. 2.44 Building with one horizontal and two vertical sections.

Fig. 2.44 shows a considerably complicated building (Cornelia Faißt) with three assigned sections. Without going to much into detail: Nobody can pretend to be able to imagine the object based on the three sections. At least *one* general view (Fig. 2.43) is required in order to get an adequate impression. Surprisingly, some architects have developed the remarkable capability of gaining deep insights into virtually incomprehensible scenes by a series of sections. Still, it is important for every architect to build physical models (Fig. 2.45).

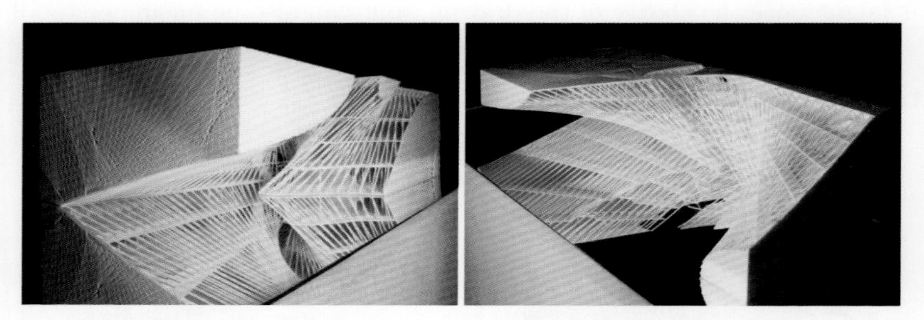

Fig. 2.45 Despite all computer simulations: The model remains very important for comprehension. The character of the outer layer is particularly apparent – it holds *two different pencils* of straight lines! The mirrored floor contributes additional effects to the presentation, but does not exactly improve one's capacity to imagine the object.

Anticipating Chapter 6, let it be said that the outer wall of the building is formed by so-called *HP-shells* (hyperbolic paraboloids), which hold two different pencils of straight lines (Fig. 6.18, 7.8). The planar sections in Fig. 2.44 of such surfaces are – as the name implies – parabolas and hyperbolas.

3 Polyhedra: Multiple faced and multi-sided

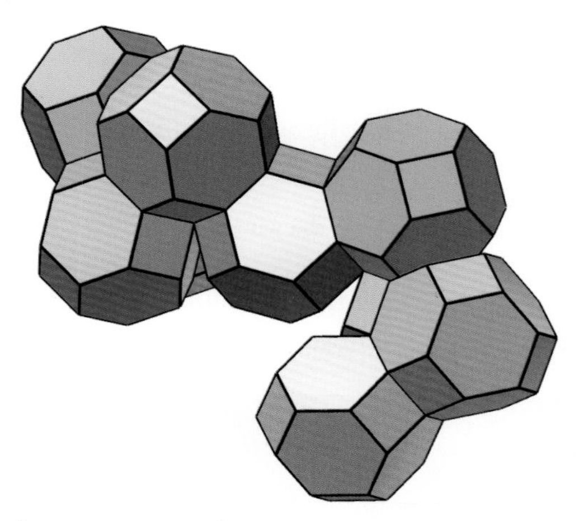

Planes are non-curved spatial surfaces. Surfaces which are composed entirely of planar polygons are called polyhedra.
Polyhedra have been (and still remain) keys to many geometric insights. In recent times, they have also become essential in dealing with computational representations of curved surfaces. These are "triangulated" or subdivided into triangles.
In any case where computers deal with arbitrary surfaces (no matter how complicated they are), an approximating polygonal surface is considered which comes the closer to the true shape of the surface, the finer the triangulation is.
As is often the case in geometry, the special cases play an important role. For example, there are a few polyhedra that are bounded by congruent polygons – most of them are the so-called *Platonic solids.* If a polyhedron contains two types of polygons, we get a further class of polyhedra, called the *Archimedean solids.* Both types of solids appear frequently in nature, and in microcosm in particular. What was investigated by the ancient Greeks more than two thousand years ago now appears under our microscopes.

Survey

3.1 Congruence transformations

Before dealing with polyhedra, we should discuss the important topic of congruence transformation. It is defined as a spatial transformation where angles and lengths (of the moving object) are preserved (Fig. 3.1).

Fig. 3.1 A congruence transformation, except for the wings

Among the congruence transformations, there are translations and rotations. Both are directly congruent, i.e., they preserve orientations.

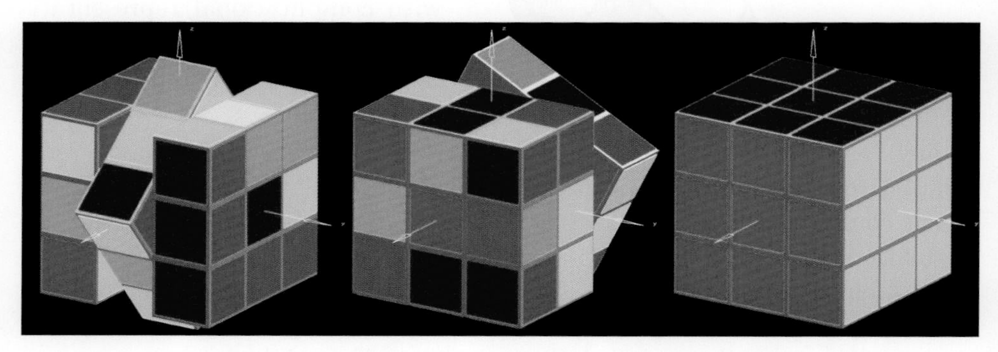

Fig. 3.2 Rotations of the magic cube according to *Rubik*

The reflections in a plane, on the other hand, are reversely congruent, i.e., they alter orientations. Fig. 3.3 shows an everyday reflection in an upright mirror.

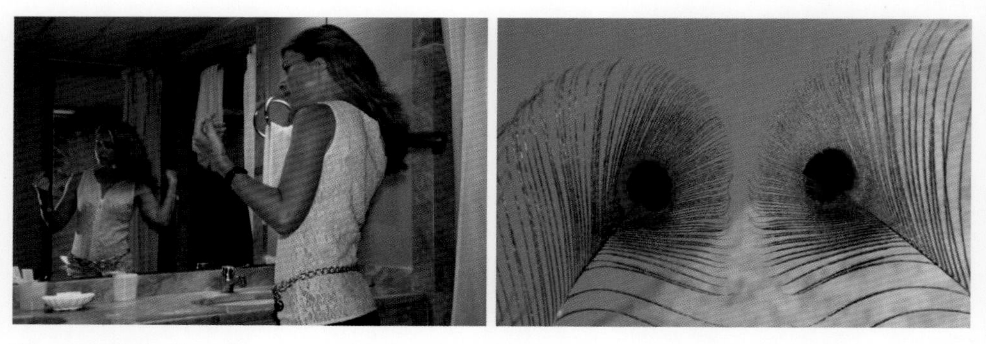

Fig. 3.3 Reflections in a perpendicular plane. The peacock's feather on the right touches the non-reflecting glass surface at the front, as indicated by the pushed in fibers.

Such a reflection does not alter left and right, as it is commonly believed, but front and back, which is especially evident in Fig. 3.3 on the right. From a

geometric point of view, this statement can be further generalized: A mirror inverts the sign of the distance of a point from the mirror plane.
Unlike reflections in a plane, reflections *in a point* are reversely congruent! In three-dimensional space, orientation can only be defined by orienting a *tripod* (e.g., three concurrent edges of a cube).

Fig. 3.4 Three (non-uniform) rotations about intersecting and parallel axes

Fig. 3.4 shows three independent rotations about intersecting and parallel axes. They are non-uniform (they are harmonic oscillations), which does not matter from a geometric point of view. By the way: The outermost right axis almost appears as a point.

Fig. 3.5 Rotation about an arbitrary axis

If one combines two rotations about intersecting axes, one gets a rotation about a third axis that runs through the intersection of the two given axes.

On the other hand, it is possible to split an arbitrary spatial rotation into several rotations about certain axes. This result is due to *Euler* and is extremely important in analytic geometry and kinematics.
Fig. 3.5 shows a rotation about an arbitrary straight line. More specifically, it is the line about which – given this specific type of bicycle – the front wheel can be rotated (let us ignore the fact that this axis slightly varies in the real world during the motion). The position of the sun in this scene is such that the shadow of the axis of rotation almost appears as a point. If one views the rotation in the direction of the axis of rotation (in the special

case, from the position of the sun), then the motion appears as a "wholly ordinary" two-dimensional rotation.
Fig. 3.2 shows repeated rotations about intersecting axes, not of the entire "magic cube", but of specific layers. In such cases, the theorem of "replacing multiple rotations by a single one" obviously does not apply, not to mention that it is quite a challenge to reverse all of these rotational steps. A computer simulation has an unfair advantage: The program simply records all arbitrary rotations taken by the user and, when the magic cube is to be "solved", simply runs the steps in reverse (with negative rotation angles).

Fig. 3.6 Left: Reflection in a plane? Right: Reflection in several planes, but not quite easy to visualize (the jellyfish are located *behind* the reflecting planes).

Fig. 3.6, at first, gives the impression of a reflection. Upon closer inspection, one notices the presence of two separate but essentially *directly congruent* animals. They can be made to overlap using a simple rotation of 180°. The axis of rotation is roughly equivalent to the midline of both parallel body axes.

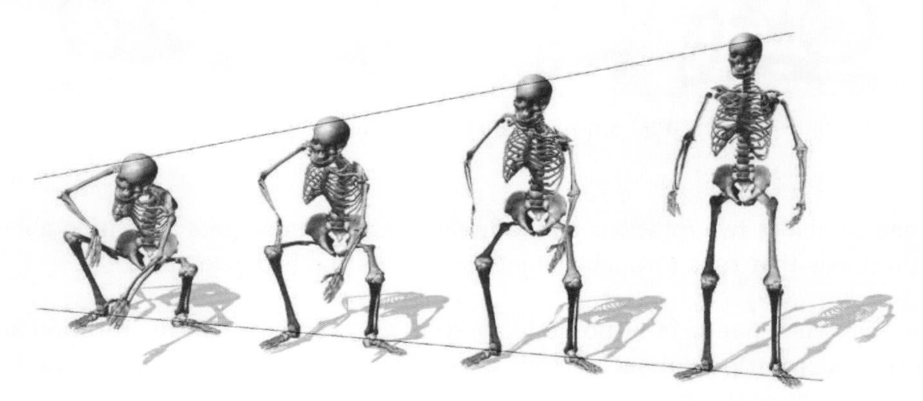

Fig. 3.7 This transformation is strictly mathematical

Fig. 3.7 shows the transformation of a skeleton from a crouching position into a stretched position. In practice, a "composite" of many rotations and translations is required. According to an idea by Hannes Kaufmann, all transitional positions can be efficiently approximated by interpreting them as linear combinations of the starting position and the final position.

Fig. 3.8 Only partly congruent transformations

Computer graphics is immensely interested in transformations that can be used to animate living creatures in a most realistic way. (the Ibises in Fig. 3.8 are real, however). Professional programs employ such techniques particularly well, and in order to understand them, some considerable knowledge about *kinematics* is required. The chapters 10 and 11 of this book are dedicated to this topic.

Fig. 3.9 Such figures result from ...

Fig. 3.10 ...reflections in a trihedron.

3.2 Convex polyhedra

Loosely speaking, a closed polyhedron is convex if it has no dents. If we connect two arbitrary vertices by a line (chord), then this segment must lie either entirely inside the polyhedron or in one of its faces.

Such solids play an important part in our world. They have geometric properties that make them particularly "simple".

Prisms and pyramids

A *prism* is formed by parallel translation of a polygon (which does not necessarily have to be planar). We can also say: One "connects" a *generating polygon* with a point at infinity. The second definition leads to a *pyramid* as a connection of a generating polygon with an arbitrary point. However, this point must not lie in the carrier plane of the generating polygon (if it is a planar polygon) and must not be a point at infinity.

Fig. 3.11 Still huge, even from far away

Seen this way, prisms are simply special cases of pyramids: The *apex* (*tip*) of the pyramid is, then, a point at infinity.
By analogy, a planar polygon is *convex* if every possible chord does not lie outside the boundary of the polygon. The following theorem should be immediately apparent:

Connecting a planar convex polygon with a fixed point (or a point at infinity) that does not lie in the carrier plane results in a convex pyramid (or a convex prism).

Fig. 3.12 From a multistage Mastaba to perfection

The ancient Egyptians have left us the most beautiful examples for regular pyramids. The construction of these colossal objects must have been as difficult as their appearance is obvious. Given a quadratic base with a side length of over two hundred meters, even the leveling of the foundation (accomplished to centimeter accuracy) is a true master work. In principle, step pyramids were built first (Fig. 3.12 on the left), after which their surface was flattened (the peak of the Great Pyramid of Giza is still very well preserved). 4500 years (!) ago, immediately after finishing construction, the appearance must have been even more impressive than it is today.

A series of simple properties

As indicated, convex polyhedra possess certain properties that can be very useful for certain geometric considerations. E.g., the following theorem is true:

> The shadow of a convex polyhedron on an arbitrary plane (as well as its contours) are convex polygons.

The proof uses the "chord-definition" of convexity and is straightforward: If we imagine such an object as being filled by tense "fibers", then these also cast shadows that cannot lie outside the shadow contours of the object.

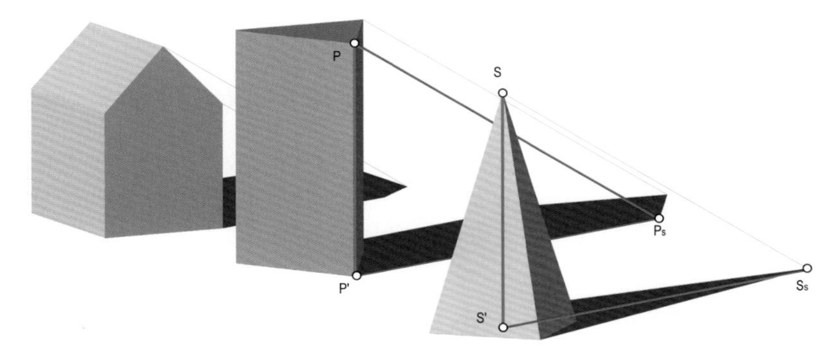

Fig. 3.13 Shadows of convex polyhedra

• Shadows of prisms and pyramids

Fig. 3.13 shows the shadows of three simple and convex objects. Shadows of prisms are often bounded by shadows of the generators of the prisms – the shadows of edges of the base and top surfaces form the rest. If the prism stands on the plane that is receiving shadow, then the edges in question are easily found: One only needs to project the light source in the direction of the edges to the shadow-receiving plane and, once there, draw the "tangents" towards the base polygon. Architecture mostly deals with straight prisms and parallel lighting (from the sun). In this case, the prism edges in question can be found by moving the plan view direction of the light rays towards the base polygon.
For pyramids, we can proceed in a similar way: In this case, all edges run through a point – this is also true for their shadows. Therefore, we get the shadow of the pyramid apex and, from this point, place the "touching straight lines" towards the base polygon. They delineate the shadow polygon. ♠

A further practical property of convex bodies that also follows from their definition is as follows:

During projection, every face of a convex body is either completely visible, completely hidden, or it appears as an edge.

By analogy – if one considers lighting instead of viewing:

Convex objects do not throw shadows on themselves. Their faces are either completely lit, completely shaded or the light rays through their edges touch along the whole face.

Computer geometry takes advantage of this fact: A normal vector $\vec{n}$ of the face of a convex body can always be oriented such that it points towards the outside. If $\vec{n}$ forms an acute angle with the projection ray (or light ray) oriented towards the projection center (or light center) $\vec{p}$, then the entire face is visible (or illuminated). It is merely necessary to determine the sign of the dot product $\vec{n} \cdot \vec{p}$. This also holds true with central projections (or central illuminations): In this case, the projection rays (or light rays) usually differ for the points of the face, but the decision whether or not the angle is acute is not affected by this.
We can derive the following rule from the aforementioned statements:

Rule of Visibility: In the projection (image) of a convex body, only fully visible or fully invisible edges go through a vertex that lies completely within the body contours. The contour polygon is fully visible.

Combination of several convex bodies

Practical situations are rarely limited to exactly one convex body. The question is: Which of the previously stated practical rules can be applied to scenes that consist of many such building blocks?
Let us discuss the procedure using a certain example. Fig. 3.16 shows a non-convex body – upon closer inspection. However, one notices that it is "harmlessly non-convex": It can easily be split into three convex building blocks: two regular eight-sided prisms and one cuboid extension.
While visibility determination for such a body is "trivial" for our imagination, the drawing of the shadows is much less so and requires a disciplined approach.

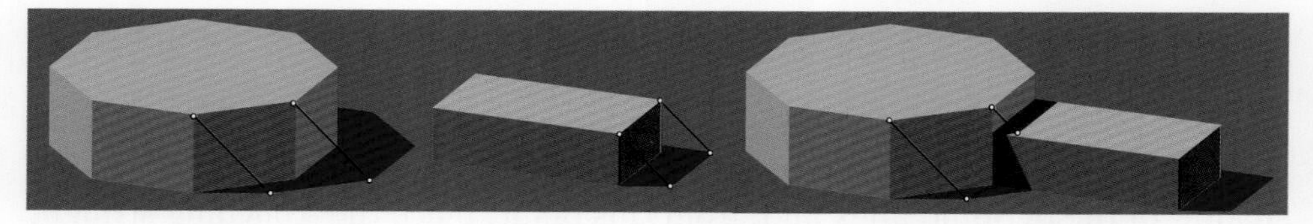

Fig. 3.14 Systematic procedure: Consider the bodies one-by-one ...

In a first step, all three partial convex bodies are considered one-by-one according to the aforementioned rules. In particular, the shadows on the base plane can be determined separately. The entire shadow on the base plane emerges from the "combination" of the separate convex shadow polygons. The thus emerging shadow polygon can easily be non-convex.

A hint for free-hand drawers (for us, free-hand drawing is an important goal to which a whole appendix is dedicated): The shadow on a surface "no longer knows to which partial body it belongs". All shadows on a surface adapt the texture of the surface on which they lie. One should also not try to highlight the separate convex shadow polygons. Shadows tend to have a "surface-like appearance". Not even the boundaries of the entire shadow polygon should be particularly stressed: Shadow boundaries are usually not sharply delimited – not even on photos. This is due to the fact that there are no true point light sources in practice. Even the sun on the firmament appears "disc-like" and is to be considered as a light cone with an aperture of half a degree.

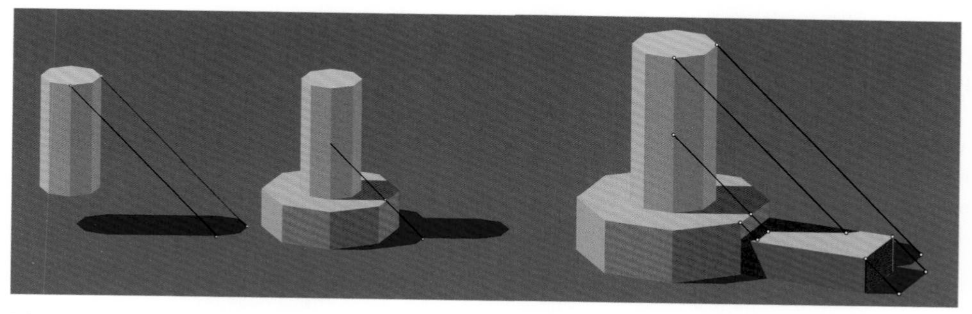

Fig. 3.15 ...and bring them into mutual relation.

Fig. 3.16 Normal view of three convex polyhedra, including the shadow

By subdividing a scene into convex parts, we can determine all surfaces that are entirely shaded ("dark"). The following important rule holds true for them:

Faces that are entirely dark cannot, in addition, receive additional cast shadows.

This simplifies our work, which now concerns the determination of the shadows of the subdivided bodies on each other. We now have to consider:

Two convex bodies never cast shadows on each other mutually. If there exist shadows in between them at all, then only one body casts shadows on the other.

This approach is a huge time-saver and can also prevent many possible errors. One can now derive a kind of *priority list* that, in most cases, is easily sortable: Disregarding certain convoluted exceptions, there is always one "frontmost" building block on which no shadows are cast by other bodies. This is followed by the building blocks on which only the first body casts shadows, and so on. Even in the rare cases where the sorting of such a list is, at first, not possible, the challenge can be surmounted by dividing single (often very long) convex blocks into two separate ones.

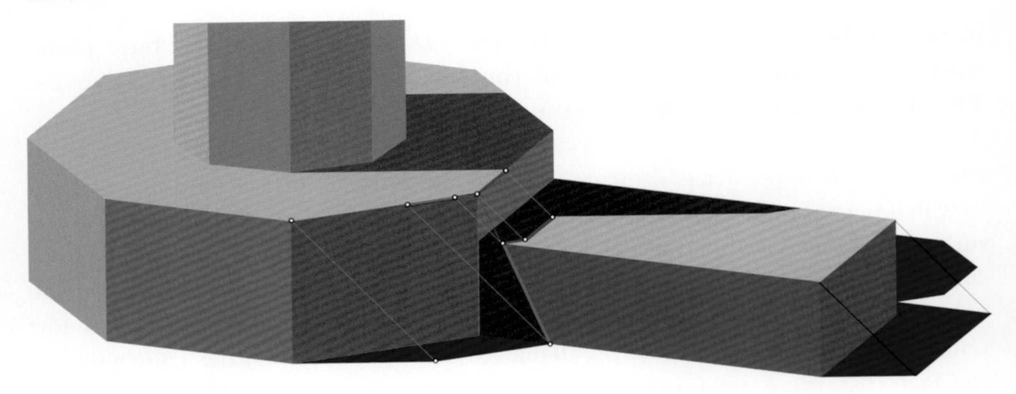

Fig. 3.17 The detail shows the inflection points of the shadow boundary.

• **Acceleration of computer programs**
If one considers all advantages that can be gained by subdividing a scene into convex partial bodies, then one can get fairly realistic images of complex scenes such as Fig. 3.18. This can prove to be important: For example, if one has to develop a "chess program", which should primarily play in an intelligent manner, but also "in addition" give the player a realistic image of the situation on the chess board, it would be a pity to waste much processing power on the highly realistic "rendering" of that image. A "real-time" visualization (at least thirty frames per second) is desirable and, in this case, easily achievable.

Fig. 3.18 Scenes composed of ... **Fig. 3.19** ... convex polyhedra

In [7], I have described many computer algorithms that make such fast rendering of realistic images possible. Despite the age of the algorithms – especially by the standards of computer graphics – they still allow computer graphics fast as lightning with correctly calculated shadows. Even the scene in Fig. 3.19 (depicting the Commodore Amiga, a groundbreaking minicomputer in its day, but completely antiquated in our time) can be animated in real-time. The algorithms are a typical example of the fruitful application of theoretical geometric knowledge in other disciplines. ♠

Fig. 3.20 on the left shows complex, "rounded" polyhedra. The shadow boundaries of this body (on which all shadow polygons are based) are still comparatively simple.

Fig. 3.20 Interpretation as full and hollow bodies

In Fig. 3.20 on the right side, the biggest polygons were removed from the polyhedron. This, of course, means that the polyhedron is no longer convex. On the other hand, one may still derive some advantages from the fact that the "hull" of the polyhedron is convex: Thus, either *all* faces potentially throw a shadow on the base plane, or *none* of them do. Furthermore, all front faces of the polyhedron are either completely illuminated or completely dark. Equipped with this knowledge, the rendering speed for the scene can be significantly increased.

Multiple light sources

Dealing with artificial light but also with regular sunlight and reflecting surfaces leads to penumbra shadows and umbra shadows: Areas in penumbra shadow are illuminated by at least one light source each, or none at all. Areas in umbra shadows are not illuminated by any light source.

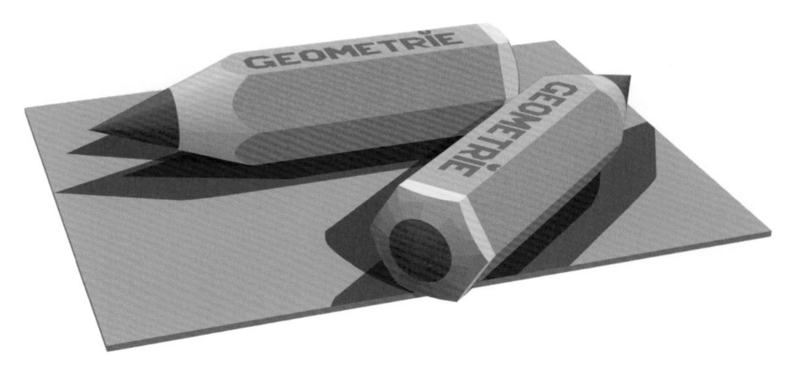

Fig. 3.21 Penumbra shadows and umbra shadows of a convex object

The result may look complicated, but geometrically speaking, it can easily be accomplished by treating each light source "separately". For convex bodies, one gets two or more convex shadow polygons. The umbra shadow lies where all polygons overlap. The other areas are more or less illuminated, depending on how many light sources are visible from their positions.

- **Shadow profiles**

One can analyze shadow casting of buildings during the day by calculating the shadows of the scene at regular intervals (e.g., each hour).

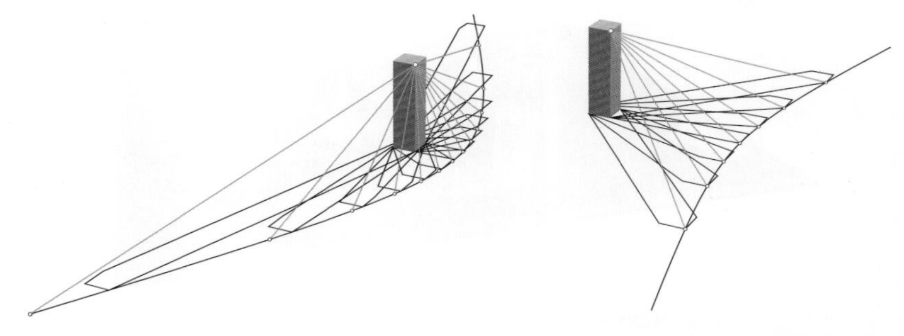

Fig. 3.22 Shadow profiles of a skyscraper (summer and winter half-year)

Fig. 3.22 shows how it can then be determined for how many hours single points remain in shadow on any given day. Such questions are highly relevant during constructional changes in dense urban areas.
By means of modern computer hardware, shadow profiles can be calculated within seconds. This is enormously useful for architects, who are by law obliged to make sure that their buildings do not cast too much shadow on their surroundings.

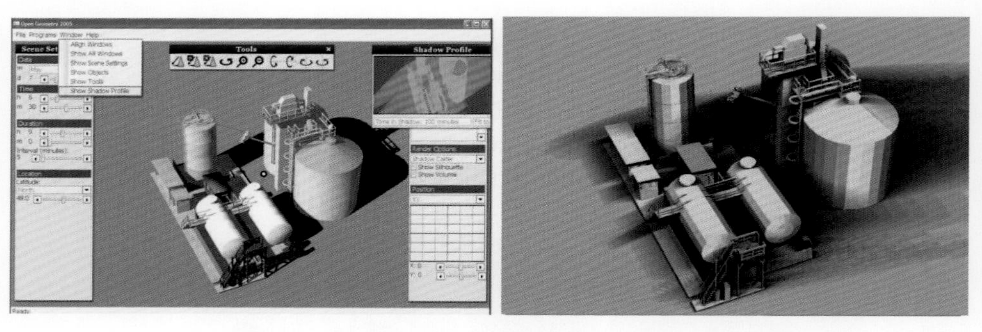

Fig. 3.23 Shadow profiles of more complex scenes

Thus, it is possible to determine the shadow duration for each visible point in the scene (Fig. 3.23, courtesy of Günter Wallner). ♠

When do we speak of a polyhedron?

Fig. 3.24 shows some objects built with Geomag (Geomag is the brand name of a toy construction system consisting of nickel-plated steel balls and connecting rods with a magnet on each end) that are not polyhedra in the strict sense: Concerning the model on the left, the rotors of the helicopter are not polygons. Concerning the object in the middle, the base and top surfaces (regular hexagons) are divided into equilateral triangles that lie in one and the same plane. By "extending" the midpoints of the hexagons from the plane, a regular six-sided pyramid, whose side surfaces are merely isosceles and not equilateral, can be constructed.

Euler characteristic

For closed polyhedra without "holes" – and therefore in particular also for convex polyhedra – , Leonhard *Euler* discovered a simple relation between the number e

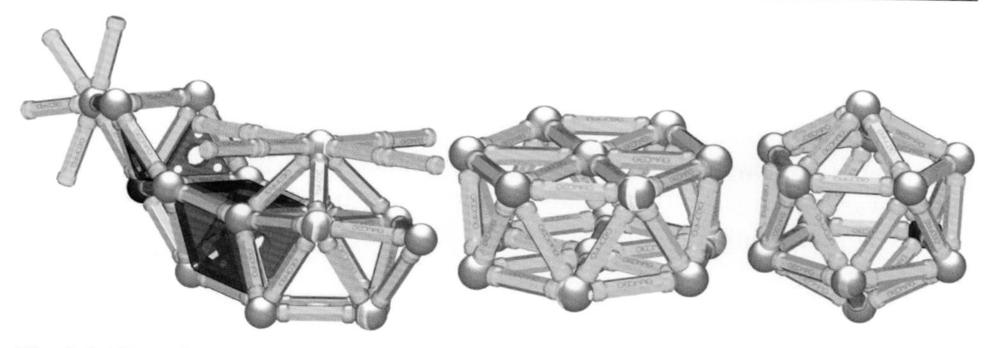

Fig. 3.24 Two of the three objects do not represent a polyhedron: For real polyhedra, all polygons must be closed, and two adjacent polygons in the plane are "merged".

of vertices, the number k of edges, and the number f of faces:

$$e+f=k+2.$$

E.g., for the cube with $e = 8$ vertices, $k = 12$ edges and $f = 6$ faces we have $8+6 = 12+2$.

Proof:

An elegant and brief proof can be shown by using the principle of mathematical induction, which is often fruitful in mathematics. The trick lies in proving the theorem for polyhedra from which we remove a face. We call this type of polyhedron "open". The number of vertices and edges of a closed and open polyhedron are the same whereas the opened one has one face less. Thus, it is to be shown that for open polyhedra we have $e+f=k+1$.

Let us first start the induction: The theorem is true for the simplest case, a single polygon ($f = 1$) with e vertices and as many edges.

Now we show that the theorem is true for $f+1$ faces if it is true for f faces: Let us assume that the Euler characteristic is correct for all open polyhedra of up to f faces. If from this it can be proven that the theorem is also true for open polyhedra with $f+1$ faces, then this conclusion can be applied again to all f.

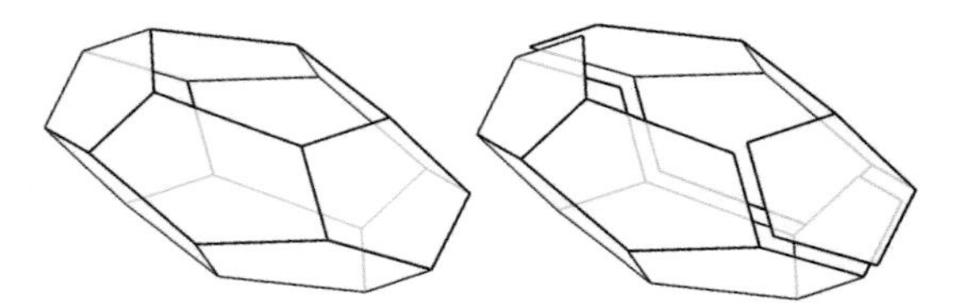

Fig. 3.25 Figure accompanying the proof

Let us consider an open polyhedron with f faces. Let us subdivide this object into two open polyhedra along an arbitrary, closed but non-planar polygon composed of m edges – exactly two of which should belong to the "top polygon" (Fig. 3.25). In sum, both parts still provide f faces, but $m-2$ vertices and $m-1$ edges more. For both partial polyhedra, according to the induction assumption, $e_1 + f_1 = k_1 + 1$ and $e_2 + f_2 = k_2 + 1$ hold true. This means $(e_1 + e_2) + (f_1 + f_2) = (k_1 + k_2) + 2$. Due to $e_1 + e_2 = e + m - 1$, $f_1 + f_2 = f$ and $k_1 + k_2 = k + m - 2$ we get $(e+m-1) + f = (k+m-2) + 2$ and thus, indeed, $e+f = k+1$. ◇

3.3 Platonic solids

About 2500 years ago, the ancient Greeks started to classify spatial geometric bodies systematically. It was obvious to ask for all polyhedra with congruent faces and equal angles along the edges. Several of these were already known, and it was soon realized that only one was missing: the *dodecahedron* with twelve regular pentagons for its faces.

As simple as possible

The *cube* – a regular *hexahedron* – is the best known example even though it not does not have the minimum number of vertices, edges, and faces: It has eight vertices, twelve edges, and six congruent square faces.
What makes the cube so simple is the fact that all of its edges are mutually perpendicular. It is, therefore, a special *cuboid*, and cuboids, from which many day-to-day objects are derived, are in any case easy to imagine.
If a sketch of a cube is to be drawn, most people first draw a square and then attach four parallel line segments of equal length (but shorter than the edges of the square). This gives the vertices in the back, behind the square. Such a sketch is quite informative, even though cubes in nature cannot be *visualized* like this. It conveys what we *know* about the cube. A technical drawer would depict two squares above or next to one another and say: This is the front view, the other is a top or a side view. This is even less ambiguous even though it is not so apparent for a novice. Every technician understands it.

Fig. 3.26 Pyrite ("Fool's gold") with cuboid crystals

Pyrite (Fig. 3.26) is one of the most frequent sulfidic minerals in nature. It crystallizes in a cubic system and often occurs in the form of well developed, cube-shaped crystals that exhibit characteristic streaks. Sometimes it also appears as a dodecahedron.
Fig. 3.27 shows a cube that is illuminated by two light sources. In general, a hexagon belongs to each light source in the shadow-receiving base plane. Both hexagons share four points in common as long as the cube stands on the plane. The umbra shadow appears where they both overlap, and the penumbra shadow where only one of the

Fig. 3.27 The shadow of a cube is generally a hexagon.

hexagons affects the surface. In Fig. 3.27 on the right side, one light source produces "grazing light". In such exceptional cases, the shadow polygon is a rectangle if the light rays are parallel.

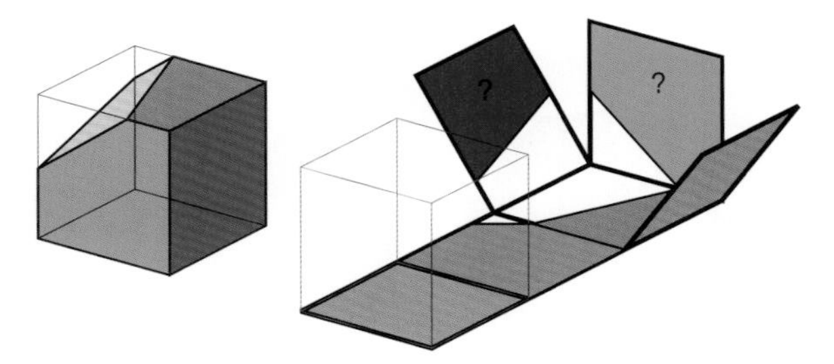

Fig. 3.28 Development of a cube including the planar section (badly chosen, since it "falls apart")

Let us unfold the model of a cube into a plane. The result is often called the *development* of the cube. We now have the need to align the six squares in such a way that they occupy the least possible space (if, for instance, we wish to build many such cube models). Furthermore, it is our intention to restrict the number of edges that have to be glued together (i.e., the number of double edges in the unfolded model). If, as in Fig. 3.28, we only wish to develop the rest of the cube after a planar section, we have to be careful not to have several parts in our hands.

Three objects from equilateral triangles

There can only be one platonic solid whose faces are equilateral quadrangles (or squares), because only three squares can meet in every vertex. If there were four, then $4 \times 90° = 360°$ would complete the full angle, and the vertex pyramid would not be spatial anymore.
This is the reason why we cannot assume equilateral hexagons as the faces of a platonic solid: At least three faces meet at each vertex, and $3 \times 120°$ (the angle between two adjacent edges of a hexagon) already adds up to $360°$. Regular polygons with more than six edges are, thus, excluded from the beginning.

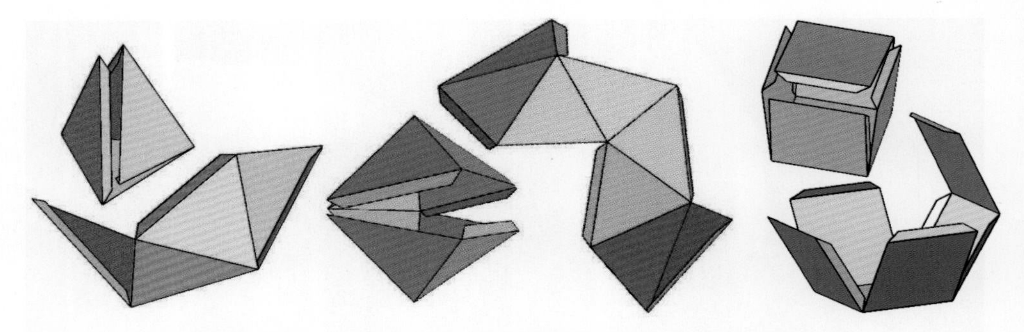

Fig. 3.29 Scenes of a development (tetrahedron, octahedron, cube)

Thus, we can only derive new platonic solids from equilateral triangles and pentagons. Equilateral triangles (with angles enclosed by edges of 60°) offer three such possibilities: A spatial corner can be formed by three, four, and even five triangles. In all three cases, the sum of all angles in a corner is less than 360°.

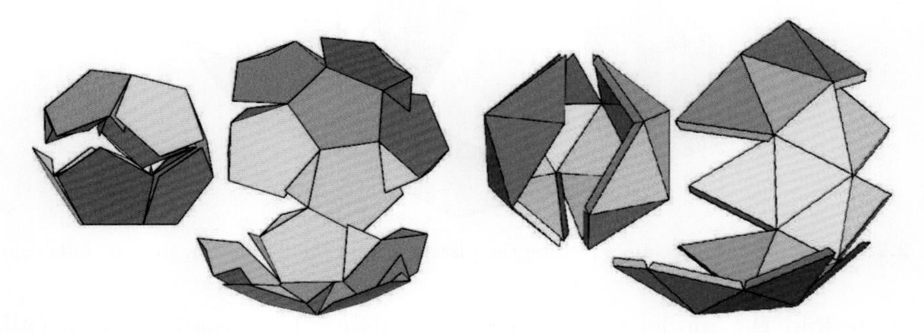

Fig. 3.30 Scenes of a development (dodecahedron, icosahedron)

If a vertex pyramid is formed by three triangles, then the polyhedron is already complete: This functions with four equilateral triangles in the form of a *regular tetrahedron*. Fig. 3.29 on the left shows how it can be constructed from its development.

If a vertex pyramid is formed by four triangles, then we get the *regular octahedron*. Fig. 3.29 in the middle shows a possible development and its folding together. The octahedron can also be interpreted as a double square pyramid with the same base. Both pyramids have half of the diagonal of the base square for their height.

If a vertex pyramid is formed by five triangles, then we get the *regular icosahedron* (from *icosi*=twenty). Fig. 3.30 on the right side shows a possible development and its reconstitution into a considerably complicated structure. This concludes the possibilities of creating regular polyhedra from equilateral triangles. If our systematic approach is used, only one body remains where three regular pentagons (angle between edges of 108°) form a corner ($3 \times 108° < 360°$).

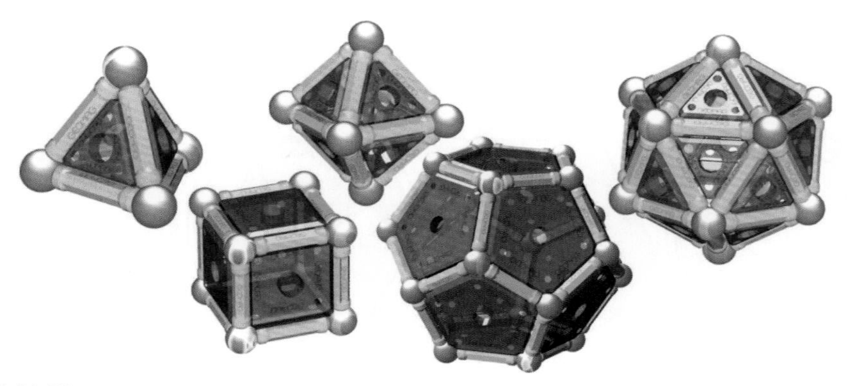

Fig. 3.31 Platonic solids constructed by using Geomag. On the right: A regular icosahedron. The spheres and rods of Geomag are uniform in size, but not in the image: The tetrahedron is slightly enlarged in comparison to the dodecahedron.

The last inexpressibly beautiful platonic solid

Half a dodecahedron (from a *pentagon*=five vertices and *dodecahedron*=twelve faces) remains if one attaches a movable, congruent pentagon to each side of a regular base pentagon, and rotates the attached objects about the edges until a polyhedron emerges (Fig. 3.32 on the left).

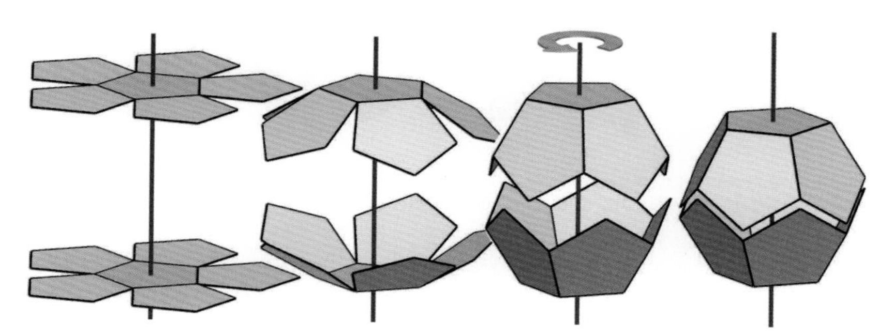

Fig. 3.32 Development of a dodecahedron

Two thus emerging, mirror-symmetrical dodecahedra fit together when one half is rotated by 36° about the perpendicular axis (Fig. 3.32 on the right). Fig. 3.30 on the left shows a practical suggestion for the construction of a dodecahedron.
The condition that all faces of a polyhedron have to be congruent is obviously very limiting. The frequent wish by architects that an arbitrary surface is approximated by small, congruent triangles (a cost-saving solution, see Fig. 6.40), thus, already fails in principle: One always gets the edges of a small number of polyhedra, among which are the regular tetrahedra, octahedra, and icosahedra (Fig. 8.32).

In practice, there exists a marginally useful cop-out: We abandon the necessity of convexity of the body and/or the faces themselves. It is then possible to obtain four additional types of platonic solids: the *star polyhedra*. Fig. 3.34

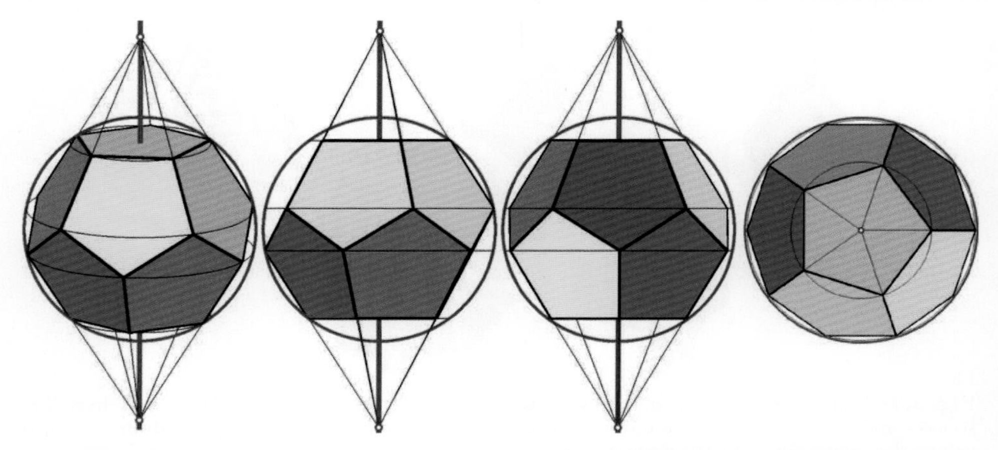

Fig. 3.33 The "finished" dodecahedron in an oblique view and in the principal views

shows these unusual structures that were discovered relatively late (by *Kepler* in 1615 and *Poinsot* in 1809).

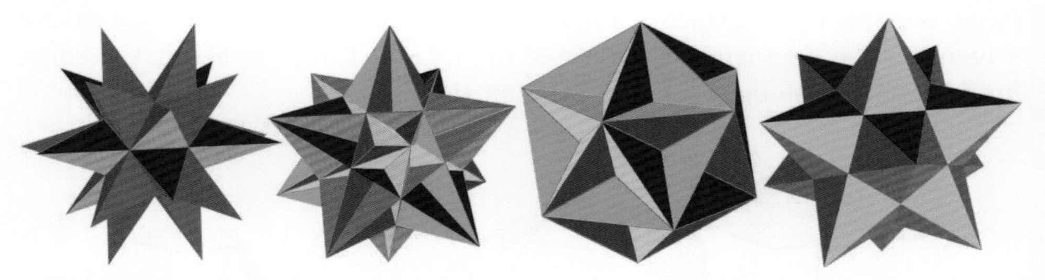

Fig. 3.34 Non-convex regular polyhedra. From left to right: Great stellated dodecahedron, great icosahedron, great dodecahedron, and small stellated dodecahedron

The "small stellated dodecahedron" – first described by *Kepler* – is constructed by twelve five-cornered stars (pentagrams) intersecting each other (Fig. 3.34 on the right). For better understanding, one of the non-convex pentagons is colored blue. In the case of the great stellated dodecahedron (Fig. 3.34 left), 20 three-sided pyramids stick on an icosahedron such that we again obtain non-convex pentagons. Twenty regular triangles are intersecting one another on the great icosahedron (Fig. 3.34 middle left), twelve regular pentagons form the great dodecahedron (Fig. 3.34 middle right).

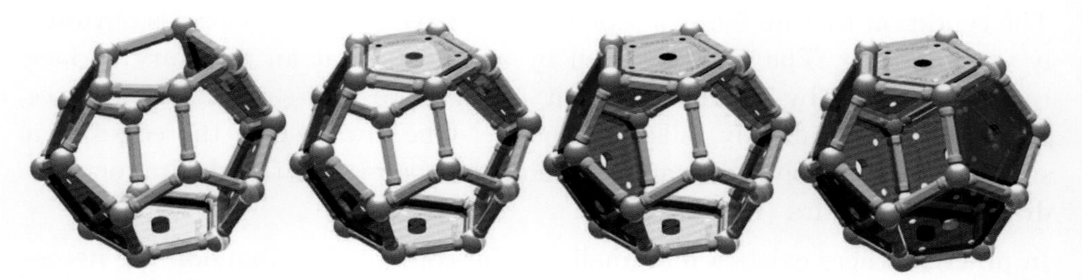

Fig. 3.35 How many panels do we need to build a stable model of a pentagonal dodecahedron?

Crafting a pentagonal dodecahedron is easy on a computer. Fig. 3.35 shows a "making of" a stable model. In practice – such as using the Geomag magnetic building blocks – it helps to "strengthen" the pentagons by inserting "panels". Only five panels have been inserted on the left, and the construction is still not stable. A sixth panel (image center left) is required for the dodecahedron to be absolutely stable.

- **Dualization of the platonic solids**

Due to their regularity, it is possible to inscribe an *insphere* and to circumscribe a *circumsphere* for every platonic solid.
If such a polyhedron is polarized at such a sphere, then the vertices correspond to the faces of the polar polyhedron (and vice versa). The new polyhedron is once again regular. If the sphere is the insphere, then the faces correspond to their points of contact with the sphere and, thus, to their midpoints. The polyhedron that emerges from these surface midpoints is once again platonic!

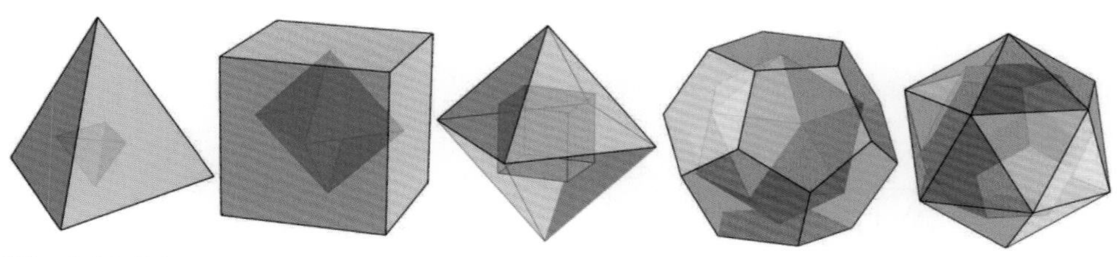

Fig. 3.36 Polarization of the platonic solid: The tetrahedron is self-dual, the cube is dual to the octahedron, and the dodecahedron is dual to the icosahedron.

The tetrahedron (4 vertices, 4 faces) is "dual to itself". Cubes (8 vertices, 6 faces) and octahedra (6 vertices, 8 faces) form a dual pair just like the icosahedron (12 vertices, 20 faces) and dodecahedron (20 vertices, 12 faces) do.

♠

Fig. 3.37 The five platonic solids, and almost a sixth

Fig. 3.37, once again, shows the platonic solids next to each other. The surface of the sixth solid (third from the left) is ideally composed of ten equilateral triangles (it can be divided into two five-sided pyramids). However, the angles of the faces (edge angles) are not all equal.

3.4 Other special classes of polyhedra

Archimedean solids

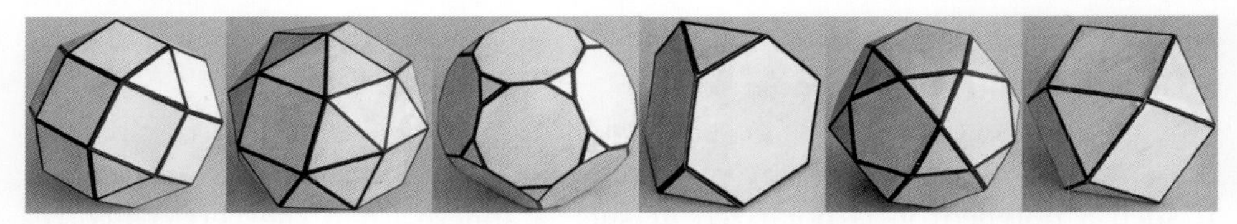

Fig. 3.38 Different Archimedean solids

A more generous condition that allows two different types of congruent surfaces presents many more possibilities (to be precise: 48). Fig. 3.38 shows several such solids that were already investigated by *Archimedes*.
The football is by far the most familiar Archimedean solid (or rather, a spherical variant of it). It can be "developed" from a dodecahedron and thus possesses 12 regular pentagons and 20 regular hexagons.

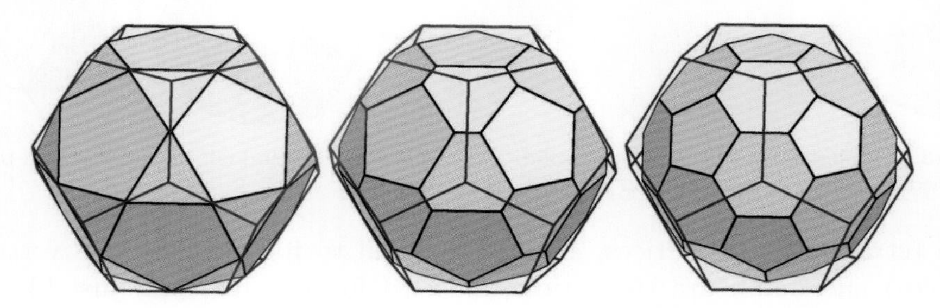

Fig. 3.39 Two Archimedean solids derived from a dodecahedron

Fig. 3.40 Variations of a football

- **Tiling of space**

If the vertices of an octahedron are cut (Fig. 3.41 on the left), a new Archimedean solid that can be used to fill up space without leaving any gaps emerges.
Such tilings of space are not at all easy to accomplish by using regular solids. It is impossible with the sole use of regular dodecahedra, see (Fig. 3.41 on the right). The proof requires a calculation.

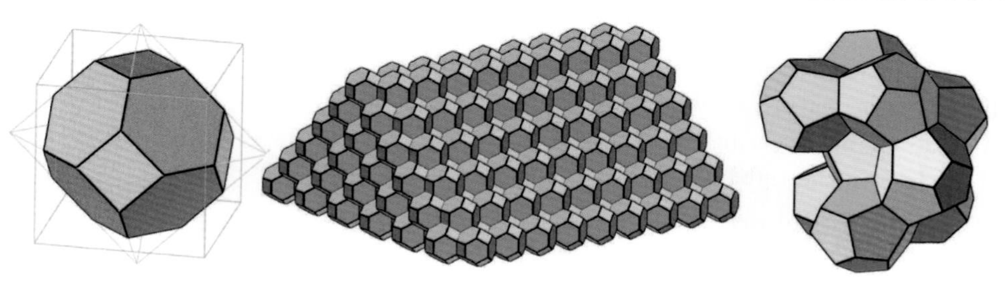

Fig. 3.41 On the left: Truncated octahedron (intersection with a cube). Center: Exact tiling using such solids. On the right: Tiling using *irregular* dodecahedra.

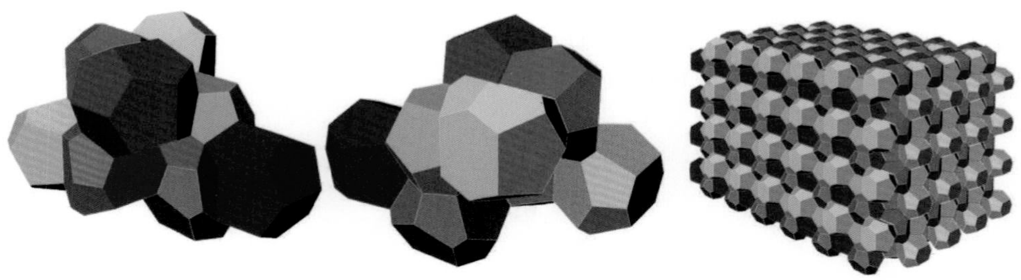

Fig. 3.42 Tiling of space with irregular 12-sided and 14-sided solids according to Weaire-Phelan. On the left and in the middle, there are two views of the basic building block, composed of eight equal-volume polyhedra.

Only recently (1993), it was discovered by Denis Weaire and Robert Phelan that irregular dodecahedra (with 12 faces) and tetradecahedra (with 14 faces) of equal volume – very similar solids possessing two hexagonal faces – can be arranged in groups of eight and used to fill space without leaving any gaps[1]. ♠

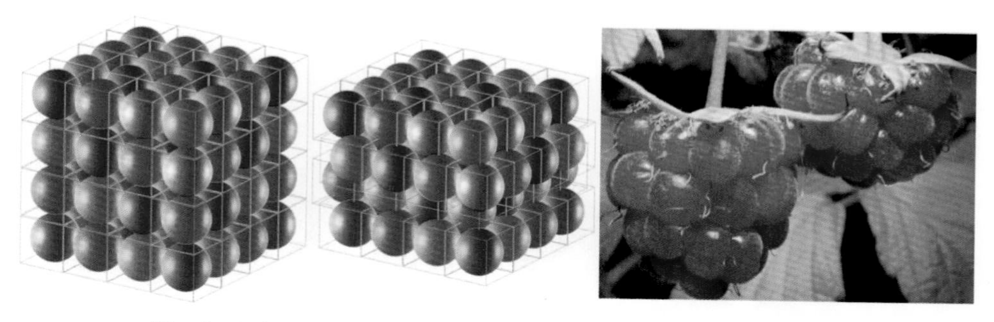

Fig. 3.43 A more or less dense sphere packing in theory and practice

The easiest way to tile three-space is, of course, using cubes. Spheres are not well suited for this purpose. The gaps caused by spheres comprise almost half of the volume of the arrangement in Fig. 3.43: The ratio of the volume of a sphere to the volume of the circumscribed cube equals $(4\pi/3) : 8 \approx 0{,}52$. The translation of the

[1] More information at `http://en.wikipedia.org/wiki/Weaire-Phelan_structure` or – with greater detail and more emphasis on mathematics – `http://torus.math.uiuc.edu/jms/Papers/foams/forma.pdf`.

positions results in a slightly better utilization of space, but even in the optimal arrangement, it is only about 3/4.

In this case, nature shows itself – as it so often does – to be rather pragmatic: A little bit of squeezing and shoving, and space is, once again, optimally utilized (Fig. 3.43 on the right).

Rhombic polyhedra

Let us take the platonic solids again, the faces of which are regular n-gons. Through each edge e, we have a plane ε perpendicular to the connection of the edge with the center C of the solid (Fig. 3.44). All planes ε that contain the sides of a face form a regular n-sided pyramid. Two neighboring pyramids touch along a normal plane ε so that the two adjacent equilateral triangular pyramid-faces form a planar rhombus.

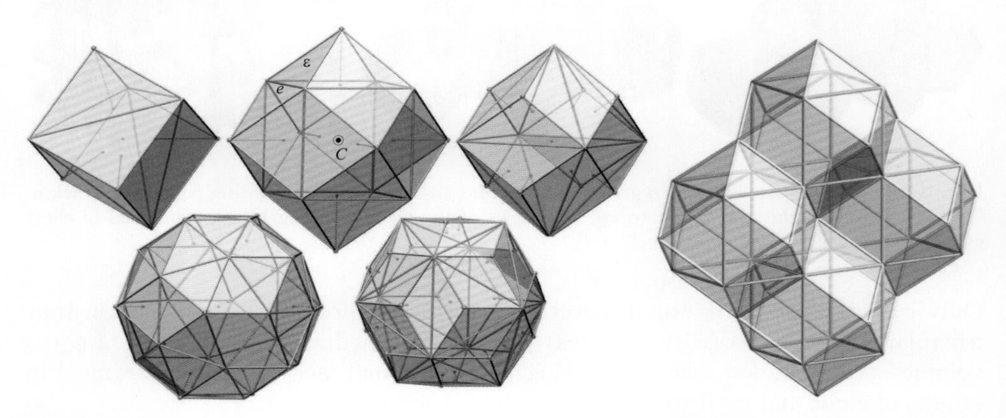

Fig. 3.44 Left block: Five platonic solids lead to two new interesting solids: The rhombic dodecahedron (12 faces) and the rhombic triacontahedron (30 faces). Right: Due to the fact that the rhombic dodecahedron can be generated by means of a cube, it follows that it is suitable for "packing space".

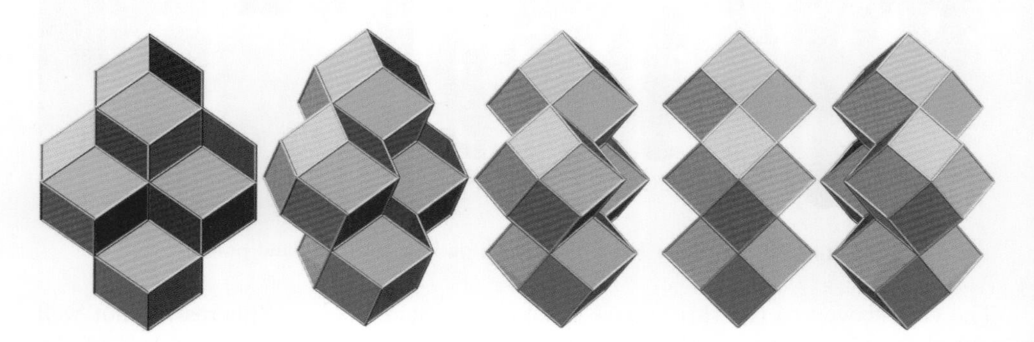

Fig. 3.45 Packing space with rhombic dodecahedra: a series of stroboscopic images during a rotation

We would now expect five new solids that consist of as many congruent rhombuses as the adjacent platonic solid has edges. However, it turns out that the cube and the octahedron (with 12 edges each) lead to the same result

Fig. 3.46 Another series of stroboscopic images: Everything is oblique and not cubic, one needs quite a few images to realize the situation.

– the *rhombic dodecahedron* (Fig. 3.44 upper middle and left). The same is true for the regular dodecahedron and the icosahedron (with 30 edges each). They generate the *rhombic triacontahedron* (triaconta=30, Fig. 3.44 lower images). Finally, the regular tetrahedron with its six edges leads to a trivial result: the cube. In this case, the rhombic faces are squares.

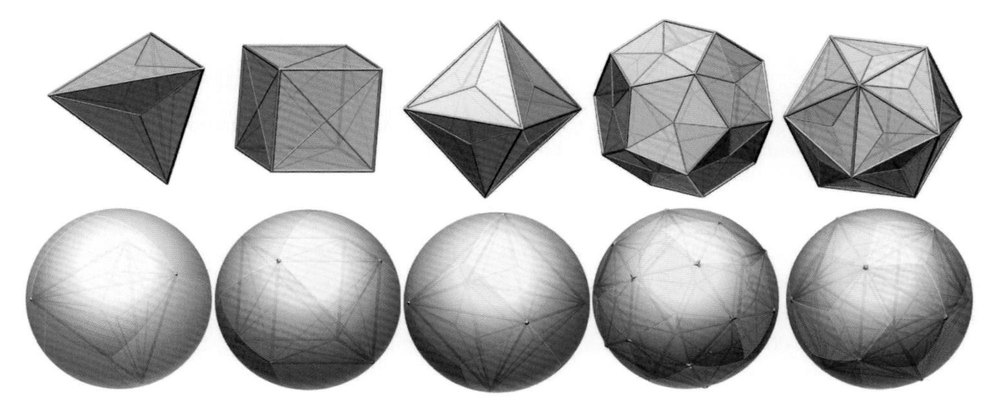

Fig. 3.47 The projection to the circumsphere of the platonic solids leads to three different remarkable rhombic tilings on the sphere.

Since a rhombic dodecahedron can be created from a cube, and cubes can be stapled to fill space, we have found another non-trivial example of tiling space (application p. 92, Fig. 3.44 right).
The upper images in Fig. 3.47 illustrate that the apices of the above mentioned regular cones through the edges of the platonic solids lie above the centers of faces. If we now project all points to the circumsphere, we get three different rhombic tilings of the sphere with 6, 12, and 30 congruent spherical rhombuses and $4+4=8$, $8+6=14$, and $20+12=32$ grid points. We will learn more about the regular distribution of points on the sphere later on.

- **More polyhedra consisting of congruent rhombuses?**

Fig. 3.48 left shows a polyhedron with 30 rhombuses (a somehow "distorted version" of our r hombic triacontahedron) in the principal views. In the plan view, the sides of the rhombuses have equal length due to the fact that they have equal length in space plus their inclination is constant. Therefore, the height differences of all sides are constant (Pythagoras), and all vertices of the triacontahedron lie in equally distributed horizontal carrier planes (to be seen in the front view).

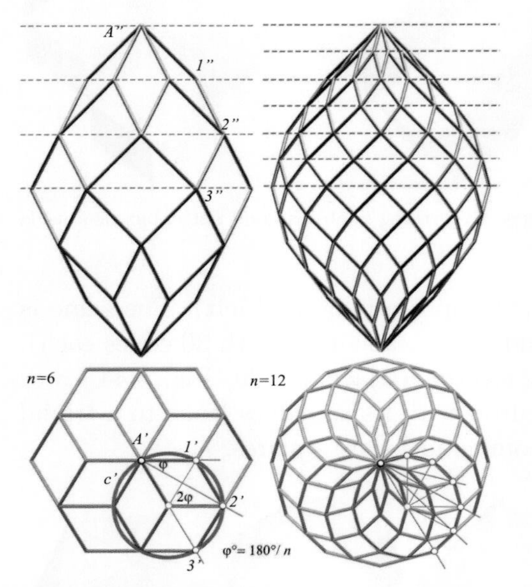

Fig. 3.48 Generalization of polyhedra consisting of rhombuses. The polyhedron on the left is a rhombic triacontahedron – its height can be varied by choosing the constant height difference of the horizontal carrier planes of the vertices.

In the plan view, the six rhombuses that appear as edges form a regular hexagon around the projection A' of the apex A. Non-trivially, the six vertices $1'$,s $2'$, $3'$, ... of three connected rhombuses are also equally distributed on a (red drawn) circle c' through A'. We could also take a circle, distribute six points A, 1, 2, 3, $\cdots$ equally, and define that these points have constant height differences. Then, we rotate the points six times about the vertical axis through A by 60° and get a polyhedron consisting of 30 rhombuses.
This method can be generalized to get more vertices (Fig. 3.48 middle). If we only distribute four points on the red circle, we get a rhombic dodecahedron. Since the height differences can still be chosen, the polyhedron's height can vary.
Fig. 3.48 right shows an ornamental smoothened realization of the polyhedron shown in the middle. Later on, when we will speak about helices, we will once again come back to this example (application p. 266). ♠

Catalan solids

In 1865, Eugéne *Catalan* described solids which are the duals of the Archimedean solids. The faces of a Catalan solid are congruent non-regular polygons

(Fig. 3.49). The vertices and faces correspond to faces and vertices of the dual type.[2]

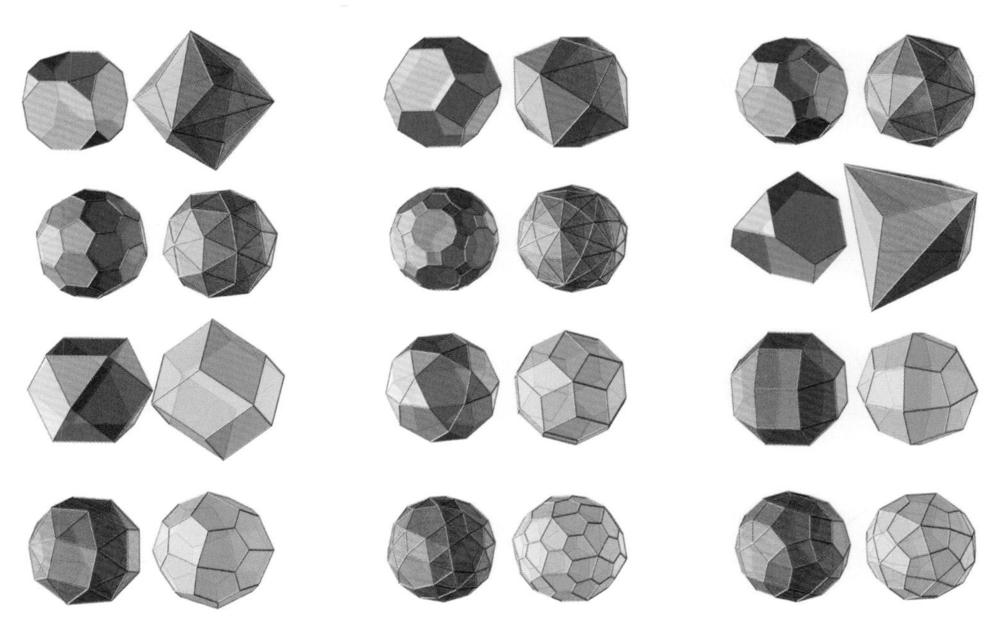

Fig. 3.49 Polarization of Archimedean solids at a sphere around the center. Six (blue-shaded) Archimedean solids result in (yellow-shaded) solids with congruent triangles, two (gray-shaded) result in solids with congruent rhombuses and the four orange shaded result in polyhedra with congruent pentagons or quadrilaterals.

Once again, we encounter pure rhombic solids: the already well-known rhombic dodecahedron and the rhombic icosahedron (20 faces).

3.5 Planar sections of prisms and pyramids

General pyramids are constructed by connecting a polygon that is not necessarily planar with a fixed point (the apex of the pyramid). If the apex is a point at infinity, then we speak of a prism.
If one intersects a prism or a pyramid with two arbitrary planes ε and $\overline{\varepsilon}$ which intersect all edges but do not contain the apex, then both intersected figures are related by *central perspectivity* (in the case of a pyramid) or *parallel perspectivity* (in the case of a prism).
When interpreting (spatial) parallel perspectivity in a parallel projection (e.g., on the drawing paper), then it represents *perspective affinity*. Accordingly, spatial central perspectivity appears in the image as *perspective collineation.*

[2] http://en.wikipedia.org/wiki/Catalan_solid provides a good survey and all the names of the solids.

With parallel perspectivity (Fig. 3.50) and central perspectivity (Fig. 3.52), corresponding points lie on parallel perspectivity rays or perspectivity rays through the apex. Corresponding sides of the section polygons intersect each other on the straight intersection line $a = \varepsilon \cap \overline{\varepsilon}$: The respective face φ of the prism or pyramid intersects ε and $\overline{\varepsilon}$ after the straight carrier line $\varepsilon \cap \varphi$ and $\overline{\varepsilon} \cap \varphi$ of the respective edges, and these have to pass through the intersection point $\varepsilon \cap \overline{\varepsilon} \cap \varphi = a \cap \varphi$. The straight line a is called the axis of perspectivity.

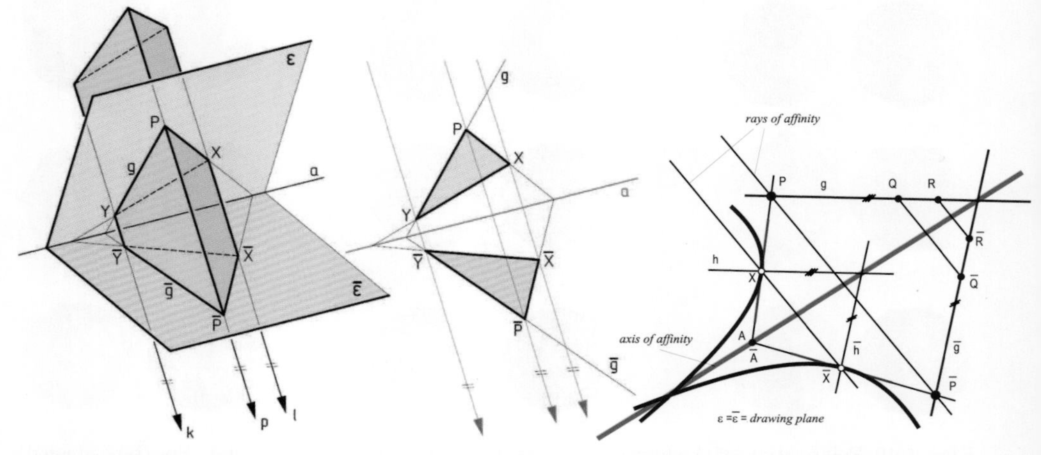

Fig. 3.50 From the spatial situation to a construction in the plane: parallel perspectivity

Fig. 3.51 Perspective affinity of two point fields

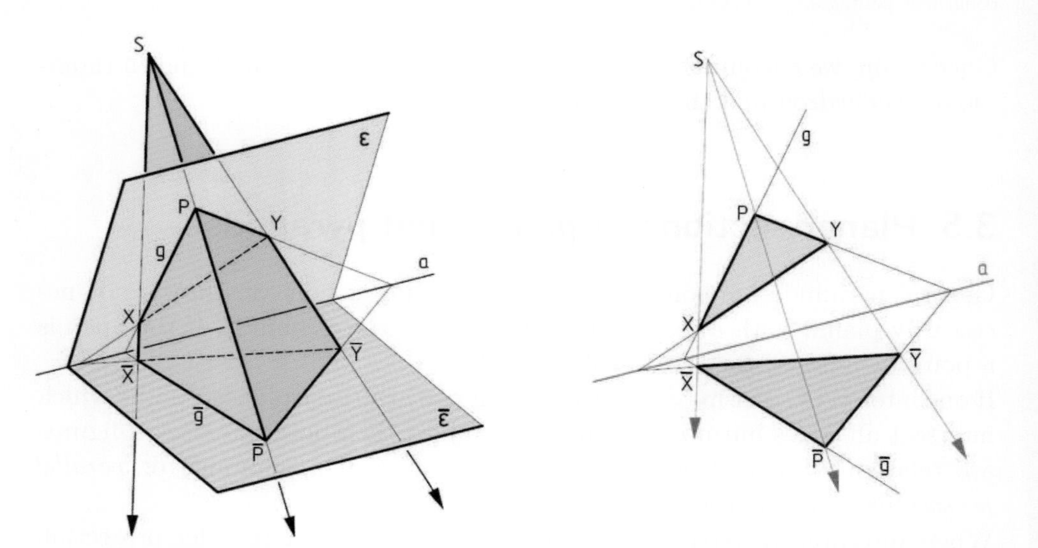

Fig. 3.52 From the spatial situation to a task in the plane: central perspectivity

If one planar section polygon of a prism or a pyramid is known, then each further section can be determined very efficiently by just using the axis of perspectivity a and only one point of the new intersection. It is then possible to restrict oneself to a "two-dimensional" construction: Let P_1Q_1 be an edge of the first section figure and P_2 the already known point on the edge through P_1. Q_2 can then immediately be determined since P_1Q_1 and P_2Q_2 intersect on a. This construction works for each projection and is truly two-dimensional.

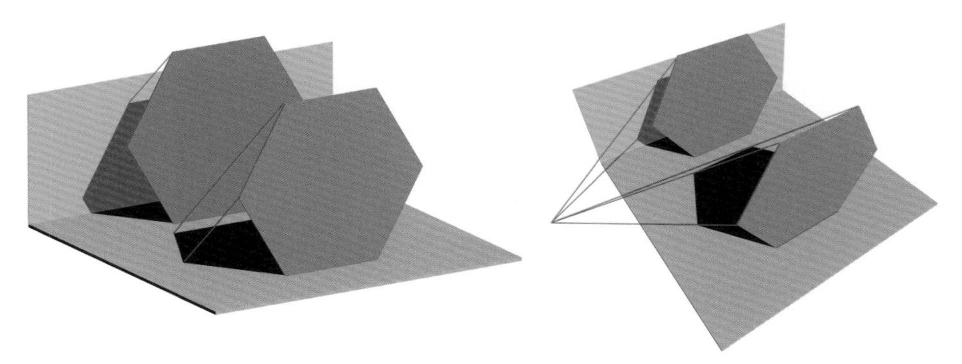

Fig. 3.53 Spatial perspectivity can be applied to all views. The parallel projection on the left and the central projection on the right show how parallel lighting is imaged. The left image contains a planar perspective affinity – the right image a perspective collineation.

It is even possible to free oneself from the spatial interpretation and to define the following:

If two planar, *linear*, and connected fields of points are related such that corresponding points lie on rays that are either parallel or pass through a fixed center, and additionally the corresponding connecting straight lines of two points intersect on a fixed axis a, then the relation is a *planar perspective affinity* or a *planar perspective collineation*. The connecting rays of the corresponding points are called the rays of affinity or rays of collineation. The axis is called the *axis of affinity* or *axis of collineation* and the center is called the *center of collineation*.

Fig. 3.54 Perspective affinity (space) $\mapsto$ perspective collineation (plane)

The connecting straight lines of respective points pass through a center, which may also be a point at infinity. The points are "mutually related via perspective". This concept was already mentioned regarding *Kiepert*'s Theorem (S. 19). One, therefore, speaks of *perspective* affinity and collineation.

The preservation of affine ratios and parallelity further applies to perspective affinity: This follows from the two aforementioned properties (assigned points lie on parallels, assigned straight lines intersect on the axis of affinity) and from the intercept theorem.
In this way, the following entities are perspectively affine / collinear in space during each parallel projection / central projection:

- Two planar sections of a prism / pyramid;
- Each planar polygon and its parallel / central projection in a plane;
- Each planar polygon and its parallel shadow / central shadow on a plane.

In anticipation of the next chapter, prisms / pyramids can also be "refined" by choosing a base curve instead of a basal polygon. In such cases, the following entities are perspectively affine / collinear:

- Two planar sections of a cylinder / cone.

Mathematically speaking, an affine relation between two fields of points $P(x/y)$ and $P^*(x^*/y^*)$ is expressed by a linear relation between the respective coordinates:

$$x^* = A\,x + B\,y, \quad y^* = C\,x + D\,y \quad (A,\, B,\, C,\, D \in \mathbb{R}).$$

In matrix notation [11], one writes $\mathbf{p}^* = \mathbf{M} \cdot \mathbf{p}$. This mapping is invertible if, and only if, M is invertible. Thus, $\mathbf{p} = \mathbf{M}^{-1} \cdot \mathbf{p}^*$, while $\mathbf{M}^{-1}$ is the inverted matrix of $\mathbf{M}$.

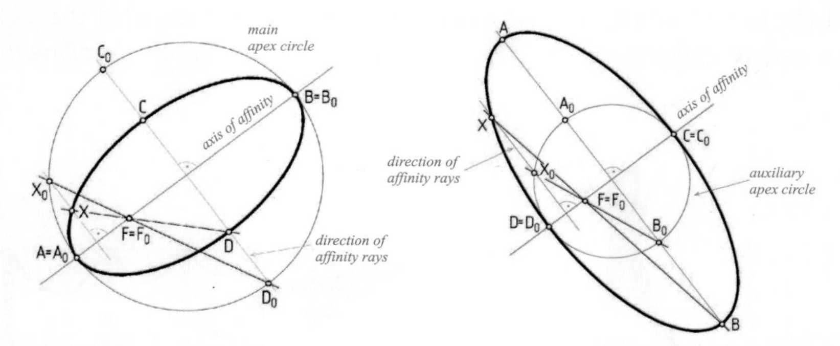

Fig. 3.55 The affine relation between a circle and an ellipse ("*de la Hire*'s construction", sometimes called the "*Proklus* construction", uses this fact and is indirectly illustrated in Fig. 11.10 on the right.)

This enormous range of potential applications explains the importance of perspective affinities and perspective collineations. We will prove, for instance (S. 136), that an arbitrary planar section of a cylinder of rotation is an ellipse. On the other hand, the base curve of a cylinder of revolution is a circle. Therefore, we have:

Circle and ellipse are mutually affine.

Fig. 3.55 on the left shows the affinity between an ellipse to its principal apex circle, and on the right its affinity to its auxiliary apex circle.

Fig. 3.56 can be seen as an anticipation of chapter 9: If circles (e, p, and h) in a plane γ are projected from a center (the eye point E) to an image plane π, one gets conic sections (e^c, p^c and h^c) after the intersection with the quadratic projection cones. These can be ellipses, parabolas, or hyperbolas.

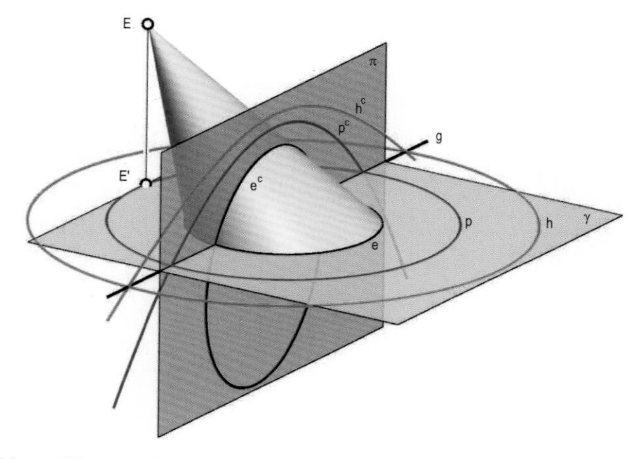

Fig. 3.56 The collinear relation between a circle and the three types of conic sections

In every case, the central perspectivity in space appears as a perspective collineation in the drawing plane. The image of the straight line g of γ and π is the axis of collineation. The image of the center E is the center of collineation. Any intersections of the circles with their projections in π lie on the axis of collineation.
Since π and γ commute, i.e., it doesn't matter who comes first, the aforementioned is true not only for the circle but for all conic sections. It is, thus, possible to "change" a hyperbola or parabola into an ellipse and even (in the special case) into a circle by means of perspective collineation. Fig. 3.57 is an example of the perspectively collinear mapping of an ellipse – after all, a photograph produces a kind of collinear mapping!
Fig. 3.58 shows two photos containing a multitude of conic sections in the form of collinear images of circles and ellipses: The left mainly contains images of circles (the circle-shaped boundaries of the horse's neck are depicted as parts of hyperbolas), but also elliptic shadows on the cylinders of revolution contained in the building. On the right there are elliptic conic sections that also appear in the photo as ellipses.

Fig. 3.57 The water basin in front of the St. Charles's Church in Vienna is elliptical. Depending on whether the plane parallel to the film plane through the lens center touches its boundary, intersects it or not, one gets a parabola (very unlikely), a hyperbola, or an ellipse.

Fig. 3.58 "Heidelberg horse": How many conic sections (or parts of which) are visible on both photos? Hint: All boundary curves on the horse, and on its neck in particular, are circular arcs. You may stop counting once you have found 3 hyperbolas and 20 ellipses ...

3.6 "Explorations" of the truncated octahedron

In this section, we will present several "explorations" of the truncated octahedron (Fig. 3.41), the only Archimedean solid that allows endlessly repeating clusters (and therefore has "space-filling potential") from an architectural standpoint. The semester project 2012 at Studio Hani Rashid was based on the premise of "The Deep Futures Expo City: Prototype 001". As an unconventional point of departure for planning a new city, the start was the making of a fully functional and spectacular worlds fair expo, the "DF EXPO 001". As a first geometrical exercise, the students were asked to explore the space-defining qualities of a single component - the truncated octahedron.

- **Splitting the volume and creating a fractal object**

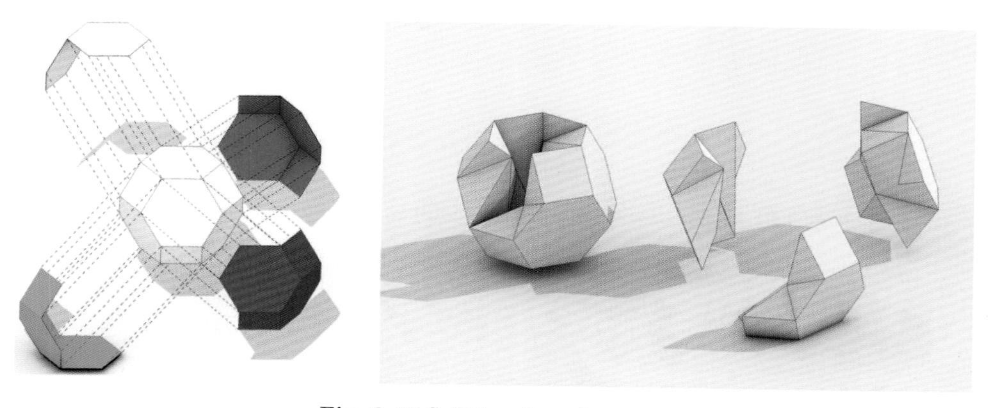

Fig. 3.59 Splitting the volume ...

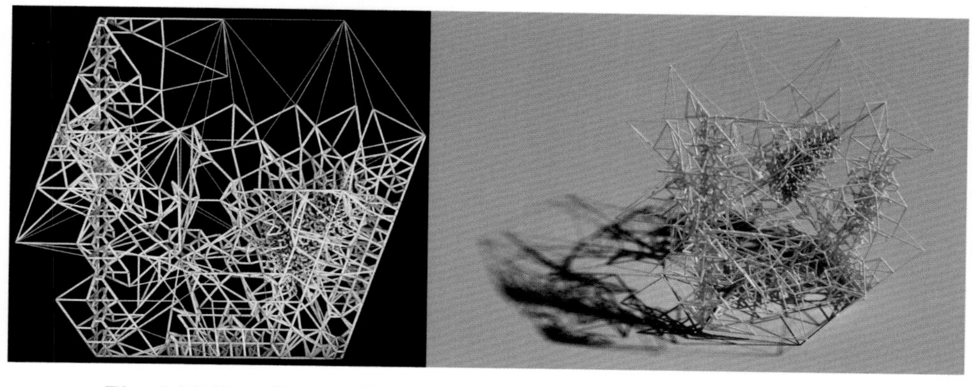

Fig. 3.60 Fractalization (Dena Saffarian, Narta Dalladaku and Peter Mears)

The truncated octahedron should be divided into four parts - three volumes and one void. Without losing its geometrical logic and with sharp mathematical precision, the solid was broken free: Divided into four parts first (Fig. 3.59), and later developed to a new geometry. When breaking open the interior of the geometry, and defining its edges, the focus was to create four equally big parts that differ in their shape. An important point was the shape of "the void", which was decided to go diagonally through the whole truncated octahedron and to touch all three parts.

It should also be used as a common ground. The places where the structures are reaching out from the three shapes, break out of their bounding box and interfere, connect or coexist with each other. ♠

• Creating a *Hamiltonian circuit*

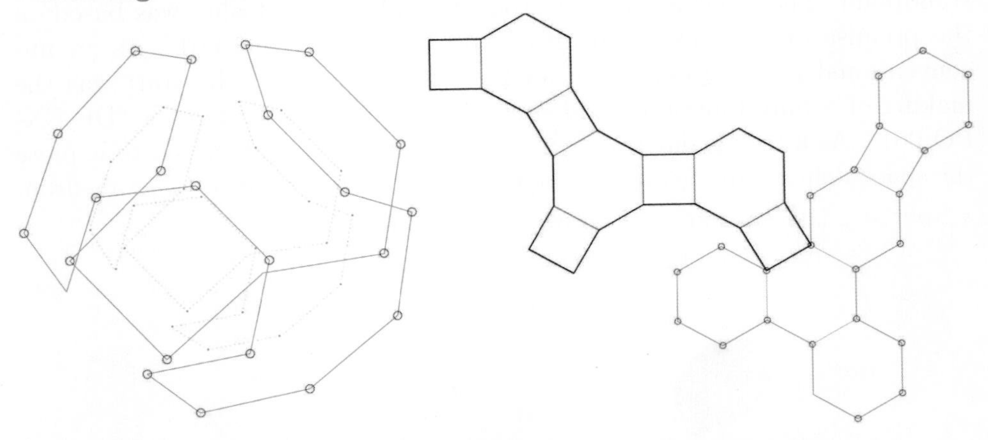

Fig. 3.61 Looking for a Hamiltonian circuit in the unfolded polyhedron

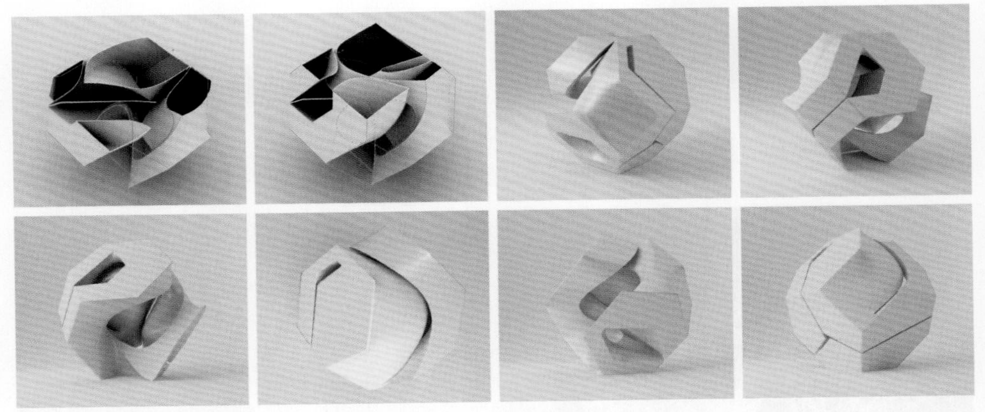

Fig. 3.62 Renderings of the circuits: Stephan Ritzer, Klemens Sitzmann, Maximin Rieder

The truncated octahedron is unfolded into the plane and a so-called *Hamiltonian circuit* ([13]), i.e., a closed polygon that visits all the vertices exactly once, is created. ♠

• Splitting and reworking

Ewa Lenart, Paul Krist, and Nora Varga also split the given truncated octahedron into four parts (Fig. 3.63-3.66). Then, each of them started to "rework" one part in their own way. The different results show fragile, pleasing structures (Fig. 3.64) that partly have the shape of hyperbolic paraboloids (see Chapter 6). ♠

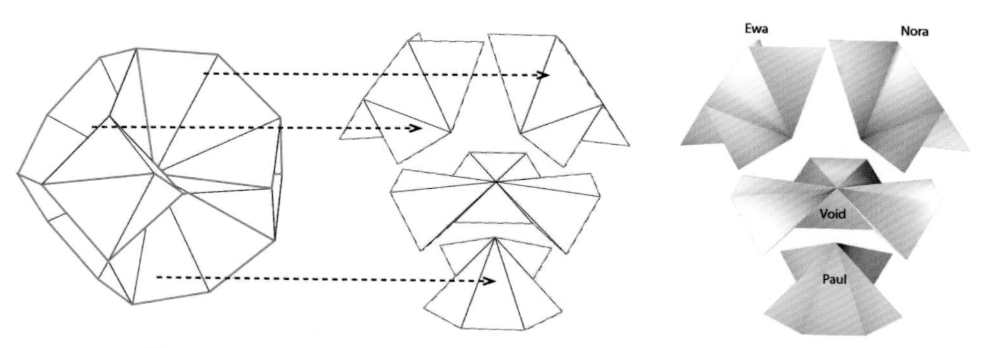

Fig. 3.63 The truncated octahedron is split "radially" into four parts.

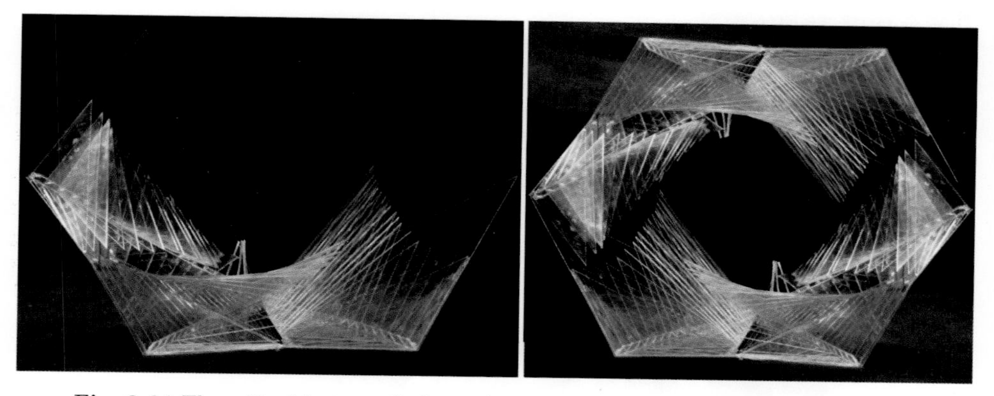

Fig. 3.64 The split objects are "refined" (on the right: a composition of two photos).

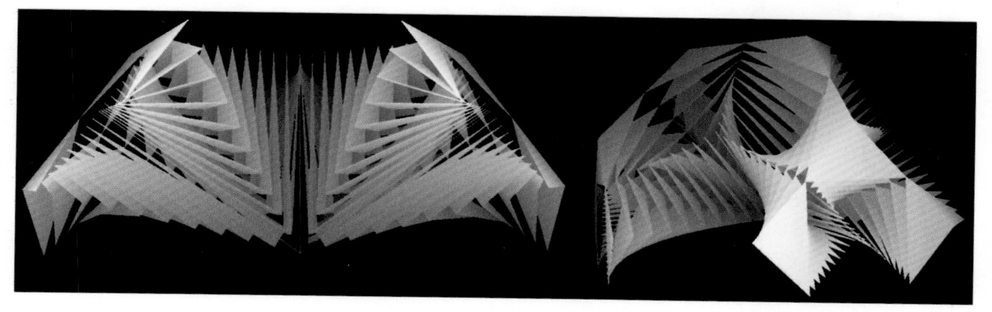

Fig. 3.65 Since regularly distributed points on skew axes are connected, so-called HP-shells (hyperbolic paraboloids) show up.

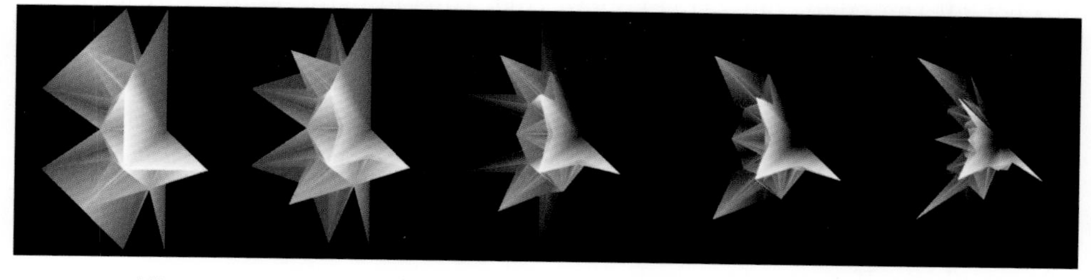

Fig. 3.66 Beautiful crystal blossom-like non-closed polyhedra are created.

• "Forced edges" lead to new shapes

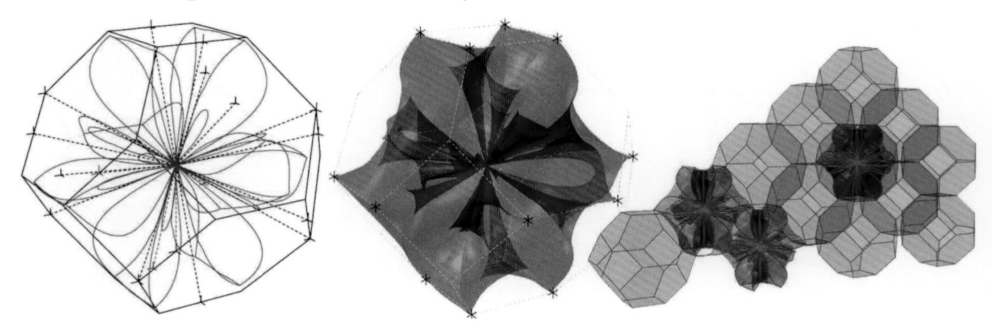

Fig. 3.67 Metamorphosis into curves (Haitham al Busafi, Arpapan Chantanakajornfung, Hulda Guðjónsdóttir, and Ivo de Nooijer)

The edges follow the active vertices towards the center in a metamorphosis into "fluid curves". Since the majority of the vertices is still attached to the original wireframe, the shape still retains considerable similarities to its parent (Fig. 3.67). Despite the drastic changes to the appearance, the contracted truncated octahedron will still fit into the spatial tessellation pattern. ♠

• Different interpretations of "the void"

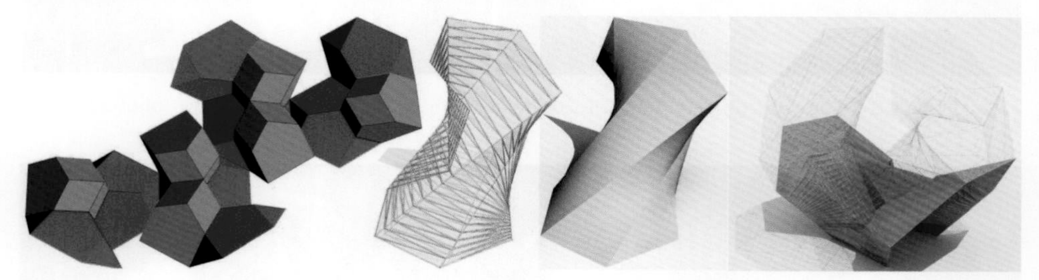

Fig. 3.68 HP-surfaces replace the void (Hessamedin Fana, Lena Kriwanek, and Kaveh Najafian)

Again, the truncated octahedron is split (Fig. 3.68 on the left). After removing the blue building blocks, the void is replaced by several HP-surfaces that span from one outer edge to an adjacent skew edge (images in the middle). The image on the right finally shows a combination of different techniques.♠

• Inversion leads to a non-Euclidean interpretations

All the 24 vertices of the truncated tetrahedron lie on a common sphere. We project all the edges onto this sphere (like Buckminster *Fuller* did with the truncated cube, Fig. 5.65). The connecting lines of the vertices thus transform into arcs of great-circles. Then, the vertices and connecting great-circles are spherically projected onto the base plane (compare Fig. 5.67), which results in a pattern of circular arcs (Fig. 3.69), since the spherical projection is circle-preserving. The plan view of the wireframe of the spherical object and the projected wireframe are inverse with respect to the equator circle e of the circumsphere. If we apply a spatial inversion to a space-filling set of truncated octahedra, we get objects like in Fig. 3.70 and 3.71. ♠

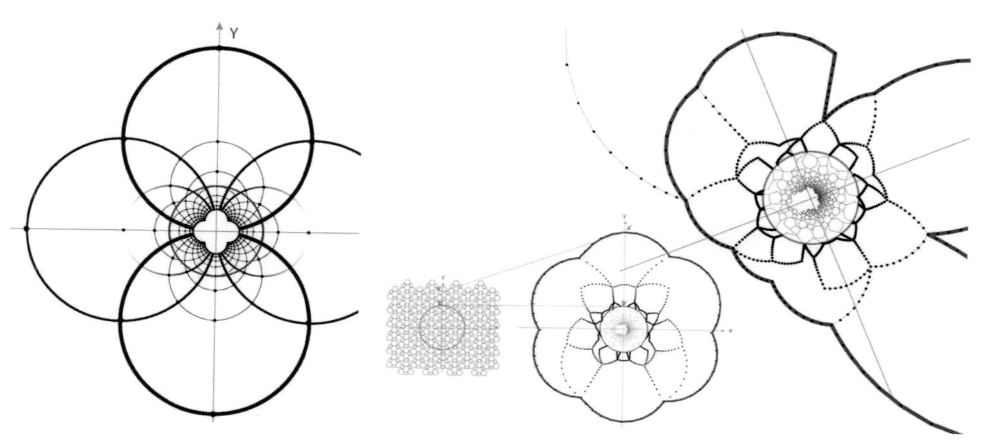

Fig. 3.69 Left: An orthogonal grid in the plane transforms into two orthogonal pencils of circles. Right: Transformations of the nets of aligned truncated octahedra.

Fig. 3.70 Left: Inversion of a system of truncated octahedra, right: A combination of the original polyhedron and the inverted scene (Abraham Fung, Robert Vierlinger, Melanie Kotz).

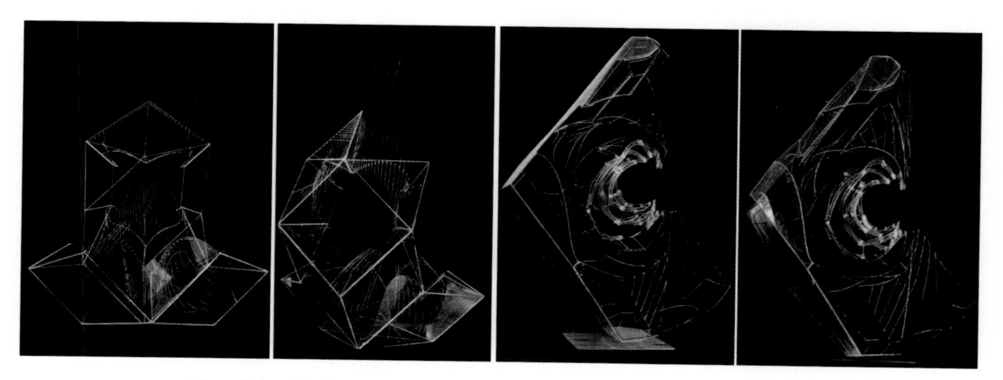

Fig. 3.71 Different models of the above (Photos: Reiner Zettl)

• **A flexible interactive installation from an Archimedean solid**

The last example in this section does not relate to the truncated octahedron, but to another Archimedean solid (Fig. 3.39 on the right): "Cerebral Hut" explores the relationship between architecture, motion, and human thought. We traditionally assume that the built environment influences our psyche. What if a kinetic architecture could establish a direct connection between the thoughts of its user and itself in order to reconfigure its physical boundaries accordingly?

Fig. 3.72 Kinetic Architecture: Guvenc Ozel (with Alexandr Karaivanov, Jona Hoier, Peter Innerhofer). The inside has a flexible triangular mesh (`http://ozeloffice.com/`).

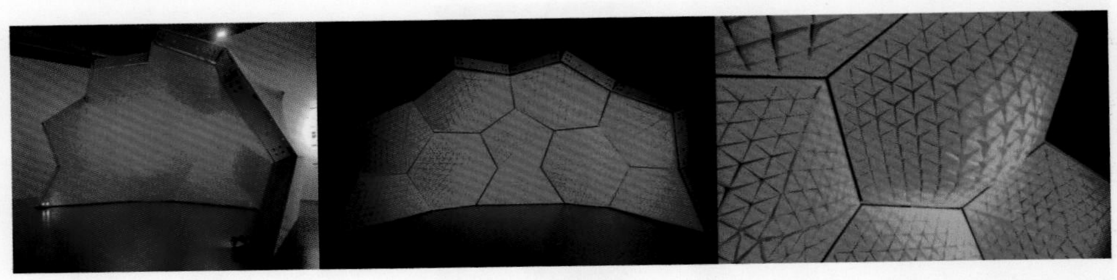

Fig. 3.73 In order to create a space that is reactive to brain activity, the team hacked and reprogrammed a commercially available device that can measure concentration levels and blinking, interpreted these data thresholds, and wrote scripts that would translate them into motion.

Research on different folding patterns and geometric structures lead to the creation of an environment that can act as a vessel for motion. Cerebral Hut plugs into a vein of contemporary research that explores kinetic environments and the relationship between technology, motion, and space, but it is the first of its kind that creates a moving architecture that directly responds to human thought.

Fig. 3.74 The biosphere by Buckminster *Fuller* (Montreal)

A remark from the geometrical point of view: The triangulation of the biosphere in Montreal (Fig. 3.74) is comparable to the (fexible!) triangulation used in the installation. The Cerebral Hut has no final or ideal design-form. Its interior and exterior is in constant transformation triggered by user participation. ♠

4 Curved but simple

In this chapter, we shall discuss the theory of curvature of planar and space curves as well as basic curved surfaces that are easy to imagine, such as spheres and cylindric surfaces. The latter differ substantially in one respect: The sphere is "doubly curved" while cylinders are just "single curved". Despite the double curvature, we will first talk about the sphere, since it is very easy to define and to imagine.

The sphere plays a central role in further geometric considerations. Every planar section of a sphere and also the intersection of two spheres is a circle.

We first need to understand the theory of planar and space curves. In space, we have two kinds of curvature, namely curvature and torsion. By using curves, surfaces can be created. The sphere may be generated by rotating a circle about one of its diameters. Cylindric surfaces can be generated by moving straight lines parallel to a certain fixed direction through space.

The images of all possible curves on a surface envelop the contours of the surface. If this contour only consists of straight lines for every projection, the surface is developable. Only such surfaces can be unfolded (developed) into the plane without distortion.

The planar section of a cylinder of revolution is an ellipse (in general), and its development is a sine curve. The intersection of two cylinders of revolution is usually an arbitrary space curve, but may in some cases split into two ellipses.

Survey

4.1 Planar and space curves

Every science tends to have its own language with which its ideas can be expressed precisely and tersely. Before discussing curved surfaces, let us, therefore, define certain terms that are especially important in differential geometry.

Basics from planar curve theory

Differential geometry deals with the limits of functions. For instance, if one considers two points P and Q of a planar curve, then its connection line forms a *secant* (lat. *secare*=to intersect). If Q now moves increasingly closer to P, then there exists an unambiguous limiting position which is called *tangent* to the curve (lat. *tangere*=to touch). One simply says: At the point of contact P, the tangent of a curve and the curve share two points which are infinitely close together. The normal to the tangent at P is called the *curve normal.* If the curve is a circle, all curve normals pass through the center. However, in general, the totality of normals to the curve envelops a curve – the so-called *evolute.*

Fig. 4.1 Tangent, normal, evolute of different curves

We can also consider *two* neighboring points Q and R to the curve point P. If the three points do not lie on a straight line, they define a circle. We now let Q and R move closer to P. This leads to a well-defined limiting position of the circle defined by three points – the *osculating circle* at the point P (a circle can also degenerate into a straight line or a point). The radius of the osculating circle is called the radius of curvature. In the vicinity of the curve point, it approximates the curve much better than the tangent. The osculating circle at P is said to have three points with the curve in common.

The locus of the midpoints of all osculating circles is the evolute enveloped by the curve normals.

Proof:
The osculating circle is determined by three neighboring points P, Q, and R. Its

center is, therefore, located at the intersection of the perpendicular bisectors of $\overline{PQ}$ and $\overline{PR}$. Let Q^* be the midpoint of PQ and R^* that of PR. The perpendicular bisectors, therefore, converge towards the curve normal in Q^* R^*. The circumcenter of PQR, thus, converges towards the intersection of two adjacent curve normals.◇

• **A circle as an evolute**

Fig. 4.1 shows curves with their evolutes. One can see that evolutes have cusps whenever the curve has an apex – a point whose radius of curvature is stationary. The right curve is especially notable. It is constructed in such a way that their evolute is a circle. It is, therefore, called a *circle's evolvent.*♠

• **Osculating circles of the ellipse**

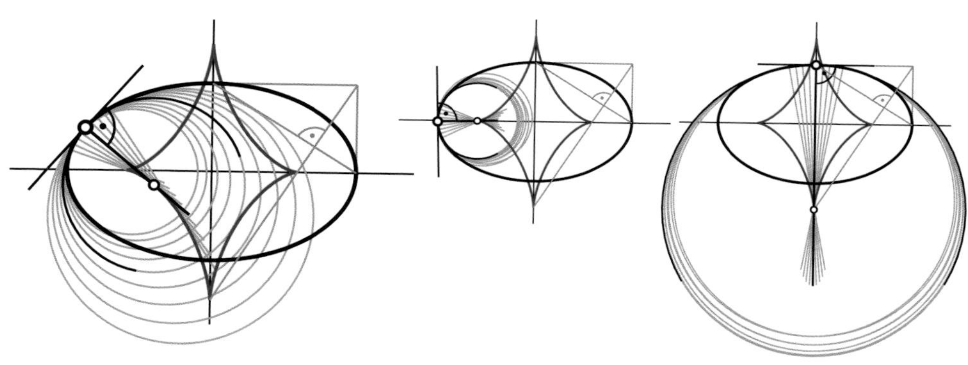

Fig. 4.2 Evolute and osculating circles of an ellipse

In Fig. 4.2, the evolute of an ellipse is visible – a diamond-shaped curve with four cusps at the centers of the osculating circles at the apexes of the ellipse. The figure also illustrates how to construct the osculating circles of the apexes, which are very useful when we draw an ellipse. ♠

• **A curve can be its own evolute**

In Fig. 4.3, one can see the logarithmic spirals with their evolutes. These are curves that always intersect a pencil of rays at a constant angle ψ. Their equation in polar coordinates (r, φ) is $r = e^{p \cdot \varphi}$. The constant parameter $p = \cot \psi$ determines how fast the logarithmic spiral wraps around its asymptotic point. The evolute of such a curve is another logarithmic spiral that is congruent to the initial curve if p is chosen appropriately. In special cases, it can even be equal to the initial curve (right image).
Logarithmic spirals play an important role in nature because they are related to exponential growth. They also have their place in technology: The photo on application p. 109 shows a paving slab lifter. Due to the weight of the object being lifted, the spirals are pressed towards the bottom, transmitting the pressure back to the slab. ♠

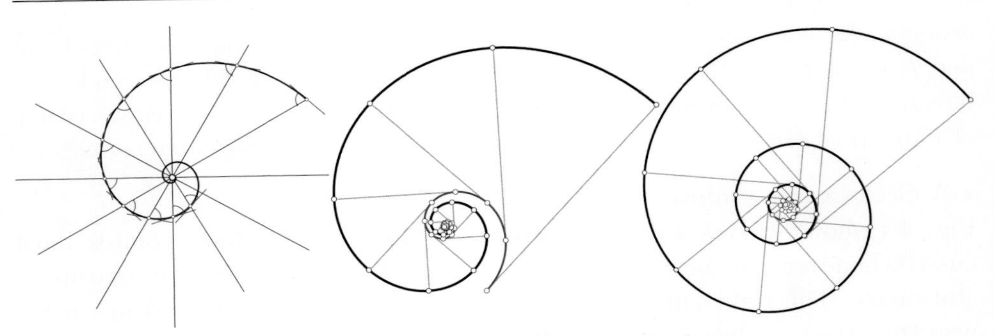

Fig. 4.3 Logarithmic spirals with their evolutes

• We smoothly rotate the steering wheel

The reciprocal value of the radius of curvature is called the *curvature* of the curve. The distance of two points, measured along the curve, is called the *arc length*. Fig. 4.4 shows a so-called *clothoid*.

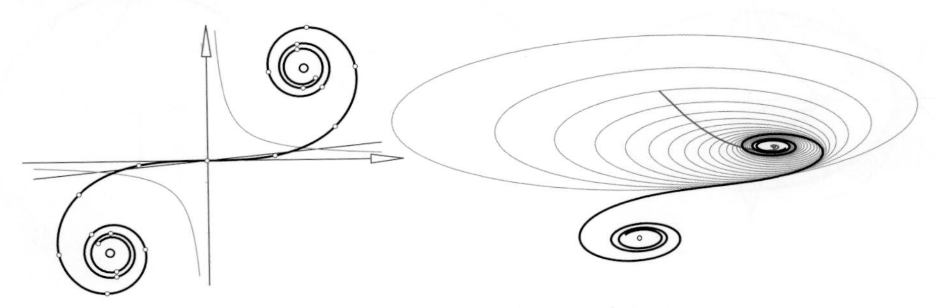

Fig. 4.4 Clothoid with winding points

It is characterized by the fact that the curvature increases linearly with the arc length. Thus, the radius of curvature decreases proportionally to the arc length (a hyperbola in the left image). Clothoids are applied in road and railway construction, where by a constant rotation of the steering wheel a straight path can be changed into a circular path with curvature continuity. The equation of the curve (it requires the numerical solution of integrals) is given in [11]. The curve wraps around two points (asymptotic points) in an infinite number of loops. ♠

• Architectural design using curves

Fig. 4.5 shows a virtual architectural design ("Intricate Surface") with a number of planar intersection curves that obey certain rules that the artist had developed ("mutations"). ♠

The theory of space curves

If the curve does not lie in a plane, then the whole issue becomes more demanding. The terms secant and tangent do not change their meanings. The curve normal, however, is no longer uniquely defined: Each normal to

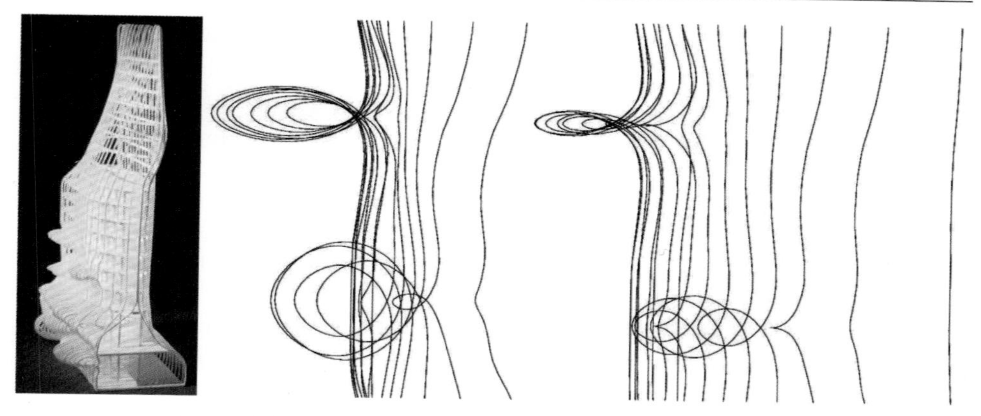

Fig. 4.5 Planar curves inspire artists (Greg Lynn, "St.Gallen").

the tangent through the point of contact is a curve normal, and lies in the normal plane ν of the curve tangent.

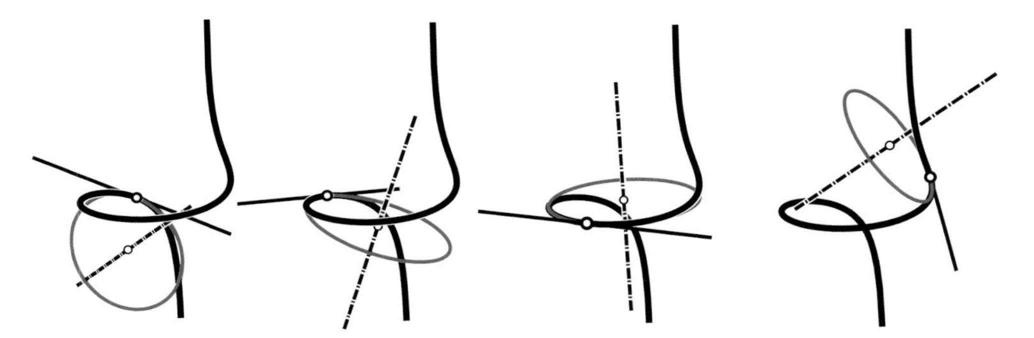

Fig. 4.6 Osculating circles of a space curve

In space, we can also consider a curve point P and *two* neighboring points Q and R. If the three points do not lie on a straight line, they define a circle in space. If we now let Q and R move towards P, the osculating circle of the curve at the point P converges to a limiting position. Its carrier plane is called the *osculating plane*. It is said: The osculating plane shares three (infinitely close) points with the curve at the point of contact.

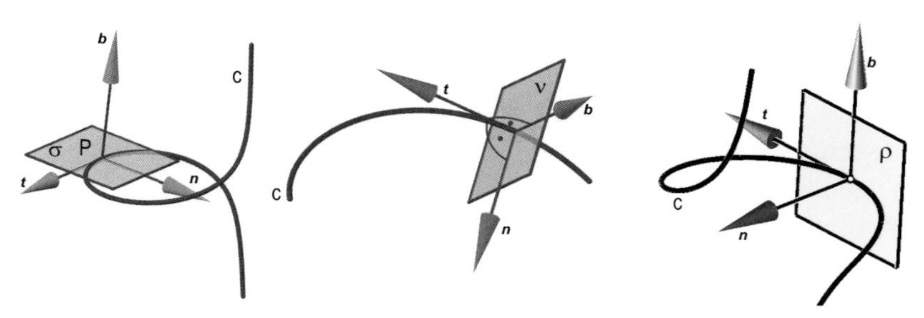

Fig. 4.7 Accompanying tripod (also called *Frenet frame*) of a space curve

Among all normals of the curve at the point P, we can now highlight one normal in particular – the one that lies in the osculating plane, also called the *principal normal.* If we now consider the normal that lies perpendicular to the osculating plane (the so called *binormal*), then we have finished the curve's *accompanying tripod.*

All osculating planes σ envelop the *tangent surfaces* of the curve, which we will discuss in detail later. All normal planes ν of a curve envelop a surface called *normal developable.* In Fig. 4.24 on the right side, we see an example of such a surface. More interesting for us is a surface that is enveloped by the so-called *rectifying planes*. The rectifying plane ϱ is spanned by the curve tangent t and the binormal b (Fig. 4.7 on the right). The thus resulting surface is called the *rectifying developable.* It occurs, for example, in twisted paper strips (Fig. 5.44).

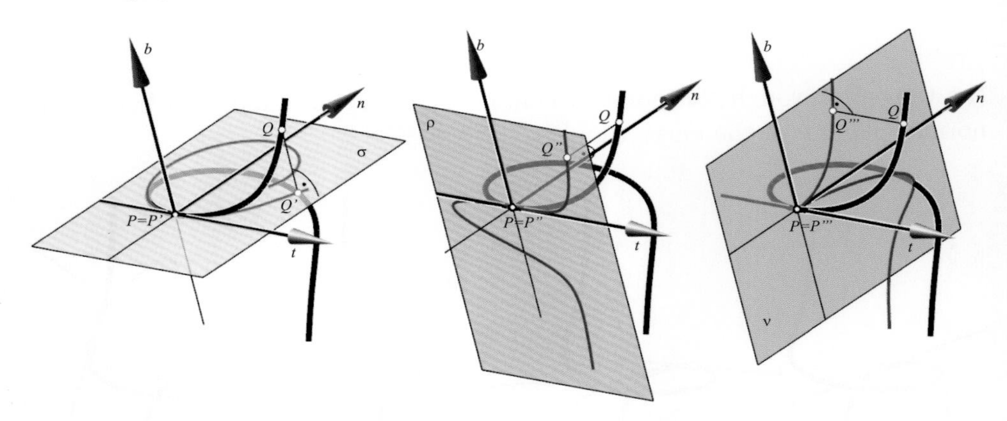

Fig. 4.8 Special projections of a curve

The normal projection of a curve to the osculating plane σ in P is a planar curve that has the same osculating circle in $P' = P$ as the space curve itself (Fig. 4.8 left). When projected orthogonally to the rectifying plane ϱ, then (in general) the curve has an inflection point $P'' = P$ (the osculating circle appears as a straight line, Fig. 4.8 middle). In the case of so-called *handle points* – points which may be found in a symmetry plane of a curve – a *flat point* occurs. When we finally project orthogonally to the normal plane ν (Fig. 4.8 right), a cusp $P''' = P$ occurs because the curve tangent appears as a point.

In the application p. 120, we will discuss the shortest curves (geodesic lines) on curved surfaces as applications of the aforementioned topics.

Next to the curvature, geometry also employs the concept of *torsion* of a curve (Fig. 4.9).

Circles (seen as space curves) have constant curvature but zero torsion. Helices as "generalized circles" also have constant curvature but additionally constant torsion.

The torsion at a curve point is a measure of the rate at which the osculating plane rotates about the tangent (by analogy to the curvature of a planar curve, which is a measure of the rate at which the tangent rotates about the

Fig. 4.9 Torsion of a space curve and something akin to an accompanying tripod

curve point). A curve with zero torsion lies in a fixed plane, which is the osculating plane (by analogy, a planar curve without curvature stays in the direction of a straight line, which is the tangent).

Space curves can create surfaces

If a curve moves through space (during which its form may change), it usually "sweeps" a surface. We call the different positions of the curve u-lines. The orbits of the curve's points form a second set of curves on the surface, the v-lines (Fig. 4.10).

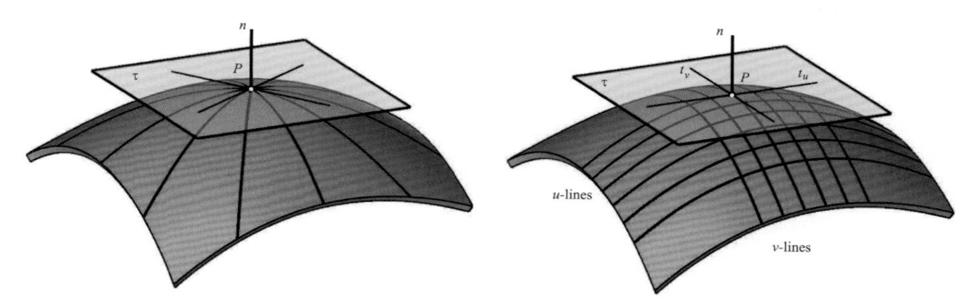

Fig. 4.10 Left: The tangent plane τ in a surface point P is spanned by all possible tangents which are all perpendicular to the surface normal n. Right: u- and v-lines of a surface. The plane τ is spanned by the respective tangents t_u and t_v.

In general, only one u-line and one v-line pass through each surface point P. The tangents t_u and t_v in P at both lines span a plane τ. It touches the surface at P and is called the *tangent plane*. The normal n of the tangent plane P is called the *surface normal*.

Contour and outline of a surface

Let us view a curved surface from a point E (in Fig. 4.11, it is a point E_∞ at infinity). Then the *contour* of the surface is usually immediately seen. It is the locus of certain points on the surface that form the *contour line* k. Let K be a contour point. It is defined such that the viewing ray EK touches the surface and therefore lies in the tangent plane τ of K. This means that it is perpendicular to the surface normal n. Let us formulate the following important theorem:

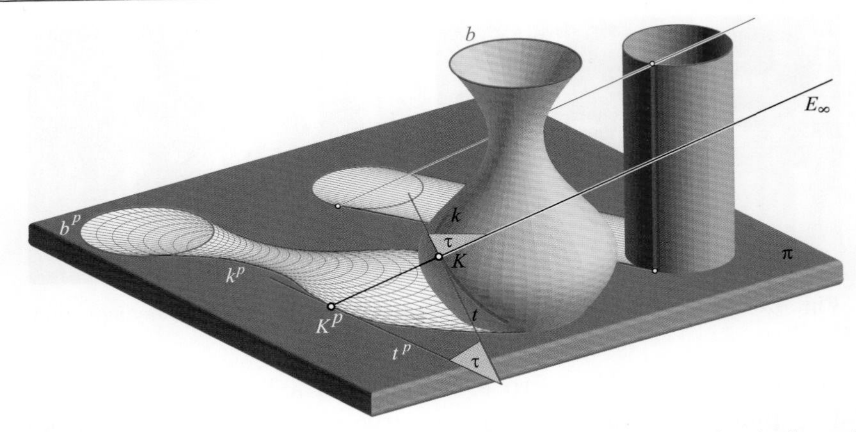

Fig. 4.11 Contour point K, contour k on the surface and outline k^p of the surface in the projection plane π. Bordering lines b (blue) are *not* part of contour / outline. Cylinder on the right: Both contour and outline are straight lines.

In a contour point, the viewing ray is orthogonal to the surface normal. The tangent plane at the contour point goes through the projection center.

In the projection to an image plane π, the surface – apart from potentially occurring boundary curves – is bounded by the projection k^p of the contour line (the *outline*).

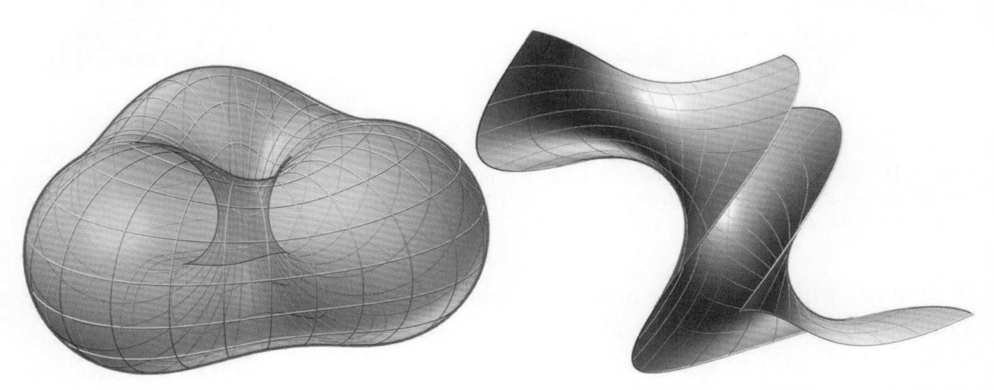

Fig. 4.12 The surface outline (red) is enveloped by the images of arbitrary surface curves (in this case, two different pencils of curves are drawn blue and green). Note that bordering lines are *not* part of the outline.

The outline k^p is also enveloped by the projections c^p of all possible surface curves c (Fig. 4.12). The following theorem is especially important for geometrically correct free-hand drawing:

If a curve on a carrier surface *intersects* the contour of the surface, then, in general, its projection *touches* the outline of the surface.

Only in exceptional cases, the projection c^p of a surface curve c has a cusp on the outline – namely, when the projection center E lies on the tangent of c at the intersection with the contour (Fig. 6.76).

• **Why do *highlights* occur on curved surfaces?**

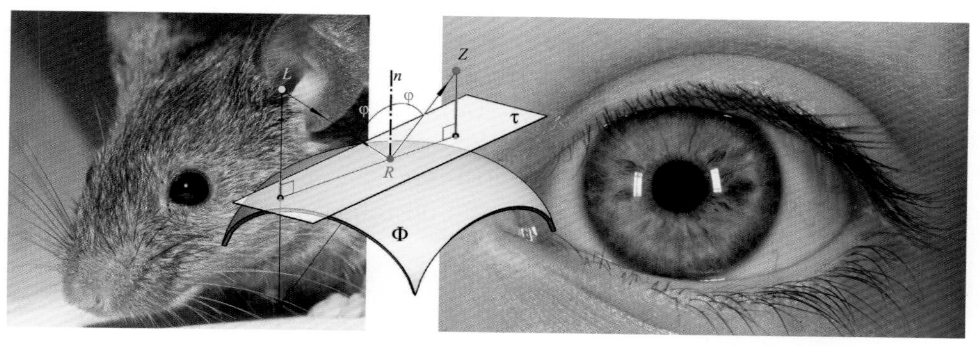

Fig. 4.13 Highlights in different surfaces. Left: One light source (the sun). Right: Reflections of different non-point light sources (flashlights).

The basic geometric assumption is as follows (Fig. 4.13): A point R of an illuminated surface is a highlight when the projection center Z and the light surface L lie in the same plane through the surface normal n in R. An additional restricting condition says that n must be the angular bisector of LR and ZR. Fig. 4.13 shows highlights on spherical surfaces. The highlights lie in planes that include the respective centers of the spheres, the position of the light source, and the lens center of the camera.♠

How can the curvature of a surface be measured?

The term *surface curvature* is much more complicated than the simple *curvature* of a curve: By merely considering the *normal sections* of the surface at a surface point P like in Fig. 4.10 on the left (these are the surface curves through P in a plane through the surface normal n), one deals with an infinite number of curvatures of curves through P.

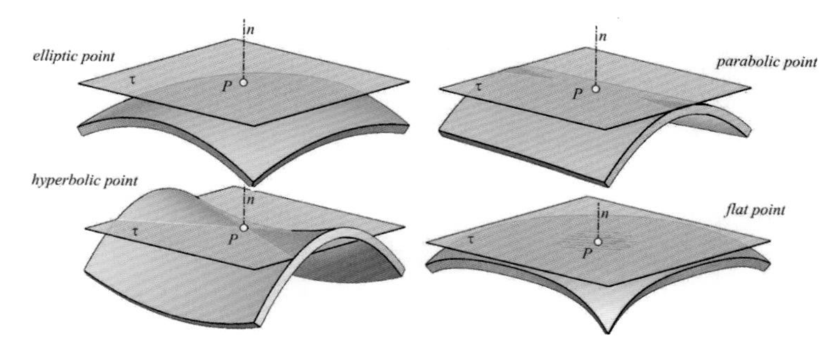

Fig. 4.14 Elliptic, parabolic, and hyperbolic surface points, flat point (bottom right)

Even among the diversity of curved surfaces, three significantly different types of surface points can be distinguished. For this purpose, let us consider the surface such that P is the contour point (the projection center lies in the tangent plane of P). If the surface now lies in *a sufficiently small neighborhood* of P *on both sides* of the tangent plane, then we speak of a *hyperbolic point* (Fig. 4.14 bottom left). Each *saddle surface* possesses points which are exclusively of this type. If the surface touches the tangent plane at one side, then then we speak of an *elliptic point* (top left; a typical example is the sphere). If there exists a certain surface curve (or tangent) through P, along which the surface, only considered locally, remains completely in the tangent plane, we speak of a *parabolic point* (in Fig. 4.14 top right we see how a tangential section "forms a cusp" at the point). Cylinders and cones represent this type particularly well. A fourth type can only happen singularly or alongside a curve on the surface: Fig. 4.14 bottom right shows a so-called *flat point.*

Practical implications

This differentiation has great practical implications. Surfaces which consist only of parabolic points (including some flat points) are the easiest to modify. What is more: Their important characteristic is that they can be unfolded ("developed") into a plane. Hyperbolically curved surfaces are much harder to modify than even elliptically curved ones.

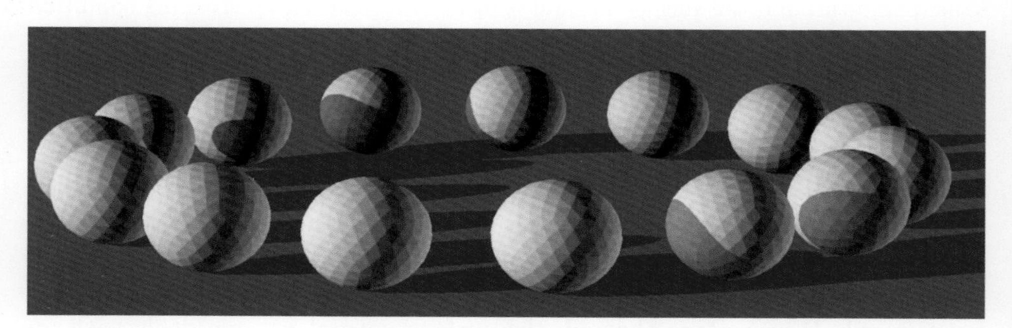

Fig. 4.15 Triangulation of the sphere

Both latter types *cannot be developed*, which is especially impractical in the case of the sphere: There is no way of mapping the surface of the Earth into the plane without distortions. While spheres can be approximated by many quadrangles or triangles more or less well, the faces will not all be mutually congruent. Fig. 4.15 shows a not-quite-perfect distribution of triangles. Fig. 4.16 illustrates how this triangulation can be derived by the "refinement" of an icosahedron.

• A moderately equal distribution of points on the sphere

The problem of evenly distributing a given number of points on the surface of a sphere is only exactly solvable for 4, 6, 8, 12, and 20 points, for it leads to patterns as on the following platonic solids: tetrahedron, octahedron, cube, icosahedron, and dodecahedron. Based on these solids, the pattern may be "refined" as in Fig. 4.16, thus allowing for multiples of the aforementioned numbers. ♠

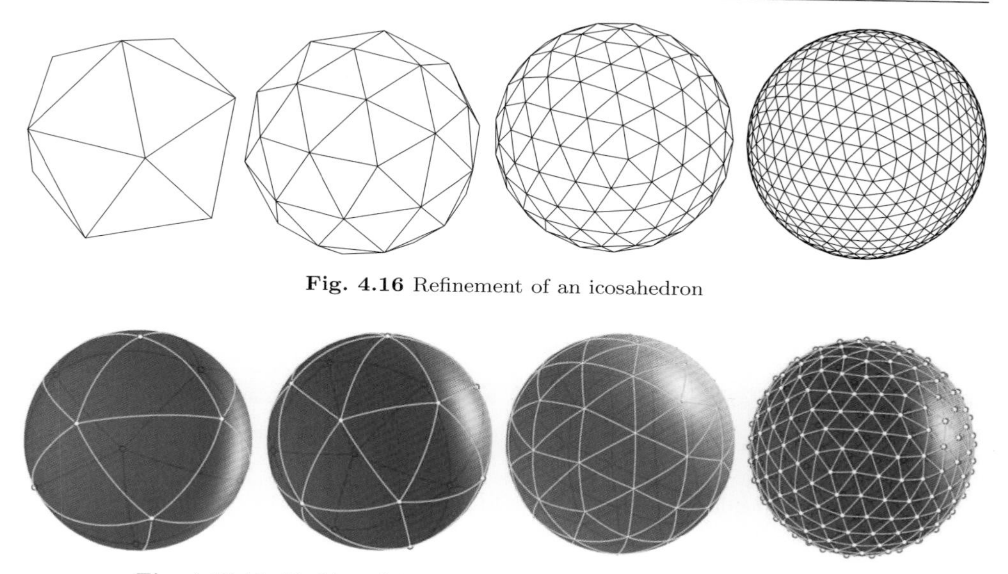

Fig. 4.16 Refinement of an icosahedron

Fig. 4.17 10, 15, 64, and 250 points distributed uniformly on the sphere.

Fig. 4.17 illustrates a possible solution for other numbers of points. The solutions were calculated by using a dynamic approach that employs quasi-physical repulsion for the points. The shortest distances between the points are parts of great-circles on the sphere (see p. 121), and the repulsive forces increase and decrease by the square of the distance. The translation of the points on the surface continues until a state of equilibrium is reached. The algorithm was developed by Franz Gruber based on the ingeniously simple algorithm by Fruchterman/Reingold [5]. This problem is also known as the *Thompson problem*, where the topic concerns the minimization of energy of n equal point charges on the sphere.

Fig. 4.18 Art work with curved approximations to spheres (Kyoto)

• Triangulation as a powerful tool

Triangulation is of high importance in the computer age: It is not only useful for the approximation of surfaces but also for the simplification of complex geometric volumes. Many particularly difficult physical problems can be solved with adequate accuracy by breaking down the situation into partial problems.

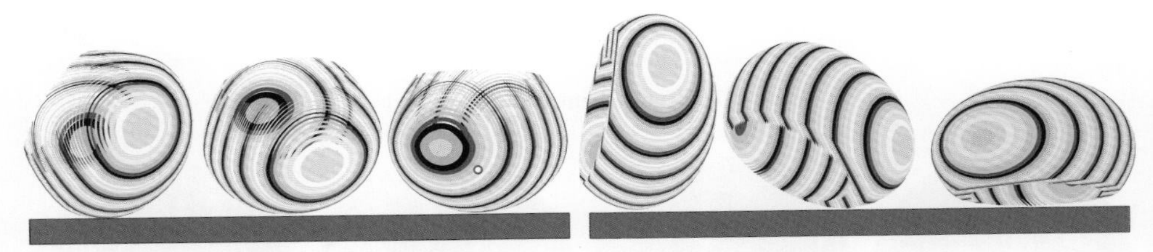

Fig. 4.19 A rolling brandy glass and a weaving ashtray

Fig. 4.19 on the left shows how a body is first cut into sections – in this case, it is a fictitious brandy glass. The emerging intersection curves can be highly complicated (closed) line strips that may also have any number of holes. In the second step, these sections are triangulated (subdivided into triangles) and can thus be interpreted as volumes (three-sided prisms with the thickness of a disc). For each prism, the centroid, the statical momentum, and the rotational momentum regarding arbitrary axes (see [11]) can be determined by using simple formulas. The sum of all centroids is called the *total centroid* (in the picture it is marked red). The sum of all rotational moments is called the *total rotational momentum*.

Fig. 4.20 Some complications arise with the introduction of fluids ...

Realistic physical simulations can be accomplished in this way. For instance, it is conceivable, given the right impulse, that the ashtray in Fig. 4.19 right can set itself upright from its seemingly hopeless situation. When simulating fluids (Fig. 4.20), however, simple geometry is no longer sufficient due to the complexity of the situation. ♠

• Geodesic lines – the shortest connection

If two points A and B on a curved surface need to be connected by a surface curve of minimal length, this will (in general) not be a straight line. We shall discuss this problem particularly in relation to the sphere, and the shortest connecting line in this special case (application p. 132).
Let us first define the following:

A *geodesic curve* on a surface is characterized by the fact that in all of its points, the respective osculating plane includes the surface normal.

This leads to the following:

The shortest connection between two points in a surface is a geodesic curve on the surface.

At this stage, let us not proceed with a mathematically rigorous proof but merely make the matter more plausible: By definition, all principal normals (Fig. 4.7) of a geodesic curve are at the same time normals of the carrier surface. Let us now consider two neighboring points Q and R of a surface point P. In the normal projection to the tangent plane in P, the surface curve is approximated by the normal projection of the osculating circle – it is a straight line because the osculating plane is its carrier plane.
In other words: Locally speaking, the curve works like a straight line – which, as we know, is the absolutely shortest path. Every other connection from Q to R would be a detour.

For two special cases, there exists a simple and easily applicable solution:

Geodesic lines on spheres are great-circles. Geodesic lines on developable surfaces become straight lines in the unfolding.

Proof:
First of all, let us consider the sphere: The osculating plane of the surface curve has to contain the surface normal and must, therefore, pass through the center of the sphere. Thus, the osculating circle at every point is a great-circle. If the geodesic curve were to deviate from the great-circle, then the osculating plane would "sway", which would lead to a contradiction.
Now to the developable surfaces: The unfolding preserves lengths, and in the plane, the shortest connection between two points is a straight line. Fig. 4.66 shows geodesic curves on a cylinder of revolution that result from the rolling-up of straight lines. These are *helices* that we shall discuss further in Chapter 7. ◇ ♠

How can we recognize a developable surface?

If a surface is parabolic at *every* point P, then by definition it also has a tangential direction in P, along which the surface stays entirely in the tangent plane. If one follows the surface along the respective tangential direction, then one always stays in the same tangent plane. This plane constantly touches the surface. It follows that parabolic surfaces are composed of straight lines and that, along these straight lines, they are touched by the same plane.
However, the condition that a surface is composed of straight lines is *not* sufficient. Quite the contrary: Most curved surfaces which are composed of straight lines, are *not* developable!

Fig. 4.21 Straight contour parts of developable surfaces must be visible from all directions.

How, then, can we find out whether a surface is parabolic throughout? If each tangent plane touches along a whole straight line, and this tangent plane by coincidence runs through the projection center, then *each* point of the straight line is a contour point. In other words: The contour of such a surface is composed of nothing but straight line segments (Fig. 4.21 and Fig. 4.22 and Fig. 4.23).

Fig. 4.22 Developable stays developable: The deformation of the cylinder creates new developable surfaces (to be recognized by the straight outlines).

The only surfaces that can be developed into the plane without distortion are surfaces without elliptic or hyperbolic points. A developable surface can be recognized by the fact that its contour is straight for any viewpoint.

Even if the contour of an object in space is merely a "harmless" curve, tips occur relatively often when dealing with the contours of *hyperbolic* surfaces: In each case the tangent plane in a contour point contains the center of the projection. This is why it is often the case that the *tangent at the contour* passes exactly through the center.

At this point, let us make an additional statement about the osculating circles of a space curve. A picture says more than a thousand words – therefore, simply "enjoy" (more information can be found in academic literature):

If a point moves along a spatial curve, the related osculating circle constantly changes. The surface that is generated by all these osculating circles can be very complicated (Fig. 4.24 left) and is always doubly curved.

Fig. 4.23 This shell has a virtual enveloping cone of revolution. Since a cone is developable, we have straight contours and shadow boundaries for any projection and light direction.

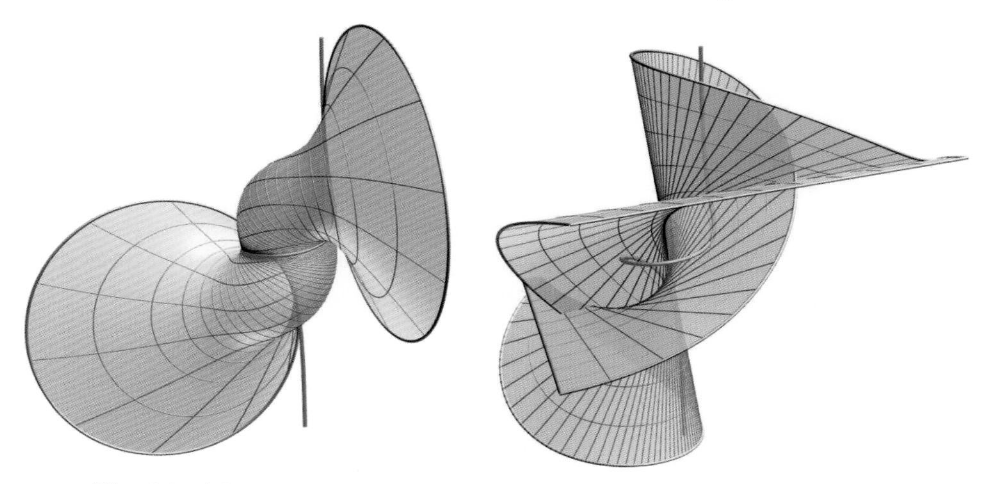

Fig. 4.24 All osculating circles of a spatial curve and their axes form a surface.

The surface traced by the osculating circles of a space curve is also doubly curved (in general). It can be seen as a spatial analogue to the locus of all centers of curvature of a planar curve – the evolute. Fig. 4.24 on the right shows the corresponding surfaces for a cubic circle (compare with Fig. 6.32).

Analogous considerations for shadows

As we already know, there is actually no difference between projection and illumination. All that we have learned so far about projection is equally applicable for central and parallel illumination.

Instead of the contour, we now refer to the *boundary of the self-shadow* (terminator), and instead of the outline of the projection, we now refer to the *cast-shadow boundary*. The cast-shadow of a surface is surrounded by the cast-shadow of self-shadow boundary and possibly by cast-shadows of the surface's boundary curves. If a surface is developable, a part of the self-shadow / cast-shadow boundary is straight (Fig. 4.25).

Just to make it more clear, we repeat it in other words: An object always produces a cast-shadow on planes, other objects and – if the object is not

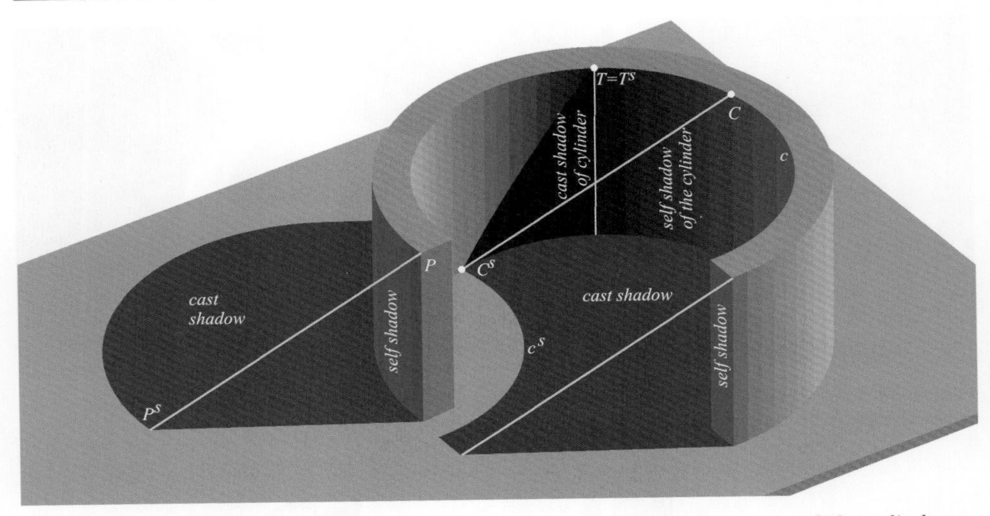

Fig. 4.25 Self-shadow and cast-shadow of curved surfaces: On the inner part of the cylinder we have a cast-shadow of the cylinder on itself and self-shadow ("dark areas").

convex – potentially on itself. However, there are parts of the surface of the object that are "simply dark" (Fig. 4.26) because they are turned away from the light source. These parts are called self-shadow.

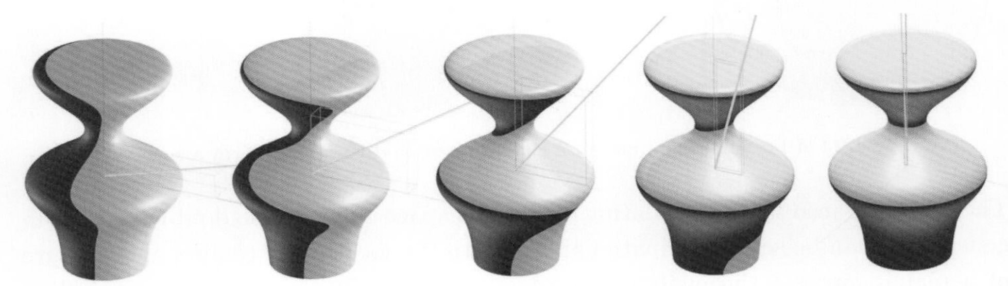

Fig. 4.26 "Dark" parts on a surface: The parts of the surface that are turned away from the light source are in "self-shadow". If the surface is not convex (like in this example), there will potentially appear additional cast-shadows of the surface on itself (not displayed in these images), especially when the light comes from higher above (images to the right).

The line that separates lit areas and areas in self-shadow is called boundary of the self-shadow or sometimes (especially in context with planets) terminator. Fig. 4.25 shows that the inner part of a hollow (and therefore non-convex) cylinder carries both types of shadow which are separated by a straight line through T. The viewer cannot distinguish between the two types, and the separating terminator is not clearly visible. Someone who draws such figures, however, or writes computer programs to render them, needs to know about this.

4.2 The sphere

Despite the fact that we have talked about developable surfaces in adequate detail, let us, nevertheless, start with a surface that is not developable. The sphere is particularly easy to imagine, occurs frequently, and has many special properties.

Fig. 4.27 Spheres as distance surfaces

The sphere is the locus of all space points that have a constant distance (the radius r) from a given point (the center M).

Fig. 4.28 Explosion plus gravitation (fireworks)

Fig. 4.27 illustrates that the flower head of a dandelion or the tips of the spines of a sea urchin define a sphere. Fig. 4.28 shows how the luminous particles of a firework travel on spheres. The same is true for sonar waves.

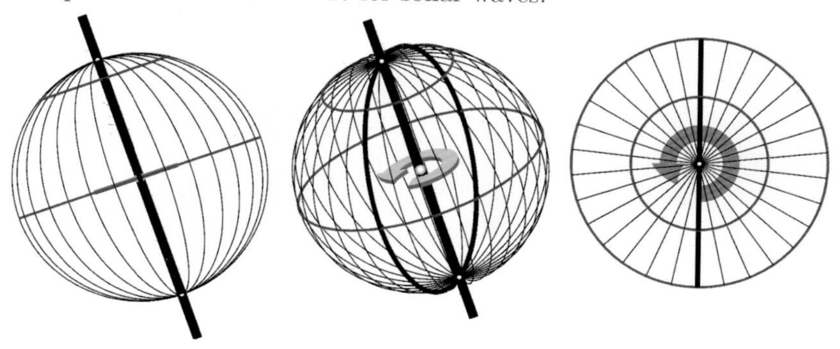

Fig. 4.29 A sphere is generated by the rotation of a circle about its diameter.

The sphere can be defined in a different way, which is advantageous for many geometric considerations:

A sphere is also formed by the *rotation of a circle about an arbitrary diameter of this circle.*

Let us consider a point P of the circle during a rotation about an arbitrary diameter d (Fig. 4.29): It describes the path circle with the axis d and the radius $\overline{dP}$. At maximum radius, the straight line MP is orthogonal to the axis d. Sphere circles with a maximal radius (the sphere radius) are called great-circles – all other circles are small circles. If P lies on d, then the small circle shrinks to a "null circle". Its carrier plane is orthogonal to the sphere's radius at the contact point (Fig. 4.30).

Fig. 4.30 Tangential planes of a sphere orthogonal to the radius at the contact point

If we now consider that d can be chosen in an infinite number of ways, we have proven the important theorem that every planar section of a sphere is a circle (Fig. 4.31):

Each planar spherical section is a circle. The intersection circle is a great-circle only if the plane contains the center. All tangent planes of the sphere are orthogonal to the radius at the contact point.

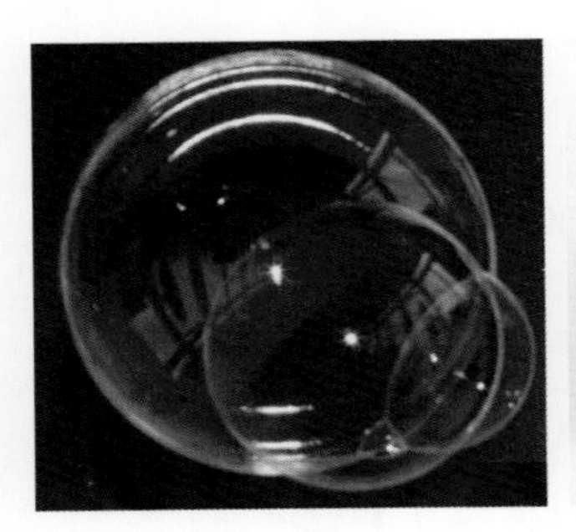

Fig. 4.31 Circles as intersections of spheres

If two circles rotate about a common diameter, we get two spheres (Fig. 4.32). During rotation, the common points of the circles generate the same circle:

The intersection of two spheres is always a circle which lies in a plane orthogonal to the straight line spanned by their centers.

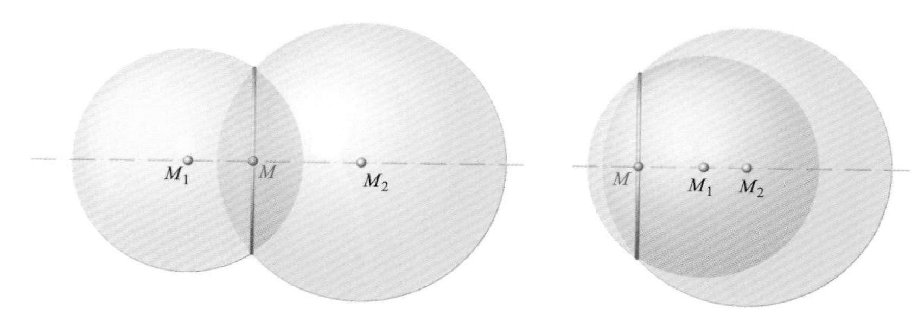

Fig. 4.32 The intersection of two spheres is always a circle.

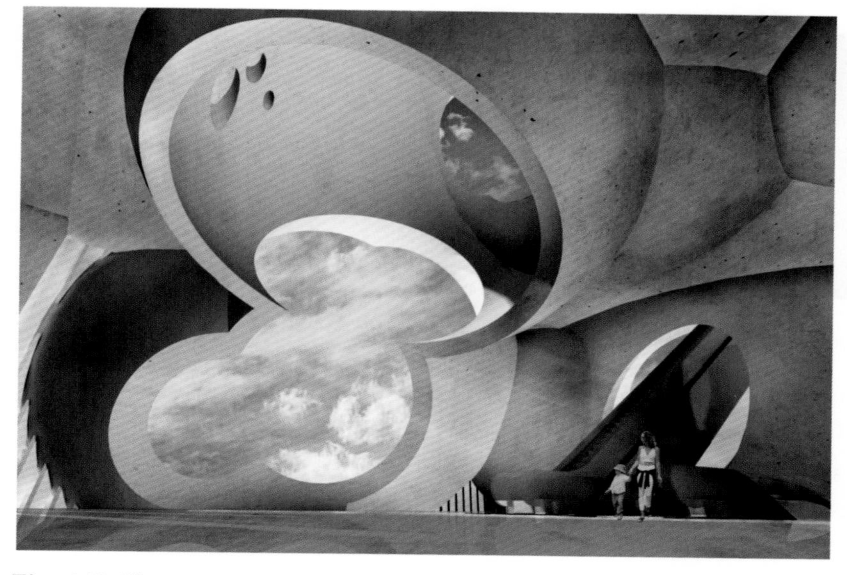

Fig. 4.33 Planar sections and mutual intersections of spheres (Lisa Sommerhuber)

An additional remark for the advanced: You may think that the theorem should be stated more precisely: Two spheres only share a circle if they are close enough to each other. In fact, it can be calculated that two spheres *always* share a circle, but it is not always a real one.

Its center is always real and lies on the straight line joining the spheres' centers. We can further say (in what may be an overstatement): Two spheres *always* share a non-real circle in the plane at infinity – the so-called *absolute circle* (compared to p. 159). Once we know this, some things are no longer surprising: Two spheres should share a space curve of fourth order, but because the absolute circle is always part of it, the intersection splits off and there only remains a planar curve of second order – a circle, which may be real or complex.

If we rotate a circle not about one of its diameters but about a chord, we get the so-called *spindle torus* (Fig. 4.34 center and right). We will discuss it further in the section concerning surfaces of revolution.

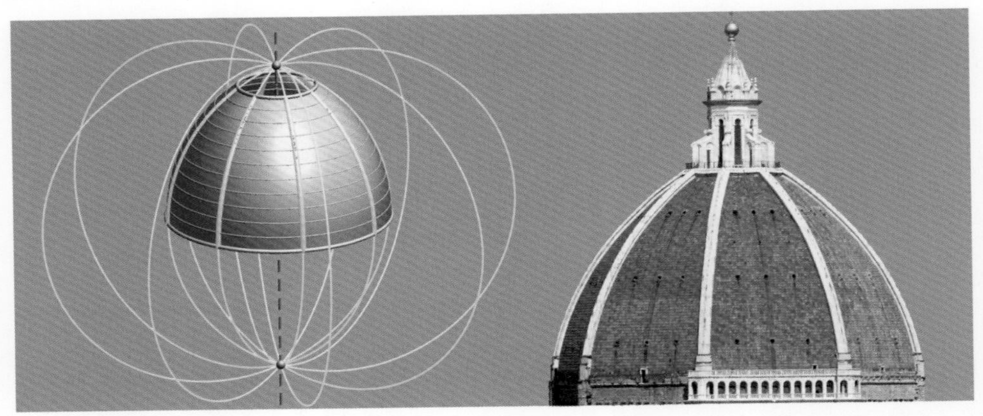

Fig. 4.34 A circle rotates about one of its chords and generates a spindle torus. Right: The roof of the dome of Florence.

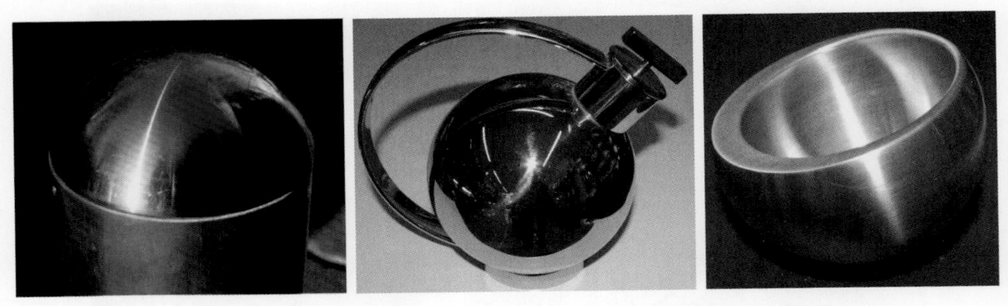

Fig. 4.35 Left: sphere cap, right: spheres in design

Parts of spheres often occur in architecture (Fig. 4.33) and technology (Fig. 4.35). One famous example is the Byzantine dome (Fig. 4.36).

• **Position of an airplane**

By using radar, the distance r of an airplane to a ground control station B can be determined. The airplane lies on the *distance sphere* around B at the distance r.

If the synchronous measurements of three ground control stations are known (whose relative positions are known), one gets three distance spheres that share two points: Two spheres intersect along a circle and all three circles

Fig. 4.36 Sphere caps in classical architecture (Santorini) and design (Munich airport)

Fig. 4.37 Byzantine domes (interior of the Hagia Sophia)

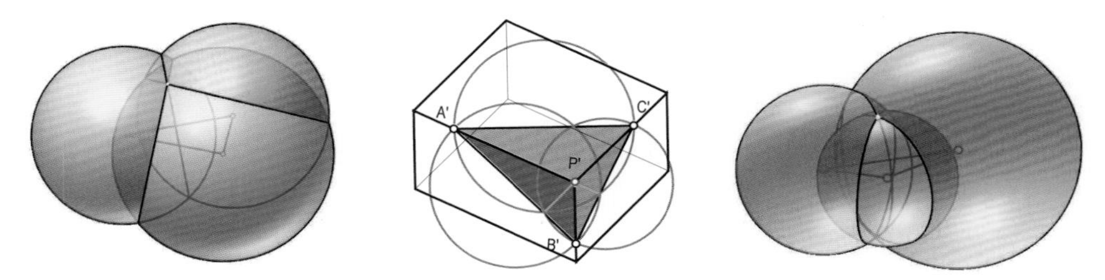

Fig. 4.38 Intersection of three spheres and cuboids through three points (center)

pass through the common points (Fig. 4.38). One of these (above the surface of the earth) is the position of the airplane. ♠

• The "spatial Pythagoras"

We are looking for a cuboid where the three edges connected to the corner P pass through three fixed points A, B, and C (Fig. 4.38 center). Due to $\angle APB = \angle BPC = \angle CPA = 90°$, the point lies on the *Thales spheres* above $\overline{AB}$, $\overline{BC}$, and $\overline{CA}$. For this cuboid corner, it can be shown very elegantly by using vector calculus that the square of the area of the triangle ABC is equal to the sum of the squares of the triangle areas APB, BPC, and CPA. The analytic proof can also be found in [11] and is given here as well due to its brevity:

Proof:

Let P be the point of origin of a Cartesian coordinate system with the points $A(u/0/0)$, $B(0/v/0)$, and $C(0/0/w)$. The triangles APB, BPC, and CPA then have the areas $\frac{u \cdot v}{2}$, $\frac{v \cdot w}{2}$, and $\frac{w \cdot u}{2}$.

The area d of the triangle ABC results from the cross product

$$d = \frac{1}{2} \cdot |\overrightarrow{AB} \times \overrightarrow{AC}| = \frac{1}{2} \cdot \left| \begin{pmatrix} -u \\ v \\ 0 \end{pmatrix} \times \begin{pmatrix} -u \\ 0 \\ w \end{pmatrix} \right| = \frac{1}{2} \cdot \left| \begin{pmatrix} vw \\ wu \\ uv \end{pmatrix} \right| = \frac{1}{2} \cdot \sqrt{(vw)^2 + (wu)^2 + (uv)^2}.$$

Thus, $d^2 = \dfrac{1}{4} \left[(vw)^2 + (wu)^2 + (uv)^2 \right] = b^2 + c^2 + a^2.$ ◇ ♠

Outline and shadow of a sphere

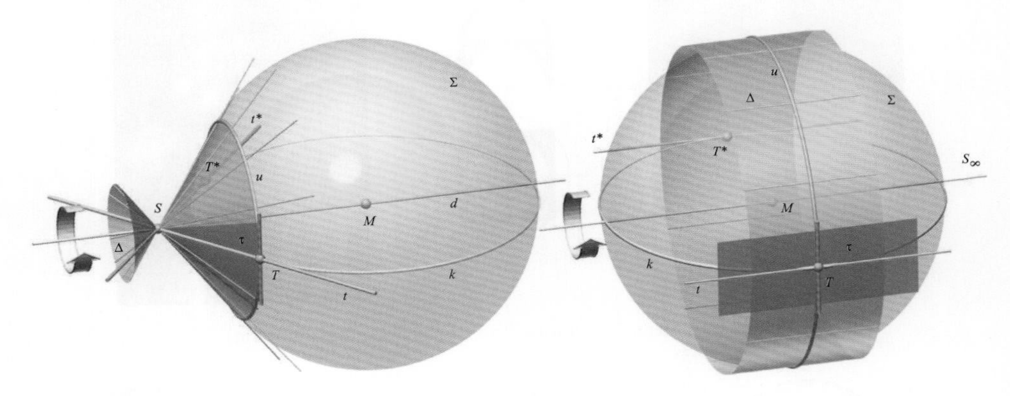

Fig. 4.39 Tangential cone and cylinder

The following theorem provides the key for all further considerations:

The tangent planes of a sphere along a small circle envelop a cone of revolution, and those along a great-circle envelop a cylinder of rotation.

Proof:
Let us imagine a circle k (center M) in the plane (Fig. 4.39) and a point S, from which two tangents t and t^* to k exist. The straight line TT^* connecting the contact points is for reasons of symmetry perpendicular to the diameter $d = MS$. Let now the carrier plane of the circle rotate about d. The circle k becomes a sphere Σ, t generates a cone Δ of revolution with the apex S, and the contact point T generates a small circle u with the axis d. The tangent of u at T and the line ST span the tangent plane τ of Δ. During the rotation τ envelopes Δ.
If we now let S pass into infinity (S_∞), then the cone of revolution becomes a cylinder of revolution, and the small circle u becomes a great-circle. ◇

Fig. 4.40 Outline and shadow of a sphere

How do we see a sphere from the center S? From S, we can consider an enveloping cone of revolution that touches the sphere.
If S is a point at infinity, then the cone becomes a cylinder of revolution. The contact curve of the cone or cylinder of revolution is a small circle or great-circle u respectively. However, this circle is already the contour of the sphere: According to construction, the tangent plane in each point passes through S. The analogue is true for central- and parallel illumination. Let us summarize:

The contour and self-shadow boundary (terminator) of a sphere is always a circle on the sphere. With parallel projection or parallel illumination, the contour and self-shadow boundary is a great-circle.

• **Tilt of the slender moon crescent – a "result of temperature"?**
Why is the slender moon crescent often "lying" in tropical regions, but more upright in northern regions?
Solution:
Whether or not the moon crescent is inclined or upright depends on the moon's terminator (see also application p. 66). It appears as an ellipse, and its apparent inclination (tilt) depends on its principal axis. It therefore matters from which direction the moon is illuminated. If it happens from the side, then the slender crescent appears to stand upright, and if the light comes from above or below, then an inclination becomes apparent. We must further be aware of the following: From the vantage point of earth, the path of the new moon follows the path of the sun, though somewhat delayed (see also p. 389). This is also approximately true for the slender moon crescent (but not for fuller moon phases).

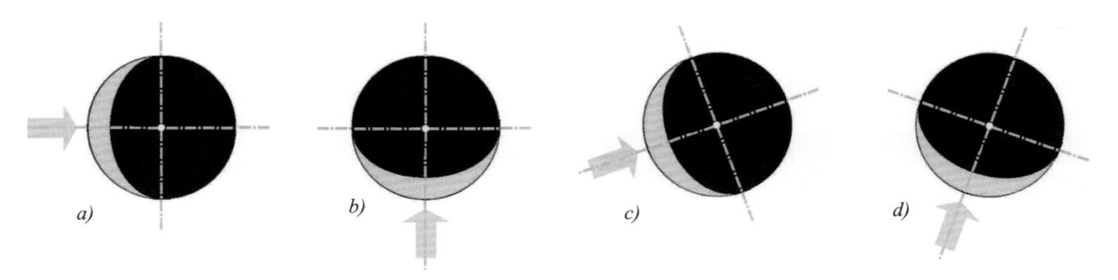

Fig. 4.41 Tilt of the slender crescent: a) at the poles, b) at the equator beginning of spring or fall, c) somewhere in-between in winter, d) somewhere in-between in summer

Let us start with an extreme: At the poles (Fig. 4.41 a), the path of the sun (if it is visible at all) appears as a circle at a fixed elevation. The slender moon crescent should also appear at a similar elevation – and not very far from the sun. When viewed from the pole, the moon is illuminated "from the side" so that the self-shadow boundary (terminator) is visible as an upright half-ellipse. In the deeper polar winter, both the sun and the moon crescent cannot be seen. (However, as is apparent on p. 389, the full moon in polar winter is very well visible. Upon closer inspection, it can be discovered that the near-full moon "appears" as in the tropics!)
Now for the other extreme: At the equator, there are two days of the year when the sun moves at a circle through the zenith (Fig. 4.41 b). The slender moon crescent

follows this path. This time the terminator appears as an inclined half-ellipse: From the perspective of the equator, in the days around new moon, it is illuminated almost precisely from the top (waxing moon, in the afternoon and evening hours) or from the bottom (waning moon, in the morning hours).
In general, path circles of the sun and the moon *in the days around new moon* are steeper in latitudes closer to the equator. The steeper they are, the flatter the ellipse appearing as the images of the moon's terminator are. It can further be said that the crescent moon will predominantly occur in summer, not in winter, because the sun's path is steeper in summer (Fig. 4.41 c and d). Therefore, it is not at all far-fetched to say that the position of the slender moon crescent is correlated to (even if not directly caused by) temperature. ♠

The shortest path on a sphere

In space, the shortest connection between two points is a straight line. On the globe, we encounter a problem when dealing with larger distances: We cannot, like Jules *Vernes*, travel through the center of the earth. The shortest connection between two points A and B on the sphere is a great-circle – a circle whose plane is defined by A, B, and the center M of the sphere. If, therefore, a ship were to travel from a certain circle of latitude to another point on the *same* latitude, the shortest path would *not* follow this circle (except at the equator or a meridian)!

The circle of latitude would be more practical for navigation purposes: It can be calculated without much trouble by measuring the angle of elevation to the peak level of the sun (see p.384). For this reason, it used to be common in old times to travel along circles of latitude – better to cross some more sea miles than to "miss" a life-saving island in the ocean: The exact determination of the circle of longitude used to be very difficult at sea. This problem was eventually solved by the use of "stormproof" clocks [11].

Fig. 4.42 Navigation systems on ships

In the age of GPS (Global Positioning System, Fig. 4.42), it may, indeed, be attempted to come close to the great-circle while crossing from A to B. Of course, one must not forget the importance of water and air streams ("jet streams").

• From Vienna to Varadero

Fig. 4.43 shows a flight path between two continents – naturally displayed on a 2D map, on which great-circles are not recognizable as such. These are photographs of a low-resolution monitor within the airplane.

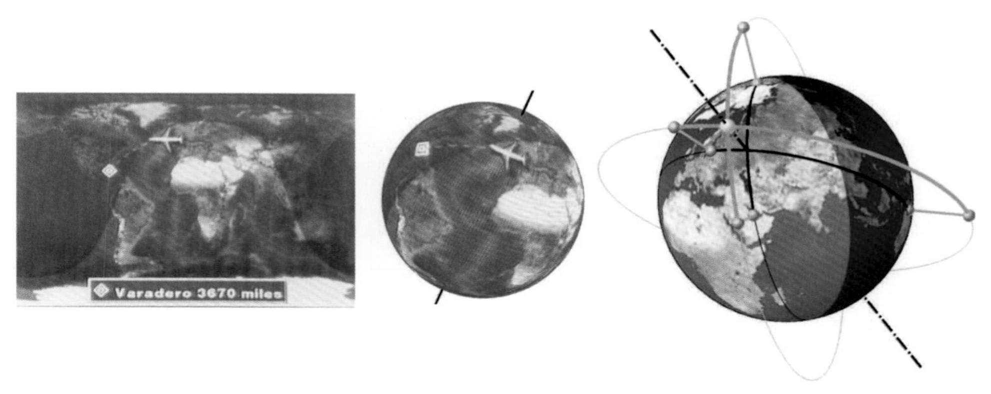

Fig. 4.43 Shortest path from Vienna to Varadero

In the middle of Fig. 4.43, the 2D map was transformed onto a sphere. The flight path should now come close to a great-circle (the offset is, indeed, negligible). The flight from Vienna (48° northern latitude, 16° eastern longitude) to Varadero in Cuba (20° northern latitude, 70° western longitude) leads (at first) somewhat into northern latitudes. From the globe shadow, which deeply envelops the north pole, it can be concluded that the flight must have taken place at the approximate time of the winter solstice. During the multi-hour flight, the shadow boundaries move westwards. The image on the right shows the ideal flight path (which is a great-circle). ♠

4.3 Cylinder surfaces

From the tried-and-true cylinder of revolution . . .

So far, we have already used the term *cylinder of revolution* because we have assumed that everybody can imagine such an object. Let us now attempt an exact definition:

A cylinder of revolution is the locus of all points in space that are equidistant to a given straight line (the *axis*) at the fixed distance r (the *radius*).

As with a sphere, a cylinder of revolution can also be explained with a motion in space:

A cylinder of revolution is generated by the rotation of a straight line about a parallel axis.

We can conclude that cylinders of revolution are "refined" regular prisms. Incidentally, computers also display them as regular prisms with a high number of edges (the eye can barely distinguish the difference between a 200-sided prism and a perfect cylinder).
Both given definitions define a cylinder of revolution as unbounded, or "infinitely long". In practice, cylinders of revolution are usually delimited by sections which are perpendicular to the axis (circles). The development of such a bounded cylinder of revolution is a rectangle with the width $2\pi \cdot r$ and the height h.

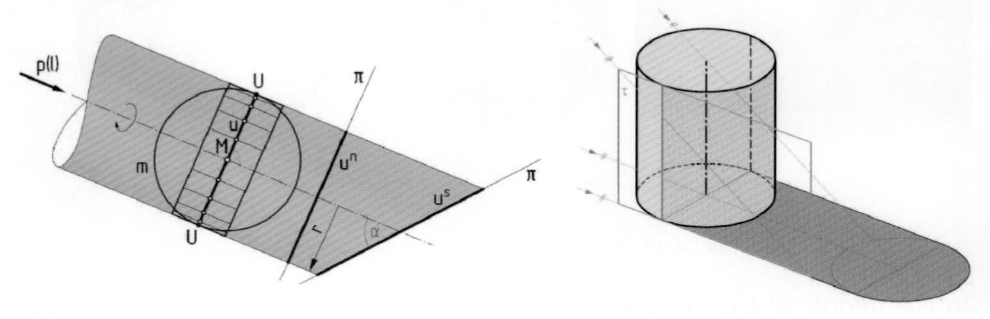

Fig. 4.44 Outline and shadow of a cylinder of revolution

Cylinders of revolution carry only parabolic points (single-curved). Their outline and their shadow are usually composed of two straight lines, including the images or shadows of the boundary circles (Fig. 4.44).

...to general cylinders

A *cylinder* is not necessarily a cylinder of revolution (as is the case in technical drawings).

A general cylinder is created by moving a straight line parallel along an arbitrary directrix.

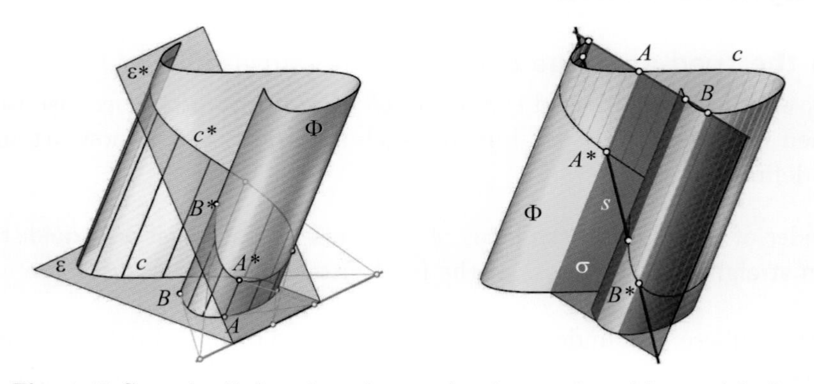

Fig. 4.45 General cylinder: planar intersection, intersection with a straight line

General cylinders (Fig. 4.45) retain important properties of cylinders of revolution, except for rotational symmetry. They are single curved (and thus

developable), which can also be recognized from the fact that their contours or self-shadow boundaries are composed of straight lines.
Two planar intersections $c \subset \varepsilon$ and $c^* \subset \varepsilon^*$ of a general cylinder Φ are perspectively affine to each other (Fig. 4.45 left, compare also Fig. 3.50). Thus, one planar intersection c contains the information about the entire surface. This can be used to get the intersection points $A^*, B^*, \ldots$ of a cylinder Φ with a straight line s (Fig. 4.45 right): We consider a plane σ through s parallel to the generating lines of Φ and intersect it with the base curve c. The intersection points $A, B, \ldots$ are the projections of $A^*, B^*, \ldots$ in direction of the generating lines.

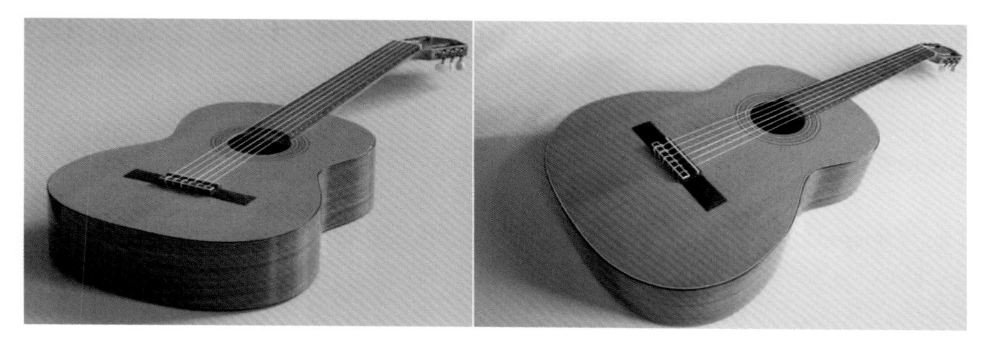

Fig. 4.46 Cylindrical boundaries of a sound board

The outline of a guitar's cylindrical boundaries is a good example. The contours are composed of straight lines, and this is true for the normal projection (the photo Fig. 4.46 left was taken with a telephoto lens), as well as the central projection (wide-angle lens used in Fig. 4.46 right). The photographs at hand were shot using light flare (indoor shot) to avoid sharp shadow boundaries. The shadow of the side wall is nevertheless recognizable as being straight.

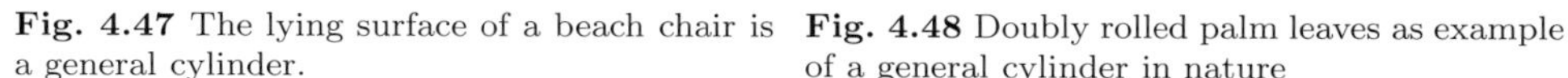

Fig. 4.47 The lying surface of a beach chair is a general cylinder.

Fig. 4.48 Doubly rolled palm leaves as example of a general cylinder in nature

A model of a general cylinder can quickly be "crafted" by everyone: One only needs to drag a piece of paper across the sharp edge of the surface of a table. This procedure changes the structure of the squashed paper and a paper roll

emerges. A phenomenon of this kind often occurs in nature, such as in plant leaves (Fig. 4.48).
General cylinder surfaces also play an important role in architecture: They can be used to shape rooftops or facades that are reasonably cheap to produce – due to the simple fact that they can be cut in the plane.
Geometrically speaking, we are often dealing with parallel projections or parallel shadows with "circumscribed" cylinders – as in the case of the sphere, where the circumscribed cylinder is a cylinder of revolution.

4.4 The ellipse as a planar intersection of a cylinder of revolution

A classical proof of spatial geometry

The "conventional" ellipse – as defined by the Greeks – occurs as a planar skew section of a cylinder of revolution.

Proof:
The proof by the Frenchman Dandelin is as genius as it is simple: A conventional ellipse is the locus of all points in the plane, whose sum of the distances from two fixed points – the focal points F_1 and F_2 – is constant. *Dandelin* noted that these points can be seen as the contact points of two congruent spheres that touch both the cylinder of revolution and the plane.

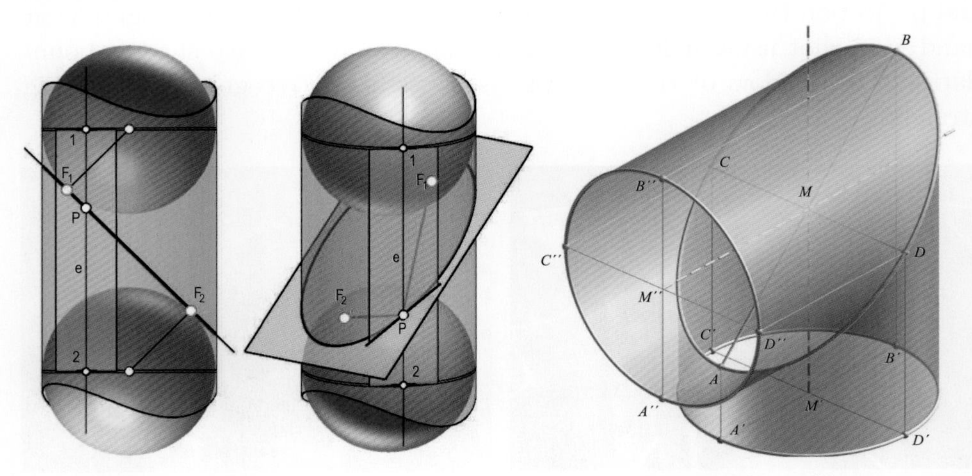

Fig. 4.49 The Dandelin spheres **Fig. 4.50** Pipe elbow

Let us call the points F_1 and F_2 (they do *not* lie on the cylinder of revolution). If we now take an arbitrary point P on the intersection curve, we notice that through the point P there exists a cylinder generatrix e, which touches the spheres in the points 1 and 2. We can say: The (identical) straight lines $P1$ and $P2$ touch both spheres, and the sum $\overline{P1} + \overline{P2}$ is constant – equal to the distance from 1 to 2. On the other hand, the straight lines PF_1 and PF_2 are likewise tangents of the spheres from P. All tangents of a sphere through a fixed point P have a constant tangent distance and form a cone of revolution. Thus, $\overline{PF_1} = \overline{P1}$ and $\overline{PF_2} = \overline{P2}$. The sum $\overline{PF_1} + \overline{PF_2}$ is also constant ($= \overline{12}$), which is precisely what was to be shown. ◇

A multitude of applications

If the cylinder is upright, the lowest point A and the highest point B on the intersection curve are the principal vertices of the ellipse (Fig. 4.50). The midpoint M is the intersection of the plane with the cylinder axis. The points C and D (which have the same height as M) are the auxiliary apexes.

Fig. 4.51 Construction of a pipe elbow: Two concentric congruent ellipses are overlapped by means of a rotation.

Both parts of an obliquely truncated cylinder can be attached to form a *pipe elbow* by rotating one of them by 180° about the normal of the plane at the center of the ellipse (Fig. 4.51). If the intersection plane forms an angle φ with the cylinder axis, then the pipe axes and the plane of the ending circles form the angle 2φ. For $\varphi = 45°$, the pipe elbow is perpendicular.

Fig. 4.52 Vaults with ellipses as intersections

Four cylinders of revolution that are cut at both sides at 45° in the manner of a wedge can be put together to create a *beam connection*. The intersection consists of a pair of ellipses that intersect each other perpendicularly in the neighboring apex (such a connection is visible in Fig. A.10).

Geometrically speaking, the same problem arises when two cylinder-of-revolution-like vaults intersect (Fig. 4.52 shows the bottom view of a *groin vault* and a *barrel vault*).

Fig. 4.53 Variations of a three-cylinder-connection

The variations of such cylinder connections (Fig. 4.53) should illustrate the enormous importance ascribed to planar intersections of cylinders of revolution.

Fig. 4.54 Intersection of two cylinders of revolution: Palmenhaus/Schönbrunn, Vienna

Proceeding with this game, one can connect six pointed cylinder of revolution wedges, producing a *spatial cross*. Six ellipses are involved, and it truly challenges our spatial imagination to sketch such an object (more in the appendix concerning free-hand drawing).

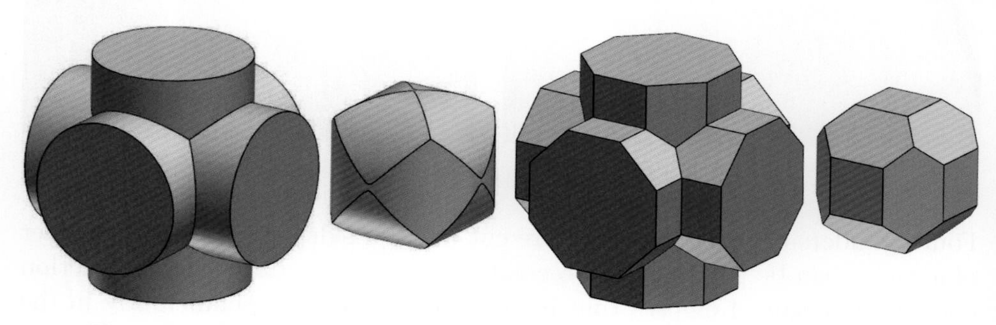

Fig. 4.55 Beam connection with three beams, a variation using eight-sided prisms

Fig. 4.55 on the left shows a beam connection with three pairwise orthogonal axes. The "interior" (center left), as a physical model, rolls well. For the computer, such intersections often lead to numerical problems ("near-identical" planes must be intersected). This is why, in Fig. 4.55, the cylinder radii were set to be slightly different, which immediately causes the intersection curves to "break off" at the double points.
In Fig. 4.55 on the right, the cylinders were approximated by regular, eight-sided prisms. The shared interior of the prisms is, then, a solid that is composed of congruent squares or hexagons.
An interesting regular four-beam connection is explained in application p. 236 (Fig. 6.97). There, the axes of the cylinders are the four altitudes of a regular tetrahedron.

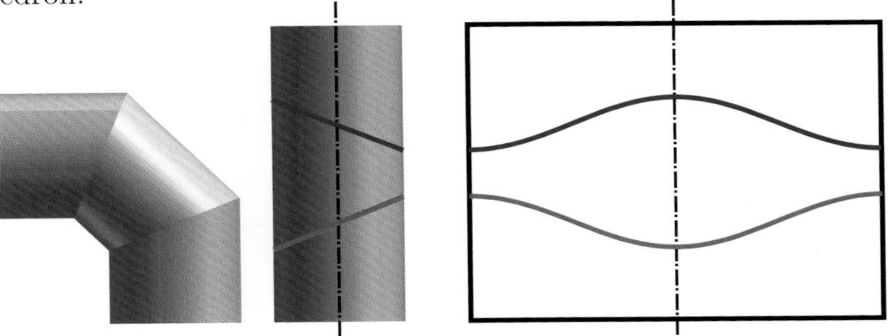

Fig. 4.56 Development of a pipe connection

Multiple cylindrical ungulas can be attached "in sequence" (Fig. 4.56), thus creating an approximation to a so-called *torus*, about which we will have much more to say.

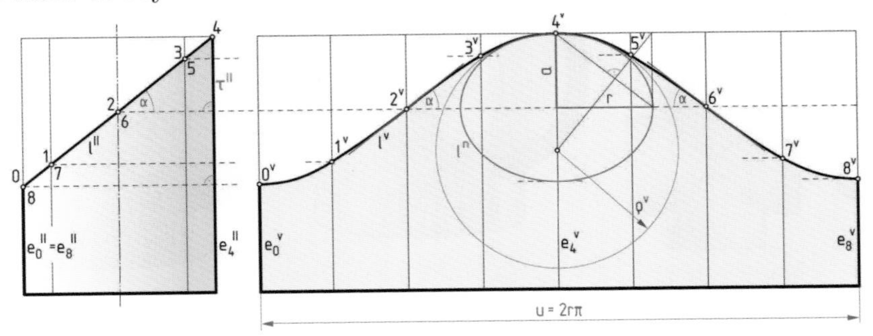

Fig. 4.57 Development of a cylindrical ungula: The intersection curve follows a sine function, of which the osculating circles and the tangents of inflection (angle β) should be specified (for example, the osculating circle at point 4 is equal in size to the ellipse that is orthogonally projected onto the tangent plane τ.

During development, a cylinder of revolution bounded by circular sections changes into a rectangle with width $2\pi \cdot r$. Which curve results from an elliptical oblique section?
This time, we need a calculation. We have to connect the arc length that occurs during development to the height of a curve point. Let us consider an ellipse point in plan view and front view. In the plan view, P' is a circle

point with the coordinates $x = r \cdot \cos u$, $y = r \cdot \sin u$ (u is the polar angle in radians). If the section plane is inclined towards the horizontal base plane (a plane at the height of the neighboring apex) at an angle of $\alpha = 90 - \beta$, then the height of P equals $z = y \cdot \tan\alpha = k \cdot y$ (k being constant). When developed, the point has the coordinates $P^v(r \cdot u/z)$, or $P^v(r \cdot u/k \cdot r \cdot \sin u)$. The developed curve is, therefore, essentially a *sine curve*.
The principal vertexes 4 and 8 of the ellipse (Fig. 4.57) correspond to the apexes of the sine curve. The auxiliary vertexes correspond to the inflection points of the sine curve. The inclination of the tangents at the inflection points equals the inclination of the plane, which can be seen in the front view.

When developing a pipe connection, it is equally valid to develop a prototype. The other parts are congruent. It is possible to proceed even more economically: By imagining the ungula as rotated by 180° *before* development, the pipe connection is stretched into a *single* cylinder of revolution (Fig. 4.56). We, therefore, only need to develop *one* cylinder of revolution with potentially multiple (though essentially congruent) oblique sections. The result is a rectangle strip, to which one or more (often congruent) sine curves are inscribed. This minimizes material wear and ought, therefore, to delight the pragmatist!

Of course, it is not always possible

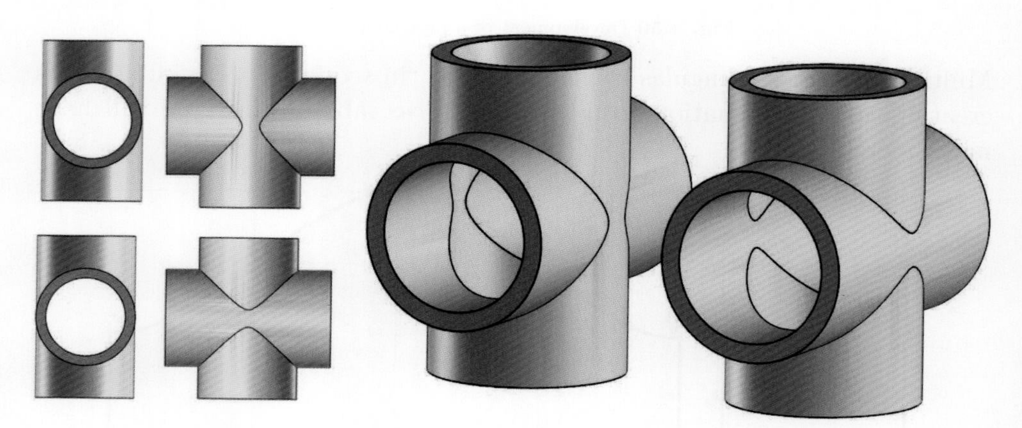

Fig. 4.58 Arbitrary intersections of cylinders of revolution

It sems to be contradictory that the intersection of two cylinders of revolution yields two ellipses since the following famous theorem holds:

Bézout's theorem: Two algebraic surfaces of the orders m and n generally share an intersection curve of the order $m \cdot n$.

Proof:
The two surfaces are intersected by an arbitrary test plane by means of two planar curves of the orders m and n. It is sufficient to show that these m have n points in common, because the intersection curve must be of the same order.

The theorem about planar algebraic curves is, as the name indicates, a classical algebraic problem and remains to be proven. The algebraic curve of n-th order is given by the implicit equation

$$\sum a_{ij} \cdot x^i \cdot y^j = 0 \text{ with } 0 \le i + j \le n.$$

The algebraic curve of m-th degree can be written in the same way

$$\sum b_{ij} \cdot x^i \cdot y^j = 0 \text{ with } 0 \le i + j \le m.$$

We eliminate one variable, say y, from both equations by computing the resultant of the two polynomials. This yields an algebraic equation of $m \cdot n$-th order. According to the fundamental theorem of algebra, this polynomial possesses $m \cdot n$ solutions (including complex and multiple solutions). ◇

Fig. 4.59 Filleting of an intersection

Fig. 4.60 Arbitrary intersection of cylinders (on the left: beaver food)

If one intersects two second-order surfaces (to which cylinders of revolution belong), then a space curve of fourth degree is to be expected.
Fig. 4.58 shows the intersection curves of cylinders of revolution with intersecting axes and slightly different radii (the curves appear as circles when projected in the direction of the axes, and as parts of hyperbolas when projected orthogonally to the plane spanned by the axes). Such intersections occur very frequently in technical applications. Fig. 4.60 shows a curious example from nature: A beaver has gnawed away at a mighty tree, which may now "fall over" in the next storm.
The intersection curve of two cylinders of revolution may also "degenerate": Both section ellipses, when put *together*, also form a space curve of fourth order (after all, they share four points with an arbitrary plane).
Fig. 4.61 illustrates how the intersection of two cylinders of revolution may degenerate if two criteria are given: First, the cylinders must have an equal radius, and second, their axes must intersect. To put it in another way: Both surfaces must share an inscribed auxiliary sphere. In the normal projection to the plane that is spanned by the axes, the hyperbola that occurs for different radii splits off into a pair of straight lines: A hyperbola cannot possess a double point.

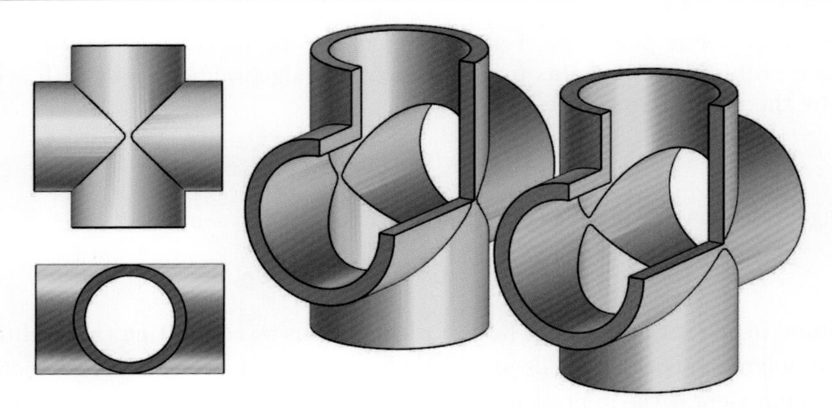

Fig. 4.61 The space curves may degenerate under certain circumstances.

The fact of degenerating intersections is not only frequently employed in technology, but also, due to a second theorem, occurs in nature with shadows:

The shadow of a planar boundary curve of a second-order surface onto itself is a conic section.

Fig. 4.62 Self-shadow of a planar intersection on a sphere or a cylinder of revolution

Proof:
The planar boundary curve is a conic section c and in the case of a sphere or a cylinder of revolution a circle (Fig. 4.63). The light rays through c form a second-degree cone Λ (an oblique cylinder in the case of parallel shadows). In both points T_1 and T_2, where the self-shadow boundary of the surface (a conic section or a pair of straight lines) intersects the boundary curve, the light cone and the surface touch one another. Due to this double tangency, the intersection curve splits off into two conic sections, one of which is the boundary curve c. The other curve c^* carries the boundary of the boundary curve's shadow onto the surface itself. ◇

• Shadows as intersection curves

In Fig. 4.64, it can be seen that the light rays surrounding the tire form an approximately oblique circular cylinder (in the photograph, it is depicted as a cone with the sun point for its apex). An intersection with the half pipe (which itself is part of a cylinder of revolution) results in a space curve of fourth order as the shadow of the tire. ♠

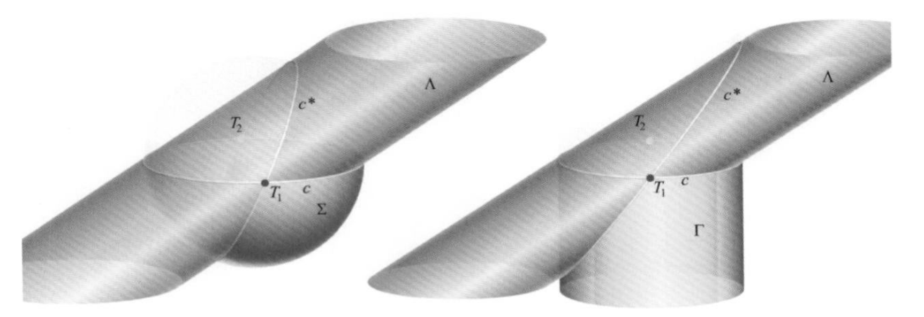

Fig. 4.63 Shadow of a sphere part or a cylinder of revolution on itself

Fig. 4.64 The shadow of a tire on a curved road is an arbitrary space curve.

The process of development and its inversion

Fig. 4.65 on the left illustrates how a sheet of brass shaped like a cylinder of revolution can be distorted into a rectangle. Mechanically, this is accomplished by continuously bending the brass at even time intervals by a small amount. The further a generatrix is away from the cutting position, the more often the bending procedure has to be repeated.

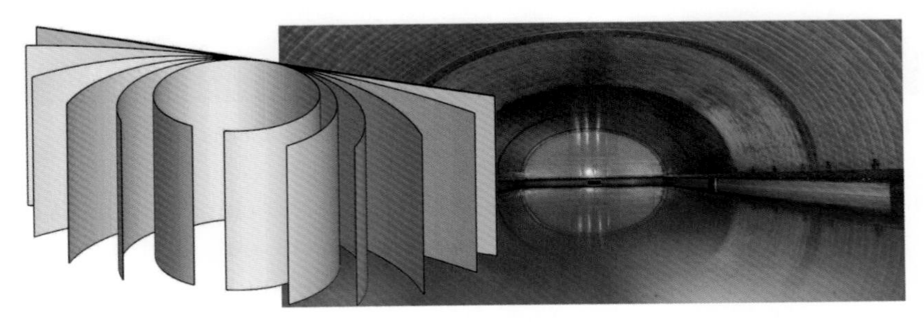

Fig. 4.65 Development of a cylinder of revolution and how it is used to roof a swimming pool.

The color-highlighted position looks approximately like half of an elliptic cylinder (though this resemblance is not exact). In Fig. 4.65 right, this position was used as a roof of a small swimming pool. The reflection in the water gives the additional appearance of a closed (and almost elliptic) cylinder.

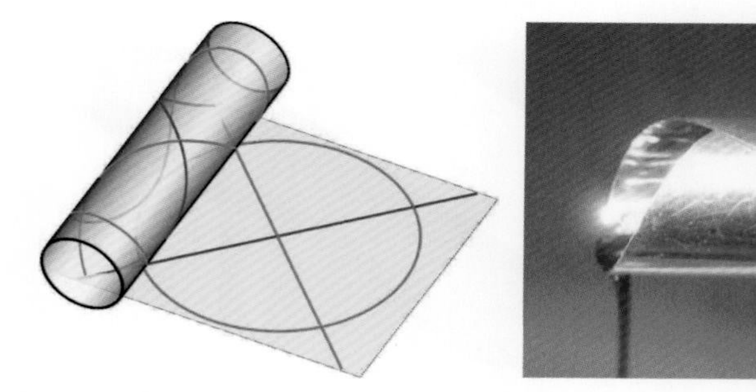

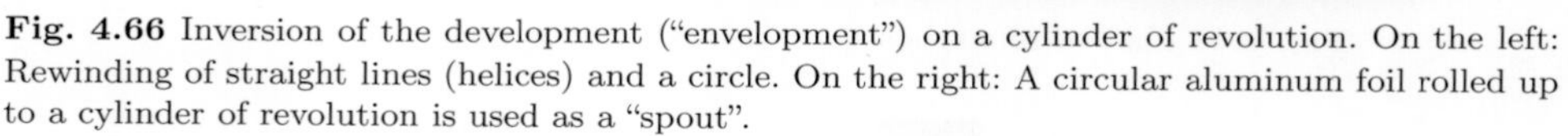

Fig. 4.66 Inversion of the development ("envelopment") on a cylinder of revolution. On the left: Rewinding of straight lines (helices) and a circle. On the right: A circular aluminum foil rolled up to a cylinder of revolution is used as a "spout".

In practice, instead of developing a surface, it is sometimes required to "envelop" it (Fig. 4.66). Simple examples include the placement of a label onto a bottle and the circular aluminum foil that is rolled up and put into a bottleneck to avoid dripping the liquid when pouring (Fig. 4.66 on the right). The "inclusion" of straight lines during envelopment produces geodesic lines on the carrier cylinder (see p.121). In the general case, these are *helices* – we shall later address these important curves in great detail.

If a circle whose diameter equals the circumference of the cylinder is included during rewinding, a curve with a tacnode, i.e., a point where the curve touches itself, is produced. Such a curve is highlighted red in Fig. 4.66 on the left. The tacnode is visible in the obscured region. In the background of Fig. 6.29, circles enveloped onto a cylinder of revolution are visible on the bamboo stools.

Fig. 4.67 Envelopment

Fig. 4.68 "Envelopment" of a circle

In sheer geometric terms, the production of potato chips in Fig. 4.68 is a tricky matter. A thin layer of pulp is obviously cut out. If it is then enveloped onto a cylinder of revolution – as with a "rolling pin" – we would not yet expect a doubly curved surface. It can be assumed that the pulp is immediately pressed onto a doubly curved stencil. The pulp is thick enough to survive the necessary stretching.

Fig. 4.67 on the left shows two ancient Egyptian bracelets where planar curves from pure gold were enveloped onto an "elliptical cylinder" (the forearm).

5 More about conic sections and developable surfaces

Cones, like spheres and cylinders, are surfaces that we can easily imagine. Like cylinders, they are simply curved and developable. Among all cone surfaces, the cone of revolution and the oblique circular cone are especially important. Their planar sections are the conic sections *per se*: ellipses (among them circles), hyperbolas, and parabolas. We can derive many common properties through spatial interpretation of these most famous planar curves.

The cones and the cylinders are developable: They can be unrolled into the plane without distortion. This characteristic is shared only with the so-called developables or torses (tangential surfaces of space curves) which are, like cylinders and cones, composed of straight lines called *generatrices*. Developables are envelopes of one-parameter families of planes where each plane touches along a certain generatrix. This makes them very simple to recognize: From any viewpoint their contours are generatrices. The shadow boundaries, too, always have straight components.

Doubly curved surfaces like the sphere cannot be unfolded into planes without distortion, but because this is such an important task, mathematicians and geometricians have spent an immeasurable amount of time on the issue of transforming the surface of the globe into a planar map such that certain geometric properties are unaffected by the process.

Survey

5.1 Cone surfaces

Just like cylinders, cones are notable due to their simplicity. We can generalize as follows (Fig. 5.1):

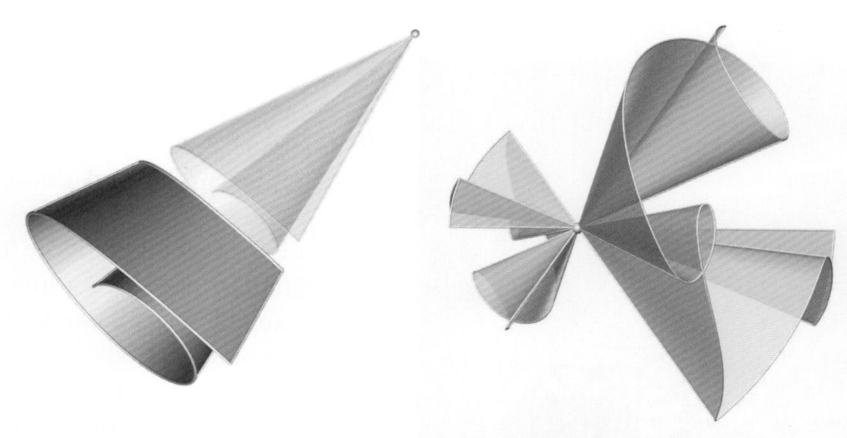

Fig. 5.1 The motion of a straight line through a fixed point (the apex)

A cone is generated by a straight line e that always passes through a fixed point S. S is the apex, e is the cone's generatrix.

The cone is not bounded and can, thus, always be interpreted as a double cone, i.e., the surface consist of both parts on either side of the apex (Fig. 5.2 on the left).

Among the cone surfaces, the *cones of revolution* are the most important:

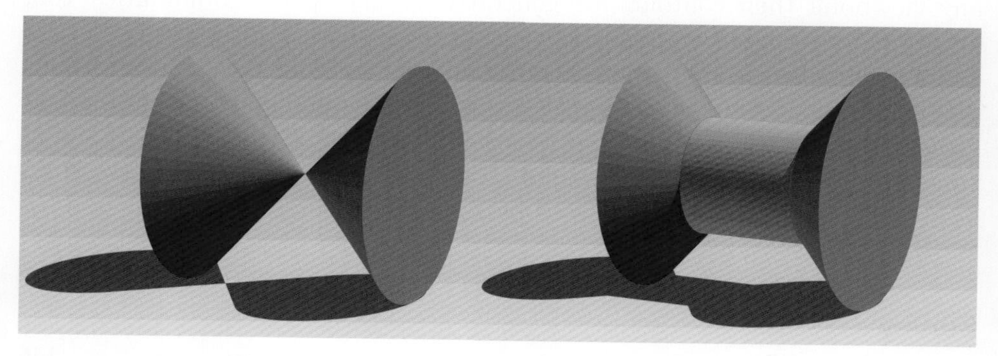

Fig. 5.2 Shadows of cones and cylinders of revolution

A cone of revolution is generated by the rotation of a straight line e about an intersecting axis.

Only parts of cones of revolution occur in practice, and these are sometimes not even easily identifiable as such (Fig. 5.3, 5.4 and 5.5).

Fig. 5.3 Cone of revolution with circles of latitude and generatrices

Fig. 5.4 Rotation of straight lines and a cone-shaped excavation

The development of a cone of revolution

All generatrices of a cone of revolution have the same length. To be more precise: all sections of the generatrix between the apex and a circular section are equally long. Therefore, if we develop a cone into the plane, it becomes a circular sector.

According to Fig. 5.6, the central angle ω of the circular sector can be derived: The length of the circular arc does not change. With the notations from the figure, we have $2\pi \cdot r = \omega \cdot s$ and thus

$$\omega = \frac{r}{s} 2\pi \quad \text{and also in degrees} \quad \omega^\circ = \frac{r}{s} \cdot 360^\circ.$$

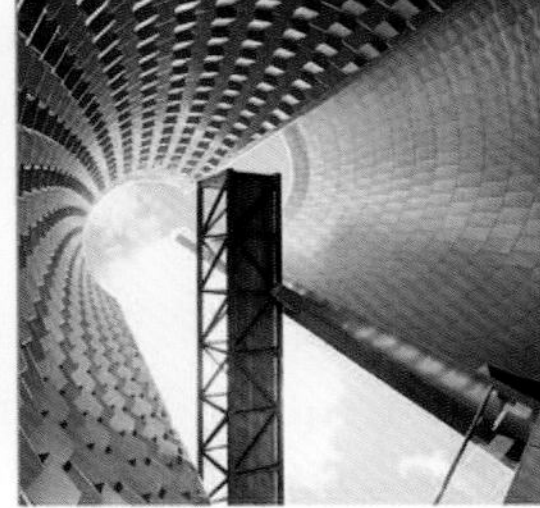

Fig. 5.5 Vulcania European Park of Vulcanism Conus by H. Hollein (parts of cones of revolution)

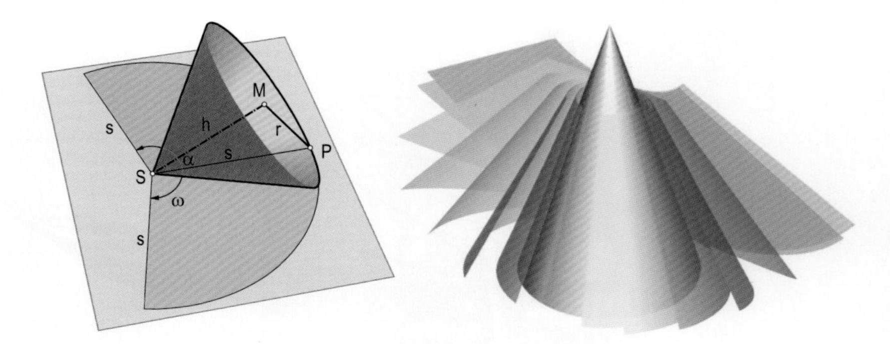

Fig. 5.6 Development of a cone

Given an equilateral cone with $s = 2r$, we get half a disc. A frustum of a cone of revolution is developed into a annular sector. Curves on the cone become curves that intersect the generatrices at the same angle as they did in space (application p. 163).

• **Envelopment onto a cone**

As with cylinders of revolution, we can also "invert" the development. We can inscribe an arbitrary curve k^v into the developed lateral area of the cone of revolution, such as a circle (as in Fig. 5.7).

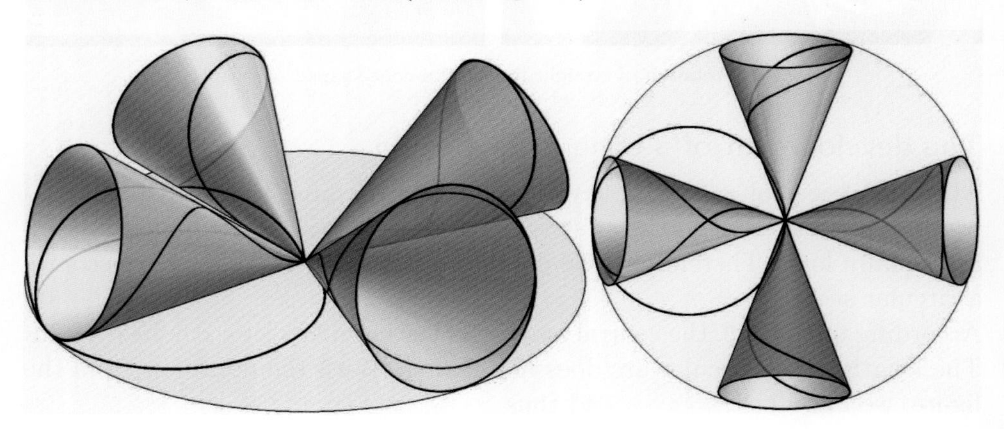

Fig. 5.7 Envelopment of a circle

The result is a space curve k on the cone that, if developed, produces the curve k^v. The curve k can be determined as follows point by point: We pick a generatrix e^v in the development. We then determine the central angle φ to the respective central angle φ^v on the base circle and thus reconstruct the generatrix e on the cone. Finally, we measure the distance $\overline{S^vP^v}$ from the development on e. ♠

General cones are, in practice, not as frequent as arbitrary cylinders. They are sometimes found as connecting surfaces of two "directrices" which often reside in parallel planes (Fig. 5.12).

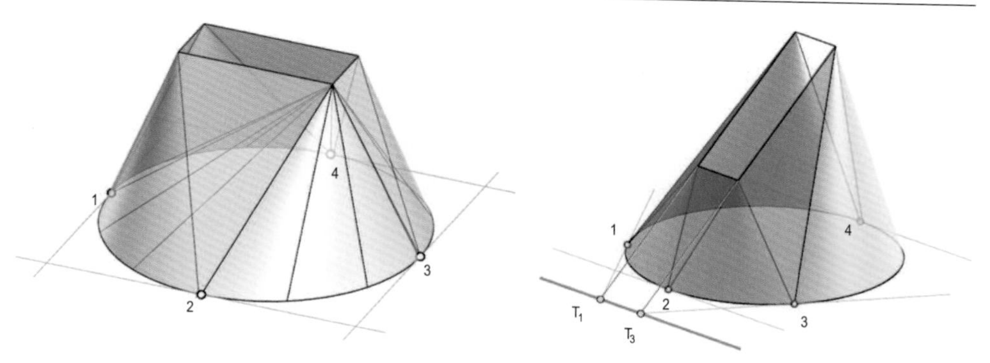

Fig. 5.8 Connecting surface between a rectangle and a circle in theory ...

• Connecting cone between a rectangle and a circle
Fig. 5.8 shows how a connecting surface composed of (planar) triangles and oblique circular cones can be pieced together in such a way that there are no edges:

Fig. 5.9 ...and practice

If, e.g., a circle k and a rectangle r in parallel planes are given as boundary curves (Fig. 5.8 on the left), then the points 1, 2, 3, and 4 on k where the tangents are parallel to the rectangle sides are to be determined. These points are then connected by triangles with the rectangle sides. Four oblique circular cones, which are touched by the triangles, remain.
If the carrier planes of the boundary curves are not parallel, then the points 1,...,4 result by intersecting the rectangle sides with the section line s and, from there, positioning the appropriate tangents to k. ♠

• The jet of a hairdryer (Fig. 5.10 and 5.11)
We want to connect a "rounded rectangle" with a circle in a parallel plane by a smooth surface.
Solution:
The following idea leads to an elegant solution: The longer sides of the rectangle can be "handled" – as in the previous example – by the use of two triangles. Two circles of different sizes in parallel planes remain (if we imagi-

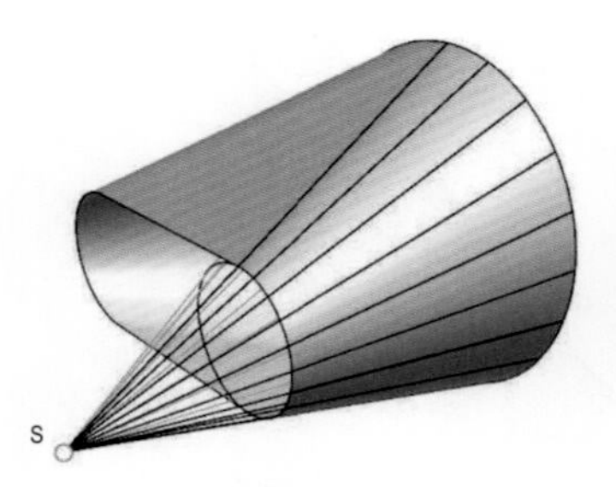

Fig. 5.10 Theory ...

Fig. 5.11 ... and practice

ne the upper half-circles as complete). There now exist two oblique circular cones with the apeces S_1 and S_2 on which the circles lie. They form the connecting surfaces that touch both triangles.
Fig. 5.12 shows an airport counter designed according to this method. The advantage, apart from an aesthetic one, lies in the fact that there are no hard edges on which to hurt oneself. ♠

• The cone by *Le Corbusier*

Fig. 5.13 depicts a design by the French architect (1887-1965) who was famous for his formally strict yet functional architecture. A more precise analysis of the sophisticated object shows its remarkably simple method of construction:
As in application p. 148 (connecting cone between a rectangle and a circle), Corbusier developed a connecting surface between a base square and a circle. He then simply and obliquely truncated the object composed of triangles and skewed circular cones. The top surface is composed of ellipse fragments (see the next section) and straight lines. ♠

Fig. 5.12 Connecting cone ...

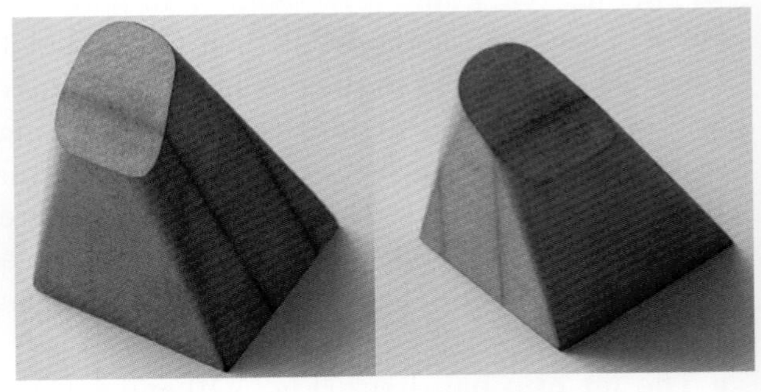

Fig. 5.13 ... of two directrices (after *Le Corbusier*)

A theorem with great technical applicability

Without doubt, geometry can be fascinating by itself, regardless of any practical applicability. It is, nevertheless, pleasing to find applications of the theorems of theoretical geometry continuously. Here is a good example:

If two second-order surfaces touch at *two* points, then the section curve splits off into two conic sections.

Fig. 5.14 Symmetrical and asymmetrical trouser parts

It can be shown that a non-degenerate intersection curve of 4-th order can only possess one double point (a self-intersection). On the other hand, it can also be proven that whenever two surfaces touch each other, the intersection curve must possess a double point precisely at the contact point.
Now, we can easily understand the following practically relevant theorem based on the aforementioned statement:

If two second-order surfaces share a *common inscribed sphere*, then the intersection curve splits off into two conic sections.

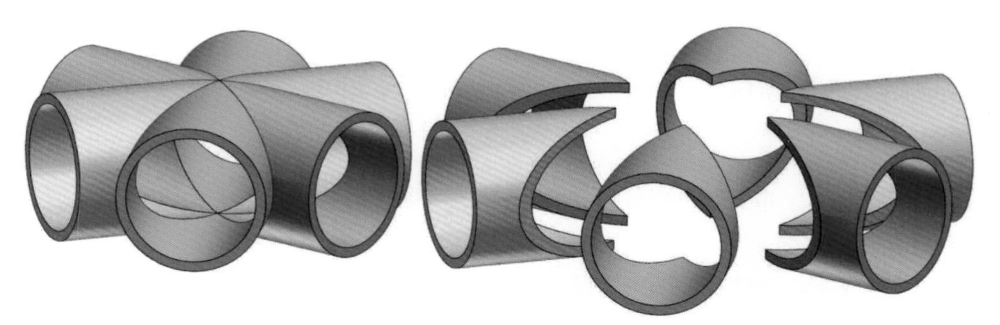

Fig. 5.15 Cylinder intersection including the explosion sketch

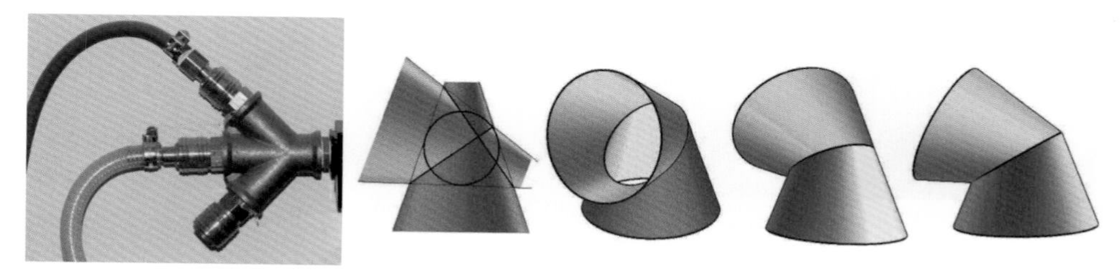

Fig. 5.16 Degenerate intersection of second-order surfaces

At first, the intersection curve of both surfaces is of fourth order. Like the carrier surfaces, it will touch the sphere twice. Both surfaces possess the same tangent plane at the contact points (which are also double points). This leads to the splitting of the curve into two conic sections.
Fig. 5.16 at the center shows two cones of revolution that have a common inscribed sphere. The section curve thus splits off into two ellipses. If only a part of the cone of revolution materializes, then only one ellipse remains.

Fig. 5.17 Exposition in Geneva

Fig. 5.16 on the right goes one step further and performs an affine transformation to the cones of revolution. Thus, they do not remain cones of revolution. In the course of it, the common ellipse changes into a new ellipse. This principle fascinated the architectural firm Coop Himmelb(l)au enough that they decided to immortalize it in multiple famous buildings (as in Geneva, where each of the towers is 40 meters tall Fig. 5.17, or in Fig. 5.18).
A spatial affine transformation is very easy to calculate: The new coordinates are simply arbitrary linear combinations of the old coordinates.

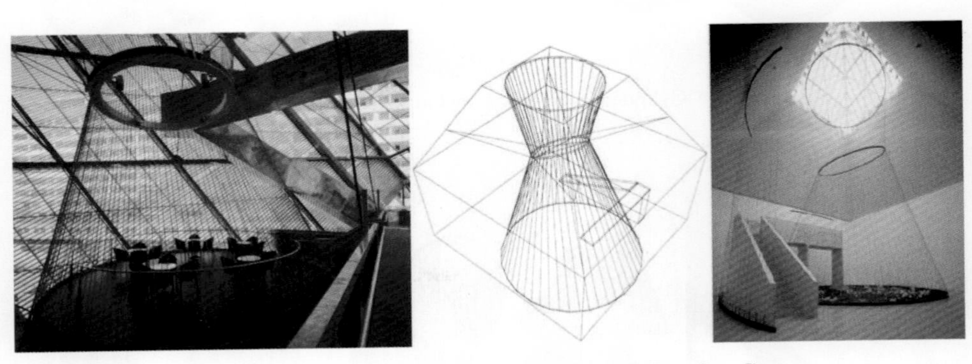

Fig. 5.18 UFA Cinema Center, Dresden and Paradise Cage

5.2 Conic sections

Planar sections of a cone of revolution

The term "conic section" (often simply called "conic") is found in every book on geometry, almost as often as "straight line" and "circle". It is also known as a second-order curve or second-degree curve. All these names imply that the curves have, algebraically speaking, two intersections with a straight line.

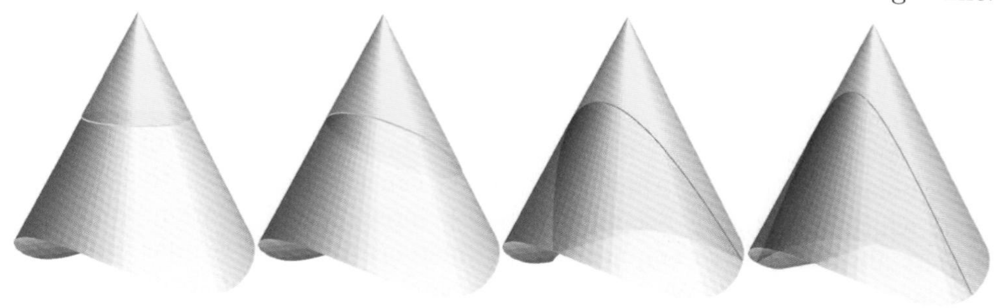

Fig. 5.19 The different conic sections

From a mathematical perspective, the curves are given by a second-order algebraic equation in the form of

$$a_1\,x^2 + a_2\,y^2 + a_3\,xy + a_4\,x + a_5\,y + a_6 = 0 \tag{5.1}$$

(not all of the first three coefficients must be zero, otherwise we are left with a linear equation). If we now insert the linearly dependent term $k \cdot x + d$ for the variable y, we *always* get a quadratic equation in x, which leads to the two intersection points. (Multiple solutions and complex conjugate pairs of solutions have to be taken into account).

What is a degenerate conic section?

We differentiate between *degenerate* and *non-degenerate* (regular) conic sections. From a geometric perspective, regular conic sections are formed by the intersection of a cone of revolution with a plane that does not contain the apex of the cone. Ellipses, hyperbolas, or parabolas, thus, emerge depending on the inclination of the plane.
A plane through the apex of the cone either intersects the cone in two generatrices, touches it along a generatrix, or does not intersect at all (more precisely: intersect along two complex conjugate straight lines with the apex as the real intersection).

For the mathematically knowledgeable: A cone of revolution with aperture 2α centered on the origin of the coordinate system and with the z-axis for its axis can be described mathematically by the equation

$$x^2 + y^2 - t^2 z^2 = 0$$

with $t = \tan\alpha > 0$. A non-horizontal intersection with a plane like

$$px + z = 0, \quad p > 0$$

through the origin (normal vector $(p,0,1)^T$) yields the condition

$$y = \pm x \cdot \sqrt{p^2t^2 - 1},$$

or a pair of straight lines which is real whenever $p^2t^2 \geq 1$, or $p \leq \cot\alpha$ is given. Otherwise, we are dealing with a pair of complex conjugate straight lines.

Focal points

If we also allow points at infinity, then all regular conic sections possess two focal points. In the case of ellipses and hyperbolas, both are regular points, and with parabolas, one of the focal points "slides" in the axis direction towards infinity.

This type of thinking has its advantages: Many properties of ellipses and hyperbolas that refer to their focal points can be transferred to the parabola, which, apparently, occupies a notable intermediate position. We either do not always need to emphasize the difference between regular and non-regular conic sections. We can imagine a pair of intersecting straight lines as hyperbolas which are simply seen from a very large distance. (A hyperbola is nearly straight at almost all points. It only changes direction very briefly in the vicinity of its vertices.) A parallel or collapsing pair of straight lines could also be imagined as an "infinitely flat ellipse".

Let us, for example, take a theorem that refers to the so-called *focal rays* (the straight lines connecting a point of a conic section with its focal points):

The tangent of a conic section bisects the angle between the focal rays.

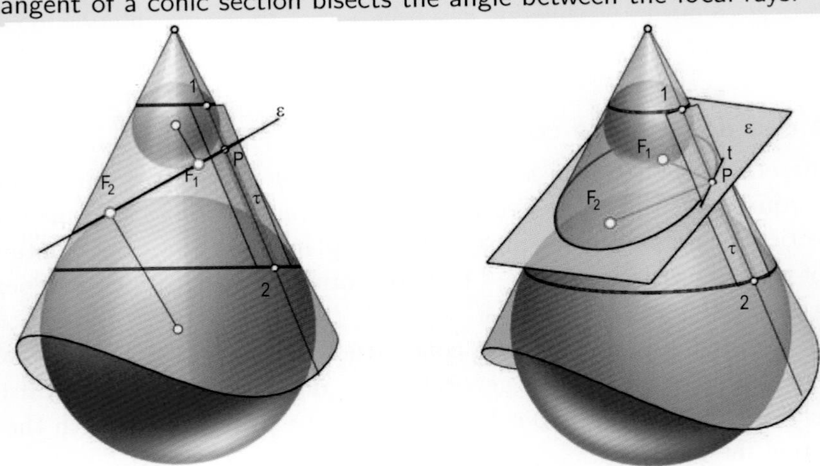

Fig. 5.20 The Dandelin spheres

Proof:

An elegant proof of this theorem is based on the proof by *Dandelin*, where a conic section is explained as a planar section of a cone of revolution. The tangent t at a curve point P can be interpreted spatially as the intersecting straight line of the cone's tangent plane τ with the plane ε. The plane τ touches the first Dandelin sphere at 1 and ε in F_1. For reasons of symmetry, t forms the same angle with the tangent segments $P1$ and PF_1. The same is true for the second Dandelin sphere, but since the carrier lines of $P1$ and $P2$ are identical, t forms the same angle with the focal rays PF_1 and PF_2. ◇

This theorem has practical applicability. More on it can be found in the section concerning second-order surfaces, where concepts such as *focal mirrors* are addressed. Furthermore, the following interesting statement can be derived:

Conic sections with two common focal points (*confocal conic sections*) intersect at right angles.

Proof:
First of all: Both conic sections are of different types – two confocal ellipses or two confocal hyperbolas possess no real intersection points. Even confocal parabolas can be disjoint (Fig. 5.23). Thus, it can only be a pairing between an ellipse and a hyperbola or parabolas opened in opposite directions. For both, it is true that the tangent is the angular bisector of the respective focal rays at the intersection. There are always two angle bisectors forming an orthogonal pair. ◇

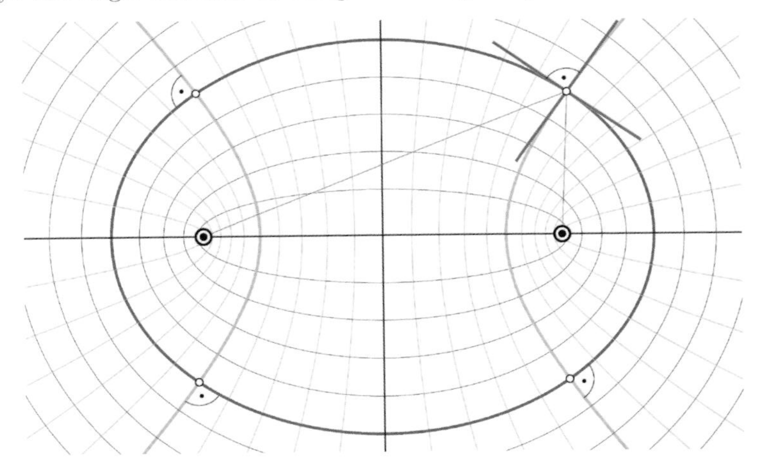

Fig. 5.21 A net of confocal conic sections

Another theorem that refers to the focal points goes as follows:

A conic section is the locus of all points at a constant distance from a fixed circle and a fixed point in its plane.

Proof:
The "proof" for the parabola actually lies in its definition: The parabola is given as the locus of all points that are equidistant to a point and a straight line. If this *directrix* is seen as a circle around the focal point at infinity, then all appears to fit once more.
Concerning the other conic sections, we shall soon prove the generalization of the theorem that includes this special case. ◇

Once again, there are applications for this theorem. For instance, let us consider the envelope of circles centered at all points of a conic section while

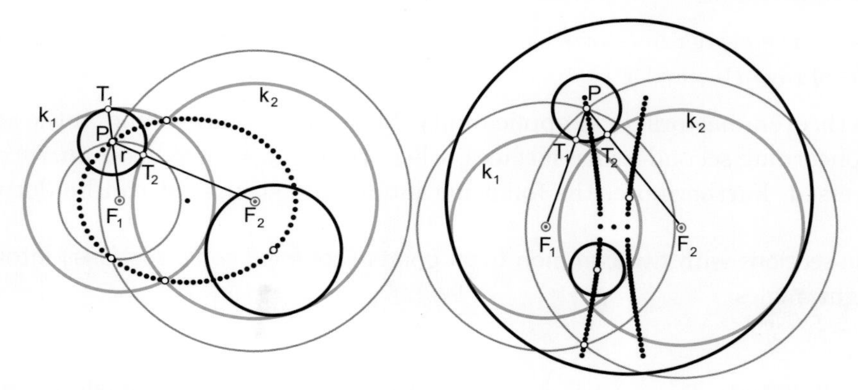

Fig. 5.22 Touching circles of two given circles and the locus of their centers

passing through a focal point. It is the "director circle" that passes through the focal point.
The above mentioned generalization of the theorem reads as follows:

Each conic section is the locus of the centers of all circles that touch two given circles in one plane.

Proof:
The parabola occurs only if one of the circles is infinitely large and thus degenerates into a straight line. We have already studied this case.
Let k_1 and k_2 be two circles with the centers F_1, F_2 and the radii r_1, r_2. We can now construct the center P of a touching circle k with the radius r if, for example, we intersect two concentric circles around F_1 and F_2 with the radii $r_1 \pm r$ and $r_2 + r$ (Fig. 5.22). However, for P we have $|\overline{PF_1} \pm \overline{PF_2}| = |r_1 \pm r_2|$. ◇

If one of the circles shrinks to a "zero circle", then the aforementioned theorem applies.

Parabolas and catenaries

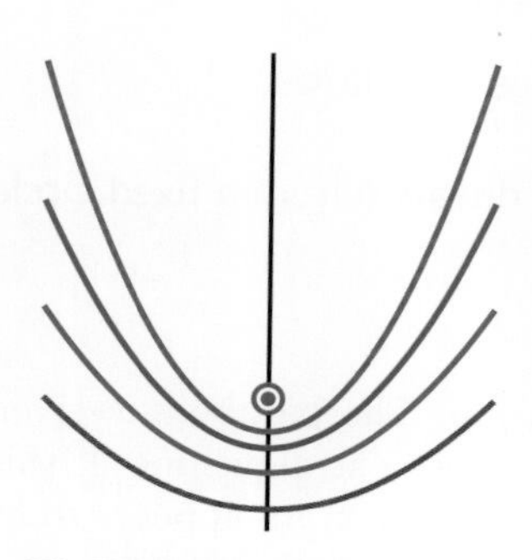

Fig. 5.23 Parabolas and ...

Fig. 5.24 ... Catenaries with different parameters

It is remarkable that there apparently exist infinitely many "significantly different" ellipses and hyperbolas, but only one prototype of a parabola. All parabolas can be derived from the prototype by a mere enlargement or shrinking. It shares this property with the circles, the catenaries (Fig. 5.25), and also with "higher-order parabolas" of the form $y = a \cdot x^n$.

Fig. 5.25 Catenaries as position of equilibrium curves (Elisabeth bridge Budapest) ...

The catenary represents the position of equilibrium of a hanging rope and is further addressed in [11].

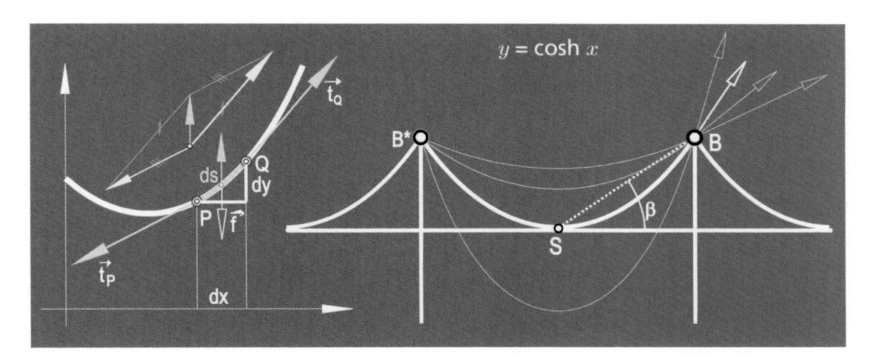

Fig. 5.26 Left: Position of equilibrium of a hanging rope, right: different tangential angles in B produce different pulling forces

We can see from its explicit equation

$$y = a \cdot \cosh \frac{x}{a} = \frac{a}{2}(e^{x/a} + e^{-x/a})$$

that the factor a is only a *factor of similarity.*

In the regions of the apex, catenaries resemble parabolas to such an extent that in practice, they can be replaced by them (though only approximately). However, if we truncate the power series of the exponential function (which is rather transcendental then algebraic) after the second-order term, we obtain the equation of a parabola. Catenaries emerge as the solutions of a differential equation: The shape of the curve emerges from the local condition that the

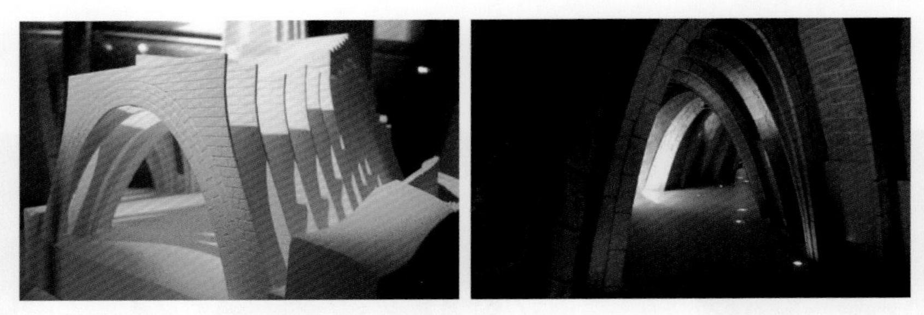

Fig. 5.27 ...and upside down catenaries for the enhancement of load capacity

tangential pulling forces of the left and right part of the rope should be in equilibrium with the mass at every point.

Fig. 5.28 A catenary turned on its head as a stable girder

A catenary that is turned "upside down" yields the optimal form of an arched bridge, where the pressure at each point is distributed evenly on both sides. Fig. 5.27 shows a model and its practical execution in the Gaudi museum in Barcelona. Fig. 5.28 shows a much bigger version.

• **Square wheels**

An interesting, though academic question: How would a road have to be constructed so that a square wheel could roll without vertical motions of its axis (Fig. 5.29)?

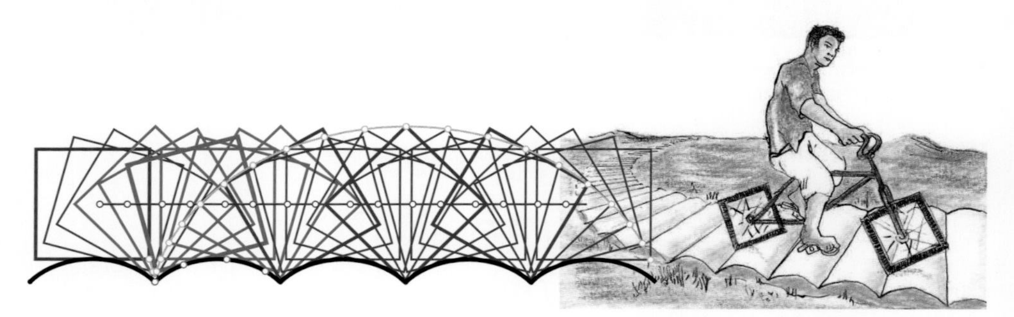

Fig. 5.29 A bumpy road and a square wheel can mutually "cancel each other".

The solution can, for instance, be found in [24] – it happens to be an upside down catenary. ♠

How many elements do we need in order to determine a conic section?

Ellipses and hyperbolas are defined by five points or five tangents – parabolas by four.

We will abandon an exact geometric proof. From an analytic standpoint, we must (in general) solve a system of linear equations with six variables – the coefficients of the conic section's equation Formula (5.1), for which only the ratio $a_1 : a_2 : \ldots : a_6$ matters.
Given five points or five tangents, it is possible to construct as many additional points and tangents as necessary precisely. The mutually dual theorems of *Pascal* and *Brianchon* are required – though we will neither prove nor even state them. *Pascal*'s Theorem is illustrated in Fig. 5.33 – a simple linear construction for an arbitrary additional point of the conic section is plainly visible.

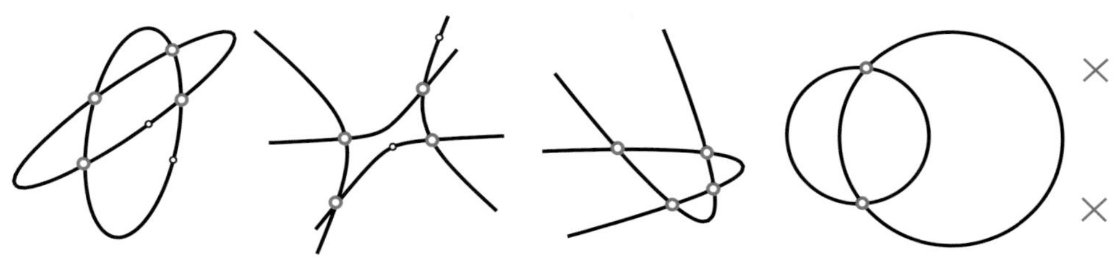

Fig. 5.30 Two conic sections (each) constructed by four shared points. Center right: A parabola is defined by four points. (Actually there may be two parabolas on four points.) Far right: Circles automatically share two complex conjugate points at infinity (marked by crosses).

Only four points or tangents are required for a parabola because the straight line at infinity is automatically a tangent and thus a fifth element. However, there exist *two* solutions which are pictured in Fig. 5.30 on the right.
A parabola "in standard position" (the axis is either parallel to the x- or y-axis) is defined by *three* elements. It is then not only possible to recognize the tangent at infinity, but also the respective contact point (in the direction of the axis).
If one point and its *respective* tangent there are known, we speak of a *line element*. In general, a parabola is, thus, defined by two line elements.
The question of how many points are required to define a conic section uniquely can be answered very quickly in the algebraic sense – by making use of analytic geometry. There, a curve of second degree is defined by six parameters, one of which can be picked arbitrarily. This means: If the coordinates of five points are inserted into Formula (5.1), we get a system of linear equations with five variables, which can be solved without uniquely.
As it is proven analytically in [11], circles always share two complex conjugate points at infinity. Despite our intuition to the contrary, this explains many apparent contradictions. It is now clear why a circle (which is a conic section) can be defined by only three points: The two missing points are the so-called "absolute points" and they always belong to the circle (see also p. 127 and p. 223).

When is the solution ambiguous?

If five elements of the same type (e.g., five points or five tangents) are given, then the conic section is defined *uniquely*. If the given element are of different types, then there exist two solutions (mixing proportion $4:1$) or four solutions $(3:2)$. Let us leave aside the exact reasons and simply solve a few of the many possible combinations.

The inverse paper strip method

An interesting possibility of "constructing" an ellipse is the so-called *paper strip method* (Fig. 5.31):

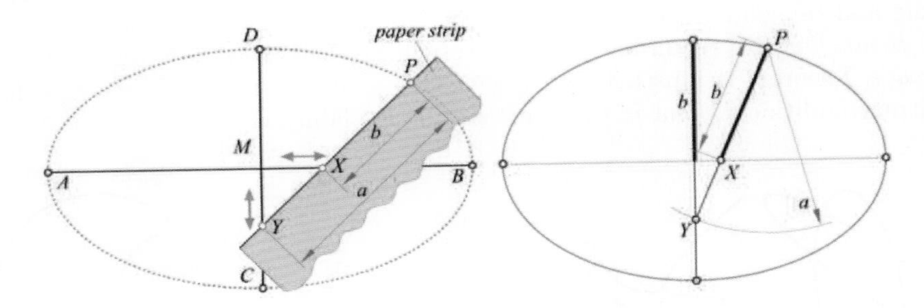

Fig. 5.31 The "paper strip method" and its turnaround

On a strip of paper we mark three mutually distinct points P, X, and Y. The paper strip is moved such that X nad Y move along the respective coordinate axes. Then, the point P describes an ellipse whose half-axes are of lengths $\overline{PX}$ and $\overline{PY}$. The proof is given in the chapter about kinematics. We can apply this knowledge in the following example: Given an axis, the apexes, and an arbitrary point P on the ellipse, the missing apex can immediately be found by fitting a paper strip by means of a compass. If, for instance, half of the major axis a is given, then $\overline{PY}$ is known, and X can be constructed. The distance $\overline{PX} = b$ is half of the minor axis.

- **Ellipse and hyperbola with axes given by a line element**

The ellipse and hyperbola are symmetric with respect to both axes. Thus, if one point is known, three other symmetric points can also be found by mere reflection. A tangent yields three additional tangents, and thus, the declaration appears to be over-determined. Due to symmetry, however, this does not produce a contradiction. Let us consider the line element in the first of four quadrants defined by the axes. For the sake of clarity, let us define the axes as the x- and y-axis without declaring which of these is the major axis.

The intersections of the tangent t with the axes shall be called X and Y. Depending on where these points lie, an ellipse or a hyperbola is to be expected.

In the "elliptic case", we can (for instance) use an affinity of the ellipse to one of its apex circles. The x-axis should be its axis of affinity. X remains static and the ray through the given point P will intersect in the "circle field" at right angle. The respective point P^* on the circle can thus, firstly, be found at the Thales circle above MX and then, secondly, at the affinity ray orthogonal to the x-axis (the affinity axis). Now, the circle around M can be drawn through P^*. The apex of the ellipse

can be found at the intersection of the affinity axis. The other apexes can be found by completion. The prinipal apexes can only be determined after that process.
In the case of a hyperbola, we should imagine a solution that may not be the shortest (when counting lines), but that represents an interesting combination of the aforementioned considerations: We first determine an ellipse that passes through the given point P and there possesses a tangent that is orthogonal to the tangent t. This ellipse is confocal to the desired hyperbola. The focal points F_1 and F_2 are thus given, and the hyperbo' is more or less finished.

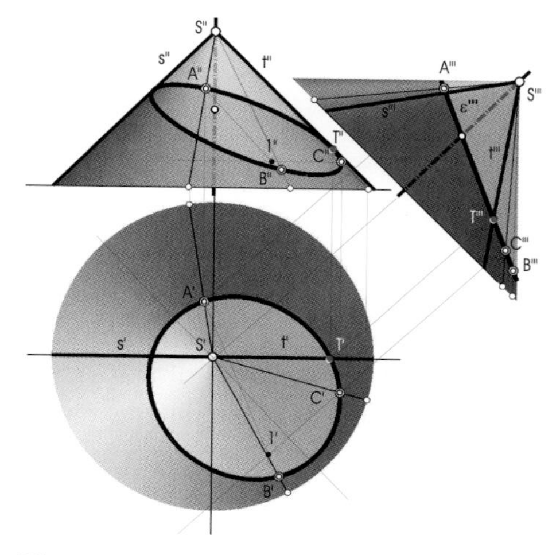

Fig. 5.32 Spatial interpretation of a conic section construction

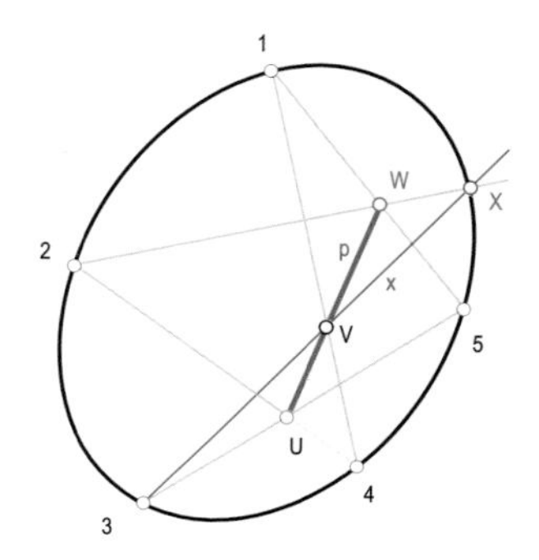

Fig. 5.33 *Pascal*'s theorem allows the linear construction of further points X – which may be located on a straight line x passing through 3 – given five points 1, . . . ,5 by means of the Pascal line p, which is defined through the auxiliary points $U = 24 \cap 35$ and $V = x \cap 14$ ($\Rightarrow W = p \cap 15 \Rightarrow X = 2W \cap 3V$).

• **Construction of conic sections through spatial interpretation**

Given some situations of initial data, the solution can be approached by spatial interpretation. The following example is already relatively sophisticated: Three points A, B, C and two tangents s and t of a conic section are given. The lines s and t can then be interpreted as outline generatrices of a right cone of revolution in a front view. One of the two angle bisectors is then the front view of the cone's vertical axis. (The initial situation is rotated according to (Fig. 5.32).)
The base plane can be "pulled up" to an arbitrary height. Now, the top views of the three points A, B, and C can be augmented by the respective generatrices. The three points thus gain a spatial significance. They lie on the cone of revolution and span a plane ε, whose intersection curve with the cone not only contains A, B, and C, but also meets the cone generatrices s and t. In the front, view we see these meets as contacts – as is the case with all other planar intersection curves whose carrier planes do not appear as straight lines. However, the augmentation of the points is ambiguous. For each point, there exist two possibilities – 8 combinations in total – of which certain pairs possess the same front views. Thus, we can expect four different planimetric solutions. An arbitrary constellation is chosen in Fig. 5.32.
In order to make the plane ε appear as a line, we augment the horizontal main straight line through C onto the straight line AB (auxiliary point 1). A side view with the viewing direction $1C$ yields the desired appearance of ε. Not only the principal vertex (as a point on the space curve) can immediately be found, but also the the contact points with the given tangents (the outline points in the front view, such as T).

♠

Construction of the conic sections

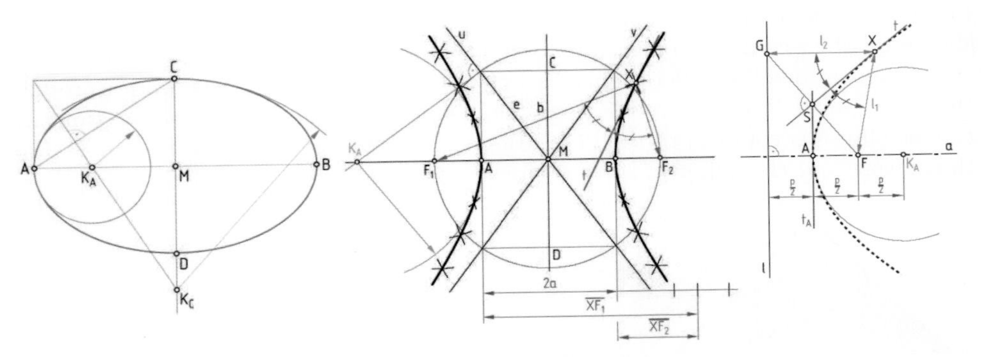

Fig. 5.34 Construction of the conic sections

When conic sections are constructed, the osculating circles in the apexes should likewise be drawn. The constructions are shown in Fig. 5.34 without proof. The ellipse always remains within the larger and outside of the smaller osculating circle. The transition of the curve from one osculating circle to another should occur “harmonically”. The plotting of arbitrary points is rather counterproductive, though, since the increased demand for drawing precision may disturb the overall balance of the curve.
The parabola can be seen as an ellipse with one apex and one focal point at infinity. As it detaches from the osculating circle, the hyperbola gradually approximates its *asymptotes*. If a hyperbola is to be sketched, the asymptotes must always be considered!
Arbitrary points are constructed by applying the definition of conic sections with the use of the focal points. In the case of ellipses, the sum of the distances from the focal points is constant. For hyperbolas, the magnitude of the difference stays constant. For parabolas, the distance to the focal point is equal to the distance from the *directrix*. The tangent at the constructed point always bisects the angle of the focal rays.

- **What does an ancient column have to do with an ellipse?**

Fig. 5.35 The production of ancient columns and what sometimes remains ...

Ancient (Greek and Roman) columns are no ordinary cylinders of revolution – they are "thicker" in the middle. In fact, instructions for the construction of the meridian were found carved in stone: A circular arc is to be constructed, then intersected with vertical lines, and finally, the vertical sections are to be multiplied (ten times, for instance). A circle whose ordinates are multiplied is, in fact, an ellipse!

Rotating an ellipse is not enough to produce an ellipsoid. Such an object is only produced by rotating the ellipse about one of its two axes. For those who are interested: The result of a rotation about an axis-parallel chord is a surface of revolution which differs from a spindle torus by an affine mapping (Fig. 4.34).

One could now argue that this is a mere splitting of hairs and that the exact shape of the column should be unimportant – as long as it is slightly thicker in the middle, who cares! But far from it! The ancient Greeks were enthusiastic "hair-splitters" in this regard, and the production of columns was quite an art of its own. For instance, columns at the corners of a building were slightly bigger – not for reasons of statics, but because they usually appear in backlight and thus give an impression of being thinner.

♠

• "Angular distorted" conic sections

If a cone of revolution that includes a planar oblique section is developed into the plane, then the section becomes a curve that, from an optical viewpoint, bears little resemblance to a conic section.

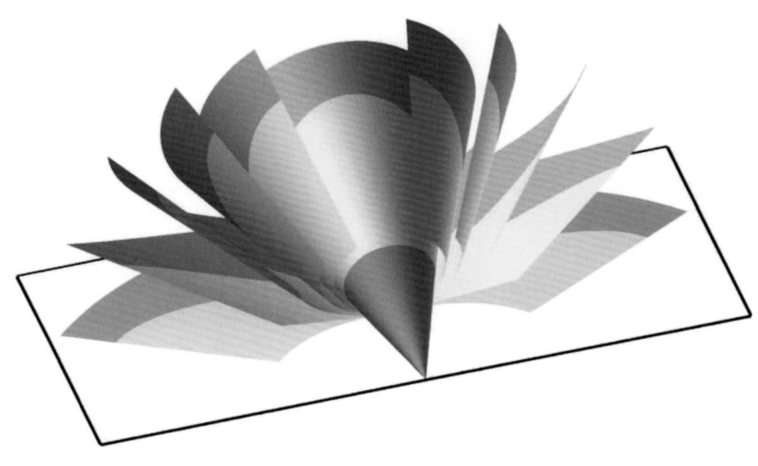

Fig. 5.36 Process of development for a cone of revolution (including a planar oblique section)

From a mathematical point of view, it is still a curve with a reasonably similar type of construction. It is produced from the top view of a conic section if the polar angle to the curve point is multiplied by a factor of $t = g/w < 1$, where g is the length on the cone generatrix seen in the top view, and w is its undistorted length. The radial distance must then be enlarged by a value of $1/t = w/g > 1$.

♠

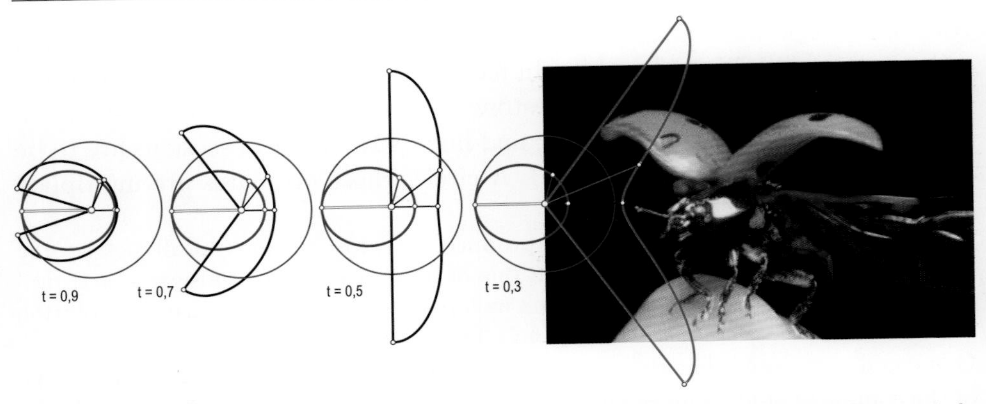

Fig. 5.37 "Angular distorted" conic sections and a resemblance, which is, nevertheless, entirely unrelated ...

• Shadows on a wall

Fig. 5.38 in the middle shows a situation from everyday life: A lamp casts a curved shadow on the wall. These are obviously conic sections, but how are they produced?

Fig. 5.38 Shadows on the wall

Solution:
To make it very clear: The conic sections have nothing to do with the cone shape of the lampshade! The circular edges of the lampshade are the important parts. For example, let us look at the bottom edge. Due to the approximately point-shaped light source (the filament wire of the lamp), there exists a projection cone of revolution that intersects the wall according to a conic section. The type of the conic section depends on the inclination of the wall with respect to the *projection cone* (the inclination of the cone-shaped lampshade is irrelevant). This is why the bottom shadow in Fig. 5.38 on the left is not a parabola (as could be spontaneously assumed), but a hyperbola. The same considerations are also applicable for the top boundary circle. In the general case, the projection cone is not the "extension" of the bottom cone (as in Fig. 5.38). We are thus faced with a second conic section that is independent from the bottom one. ♠

• Parabolic trajectory

Fig. 5.39 shows a game using approximated parabolic trajectories. If an

Fig. 5.39 Approximated parabolic trajectories

object is thrown in the air at an angle, then its trajectory (assuming no air resistance) will resemble a parabola. The special view Fig. 5.39 gives the impression of a "jumping jet of water". Irregularities in the parabolic arc arise due to several factors invisible to the naked eye (the image was taken using a very short exposure time). ♠

• Why are the orbits of planets ellipses? (Fig. 5.40)
Kepler discovered this fact by means of very accurate measurements. It is hard to find a geometric explanation, since we therefore need to solve differential equations[1]. More about the orbit of our planet in Chapter 11. ♠

Fig. 5.40 The orbits of planets are ellipses.

[1]`http://user.gs.rmit.edu.au/rod/files/publications/Satellite{\%}20Orbits.pdf`

5.3 General developables (torses)

The third and last class of simple curved surfaces (next to the omnipresent cylinders and cones) is not very well known. No wonder: These seldom spectacular surfaces often occur in a very hidden form – their outline is always straight.

As we have already said, developables are tangent surfaces of space curves. They are enveloped by planes – the osculating planes of the space curve.

In the course of any type of spatial motion, each plane envelops a developable, a cylinder, or a cone. It touches the emerging surface at each position along a whole generatrix.

Fig. 5.41 Sand dunes as slope developables

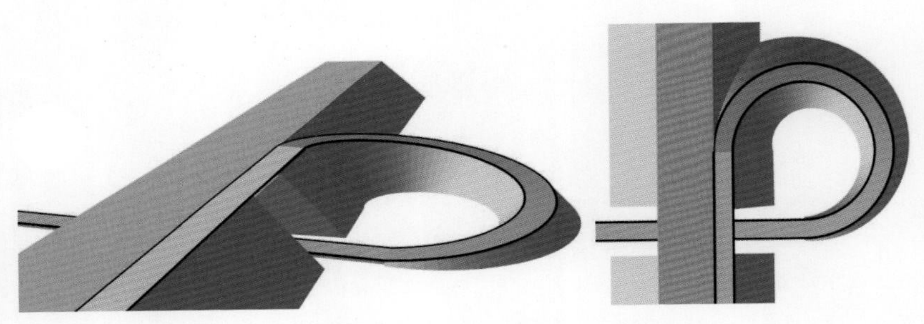

Fig. 5.42 Surfaces of constant slope in road construction

Slope developables are an important class among the developables. They are characterized by the fact that all their tangent planes are equally "sloped" and thus possess the same angle of inclination to a base plane. Such developables are found in sand dunes, contributing to their equilibrium: Sand is "trickling" downwards until a certain angle of inclination is reached. At a certain angle, depending on the grain size, the friction of the grains is large enough for a state of equilibrium to set in. Let us think of the ridge of a sand dune. From each of its points, sand trickles down in all directions and forms a cone of revolution (slope cone). The dune is the enveloped surface of all slope cones (Fig. 5.41 left). Pairs of adjacent cones intersect along a hyperbola that, in extreme cases, becomes a pair of straight lines passing through the tip of the cone.

When planning roads through arbitrary terrain, the engineers rely on this insight: For the used material (sand or gravel), the critical angles are known. The engineers now consider corresponding developables through the edge of the road and solve two problems at the same time: First, the developables are stable and there will be no landslides, and second, the necessary motions of material are minimized.

How to create developable strip?

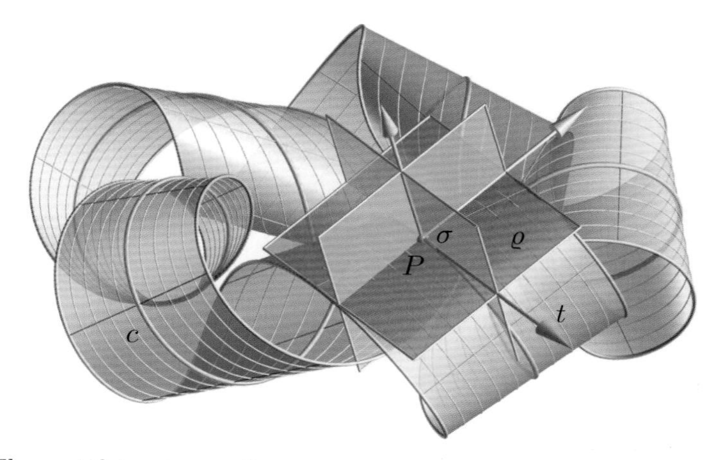

Fig. 5.43 The rectifying torse of a space curve can be parameterized such that the whole surface develops into a rectangular strip with the midline as the rectified curve.

We know that, next to the "trivial" cylinders and cones, only a class previously called torses, but now called developables, can be developed. Let us take a rectangular strip of paper and attach the narrow edges to each other. The hereby generated ribbon is, of course, developable, since we cannot stretch or compress the paper without tearing it. If we do not twist the paper strip, then we get a cylinder. The midline of the strip is, thus, its normal section. Twisting produces a developable with the midline being a curve in space.

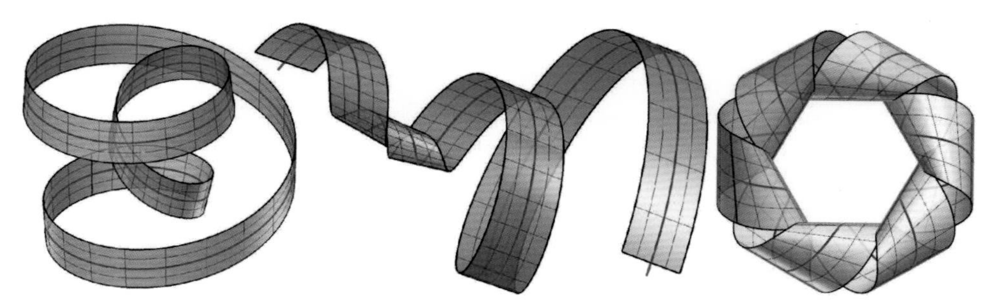

Fig. 5.44 Rectifying developables of three space curves (blue)

• **Deliberate creation of paper strips**

Is there a "paper strip" for an arbitrary space curve c so that c is its midline (Fig. 5.43,Fig. 5.44)?

Let us consider three developables that are connected to the space curve: the *tangential surface* as the envelope of all osculating planes, the *normal developable* as the envelope of all normal planes, and the *rectifying developable* as the envelope of all rectifying plane (planes orthogonal to the principal normal).

The midline c becomes a straight line when spread out in the plane and is, thus, a *geodesic line* on the developable. We know of these lines that their osculating planes must be perpendicular to the tangent plane at any point (p. 121). The tangential surface, whose osculating plane is also its tangent plane, is, thus, not qualified for our purpose. We can see from Fig. 4.24 on the right that the normal developable does not even carry the curve. This only leaves the rectifying developable, for which all the necessary conditions are met: The curve lies on the developable and is also its geodesic line.

Despite the fact that the development preserves angles, the generatrices of the developable do not meet the midline orthogonally. Quite on the contrary: The only straight line in the rectifying plane which is orthogonal to the tangent of c is the *binormal*, and the locus of all binormals is not a developable. Thus, the method of intersecting two neighboring normal planes of the curve does not lead to the solution. In this case, it results in the binormal developable which is irrelevant here. ♠

Fig. 5.45 "Development of a developable": How is it done? Center: Two congruent parts, one inside the other. On the right, one of these parts was developed.

• **"Development of a developable"**

Every developable can be seen as the tangent surface of a space curve g. The curve g is known as the developable's *edge of regression*. The tangents g are the generatrices of the developable. The development makes the edge of regression a planar curve g^v. The generatrices of the developable are tangents of g, and thus, the corresponding lines in the development are the tangents of g^v.

The edge of regression is seldom relevant in practice because, out of the infinitely large tangent plane, only the parts that do not border on the edge of regression are normally used. The edge of regression is meaningless in regard to the windmill in

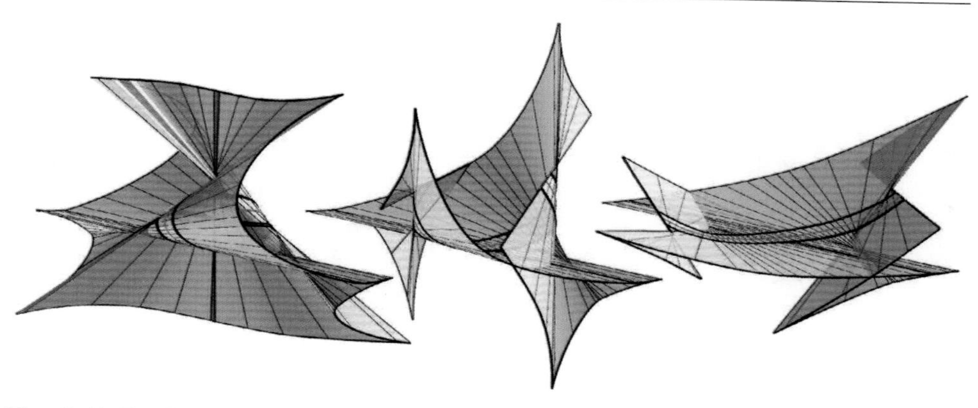

Fig. 5.46 Development of a developable which contains the Möbius strip. Each intermediate position is a developable, on which the jointly transformed strip is bounded by geodesic lines.

Fig. 5.45. Something similar is true for the oloid, whose development is depicted in Fig. 6.49, and for the development in Fig. 5.46. ♠

• Approximation of a curved surface through rectangular strips
A common technique (especially in shipbuilding) is to approximate moderately curved surfaces by means of developable strips.

Fig. 5.47 A curved surface is "modeled" from planks.

Wooden planks – but also aluminum planks in case of tensile loading – are, for example, well suited for "twisting and bending". This system of modeling can be seen in Fig. 5.47. ♠

Let us generate arbitrary developable surfaces

Architecture often demands the use of developable surfaces, and no wonder: The creation of such surfaces is many times easier and cheaper than that of doubly curved surfaces.

Some good news and some bad news: While it is always possible to find a developable surface connecting two arbitrary boundary curves, developable surfaces are characterized as always having straight-line contours. The connecting surfaces could, therefore, not have anything "curvy" or "spherical" about them. In other words, they would, in many cases, not be curvy

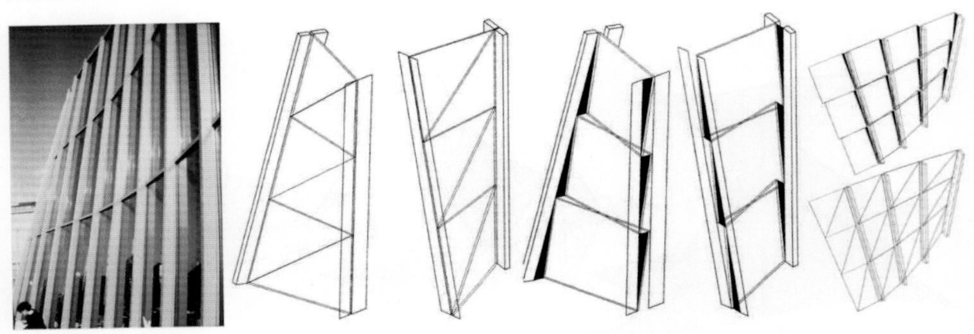

Fig. 5.48 How architects construct curved and reasonably priced glass facades ...

enough. A program which allows the direct modeling by using developable surfaces is introduced on p. 289 – and can, nevertheless, achieve good results.

- **Developable connecting surfaces**

Let c_1 be an arbitrary curve in space and c_2 a curve in a plane ε_2 (Fig. 5.49). We will now introduce a procedure for finding the developable connecting surface Φ of c_1 and c_2:

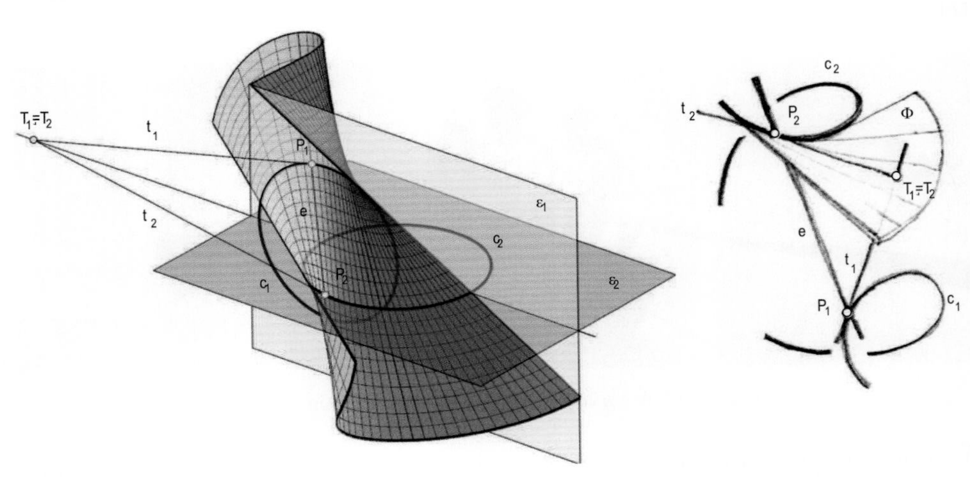

Fig. 5.49 Construction of a connecting developable (on the left: Oloid)

We determine the tangent t_1 for an arbitrary number of points $P_1 \in c_1$, intersect it with the carrier plane ε_2, and draw a tangent t_2 towards c_2 from the intersection point $T_1 = T_2$. The respective contact point P_2 determines, along with the initial point P_1, a generatrix e of the surface Φ. The tangent plane of Φ is determined along the whole generatrix by the connection of e with t_1. Due to our construction, t_2 also lies in the tangent plane – which is no contradiction. The thus defined series of tangent planes envelops a developable. This construction also applies if both c_1 and c_2 are planar curves. If both curves c_1 and c_2 are really space curves, the problem is more difficult, but still solvable: Instead of the carrier plane ε_2 of c_2, we now consider the

Fig. 5.50 Centrum Bank of Vaduz (Hans Hollein) with cylinders as surfaces

tangent plane Φ_2 of c_2. The tangent t_1 at c_1 should not be intersected with the plane ε_2, but with Φ_2 – a task for which we need a computer!
An interesting example of a special connecting developable is pictured in Fig. 5.50. Here, the directrices lie in parallel planes. In such cases, the generatrices of the connecting developable are the straight lines connecting the points on both generatrices in which the tangents are parallel to the directrices. However, the image on the right indicates that all generatrices are parallel – and thus, a *cylinder* is given. The image is based on a cylinder, which was then intersected with the side surfaces. ♠

Reflection in a developable surface

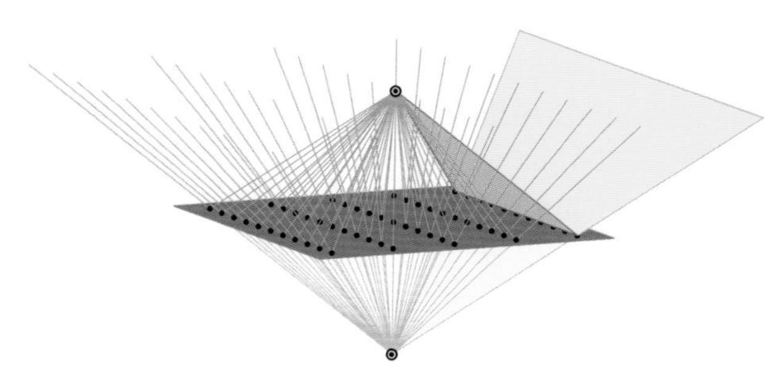

Fig. 5.51 The theory behind the reflection of light rays in a plane

Reflections in curved surfaces belong to very sophisticated geometry. If all light rays that emanate from a point-shaped light source are reflected in such a surface, we get a jumble of rays which envelop a new surface: the so-called *focal surface* or *caustic*. In general, such focal surfaces are very complicated and full of "singularities". In application p. 184, we shall discuss the reflection in a sphere.
The problem is simplified significantly if the reflecting surface is only single curved: Cylinders, cones, and developables are touched by a fixed plane along a whole generatrix.
Fig. 5.51 shows what happens during reflection in a plane: All light rays from a point-shaped light source (which – like the sun – may be infinitely far away)

through an arbitrary straight line of the plane form a plane. The reflected light rays are restricted to the reflected plane, which is itself determined by the straight line and the reflected light source.
If the reflecting surface is single curved, we can replace it along each generatrix by its tangent plane. As with the plane (Fig. 5.52), the simple situation remains:

All light rays that are reflected in a generatrix of a simply curved surface lie in a plane (the *focal plane*) which passes through the point that is reflected in relation to the tangent plane along the generatrix.

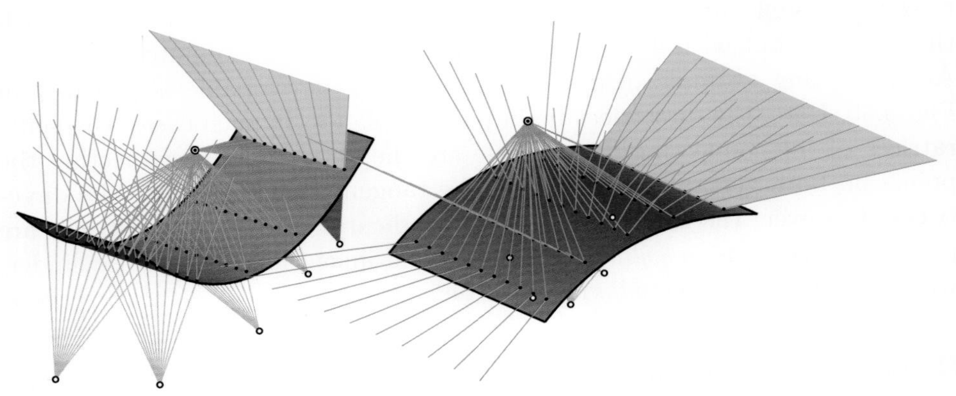

Fig. 5.52 Reflection in a simple curved surface

If we now let the generatrix move along the surface Φ, we get a pencil of focal planes which envelop the developable. The focal surface of a developable surface is, thus, itself developable. To give a more concrete example, the focal surface of a cylinder is again a cylinder. In application p. 184, we will find out more about the reflection in a cylinder of revolution. The locus of all light sources reflected in the tangent planes is a curve – the *focal line.*

• **Shiny generatrices on aluminum cylinders**
Let us consider Fig. 5.52 on the right: One of the rays that are reflected along a generatrix appears as a point. We can now imagine parallel light and picture the surface from a greater distance, which causes the viewing rays through the points of the generatrices to be approximately parallel. Thus, every point on the generatrix "shines up", and the whole generatrix glows brightly. Such effects are often visible on turned parts of aluminum (cylinders of revolution or cones of revolution) and also with glass bottles in sunlight. ♠

• **Glowing curves**
We encounter the following situation with reflections in a plane: A developable surface reflects a light source (such as a spotlight or the sun). All focal planes intersect the plane in a straight line. The focal lines envelop a curve – the intersection curve of the focal surface with this plane. We thus have: *The reflection of a developable surface on a plane is a glowing curve.* The curve is glowing because it focuses all light that interacts with the surface (and sometimes even causes heat effects). ♠

• **Light games in the water** (Fig. 5.53):
Everybody knows the fascinating, constantly changing reflection patterns of sunlight on the sides of boats or buildings. The focusing effect of the light rays is evident here.

Fig. 5.53 Light games above water (left) and below water (right)

This permits the speculation that the surface of moving water is only single curved. This is, in fact, true (though only approximately so) if the motion of water is caused by wind. In such situations, the emerging wave surfaces are (ideally) cylinders – thus, single curved. But we are also dealing with the reflections of the waves on harbor moles and boats. This produces new, cylindrical waves which overlap the original ones. In any case, these are the causes of a beautiful spectacle ... ♠

• **Reflecting compact discs – and curious effects**
The curious reflections on CDs must have aroused the interests of many. Two things are apparent: Firstly, these reflections seem to be glowing straight lines that can swiftly change position (and essentially rotate about the midpoint of the CD). Secondly, all rainbow colors seem to be visible among the reflections. This cannot be caused by the tracks (of which there is actually only one): The same effect is visible on blank CDs.

Fig. 5.54 A CD reflects all colors of the rainbow.

Other properties of light are at play here – apart from the mere reflection. Roughly speaking, we are dealing with interferences. CDs are essentially manufactured from plastic polycarbonate, whose one side is vaporized with an aluminum layer (the aluminum reflects the laser beam which reads the data). In order to prevent chemical reactions, the disc is then coated with transparent plastic.
When a light ray passes through the plastic coating, a part of it reflects in the surface. What remains is refracted at the surface based on to the angle of incidence

(and also splits up into its color components). The refraction occurs again on the other side of the plastic coating – and no later than at the aluminum layer, reflection occurs again, as does "refraction", and the ray exists parallel (but slightly shifted) to the ray reflected in the first surface. Diffraction of the light rays also takes place due to the presence of an optical diffraction grating (many thousands of openings per inch). The reflected and (following the many consecutive reflections) slightly shifted light rays now interfere with one another. This may lead not only to the augmentation but also to the cancellation of whole color components. Similar effects occur on butterfly wings and fish scales. ♠

- **Artistic inspiration**

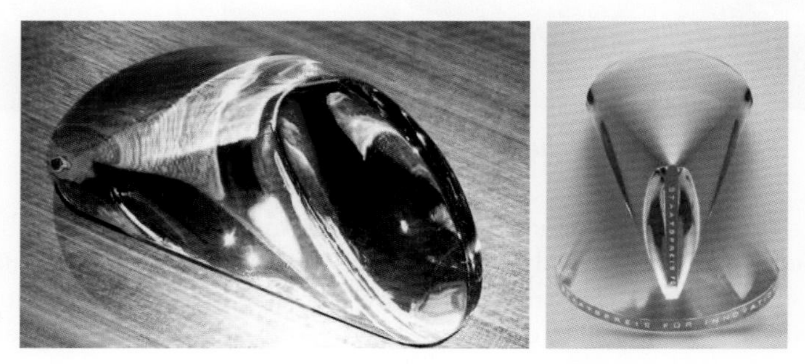

Fig. 5.55 The oloid as a piece of art

Fig. 5.55 shows a glass oloid by Konstantin Ronikier. Here, reflection and refraction effects are incorporated into an entire artistic appearance. ♠

5.4 About maps and "sphere developments"

The dilemma about the development of doubly curved surfaces

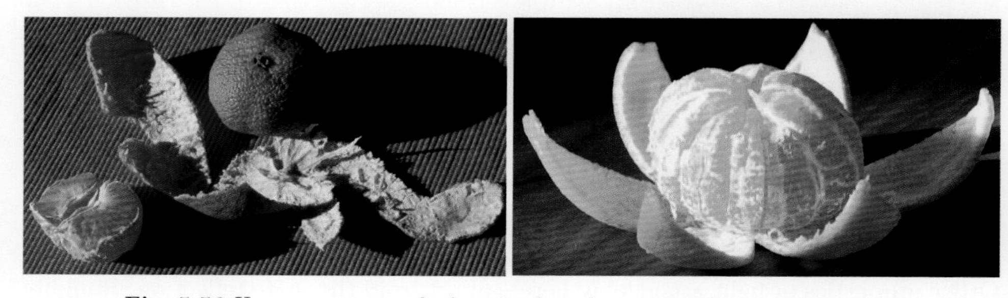

Fig. 5.56 However you may look at it, the sphere is simply not developable!

We know at this point that doubly curved surfaces cannot be developed – let alone a sphere! Every attempt to press the surface of a sphere into the plane ends with expansions and compressions, or with the tearing apart of the whole object altogether. The method of peeling tangerines pictured in Fig. 5.56

comes close to Walter *Wunderlich*'s approach in Fig. 5.57. The method in Fig. 5.56 on the right exhibits certain similarities to the approximation of the sphere by cylindrical strips in Fig. 5.57 (center). Nature occasionally solves the problem abruptly (Fig. 5.58).

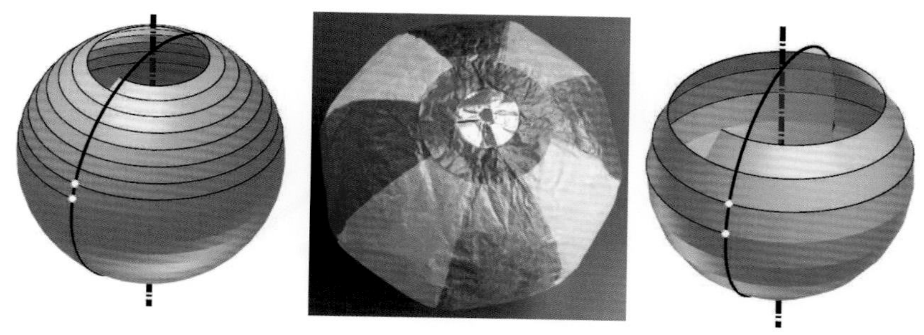

Fig. 5.57 The "apple peeling method" by *Wunderlich* through approximation by a cylinder

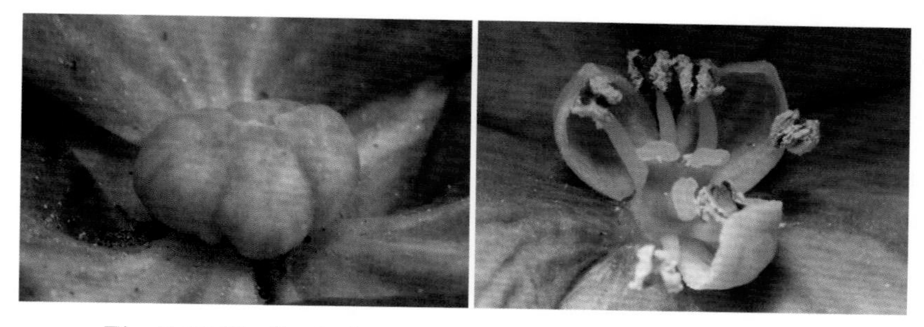

Fig. 5.58 The "brute-force approach" sometimes works as well ...

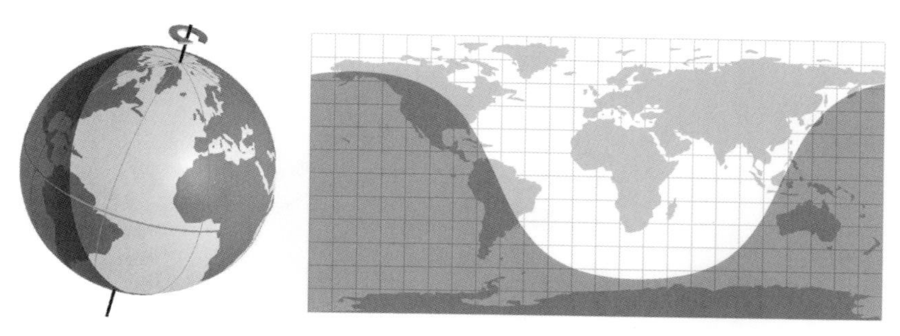

Fig. 5.59 The shadow boundary in the distorted coordinate grid

All of this despite the fact that the development of a sphere would be important for so many applications – especially for cartography! It is obvious that the maps with which we constantly deal must have been produced by using certain tricks that accommodate the curvature of the sphere. This must necessarily have occurred at the expense of geometrical properties that are well known to us. See Fig. 5.59, for example, where those parts of the globe are shown that are currently illuminated by the sun.

The situation in the left image is very clear: The shadow boundary of the globe is a great-circle. The right image shows the respective situation in a relatively common world map where geographic longitudes and latitudes are simply used as Cartesian coordinates. The great-circle becomes a wave-like curve (but not a sine wave). With such transformations, it is very simple to find a geographical position. But that's about it: Angles, lengths, and areas appear greatly distorted. This is especially evident in polar regions – around the equator, distortions tend to be the smallest.

Fig. 5.60 A slightly unusual "development" of a sphere ...

• Is it developable after all?

Architects are usually keen on building models of their sketches. Nowadays, they mostly employ state-of-the-art modeling software for this purpose. If they then activate the "develop"-function of such a software package, the computer will not hesitate. In the case of a sphere, a strangely frayed object will emerge that, upon closer inspection, turns out to consist of ring sectors. Why is this?

Fig. 5.61 ... with different qualities

Fig. 5.61 and Fig. 5.60 show the spatial situation: The sphere is more or less accurately approximated by frustums of cones of revolution. Each single frustum is developable, producing a ring sector. The better the approximation, the more circular rings are produced.

As original as this idea is, it is not applicable to cartography: One needs only imagine a world map where all the continents are "dissected" on separate strips ... ♠

An angle- and circle-preserving projection

We now consider the so-called *stereographic projection* of the sphere. It is a central projection with the center of the projection on the sphere, and the image plane is (e.g.) the opposite tangent plane. This kind of projection plays an important role in *cartography* due to the following properties:

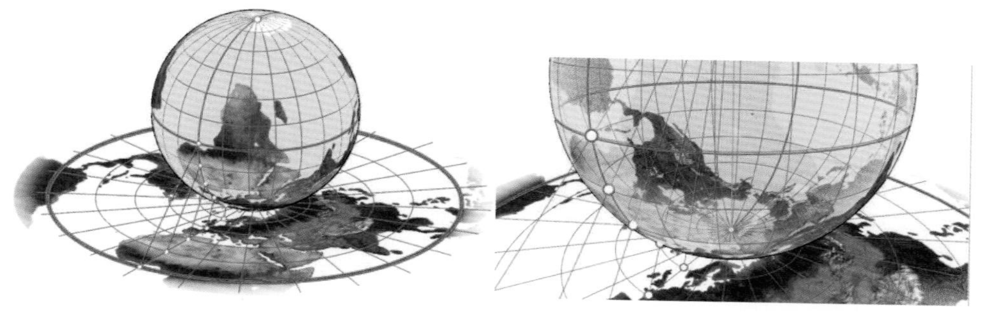

Fig. 5.62 Stereographic projection of the globe from the south pole

After stereographic projection, section angles and circles remain preserved.

Proof:
Let Z be the center of the stereographic projection from the sphere to the plane and π be the tangent plane through the antipode of Z. Without loss of generality, we may assume π to be horizontal. The straight lines of π correspond to circles through Z that result from the intersection of their connecting planes with the sphere. Let us take a closer look at two such straight lines a and b with the intersection P and the enclosed angle φ. They correspond to two circles a_k and b_k on the sphere through P_k and Z.
Due to symmetry, the angle enclosed by both circles on the sphere is the same in both common points. However, the angle enclosed by a_k and b_k at Z equals φ, because the respective circle tangents lie in the horizontal tangent plane of Z and are, therefore, parallel to a and b. Therefore, a_k and b_k also intersect in the point P^k at an angle of φ. Whatever is true for the angle between two straight lines is also true for the angle enclosed by two curves, since the angle between curves is defined as the angle of the respective tangents. We have thus showed that the stereographic projection preserves angles. It is also said that the stereographic projection is conformal.
We must briefly elaborate on the proof that the stereographic projection also preserves circles: Stereographic projection on π is basically no different from a central projection on π – under special consideration of the points on an auxiliary sphere, which was placed "between" the center and the image plane (more precisely: the restriction of a central projection to the sphere). It is not forbidden to project points

Fig. 5.63 Stereographic projection of the globe from the antipode of Vienna

or straight lines that do not lie on the sphere. However, the preservation of angles and circles, in general, only applies to the sphere elements.
Let k_k be a circle on the sphere. We now observe a cone of revolution that touches the sphere along k_k. It has the apex S_k and nothing but generatrices that intersect the circle *on the sphere* at right angles. Let us project S_k from Z towards $S \in \pi$. Then, the generatrices correlate to straight lines through S. On each such straight line, there exists the image of a point of k_k. The preservation of angles applies there (and only there). As on the sphere, the tangent of the projection $k \in \pi$ of k_k must intersect the projected cone generatrix orthogonally. However, there exists only one type of curve that intersects a pencil of rays orthogonally: the circle. Moreover, S must be the midpoint of this circle!
If the sphere circle k_k passes through the center Z, then the apex S_k of the "adjacent" cone of revolution is located at the same height as Z. Its projection S is, therefore, a point at infinity. The pencil of rays of the projecting cone generatrix becomes a pencil of parallels. The projected curve k becomes a straight line that intersects this pencil orthogonally – but also, if you like, to a circle with its "center at infinity".◇

A practical map

If the points on the globe are projected from the center of the globe onto an arbitrary tangent plane, we get a world map where all planes – including the great-circles – appear as straight lines through the center. However, great-circles are (as we will soon discover) the shortest connecting paths between two points on the sphere surface. On such a map, it would, thus, be sufficient to connect two points with a ruler in order to draw the shortest path (Fig. 5.64).
As remarkably simple as this principle may be, it produces a very significant map distortion, because the center of projection is relatively close to the projection plane. If, for instance, the northern hemisphere was projected onto the tangent plane of the north pole (many airplanes choose routes far up north on long-distance flights), then the entire southern hemisphere (including the

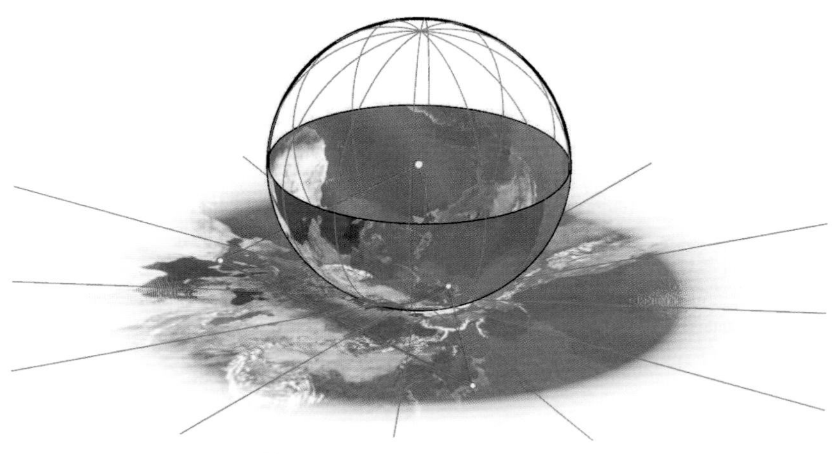

Fig. 5.64 From Tunis to Tokyo

equator) would not even appear on the map. Even the latitudes south of the 30th parallel would lie very far outside.

• From Bangalore to Buenos Aires

The pragmatic American architect Richard Buckminster *Fuller* solved the problem of flight paths in his own way (Fig. 5.65): He projected not onto the plane, but onto the side surfaces of a cuboctahedron ingeniously fitted onto the Earth, on which eight equilateral triangles and four squares are located (the name derives from a combination of a cube and an octahedron). By complete accident, the continents turned out to fit relatively well on the surfaces of the polyhedron. If such a cuboctahedron is developed in the plane, one gets a map that isn't very distorted (but not connected either). On it, the shortest flight paths appear as straight lines as long as one keeps to the side surfaces. The kink happens upon transition to the next surface. However, within continents, the flight routes are very easy to find.

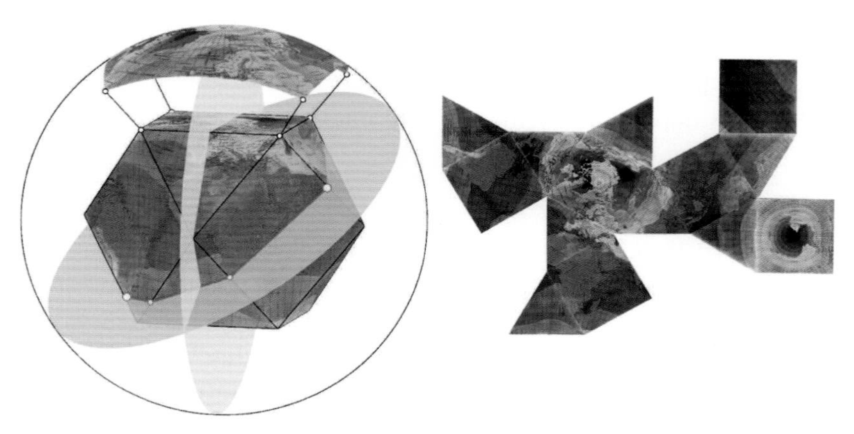

Fig. 5.65 With Buckminster Fuller on the Cuboctahedron from Bangalore to Buenos Aires

The stereographic projection offers a theoretically interesting solution to the flight path problem. On a map constructed by such a projection of the globe, the longitudes and latitudes appear as circles due to the preservation of angles and circles – with each longitude intersecting each latitude perpendicularly. The shortest paths between two points are, thus, also circles that must fulfil a certain condition that we will work out shortly. By defining a circle, one also knows the course angle in every point of the flight or ship path.

• **Shortest connection**

In a stereographic projection, the shortest path between two distant globe points A and B is to be determined.

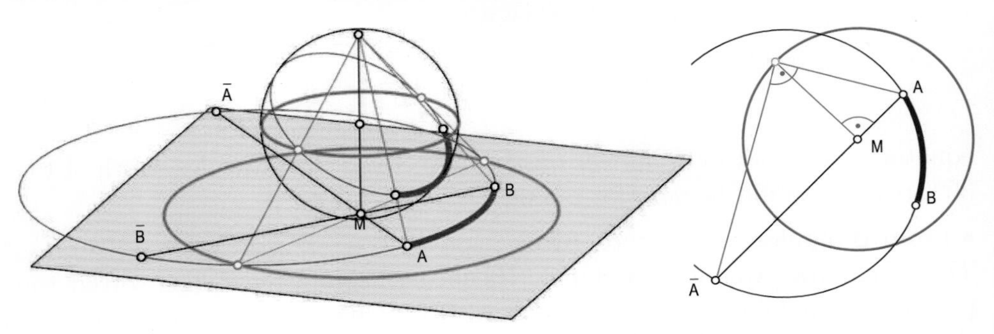

Fig. 5.66 The shortest path in a stereographic projection

Solution:

The shortest path from A to B on the sphere is a great-circle (Fig. 5.66). It is unambiguously defined by the carrier plane ABM (M is the midpoint of the Earth). If two arbitrary great-circles are intersected, then the intersection points lie diametrically (relative to M). In stereographic projections, great-circles are, thus, circles that pass through two opposite points on the inversion circle (which is the image of the equator). According to the theorem concerning the constant power of a point regarding a circle (p. 14), we, therefore, know the product of the distances on the secants through M (through A as well as B). This way, two additional points $\overline{A}$ and $\overline{B}$ on the great-circle in question can be determined (though only one is sufficient for construction).

♠

The inversion at a circle from a "higher perspective"

Let us once again return to the inversion in the plane. As interesting as it might have been theoretically, it still seemed a little bit arbitrary. Does the transformation $\overline{MP} \cdot \overline{MP^*} = r^2$ actually have a "higher" significance?

Let the equator plane π be the image plane of the stereographic projection and the north pole Z be the center of projection (Fig. 5.68), and let S denote the south pole. P_k and P_k^* are two points on the sphere with the same orthogonal projection (top view) in π. The points Z, P_k, and P_k^* now lie on

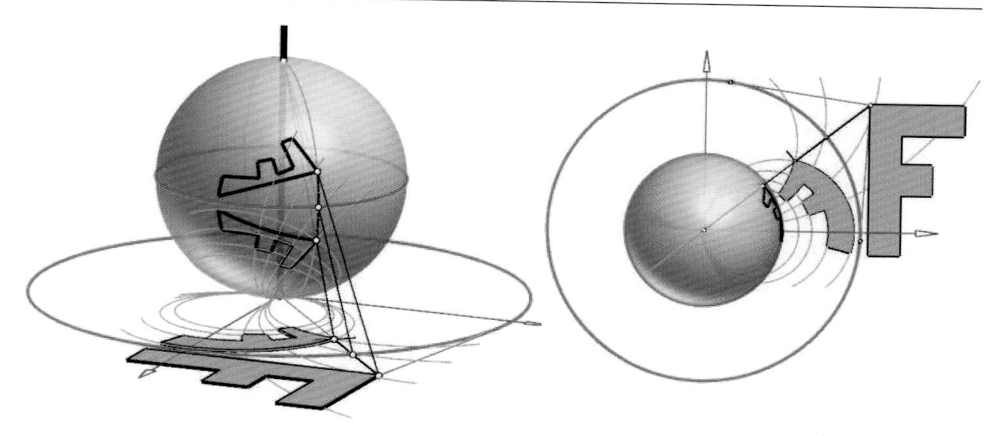

Fig. 5.67 A beautiful spatial interpretation of the inversion

the same great-circle, which we shall, from now on, imagine as the sphere outline.
Let P and P^* be the respective stereographic images of P and P^*. The points P_k, P^*, and S thus lie on a straight line (reflected in π) which, according to *Thales*' Theorem, is perpendicular to PZ.

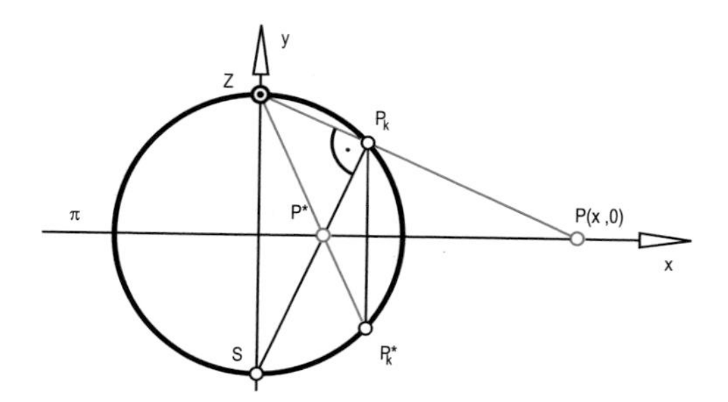

Fig. 5.68 Inversion and stereographic projection

For the next simple calculation, let us imagine a two-dimensional coordinate system with the center M of the sphere as its origin, the x-axis in the direction of P and the y-axis in the direction of Z. If x_0 denotes the x-coordinate of P, then the straight line PZ has a slope of $-\frac{r}{x_0}$ ($r=$ sphere radius). The slope of the straight line P_kS is, thus, $\frac{x_0}{r}$ and its equation is, therefore, $y = \frac{x_0}{r} \cdot x - r$. In order to preserve P^*, we only need to intersect it with the x-axis:

$$\frac{x_0}{r} \cdot x - r = 0 \Rightarrow x_0 \cdot x = r^2.$$

This already shows that P and P^* are inverted at the circle around M with the radius r.
If the tangent plane at S is chosen instead of π as the image plane of the stereographic projection, then all image lengths are doubled, and we get the following theorem:

Let π be the horizontal drawing plane in whose origin a sphere with diameter 1 is located. If overlapping points on the sphere are projected from the topmost point of the sphere onto π, we get points which form an inverse pair with respect to the unit circle in π.

Strictly speaking, the theorem would later have to be formulated as follows: Let Z be a point on the sphere, π be the tangent plane of the sphere in the opposite point, and k_0 be the great-circle of the sphere parallel to this plane. Then, both hemispheres (bounded by k_0) are mapped via the central projection from Z to π to point fields which are related by an inversion with respect to k_0.
This admittedly far-fetched inversion of the plane is considerably easier for a three-dimensionally thinking being – it is merely the reflection of a sphere half in a plane. The "missing link" is a stereographic projection. Of course, we could now think about whether or not a four-dimensionally thinking being would find it equally easy to invert three-dimensional space on a sphere – such as using the generalized stereographic projection of space on a "hypersphere". It is, indeed, possible to extrapolate the line of reasoning, and from a mathematical standpoint, it works! The only catch: Calculating such a phenomenon is perfectly feasible, but *imagining* it is no longer possible.

From a higher perspective, the most important properties of inversion are very easily provable:

The inversion maps circles to circles and preserves angles. Straight lines correspond to circles through the midpoint of the inversion circle.

Proof:
Preservation of angles (conformity) and preservation of circles alike are deducible from the respective properties of the stereographic projection, which is applied twice (in opposite directions).
Likewise, the reflection in the sphere does not affect the section angles or the circle property. In the case of stereographic projection, straight lines correlate to circles through the projection center (the intersection circle of the image plane with the sphere). During reflection, they become circles through the "receptor point" M, which likewise become circles through M in the case of the inverted stereographic projection. ◇

Another proof for the preservation of circles can be found in [11] and is therein conducted in a pretty elegant way by the use of complex numbers. The advantage of the geometric proof is, once again, the simple fact that its "residual products" promote geometrical comprehension. One insight somehow always produces the other, leading eventually to a rounded geometrical "worldview", in which many conclusions become clear whose proofs would otherwise have to be "derived" mathematically.

5.5 The reflection in a circle, a sphere, and a cylinder of revolution

So far, we have often mentioned the reflection in a circle (such as with inversion), even though the reference to a physical reflection was not always intended. The reflection in a surface which we know from nature follows two rules:
Firstly, the incoming and the outgoing light rays lie in a plane which also contains the surface normal. Secondly, the angle enclosed by the normal and the incoming ray equals the angle enclosed by the normal and the outgoing ray. ("angle of incidence = angle of reflection").

- Reflection in a plane

Let E be the eye point and S the point that emits light (Fig. 5.69 on the left). If we look into the reflection plane τ, we see a reflecting point R. We can geometrically determine R by reflecting S in τ and connecting this point S^* with E. The eye can no longer distinguish if it is looking at the reflecting point R or the reflection point S^*.
In fact, both rules are observed: SR and RE lie in a plane which includes the normal n, just as the angle of incidence ζ_1 and the angle of reflection ζ_2 are identical. The reflection in a plane is, therefore, a simple (linear) problem.♠

- Reflection in a curved surface

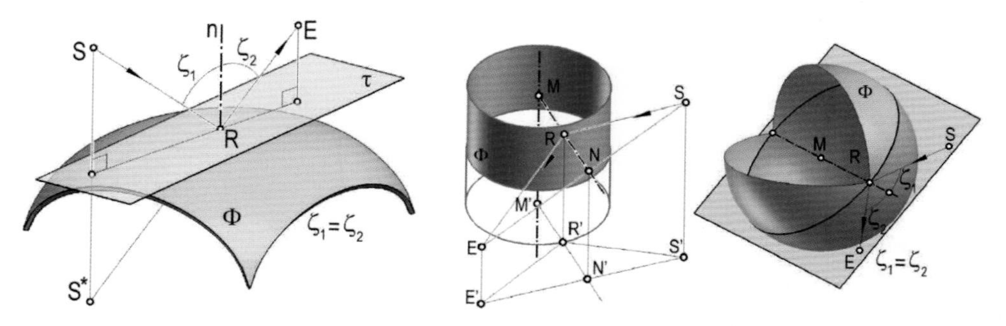

Fig. 5.69 Reflection in curved surfaces

The reflection in a curved surface is, of course, a remarkably difficult undertaking in the general case. Let τ be a tangent plane in an arbitrary surface point R (Fig. 5.69 on the left). The point S in space can thus be reflected in τ. If, by coincidence, the straight lines ER and ES^* happen to be identical, then we have found the reflecting point of S in R (complicated surfaces may contain multiple "reflecting points").
In practice, we can find all reflecting points by conducting the aforementioned test for a high number of surface points. In cases where such reflection points are "nearly" found, the immediate vicinities can further be scanned by using approximation techniques.
♠

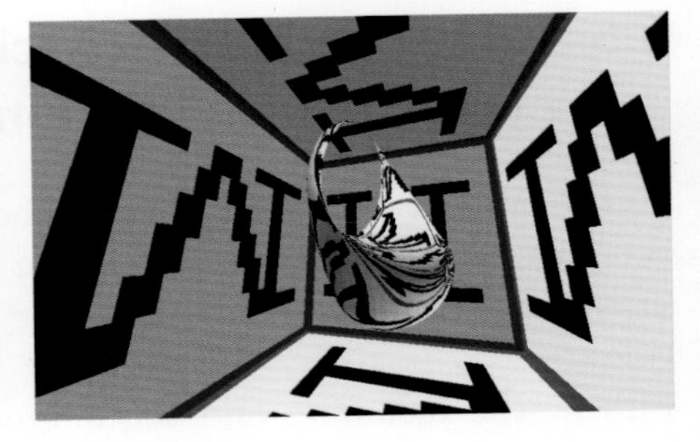

Fig. 5.70 Reflection in a curved surface

This generalized search method is, of course, computationally expensive – but we will not further discuss it here. In his diploma thesis, Franz Gruber developed a method for determining reflecting points as in Fig. 5.70 in "real time" (multiple images per second).
Two special and somewhat frequent cases are interesting for as.

- **Reflection in a cylinder of revolution**

Let Φ be a cylinder of revolution (Fig. 5.69 at the center), which for the sake of simplicity, we shall imagine as being positioned vertically. If we project all points normally into the base plane, the problem of reflection is reduced to the two-dimensional reflection on the base circle. We now just have to find a point R', which, if seen from E', is the reflecting point of S' in the base circle: The same angles in the base plane cause an equality of angles in the plane that is formed by E, R, and the cylinder normal $n = MR$.

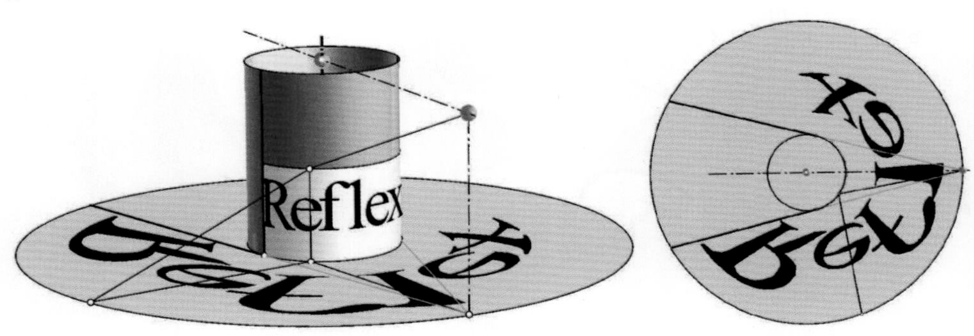

Fig. 5.71 Anamorphoses: Reflection in the cylinder of revolution

The reflection in the cylinder of revolution, which is familiar to us from many halls with reflecting columns, was historically used to create so-called *anamorphoses*. The observer sees a bizarrely distorted image that appears "normal" when seen in a cylindrical mirror (Fig. 5.71). ♠

- **Reflection in a sphere**

The reflection in a sphere can likewise be traced back to the reflection in a great-circle: The sphere normal always contains the center M, and the light rays ER and RS in question lie in the great-circle plane MES. ♠

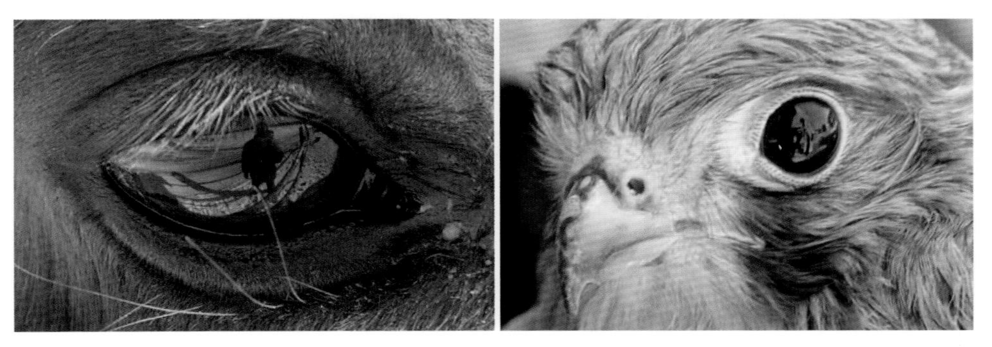

Fig. 5.72 Reflection in spheres

Thus, the problem of trying to get a grip on the physical reflection in a circle is becoming more urgent. To put it up front: The problem is not easily solvable because it can be shown that it would require the solution of a fourth-order equation. This is no problem for the computer (there are formulas for solving fourth-order equations), but it is in general unsolvable with just a ruler and a compass alone.

Fig. 5.73 The physical reflection in a sphere

A simple special case can be answered relatively fast: If the point S (which is to be reflected) happens to have the same distance from the circle (in practice, it would be from the sphere or the cylinder of revolution) as the eye point E, then the reflecting points R_i will be located precisely on the angle bisector of MS and ME. Armed with this knowledge, the specular points in the eyes (Fig. 5.74), which emanate from a light source, are easy to localize. The other way: The location of the light source is equally easy to reconstruct, given the location of the specular points.
Even in the general case, we should not immediately give up. Let us illuminate its "trimmings" somewhat:

• **Conic sections with given focal points which touch a given circle**
We know that in the case of conic sections, the curve tangents and curve normals bisect the angle of the focal rays. A light ray through the focal point E is thus reflected in another focal point S.

Fig. 5.74 Specular points on sphere caps

If such a conic section touches a given circle k at the curve point R, then the tangent to the circle equals the tangent to the conic section at R, and we have thus reduced the problem to the reflection in a circle.

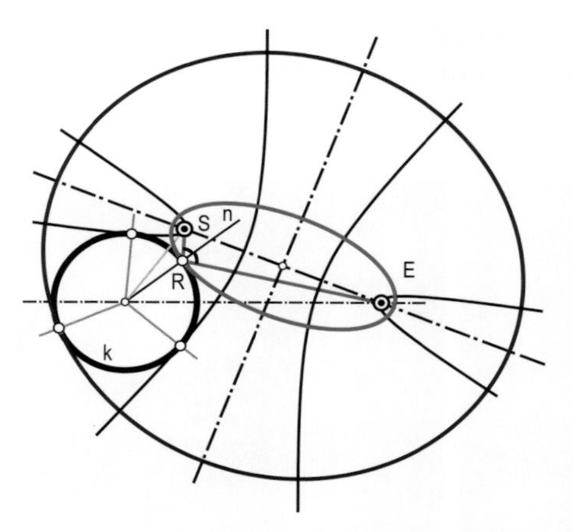

Fig. 5.75 Four confocal conic sections which touch a given circle

In order to find an additional point on the conic section in question, we now search for the point R on k, which is the reflecting point of S on the circle k with respect to E. In the case of the outer reflection, the solution seems to be unambiguous, but it still cannot be done without the solution of a fourth-order equation. Not all of the four possible solutions have to be real. We are dealing with four confocal conic sections (Fig. 5.75) which, by the way, mutually intersect orthogonally, as we have already proven. ♠

The question still remains how the four "reflecting points" of S on the circle k can be geometrically determined for the given eye position (Fig. 5.77 on the left). The answer can merely be: by trial and error. By moving R along the circle, the light ray ER is varied. The respective incoming light ray is always the reflection in relation to the circle normal and can easily be constructed. At some moment, S will change sides relative to this reflection – and the solution lies precisely there, with graphical accuracy.

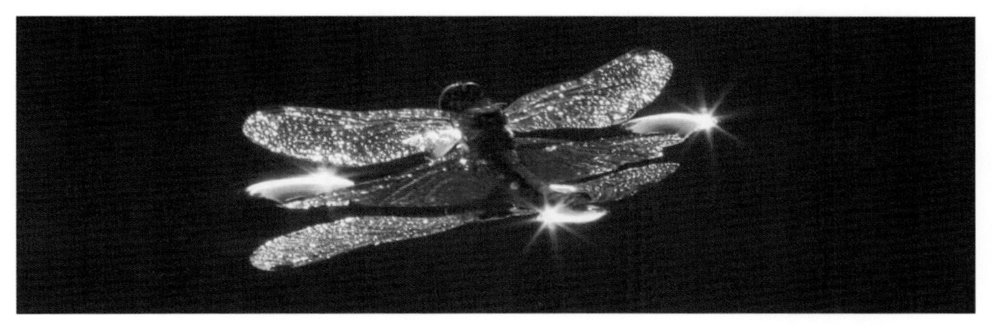

Fig. 5.76 The dragonfly impresses the water surface in three points, which are immediately visible through the specular highlights.

• Focal lines of a circle

The aforementioned method of "trial and error" produces a family of reflected rays which envelops a curve. The form of this so-called *focal line* (*caustic*) may vary. For points outside of the circle, it will look approximately as in Fig. 5.77 on the right.

The term "focal line" is derived from the fact that instead of the eye point E, a point light source can be imagined, whereupon such brightly illuminated curves, indeed, appear on cylindrically reflecting walls. A point E inside the circle k produces focal lines of the same type as in Fig. 5.78.

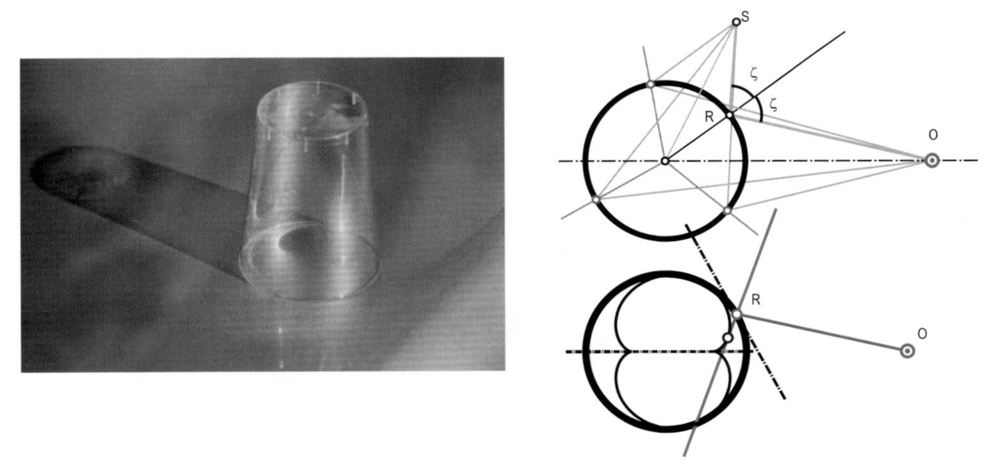

Fig. 5.77 The reflected rays envelop a pointed curve.

In each case, from E, there exist multiple (at most four) tangents of the focal line which lead to the desired reflecting points.

During reflection in the outer wall of the circle, *only one* of the four solutions is practically relevant. It can also be found by trial and error. If E lies inside the circle, then *all* solutions may actually be visible.

Fig. 5.79 serves to illustrate why spatial inversion is often referred to as reflection in a sphere despite the fact that a real physical reflection in a sphere has nothing to do with it: Inversion gives us a well defined spatial

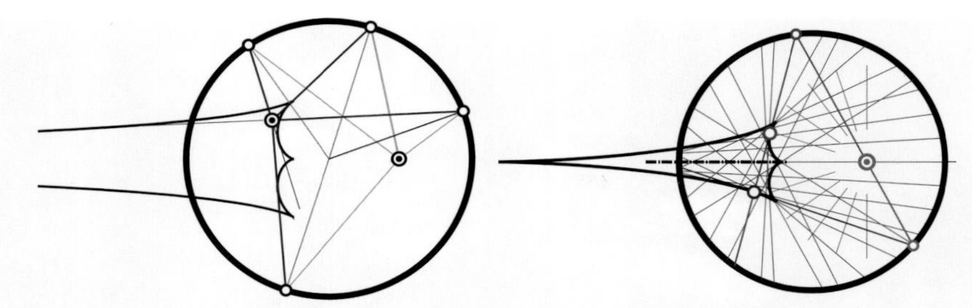

Fig. 5.78 The point may also be located inside the circle.

object, which is generated regardless of the observer (in the particular case, it is a Dupin *cyclide* which is further discussed on p. 232). The reflected image which appears on the sphere changes its appearance along with the location of the viewer.

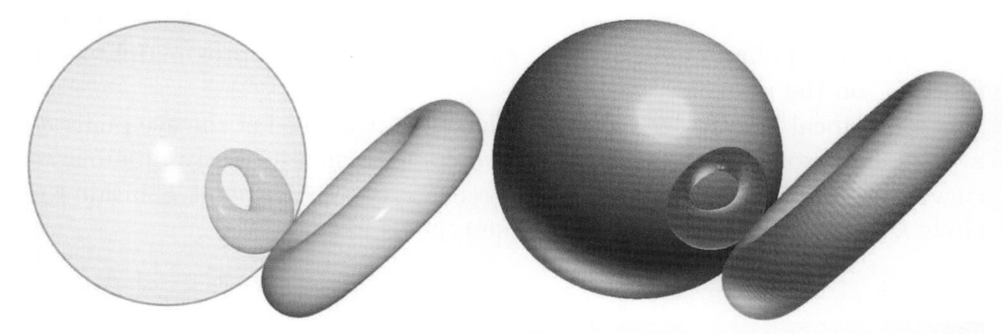

Fig. 5.79 Inversion and physical reflection in a sphere

When put this way, the optical similarity of both images in Fig. 5.79 is only apparent because the inverted object is observed from the right vantage point. By the way: M.C. *Escher* has approximated his spherical reflections by means of inversion.

Fig. 5.80 A torus as an air bubble, encased in glass

By comparison, Fig. 5.80 shows the image of a torus distorted due to refraction. ♠

6 Prototypes

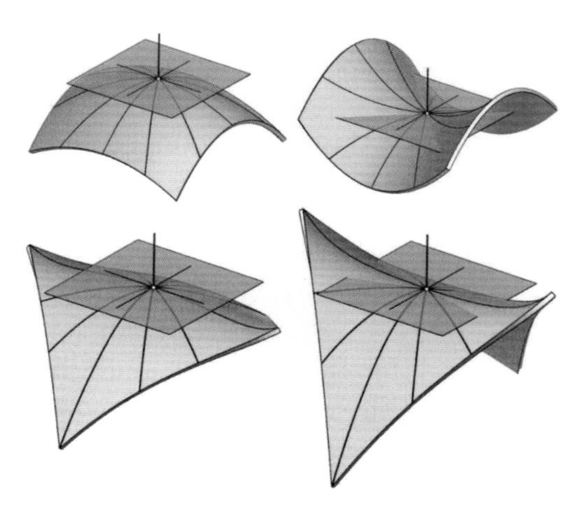

The majority of all curved surfaces cannot be developed into a plane. "In the neighborhood of a surface point" doubly curved surfaces can bend in one direction, lying on one side of the tangent (as with the sphere) or – as with the saddle surface – bend in two directions, lying locally on both sides of the tangent plane.

By examining the simplest cases of doubly curved surfaces, i.e., second-degree surfaces, from the perspective of their behavior regarding the tangent plane, far-reaching statements about local curvature characteristics of all surfaces can be made.

Surfaces that consist locally only of elliptic or hyperbolic points cannot be flattened out in the plane without deformation – they would need to be parabolic in all points.

However, second-order surfaces are not only interesting as prototypes. They possess a multitude of special properties with technical applications. A classical example is the parabolic headlamp.

Surfaces of revolution shall serve as a first class of more general surfaces to be discussed in one of the following sections. Aside from the second-order surfaces of revolution, the torus also plays an important role. This fourth-order surface has numerous applications, and can not only be generated by the rotation of a circle about a coplanar axis but also by the rotation of a sphere.

Survey

6.1 Second-order surfaces

The simplest type of surface is a plane, which in space possesses one intersection point with every (general) straight line. It is therefore said: Planes are first-order surfaces (and there exists no other first-order surface).
Each surface that possesses two intersection points with a (general) straight line (these must not always be real and may also coincide) is called a *second-order surface*. We are already familiar with several types of them: The *sphere*, the *cylinder of revolution*, and the *cone of revolution*.
By applying *affine mappings* (which do not change the aforementioned number of intersections) to such surfaces, *ellipsoids*, *elliptic cylinders*, and *quadratic cones* can be produced.
Through *collinear mappings* (they are also linear mappings that do not change the order of the surface), *paraboloids* and *hyperboloids* can be produced – as well as the *parabolic and hyperbolic cylinder*. As we shall soon discover, it is impossible to distinguish between different types of cones produced by collineation, which is why they are generally referred to as *quadratic cones*.

The doubly curved second-order surfaces – ellipsoids, paraboloids and hyperboloids (two-sheeted) – differ with respect to affine mappings but there are collinear mappings that can transform them into a sphere. The single curved second-order surfaces – elliptical, parabolic and hyperbolic cylinders and quadratic cones – are affine and collinear with respect to the cylinder or the cone of revolution.

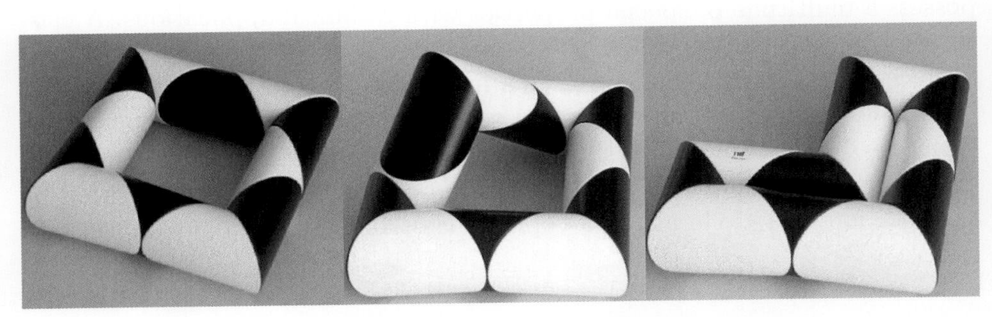

Fig. 6.1 Elliptical cylinders carry circular intersections.

Fig. 6.1 shows an interesting "experiment" with elliptic cylinders which are actually oblique circular cylinders: in contrast to ellipse-shaped sections of the cones of revolution, the single "hooves" can be steplessly twisted due to their circular shape.

Ellipsoids

As we already know, all cylinders and cones are developable because they are single curved. Ellipsoids, on the other hand, are doubly curved – just like spheres. In general, ellipsoids possess "three axes" but are not rotationally symmetric – they possess three pairwise-perpendicular planes of symmetry.

By examining the ellipsoid that is produced by the *rotation of an ellipse*, one can differentiate between an *oblate* and an *egg-shaped* ellipsoid of revolution.

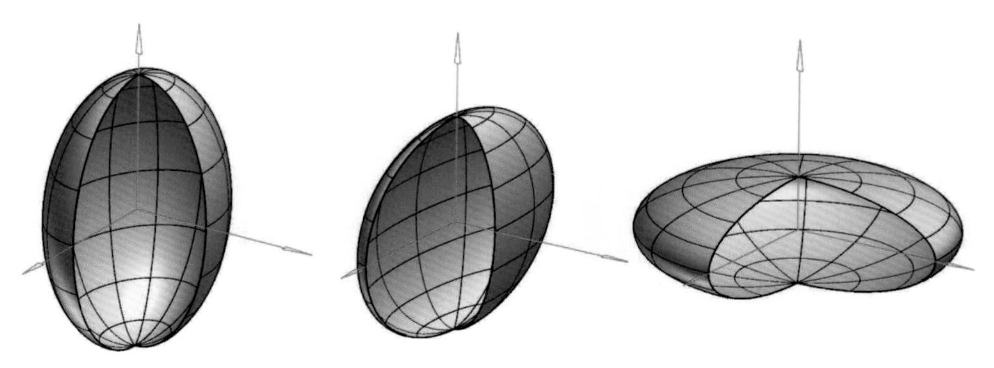

Fig. 6.2 Left and center: ellipsoid with three axes; on the right: ellipsoid of revolution

• The shape of the Earth

As is well known, the globe is not precisely a sphere but rather approximately an oblate ellipsoid of revolution, also called a *geoid*. The exact shape of the Earth corresponds to an ellipsoid with many tiny indentations and protrusions. These "blemishes", including the oblateness at the poles, are highly relevant for physicists and geodesists. From a purely optical perspective, the Earth is *practically indistinguishable from a sphere.* ♠

Due to their affinity, ellipsoids "inherit" important properties from the sphere:

Each planar section of an ellipsoid is an ellipse. The outline of an ellipsoid is an ellipse in cases of normal projection and a conic section in cases of central projection.

• Ellipsoids in dome construction

Fig. 6.3 An exact elliptical dome and "something similar" (Karlskirche, Vienna)

Fig. 6.3 on the left shows the dome of the Reichstag in Berlin, which has the exact shape of a half-ellipsoid of revolution. In Fig. 6.4, both normal projections of the dome as well as one central projection (perspective) can be seen.

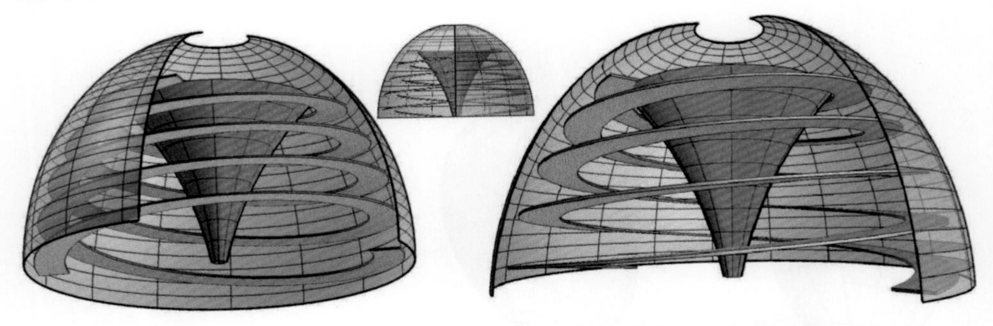

Fig. 6.4 The Reichstag dome in Berlin is a half-ellipsoid of revolution.

The cathedral in Florence is duly famous, not the least for its dome (built by *Brunelleschi*, Fig. 9.2). However, the dome is not an exact ellipsoid – not even roughly! In fact, it is approximated by cylinders, which mutually intersect along ellipses. These ellipses do not have their apexes on their respective rotational axes. ♠

• Ellipsoid in industrial design

Fig. 6.5 Spherical and ellipsoid-shaped teapot

Fig. 6.5 on the right shows Boris Sipek's teapot in the form of an ellipsoid. Strictly speaking, the teapot resembles a "magic egg" in that it raises the question of how the separate pieces are assembled. Sipek deliberately uses affine transformations in much of his work. A spherical counterpart to his teapot is also available (on the left). ♠

• Ellipsoids with three axes serve as aesthetic lamps

The lamps in Fig. 6.6 are perfect ellipsoids with three axes. The aesthetics of ellipsoids definitely inspire artists. ♠

Fig. 6.6 Aesthetic lamps "Queen Titania", designed by Alberto Meda ans Paolo Rizzatto

Mathematically speaking, second-order surfaces are expressed by a so-called *implicit equation* of the form

$$a_1\,x^2 + a_2\,y^2 + a_3\,z^2 + a_4\,xy + a_5\,yz + a_6\,xz + a_7\,x + a_8\,y + a_9\,z + a_0 = 0.$$

If a straight line with the parametrization

$$x = u_1 + t\,v_1,\ \ y = u_2 + t\,v_2,\ \ z = u_3 + t\,v_3$$

is intersected with the surface, an algebraic equation of degree two is found for the paramter t. Each of the two solutions corresponds to an intersection point.

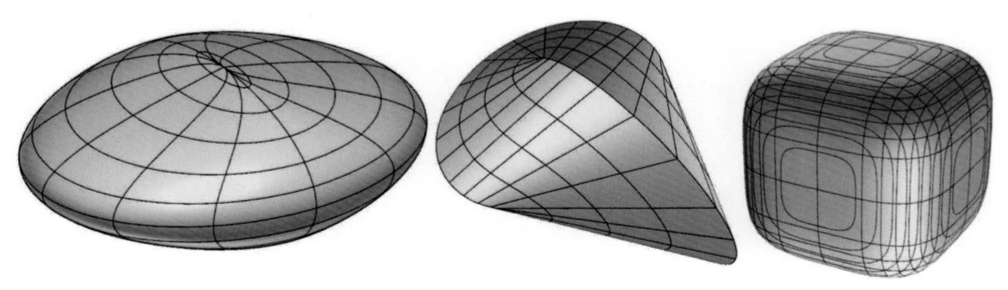

Fig. 6.7 Convex surfaces but not of second order

When counting intersection points, it would be unwise to rely on mere visual appearance. The convex surfaces in Fig. 6.7 (which possess neither indentations nor constrictions seem to have at most two intersection points with any straight line. However, not every surface of a convex solid is of second order. The surface on the right in Fig. 6.7 is, in fact, a *supersphere* with the implicit equation $x^4+y^4+z^4 = r^4$. Therefore, it is a fourth-order surface. The surface tangents in the six flat points are in "higher order contact" with the surface.

In this case, a second-order surface has enormous advantages for our deliberations: For instance, we then know that each planar section with this surface is a second-order curve – and therefore a real or complex conic section (in general). This conic section may also degenerate to a pair of straight lines (real or conjugate complex) or a single straight line (always real) that should be counted twice.

Listening and focussing ...

Due to their elegant form, egg-shaped ellipsoids of revolution are often found in design. By making use of the ellipse's focal point property, a focusing lamp may be produced, whose reflecting interior resembles an egg-shaped ellipsoid of revolution. It focuses all light that is created at the focal point of the lamp onto the second, not materialized focal point.

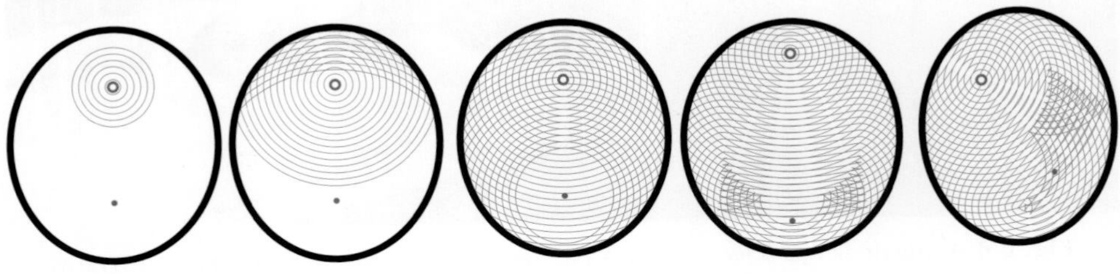

Fig. 6.8 Wavefronts emanating from points inside the ellipse

Aside from light, other types of waves may also be focused this way, such as thermal radiation (burning mirror) and sound waves (Fig. 6.8). The focusing works best if the source is located exactly at one focal point (left to center) – however, the phenomenon does not break down given slight deviations (right images). The figure shows the progression of wavefronts and enables us to observe interferences.

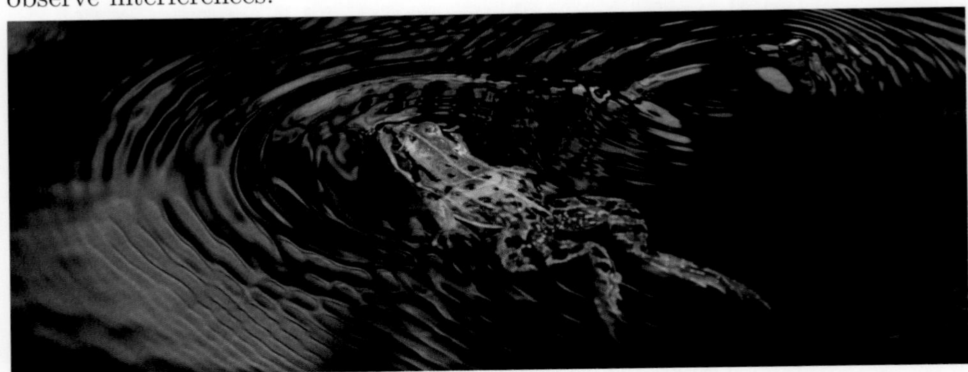

Fig. 6.9 Visible wavefronts

• Whispering bowl and odd acoustic phenomena

Fig. 6.10 shows two opposing bowls in a library of Vorau Abbey (Styria) which resemble the cut-off caps of ellipsoids of revolution. Prior to precise measurement, one may also have thought that the bowls are spherical: The difference between the ellipsoid cap and the respective *osculating sphere* at the apex is practically negligible.

By placing one ear close to a focal point, it is possible to hear words which have been whispered over 20 meters away near the other focal point.

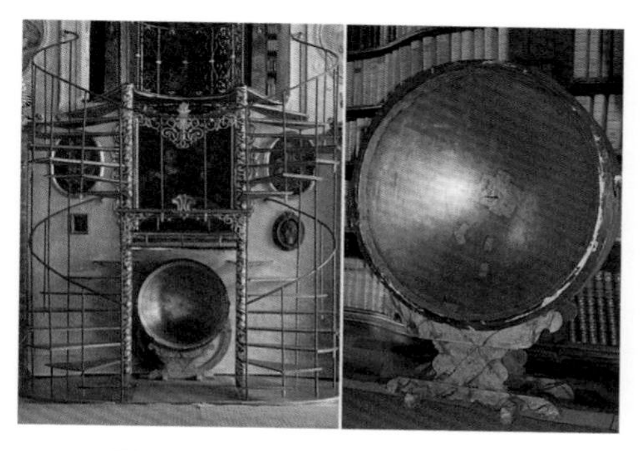

Fig. 6.10 Whispering bowls ...

Fig. 6.11 ...and sound worlds

Fig. 6.11 shows an artistic version of the acoustic effect by (Bernhard Leitner). Paraboloids were used in that example, which, as we shall soon see, can be regarded as extreme cases of ellipsoids. Sound is being projected towards the bowl from the focal points and is then being focused parallel and directed towards a wall, where, after being reflected, the acoustic sensation appears at the angle of reflection. ♠

Paraboloids

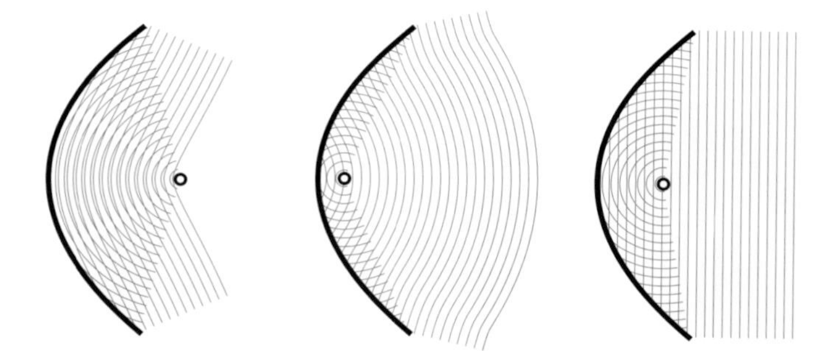

Fig. 6.12 Wavefronts of a parabola

Fig. 6.12 illustrates the reflection of wavefronts when using an ellipse instead of a parabola. It can be seen that a wavefront from planes occurs whenever the starting point is also the focal point.
A *paraboloid of revolution* can be constructed by rotating a parabola about its axis – Fig. 6.12 further illustrates the many technical applications of such surfaces as reflectors (radar screens, satellite dishes, radio telescopes, etc.).

- "Receiving dishes" (Fig. 6.13 left)

In theory, a satellite dish has to be an exact paraboloid of revolution, and the receiving device has to be located exactly at its focal point. As with the whispering bowl previously mentioned, however, it is possible to approximate it pretty well through an *osculating sphere.*
It is only important that the receiver is not positioned at the center of the sphere but half-way between the center and the surface: The radius of the

Fig. 6.13 Paraboloids of revolution as receivers or effective search lights

parabola's osculating circle at the apex is twice the distance of the focal point to the apex. Since the receiver is not point-shaped to begin with, the devices allow for the axis of the bowl to be slightly different from the signal direction. In such cases, it is required for the sensor to be positioned off-center (compare it to Fig. 4.42)! ♠

• **Parabolic headlights and sun collectors**

By positioning a point light source at the focal point of a reflecting paraboloid of revolution, it can be ensured that all light rays that hit the mirror will be reflected parallel to the axis. Such lights are called "long distance beams", Fig. 6.13 on the right).

Fig. 6.14 Parabolic headlights ...

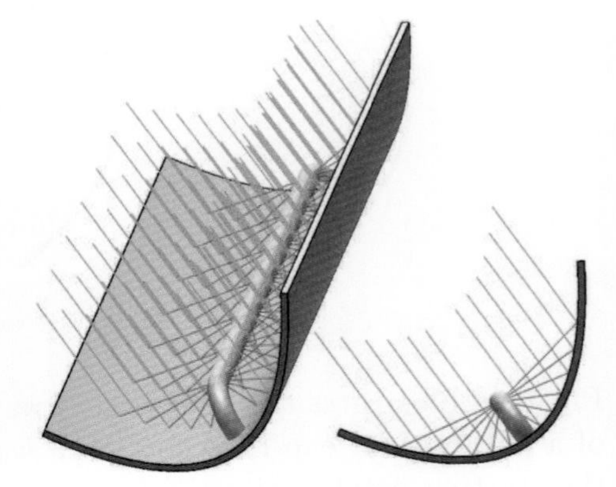

Fig. 6.15 ... and their inversion

In practice, rod-shaped light sources are often used. *Parabolic cylinders* are utilized for such cases (Fig. 6.14), which are also useful for focusing sunlight on a pipe. This pipe may contain liquid (e.g., thermo oil) which is first heated and then diverted. The screen is oriented in an east-west-direction and is constantly moved to ensure that the sun stays in its symmetry plane. ♠

Parabolic surfaces of translation

Paraboloids of revolution which have undergone affine transformations – in other words, *elliptic paraboloids* – are also second-order surfaces. It is possible to get *hyperbolic paraboloids* by means of certain collineations. We shall prefer another definition:

Fig. 6.16 Paraboloids as translational surfaces

By moving a parabola along a second, fixed parabola with a parallel axis and a different carrier plane, the moving parabola sweeps a paraboloid.

If both parabolas are curved on the same side, we get an elliptic paraboloid – otherwise, a hyperbolic paraboloid (Fig. 6.16). If one of the parabolas is not curved at all (a straight line), we get a parabolic cylinder. The two parabolas are interchangeable in their function. Thus, the paraboloids carry at least *two pencils of parabolas of equal importance* (as we shall soon show, there exists an infinite number of pencils).

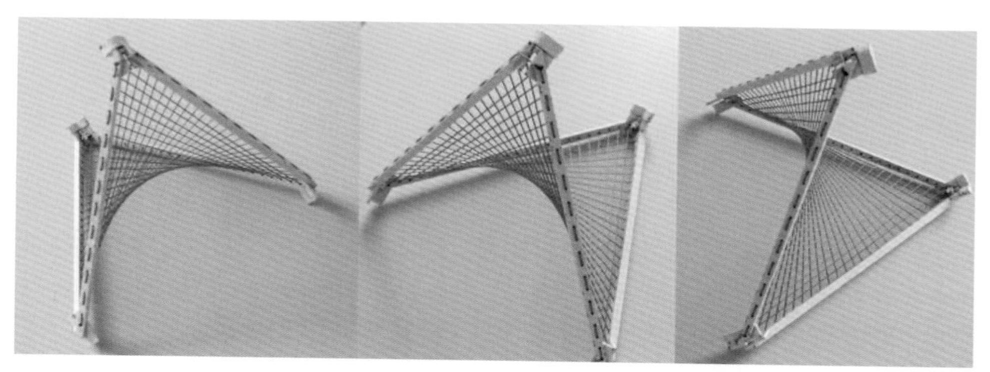

Fig. 6.17 Hyperbolic paraboloid as the result of congruent rows of points

A paraboloid in its standard position (includes the origin, the axis is the z-axis) can be described analytically by the equation

$$z = A \cdot x^2 + B \cdot y^2.$$

If A and B are of equal sign, then the paraboloid is elliptic – otherwise it is hyperbolic. If A or B disappears, then we are dealing with a parabolic cylinder.

If a paraboloid is intersected with an arbitrary plane $y = k \cdot x + d$ parallel to the axis, then the intersection can be given an equation by inserting this linear equation into that of the paraboloid. One gets the equation of a parabola

$$z = A \cdot x^2 + B \cdot (k \cdot x + d)^2 = (A + B\,k^2) \cdot x^2 + 2Bkd \cdot x + Bd^2.$$

This means: *Each intersection with a plane parallel to the axis yields a parabola.* Its curvature depends only on k, not on d. Thus, on this paraboloid, there exist *infinitely many congruent parabolas that can be translated into each other* for each stance of the planes (as expressed by k).
In the case of hyperbolic paraboloids, A and B have different signs, and A/B is negative. With $k = \pm\sqrt{-A/B}$, the coefficient of the quadratic term disappears and the section parabola becomes a straight line. This means that a hyperbolic paraboloid carries two families of straight lines which form a grid of parallel straight lines if seen from the top. Let us summarize:

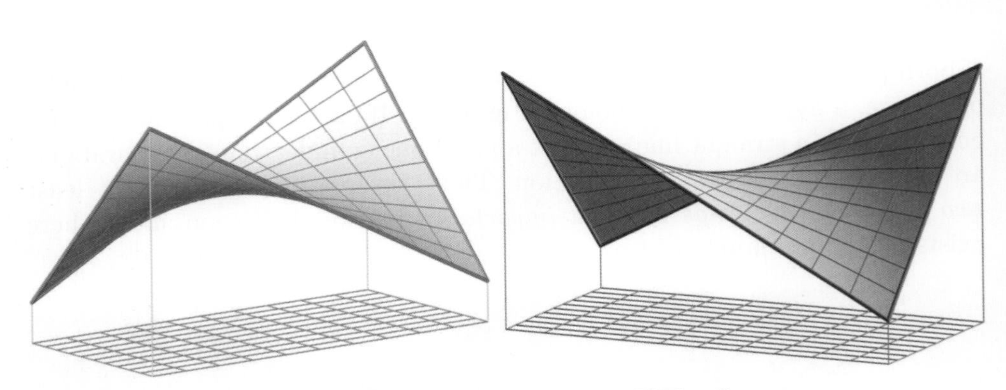

Fig. 6.18 Hyperbolic paraboloids (HP-bowl)

Each planar section of a paraboloid which is parallel to the axis is, in general, a parabola. Concerning the hyperbolic paraboloid, the parabolas degenerate to lines for two particular stances of planes. Hyperbolic paraboloids, thus, carry *two different families of straight lines which appear parallel when projected in the axis direction.*

The boundaries of the surface in Fig. 6.17 are the generatrices that form a *skew quadrangle*. The rows of points formed by correlated points, i.e., these points are joined by generatrices, are congruent if the boundary generatrices appear symmetric when viewed from above – otherwise, they are only similar. The following can, thus, be said:

A hyperbolic paraboloid is also generated by congruent or similar rows of points.

Classically beautiful and very useful

Paraboloids play an important part in surface theory, as all curved surfaces behave like paraboloids on a local level. We will discuss more about this in the next section.

Hyperbolic paraboloids are even important in architecture, where they are simply called *HP-bowls*. For instance, they can be used to roof rectangular areas in an elegant and very stable manner (Fig. 6.18). Construction is especially simple as the wireframe can be constructed from straight rods.

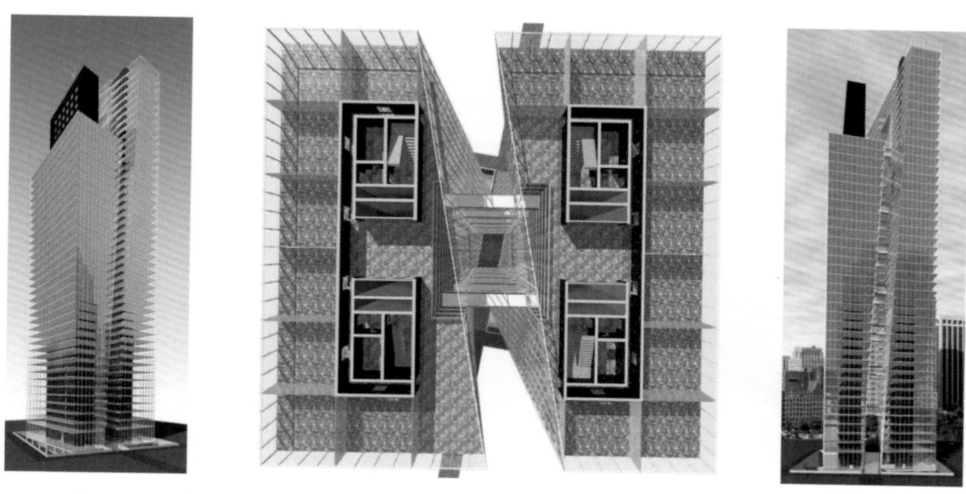

Fig. 6.19 Skyscrapers with HP-bowls as interior walls (in the middle: the top view)

Fig. 6.19 shows a sketch of twin towers by Paolo Piva where the lateral surfaces are derived from HP-bowls.

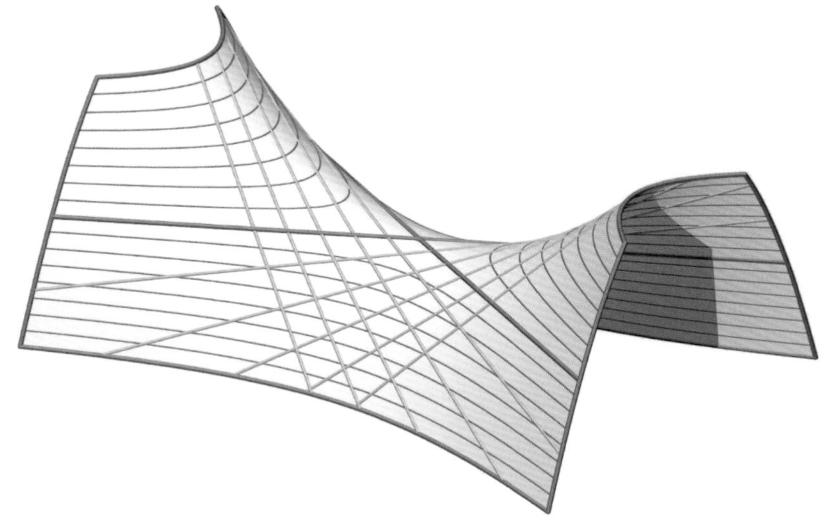

Fig. 6.20 A hyperbolic paraboloid carries two families of straight lines, hyperbolas, and parabolas.

Hyperboloids

HP-surfaces share the property of being able to carry two different families of straight lines with only one other type of surface: The one-sheeted hyperboloids, which are also second-order surfaces.

A *hyperboloid of revolution* is formed by rotating a hyperbola about one of its axes. Depending on the axis of rotation, the result is either one-sheeted or two-sheeted.

The *two-sheeted hyperboloid of revolution*(Fig. 6.21) is not as practically important as the one-sheeted variety. It can be used as a lampshade for a precisely point-shaped light source (Fig. 2.5) because, due to the fundamental properties of conic sections, light that shines at a focal point is reflected in such a way that it appears to emanate from the second focal point (and, in the case of the hyperbola, in the extension). However, direct incidence of light would have to be prevented by local coverage, which would prevent any lighting to occur near the light axis.

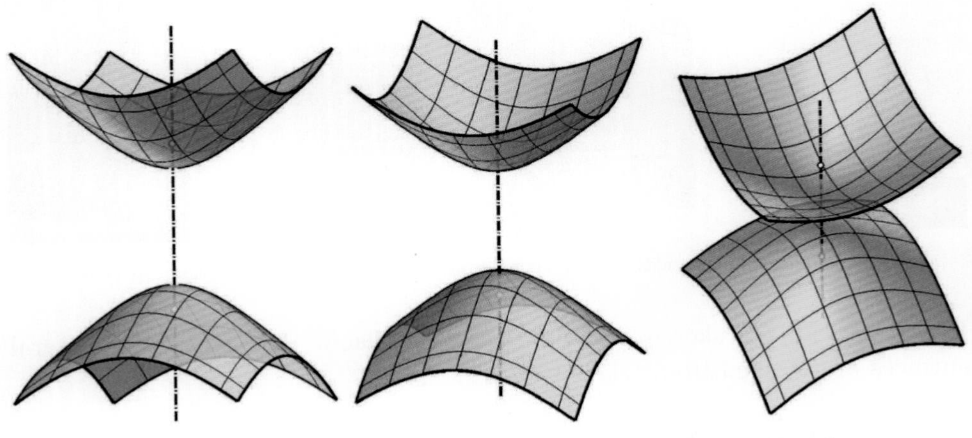

Fig. 6.21 Two-sheeted hyperboloid of revolution from different perspectives

The one-sheeted hyperboloid of revolution is much more important in practice. It is generated not only by rotation of a hyperbola about its conjugate axis but also by the rotation of a straight line about a skew axis. This not-at-all-trivial insight is due to Christopher *Wren*, the constructor of St. Paul's Cathedral in London.

A hyperboloid of revolution with the axis z is described by the equation

$$\frac{x^2+y^2}{a^2}-\frac{z^2}{c^2}=1.$$

If it is intersected with a tangent plane $x = a$ (parallel to the z-axis), then the equation of two straight lines can be found as

$$\frac{y^2}{a^2}-\frac{z^2}{c^2}=0 \Rightarrow z=\pm\frac{c}{a}y.$$

A rotation gives us both families of straight lines which lie in the plane.

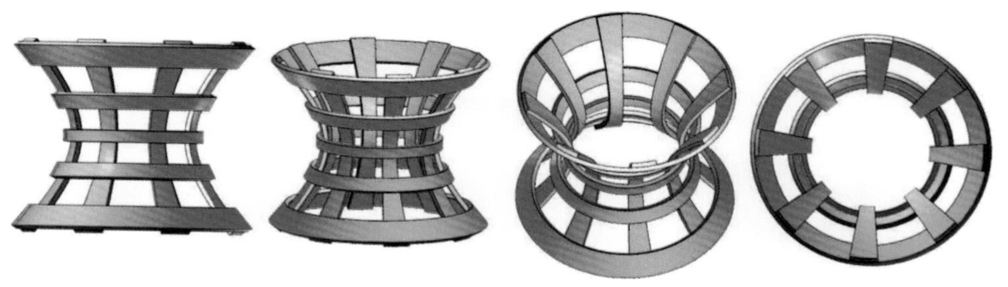

Fig. 6.22 Generation of a hyperboloid by means of a hyperbola

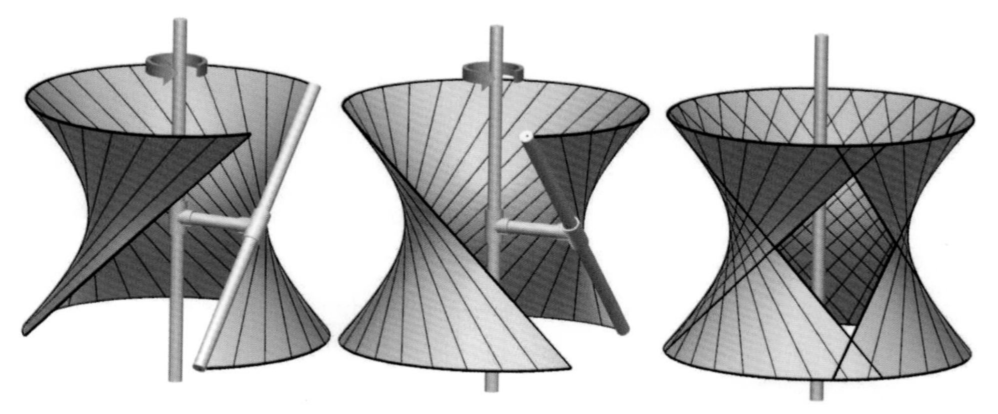

Fig. 6.23 Double generation of a hyperboloid of revolution by means of rotating straight lines

Let us now consider an arbitrary point P in the plane. One straight line e of the first family and another f from the second family pass through it. The straight lines e and f span a plane τ. The section of a plane with a second-order surface is always a second-order curve. This curve obviously degenerates into two intersecting straight lines. Therefore, it could be said that e together with f forms "degenerate" hyperbola. We now know two curves e and f through the surface point P (straight lines are first-order curves which can, for instance, be deduced by their linear parameter representation). Their tangents in P – the straight lines themselves – by definition span the tangent plane, and thus, $\tau = ef$ is the tangent plane at P.
Nothing changes about this statement even if we apply an arbitrary spatial affinity to the hyperboloid of revolution. We therefore have:

Every tangent plane τ of a one-sheeted hyperboloid intersects at a pair of straight lines e and f. Within each of the four sectors defined by e and f, the surface always stays on the same side of τ.

It can be shown, using similar modes of thought, that every non-degenerate second-order surface that carries *one* family of straight lines $\{e\}$, must also carry a second family $\{f\}$: Each planar section through e yields a second-order curve which, nevertheless, possesses a "straight-line component" through e and must, thus, degenerate into a pair of straight lines.

Fig. 6.24 Barcelona and its affinity to hyperboloids: Arriving at the airport, the control tower Barcelona (*Ricardo Bofill Levi*) is the first hyperboloid to be seen. To the right: Antoni Gaudí's model for the hyperboloids in the roof of the *Sagrada Família* (compare Fig. 6.25).

Fig. 6.25 Roof of the *Sagrada Família* in theory (middle) and practice

Thus, it is clear that two pencils of straight lines must exist on the HP-surfaces – just like on the hyperboloids. But what about cylinders and quadratic (elliptic, hyperbolic, and parabolic) cones? The tangent plane τ not only touches in a single point, but along a whole generatrix. Thus, two "neighboring generatrices" (infinitely close together) lie on τ: The second-order curve degenerates into a pair of identical straight lines.

- **Stable and light-weighted**

The cooling towers of many power plants have the shape of hyperboloids of revolution. The ones depicted here (Temelin) are 155 m high and have a ma-

Fig. 6.26 Cooling towers of a nuclear power plant. The geometry would be perfect, examples of the past, however, have provided proof of sometimes uncontrollable reactions ...

ximum diameter of 130 m. The diminution towards the top causes an increase of speed of the hot water vapor. This leads – according to the aerodynamic paradox – to diminution of pressure which chills the steam. ♠

• Generalization of the generation

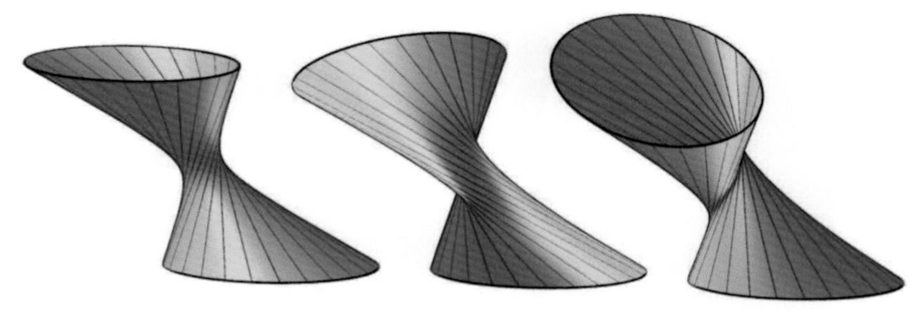

Fig. 6.27 Generalization does not always work.

It may stand to reason that the double generation of a hyperboloid of revolution can be generalized by two rotating straight lines. In fact, the double generation works as long as the generating circles lie in parallel planes. The result is then a general hyperboloid (Fig. 6.27 left) that carries two families of straight lines. By "tilting" the circles, the two families of straight lines produce two different surfaces that have nothing to do with a hyperboloid (Fig. 6.27 center and on the right). ♠

• Result of rows of points on circles in parallel planes

If one point moves on a circle and the second point moves, at the same angular velocity, on a circle in a parallel plane, then the connecting straight line generates a one-sheeted hyperboloid of revolution (in general). If the respective circle tangents are parallel, a cylinder is produced (this also works if the circles are not positioned directly above each other – unlike in Fig. 6.28). It is also possible to consider a one-sheeted hyperboloid as the *product of congruent rows of points on parallel circles.* If the points move through their

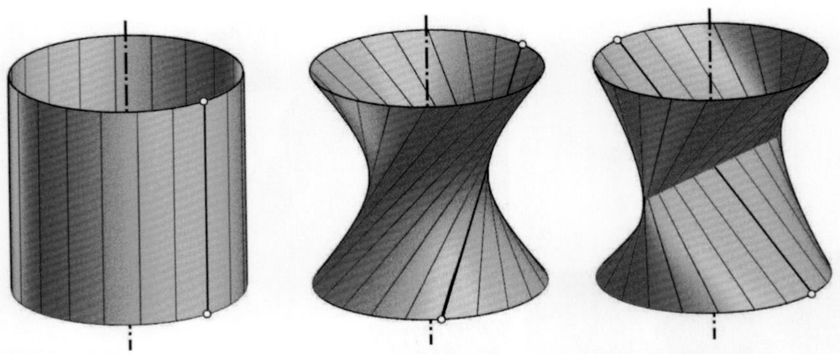

Fig. 6.28 Generation of directly or reversely congruent rows of points

carrier circles in opposite directions, a new surface is created that is no longer of second order. In the special case pictured in Fig. 6.28 on the right, a so-called *circular conoid* is generated, which we shall discuss in a later chapter (Fig. 7.3). ♠

Fig. 6.29 Hyperboloids of revolution and – in the background on the bamboo stools – interesting "envelopments" on cylinders of revolution (see also p. 143).

The fact that two independent families of straight lines lie on a one-sheeted hyperboloid of revolution has practical applications. Stools of this shape (Fig. 6.29) have an enormous carrying capacity. The gargantuan cooling towers of power plants are also shaped like hyperboloids of revolution. This allows them to be stable despite having relatively thin walls.
Two congruent hyperboloids of revolution can, under some circumstances, fit together (Fig. 6.30). The surfaces touch along a generatrix. If one imagines both axes as fixed and allows one of the hyperboloids to rotate about its axis, then the other one would also have to rotate about its axis – given "glide-free friction".
This way, the rotation about its own axis can be transferred consistently onto a rotation about a second skew axis. In practice, this is used for the design

of so-called *hypoid gears*, despite the fact that such gears are not optimally suited for transferring rotations onto skew axes. This is the case because the hyperboloids do not roll but they, in fact, glide simultaneously along each other in so-called *grinding* motions. *Involute gears*, which we shall discuss in a later section, are much more frequently encountered. Their gear flanks are developables, which allow "instantaneous rolling" without gliding. Another highly original solution to this problem can be found in Fig. 11.14.

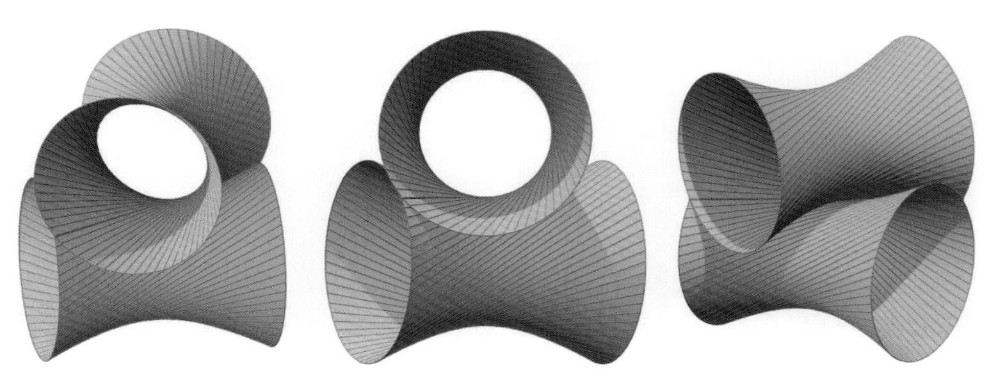

Fig. 6.30 Grinding hyperboloids of revolution

We already know that every second-order surface intersects an arbitrary plane along a second-order curve. It can also be shown that:

The contour of a second-order surface lies in a plane and is, therefore, a second-order curve. In the case of doubly curved surfaces, it is a conic section, and in the case of single curved surfaces, it is a pair of straight lines (in general).

Proof:
A doubly curved second-order surface arises from a sphere by means of collineation. The inverse of the collineation allows us to transform such a surface back into a sphere. The center of the projection transforms into a point. At this point the cone (of revolution) built by projecting surface tangents is centered. In the collineation, the circle of contact corresponds to the contour of the second-order surface, which is, therefore, a conic section.
By analogy, it is possible (using collineation) to transform a single curved second-order surface back into a cylinder or cone of revolution. ◇

In Fig. 6.30, the contour of the hyperboloid of revolution corresponds (among other possibilities) to a hyperbola four times, to an ellipse once and to a circle once as well.
Fig. 6.21 illustrates that the contour of a second-order surface does not always have to be real: In the image, no outline seems to exist for the two-sheeted hyperboloid of revolution. Mathematically speaking, a contour, indeed, exists – but merely an imaginary one. Imaginary contours also occur if center of the projection is inside the second-order surface.

Whatever holds true for the contour is also applicable to the self-shadow boundary. It can be shown that any self-shadows of the surface's boundary curves must also be conic sections:

Self-shadow boundaries and shadows of a planar section of a second-order surface onto itself are second-order curves.

Fig. 6.31 The ellipse as a shadow on the inner side of a second-order surface

The proof for this statement shall now be sketched: Let L be a light source and c the shadow casting planar section of the surface Φ (a second-order curve). Thus, the projection cone Λ from L is a second-order cone (in the case of parallel light rays, it is a second-order cylinder). $\Lambda \cap \Phi$ is to be determined. Firstly, according to *Bézout*'s theorem (p.140), a spatial *fourth-order* curve is to be expected. However, this curve must already contain the planar section c – a second-order curve. Thus, the remaining section, which is the shadow of c, may only be of second order.
Fig. 6.30 illustrates this remarkable theorem. Shadows of hyperboloids *on other hyperboloids* are also visible, which are indeed more complicated (fourth-order) spatial curves.

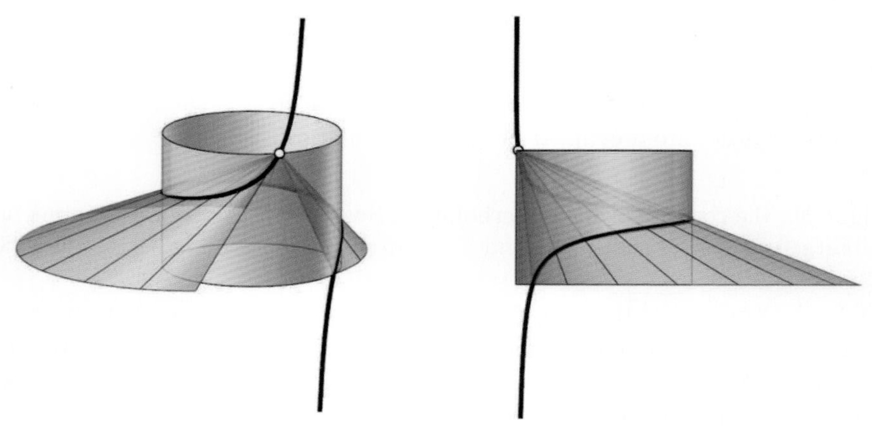

Fig. 6.32 Cubic circle as an intersection curve

• Cubic circle: By intersecting a second-order cylinder with a second-order cone, a spatial fourth-order intersection curve can be predicted for the general case. If, however, the cylinder and the cone share a generatrix, then this straight line is part of the curve. It is said to separate itself from the intersection. If both surfaces touch along this generatrix, then the remaining section is a conic section – otherwise, it is a spatial third-order curve. Such a curve is visible in Fig. 6.32, where a cylinder of revolution and an oblique circular cone were chosen. The result is called an *upright cubic circle.* When viewed from above, the curve appears as a circle, and in other projections as a third-order curve. A cubic circle belongs to the simplest nonplanar algebraic spatial curves and is, thus, frequently employed for purposes of illustration. For interested readers: A possible parameter representation of the curve is given by

$$x = r \cdot \cos u,\ y = r \cdot \sin u,\ z = h \cdot \tan u/2\ (u \in [-\pi/2, \pi/2]),$$

where r is the cylinder radius and h is a scaling factor for the height.
A spatial curve with an even simpler parameter representation is the *cubic parabola*

$$x = t,\ y = t^2,\ z = t^3.$$

6.2 Three types of surface points

In the last section, we have discussed the topic of second-order surfaces extensively. This was for two reasons: Firstly, these surfaces have numerous practical applications – as we have seen. The second reason should soon become clear:

Curvature of a planar curve

Fig. 6.33 Curves with intricacies

Fig. 6.34 Osculating cylinder of rotation ...

Let us start with a two-dimensional comparison and imagine an arbitrary curve in the drawing plane (without intricacies such as tips, kinks, or jumps as in Fig. 6.33).

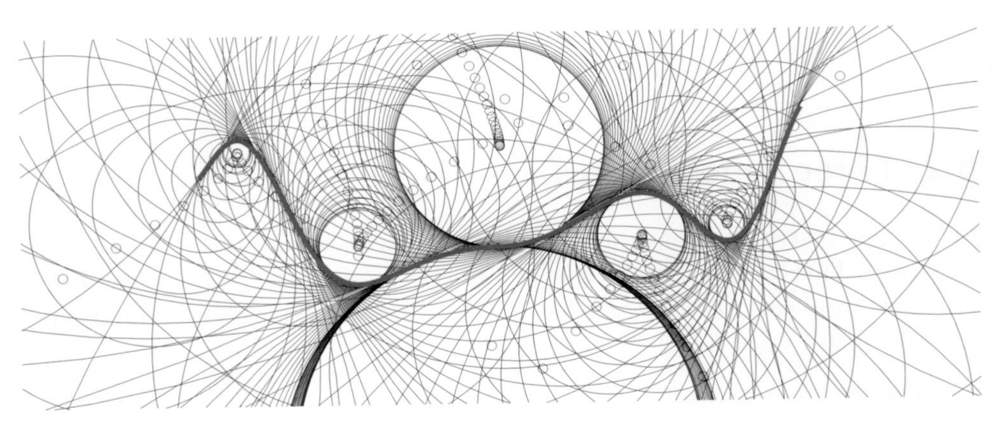

Fig. 6.35 Osculating circles of a planar curve

At each point P of this curve, there exists a tangent. It can be interpreted as the "linear approximation" of the curve in the vicinity of the point P. There also exists an *osculating circle* at each point, which does a much better job at approximating the curve. It could be said that it represents a quadratic approximation. Depending on the side on which the osculating circle is located, it is already possible to distinguish between positive and negative curvature. The smaller the osculating circle, the larger the curvature (Fig. 6.35).
Fig. 6.36 shows another kind of quadratic approximation of a curve. This time, osculating *parabolas* were used instead of osculating circles. Like osculating circles, they are also each defined by three neighboring points as well as the additional specification of the direction of the axis.

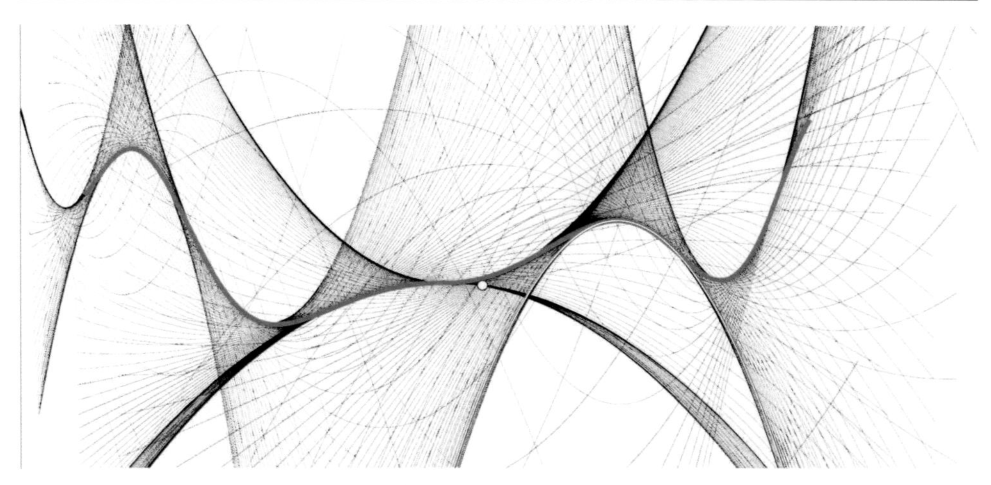

Fig. 6.36 Osculating parabolas of a curve

In space, curvature is not at all easy to define!

Let us now try to extrapolate the procedure into space: If P is a surface point, then the corresponding tangent plane $\tau \ni P$ is a *linear* approximation of the surface. However, what corresponds to the osculating circle? Let us consider an arbitrary surface tangent $t \ni P$. There exist infinitely many spatial curves c in the surface whose tangent is t. The osculating circles may be different for various curves. *Meusnier* has shown that all the osculating circles of the surface curves through P tangent to t form a sphere – the *Meusnier* sphere (Fig. 6.37) – defined by the line element (P, t). The center of this sphere is also the center of curvature for the *normal section* of the surface through t.

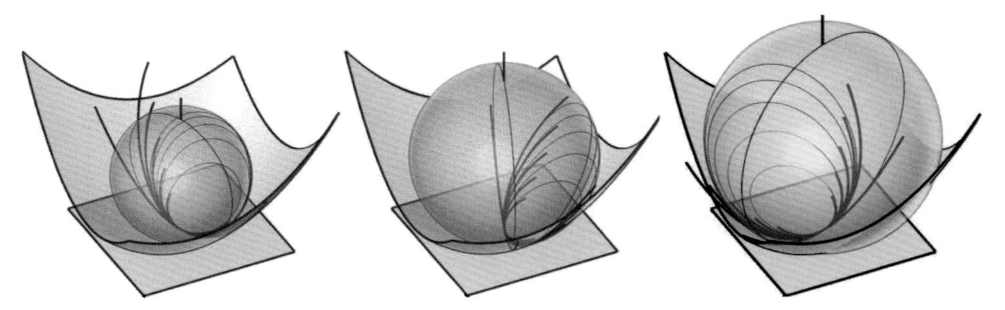

Fig. 6.37 *Meusnier* spheres for different tangent directions

Principal curvatures

The existence of the *Meusnier* sphere is remarkable in and of itself. Yet, what happens if we change t? Of course, this problem was already investigated in great detail. However, the issue is very elaborate and does not yield any immediately visual insights – it, thus, falls outside the scope of this book.

The following shall be presented without proof: At each surface point, there exists a greatest and a smallest *Meusnier* sphere. They belong to a pair of orthogonal surface tangents, so-called *curvature tangents*. Let us intersect the

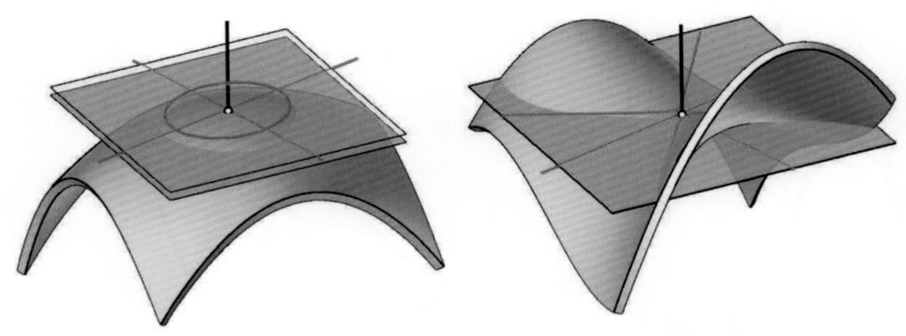

Fig. 6.38 Planar intersections parallel to the tangent plane, Dupin's indicatrix

surface with a plane which is parallel to the tangent plane but only "infinitesimally" distant to it (Fig. 6.38). In an infinitesimal small neighborhood of the surface point, the intersection curve with the surface can hardly be distinguished *locally* from an ellipse, or a hyperbola, or a pair of parallels – known henceforth as *indicatrices*. The axes of the indicatrix are the curvature tangents. This also means that the *curvature of all normal sections is known as soon as the indicatrix is defined* – in particular, the knowledge about the curvature along the curvature tangents is enough. This insight dates back to *Dupin*.

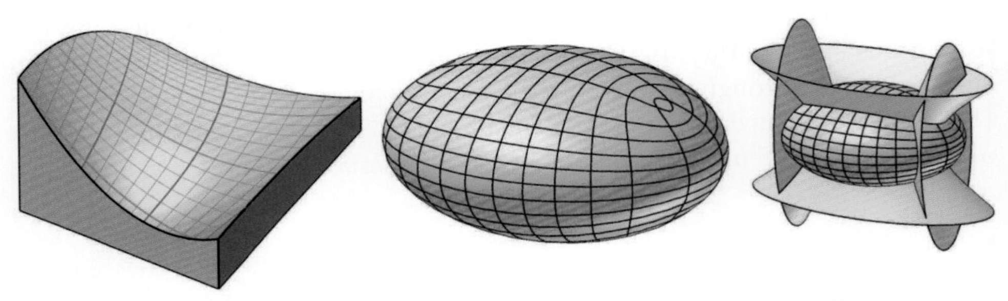

Fig. 6.39 Left: curvature lines on a general surface, middle and right: curvature lines on an ellipsoid with three axes.

Curvature lines are the surface curves where all tangents are curvature tangents and which, therefore, follow the direction of the strongest and weakest curvature.

There are two families of curvature lines. The curves from one family intersect all curves from the other family orthogonally (except at umbilics). If we consider a surface point and intersect the surface orthogonally to its tangent plane τ, we get ellipses or hyperbolas whose axes are parallel to the tangents of the curvature lines in τ. Fig. 6.39 in the middle shows curvature lines on a three-axis ellipsoid. It can be shown that the curvature lines are intersection curves with "confocal" hyperboloids (Fig. 6.39 on the right).

Fig. 6.40 in the middle and on the right shows the design and the construction phase for the project "hangar8" (Volkmar Burgstaller). A three-axis ellipsoid is triangulated by being first cut into slices, after which the points of an

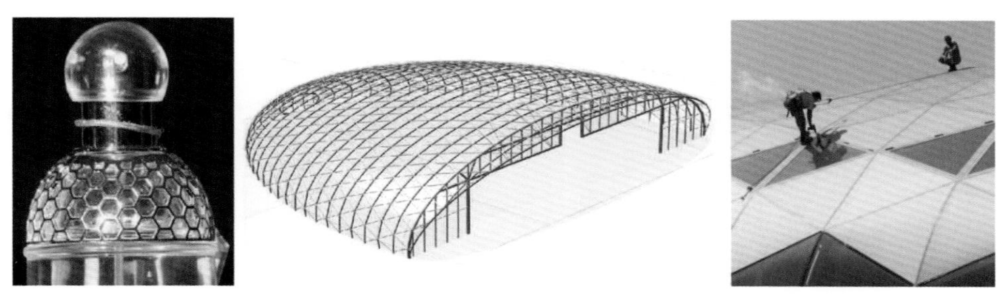

Fig. 6.40 Three-axis ellipsoid with a clever triangulation

intersection line are connected to the points of the next intersection line in a way that produces approximately equilateral triangles. We know that this can, indeed, only work "approximately" – unless we are dealing with very special polyhedra, no surface can be approximated only by congruent triangles. Fig. 10.25 shows the same surface but from the inside. Fig. 6.40 on the left shows a perfume bottle by Guerlain (an ellipsoid of revolution) where a similar method of triangulation was applied, and groups of six triangles were combined into hexagons.

Osculating parabolas lead to an important theorem

The other approach would consist in using osculating parabolas instead of osculating circles. They agree in curvature with any intersection of a plane through the surface normal. Like the osculating circles of the normal sections, they should be defined by three neighboring points on the normal sections. This time, the result – the surface formed by all possible parabolas – is simple: It turns out to be either the tried and true paraboloid (elliptical or hyperbolic) or a parabolic cylinder!

If we attempt to approximate the surface quadratically by using parabolas, we always get the same, previously noted three surfaces. The approximation quality will improve, the smaller the chosen area where the comparison takes place. Thus, from a *local* perspective, there exist only three types of surfaces:

Essentially, every surface of any conceivable complexity is locally, i.e., in a sufficiently small neighborhood of a point, curved like an elliptic or hyperbolic paraboloid or like a parabolic cylinder (Fig. 6.41). The surface point is, therefore, called either elliptic, hyperbolic, or parabolic.

Paradoxically, the two-sheeted hyperboloid consists only of elliptic points even though it carries a multitude of hyperbolas (Fig. 6.21). Even the contours (if real) appear as hyperbolas!

If we consider not only a "microscopic fragment" but the whole surface, then the issue becomes much more complicated. At the level of the whole surface, there are no limits to the possible variations. Nevertheless, it is still possible to speak of elliptic, hyperbolic, or parabolic curved parts of a surface. In general, the local properties do not change arbitrarily, but continuously.

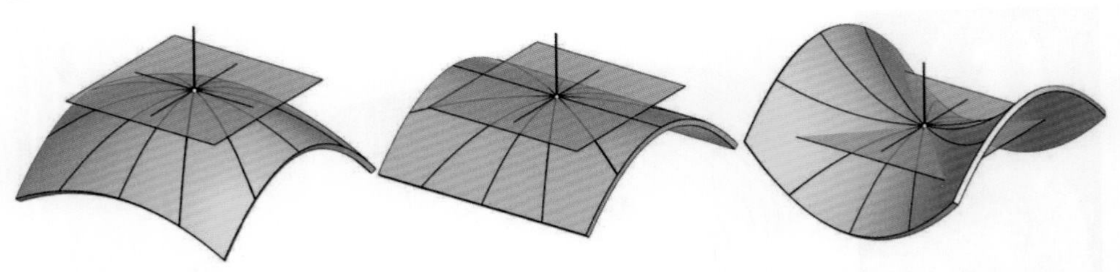

Fig. 6.41 The three different types of surface points

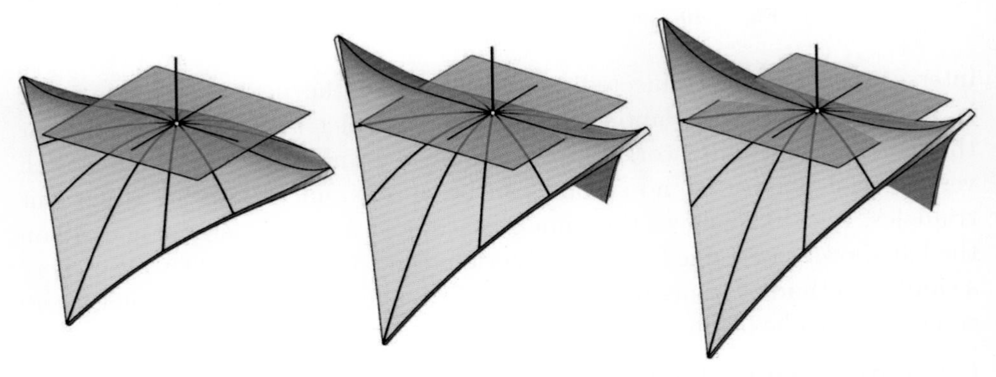

Fig. 6.42 The "saddle point" is always of a different type.

Fig. 6.42 shows three surfaces which, at first glance, resemble an HP-surface (hyperbolic paraboloid). At the supposed *saddle point*, the surfaces are differently curved (elliptic on the left, parabolic in the middle, and hyperbolic on the right). *In the neighborhood* of the point in question, only the right surface is actually related to the HP surface. This can also be seen due to the fact that the tangent plane "penetrates" the surface in more than just the contact point (just locally). In the case of the middle surface, there also occurs an intersection with the tangent plane, but not in the immediate neighborhood of the point.
The exact HP-surface and the one-sheeted hyperboloid of revolution carry only hyperbolic points, while convex surfaces (surfaces which envelop a manifold without indentations or neckings) are always either elliptically or parabolically curved.

• **An exciting experiment**

Can a two-dimensional being on a surface find out, to what extent his "carrier surface" is curved, without leaving the surface and thus entering three-dimensional space?
The mathematician's answer is: Yes! The being would merely have to draw a circle with the radius r (Fig. 6.43) and measure its circumference. If this circumference is $2\pi \cdot r$, then the being lives on a parabolic surface – if greater, then on a hyperbolic surface. A being on an elliptic surface – just like ourselves on the globe – would conclude that the circle is smaller than expected.
Let us find out based on a specific example: Let us draw a circle around the north pole with a radius of $r = 10\,000$ km (a quarter of the Earth's circumference). Thus, all points of the circle lie on the equator whose circumference is 40 000 km. In fact,

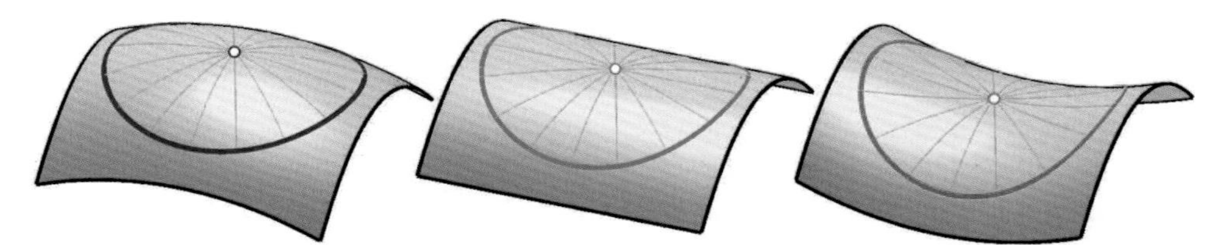

Fig. 6.43 The test using the circle

it is $2\pi \cdot r \approx 62\,800$ km, which means that our carrier surface must be elliptically curved – and quite strongly so. This everyday fact of contemporary science was assumed to be true for many centuries (and was prefigured by several great minds of the ancient world such as *Eratosthenes*, who knew the circumference of the Earth with remarkable accuracy 2200 years ago). The "rigorous proof" was not provided until the age of rocket science.
This experiment has an additional, particularly exciting extrapolation: Mathematicians can prove that it also works in any dimension. We can detect whether space is curved without entering the fourth dimension – it is just not as simple. Our space does not seem to be curved on a large scale. *Locally*, however, there are lots of noticeable indentations in the neighborhood of large masses (such as black holes).♠

• Light edges

Fig. 6.44 shows two surfaces of translation which differ only slightly: In the right image, a cubic parabola is being translated along a second cubic parabola. The surface is continuously differentiable to the second degree – even at the critical "midline" where the color changes. In the left image, a translated cubic parabola is being approximated by two regular parabolas which do not have a transition of constant curvature at their apex. A layman has to examine the images many times to notice that in the left, there are bends at the midlines next to the shadowing. Nevertheless, such subtleties play a great role in practice: In the construction of car bodywork, light edges are mostly undesirable. Due to the reflective paint, car bodies are visually very sensitive to such details. It bears a likeness to detecting indentations in a fly's eye (Fig. 6.46).

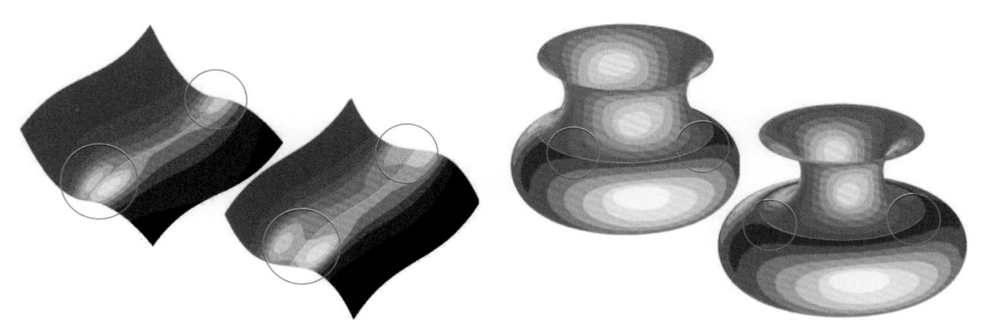

Fig. 6.44 Curvature jump ...

Fig. 6.45 ... or not?

Fig. 6.45 shows two surfaces of revolution where two differently curved meridian circles are combined. The transition occurs on the left at an arbitrary point, and on the right at the exact location where the tangents of the circle are horizontal. On the first surface, elliptic and hyperbolic points coincide unexpectedly. On the right, the horizontal tangential surface touches both partial surfaces along the mutual circle. Its points are parabolic for both partial surfaces. This is why the curvature jump is not noticeable as a "light edge" – as opposed to the left image.

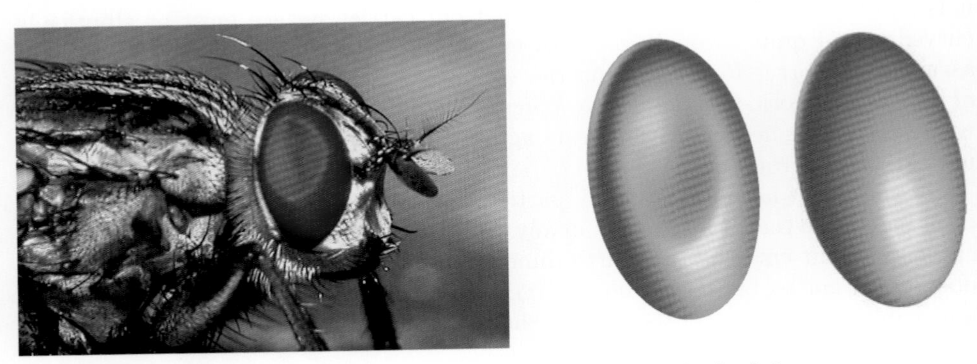

Fig. 6.46 The light reflection shows an indentation in the fly's eye.

Fig. 4.35 shows a sphere cap adjacent to a cylinder of revolution. At this point, there is not only a noticeable edge but the light reacts completely differently in the reflection! ♠

If a surface is composed exclusively of parabolic points, then the surface is only single and not doubly curved. There only exist a few classes of surfaces with this property, and they play an important role in practice: There are the developable surfaces which include the cylinder and the cone (see also chapter 5). They can be manufactured relatively easily by bending the "blank cut" without elongation or compression (Fig. 6.47 on the left).

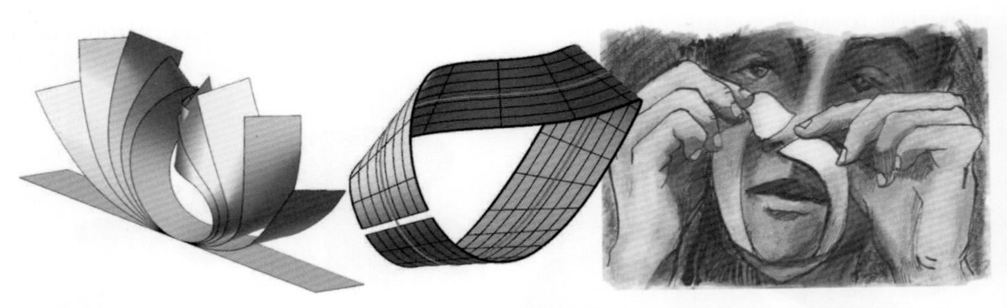

Fig. 6.47 Developable surface: Möbius strip (on the right: instructions for its creation)

Developable surfaces can be visually recognized by their contours, which from every direction are composed entirely out of straight-line segments. Here, one ought to be reminded of the contour generatrices of cones and cylinders (boundary curves are not counted as part of the contour – only the surface curves along which the tangent planes appear as a line segment). Some lesser-known examples include the developables of constant slope, which are used during street construction, the oloid (Fig. 6.49), and the Möbius strip (Fig. 6.47).

Fig. 6.48 Developable surfaces from different perspectives

• Revolvability with practical consequences

The oloid (compared with Fig. 5.49 and 6.86), invented by the Swiss engineer Paul Schatz, is one of the developable surfaces with practical application. It can be rolled along a table surface in a wobbling-like motion, which must have given it its onomatopoetic name "wobbler" in the English-speaking world (Fig. 6.49 on the right).

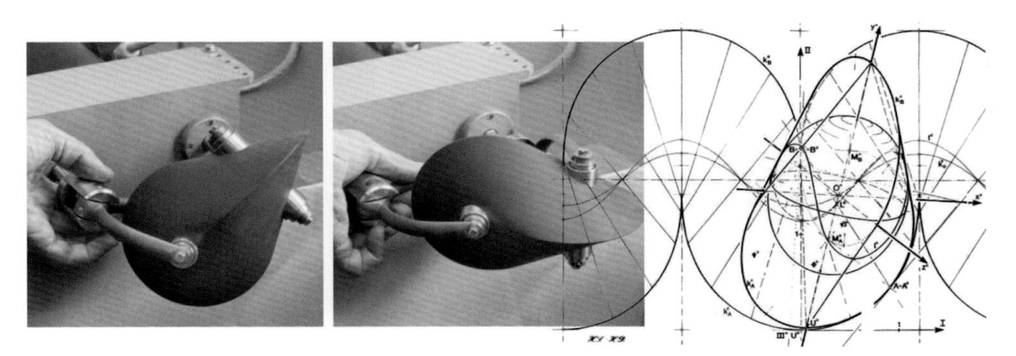

Fig. 6.49 Left, center: Revolution of the oloid according to Paul Schatz (`www.oloid.ch`). Right: Development according to Stachel/Dirnböck (`www.geometrie.tuwien.ac.at/stachel`).

As a consequence, the oloid can be revolved in a rather unusual way (Fig. 6.49). If this revolution is conducted in water, it produces remarkably widely radiating currents which can be used, for instance, to treat sewage water. ♠

6.3 Surfaces of revolution

Rotationally symmetrical surfaces play an important role in nature and technology.

A surface of revolution is created by rotation of an arbitrary curve about a fixed axis. If the generating curve lies in a plane through the axis (meridian plane), it is called the *meridian* of the surface.

Fig. 6.50 Surface of revolution in theory and nature

Each point of the generating curve if rotated creates a *circle of latitude* (sometimes also called a *parallel circle*) of the surface. If the curve does not lie in the meridian plane, the axially symmetric meridian can be determined by intersecting all circles of latitude with a fixed meridian plane.

Fig. 6.51 Surfaces of revolution with meridians and circles of latitude (oblique axis)

If the meridian has a tangent parallel to the axis of rotation, then a rotation of the associated point produces a circle with a locally minimal or maximal radius (e.g., *equator circles*). In general, *boundary circles* do not belong to this group of circles. If the meridian tangent is orthogonal to the axis, the circle's plane touches the surface along the entire circle.

During the rotation, the tangent of the meridian generally traces a cone of revolution with its apex on the axis of rotation. It can also be said that the surface is enveloped by a family of tangent cones. The surface can also be interpreted as the envelope of all cones, whose apexes lie on the axis and which also touch the meridian (Fig. 6.50 on the right).

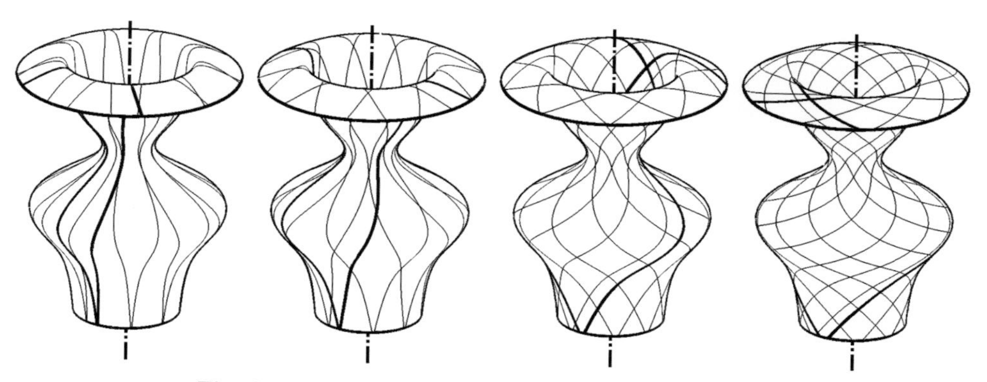

Fig. 6.52 Two families of curves on a surface of revolution

Surfaces of revolution can cope with any rotation about the axis and reflections in an arbitrary meridian plane. If the generating curve is not a planar curve (Fig. 6.52), then, in addition to the curve itself, any curve reflected in an arbitrary meridian plane is "generating". (*Any* curve on the surface of revolution generates at least a part of this surface.)

Fig. 6.53 Along each parallel circle of a surface of revolution, one can inscribe a sphere (left and middle) and a cone of revolution (right).

• Vase with an ellipse as its finishing part

The lower part of the vase pictured in Fig. 6.54 is created by the rotation of a sine wave. In the upper part, it is designed so that its finishing part is an ellipse inclined towards the axis.

Now the question arises which surface of revolution allows this simple oblique intersection. Locally, it must be a second-order surface of revolution! In

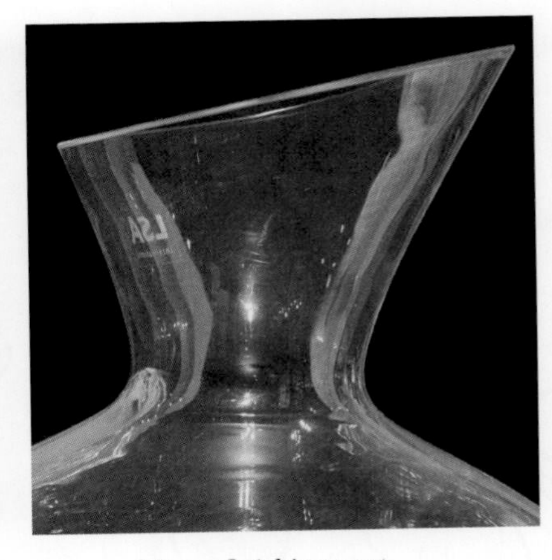

Fig. 6.54 Vase with an ellipse as an oblique finishing part

this particular case, the surface is hyperbolically curved. In its upper part, the surface is a hyperboloid of revolution, and in the part near the oblique intersection, it is nearly a cone of revolution. ♠

- Distorted wine barrel

A wine barrel is a well-known surface of revolution and used to function as the starting point of many important calculations. *Kepler*, for instance, used it to develop a simple formula which is of high significance in mathematics ("Simpson's rule").

Fig. 6.55 Affinely distorted surfaces of revolution (on the left: the original)

Fig. 6.55 shows a "poster" of a visitor's lounge: The wine barrel is affinely distorted. The visitor would only notice this when coming close to the object and being able to view it from its side. On the right side, the same photo has simply been stretched planimetrically in its horizontal direction – which re-establishes the appearance of a surface of revolution. ♠

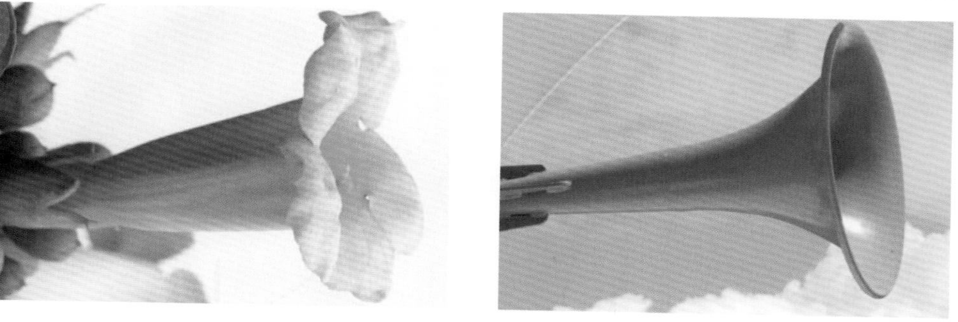

Fig. 6.56 Contour of a surface of revolution in nature and technology

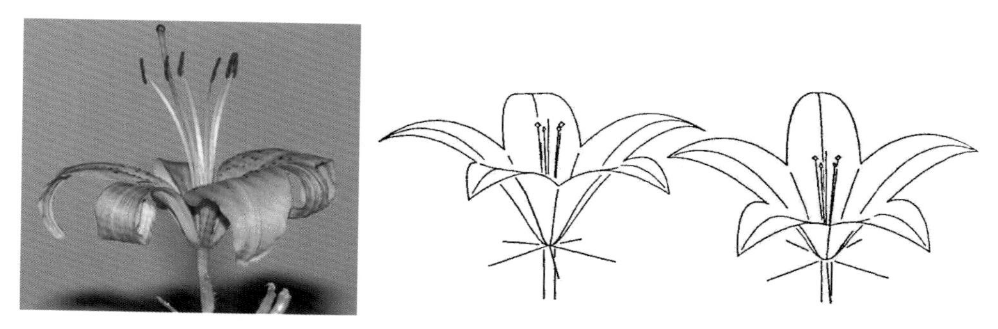

Fig. 6.57 Broken surface contours in nature (fire lily) and computer graphics

The torus is a frequent building block of higher order

A sphere is constructed by rotating a circle about an arbitrary diameter. If the axis of rotation is translated within the circle plane into an arbitrary position, then a much more complicated surface is produced by subsequent rotation. It is called a *torus* or *ring surface*, and is, as can be calculated, an algebraic surface of fourth order. Geometrically speaking, it can be reasoned that a pair of circles lies in each meridian plane (Fig. 6.58) which is intersected by an arbitrary straight line in four points at most.

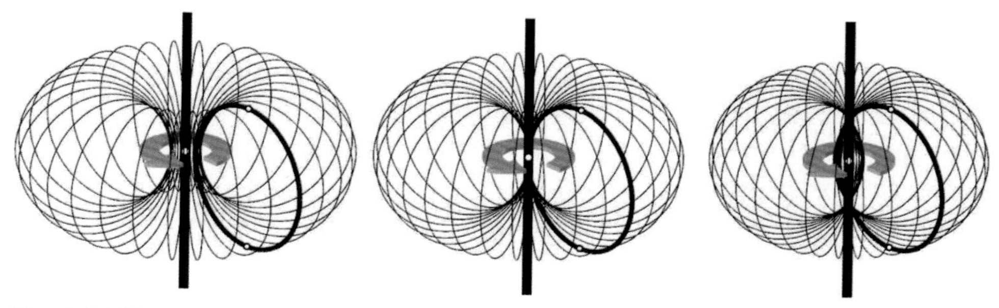

Fig. 6.58 The torus as a surface of revolution: A circle rotates about a straight line in its plane.

The same surface is also enveloped if a sphere for which the circle in the meridian plane is a great-circle rotates. It is, thus, possible to "roll around" a sphere inside a torus – which has many implications for practice.

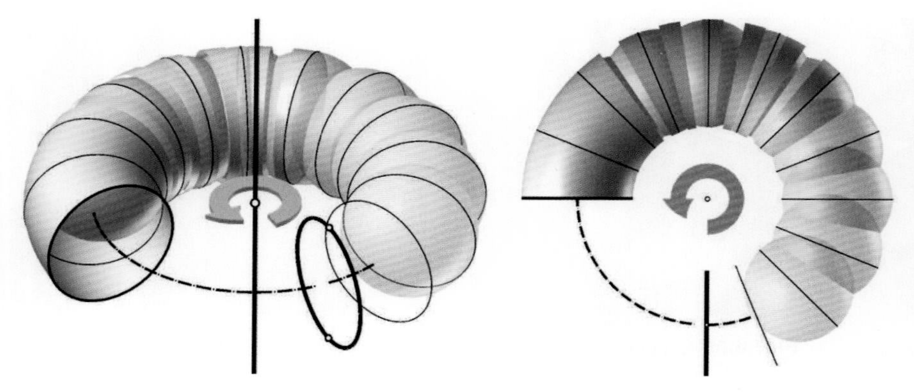

Fig. 6.59 Torus as an *envelope* of spheres

A torus is a fourth-order surface of revolution which is created by the rotation of a sphere or a circle about an axis located in its plane which does not contain the center.

The outer ring of a torus is composed of elliptic points and the inner ring of hyperbolic points. The circles, on which the torus can be placed onto a plane, carry parabolic points. Each circle of this kind is entirely contained in the tangent plane of each such point.

Fig. 6.60 Torus chains: Each link is made of half ring surfaces which are joined cylindrically.

A standard steel chain (Fig. 6.60) is composed of links made from two half ring surfaces which are connected by cylinders of revolution. The radius of the circle which generates the torus is slightly smaller than the radius of the inner equator circle.
Rotating only parts of the generating circle (meridian circle) produces open surfaces of revolution which are, geometrically speaking, still labeled as tori. The same is true if the rotation is applied to the whole circle but does not proceed by 360° (Fig. 6.61 on the left).
Fig. 6.62 shows the interior perspective view of a torus – a challenge to the light sensitivity of the photo film and to the nerves of the "participants".
Fig. 6.63 shows two surfaces which are closely related to the torus. More precisely: The left surface is a half-torus, or is bounded by surface parts

Fig. 6.61 Torus part and affinely distorted tori

Fig. 6.62 A light at the end of the torus-shaped tunnel

which belong to a half-torus. The right surface is constructed in a way that resembles a torus but the meridian circle is continuously becoming smaller. Thus, this surface is no longer a surface of revolution.
Fig. 6.64 shows further variations. The left image is a computer graphic. If a doubly curved torus was to be modeled using a loop, it would only result in an approximation because the surface thus generated is only single curved. The model on the right dates back to M.C. *Escher* and is often called an "Escher torus" – even though this is not quite correct.

Non-trivial sections of the torus

Fig. 6.63 One of the two surfaces is not a torus.

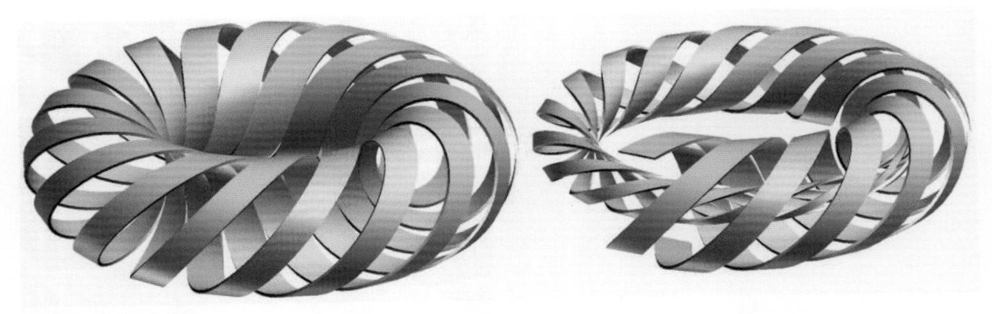

Fig. 6.64 Other variants ...

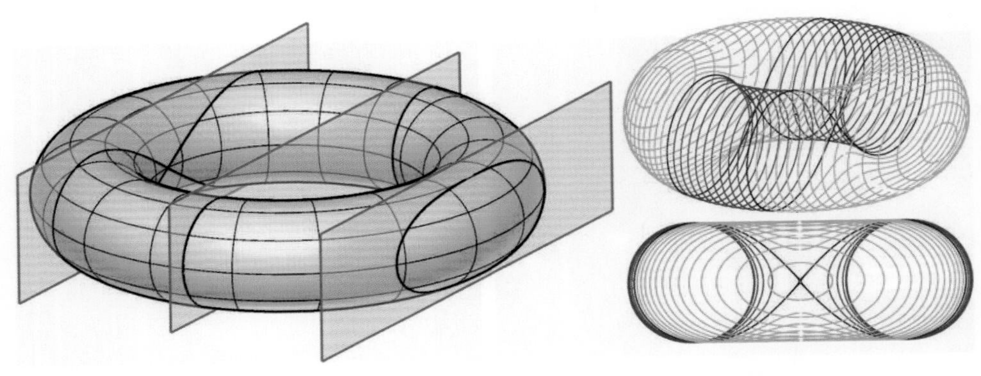

Fig. 6.65 Intersections of a torus with planes parallel to the axis

Algebraically speaking, the torus is a fourth-order surface. Each planar intersection curve is therefore, in general, a fourth-order curve. Among these are the so-called *Cassini ovals* which are produced if the plane that intersects the torus is parallel to the axis of the torus. These curves have already caught our attention (Fig. 1.31, Fig. 6.65 on the right). If the intersection plane includes the axis, then the curve splits off into two circles.

There exist two other positions of the intersection plane where the intersection curve splits off into two circles: Each plane orthogonal to the axis intersects along two concentric circles. These two circles coincide if the plane (orthogonal to the axis) touches the torus.

Fig. 6.66 *Villarceau* circles in theory, as a wooden model and in art (Benno Artmann)

There even exists a non-trivial case where the section curve splits off into two circles: The plane has to touch a pair of axis-symmetric meridian circles in hyperbolic surface points. We now have the case that two surfaces (the torus and the plane) share a fourth-order curve, which – due to double contact – has to possess *two* double points. However, this is only possible if the intersection curves degenerates. In this case, the curve would have to split off into two circles but given another fourth-order surface, they might be arbitrary conic sections.

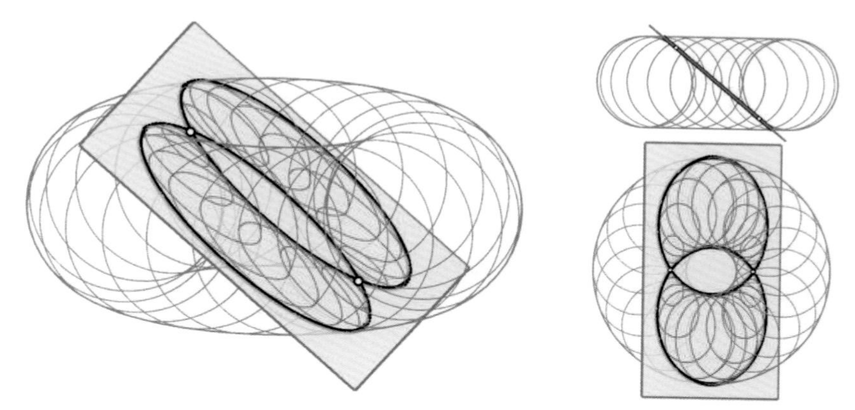

Fig. 6.67 *Villarceau* circles as envelopes

Both conic sections are envelopes of the pencil of circles that is cut out from the torus generating spheres (Fig. 6.67). Thus, the envelopes must be *circular* (i.e., they are passing through the absolute points of the intersecting plane – see p. 159). This remarkable discovery is due to the French geometrician *Villarceau*.

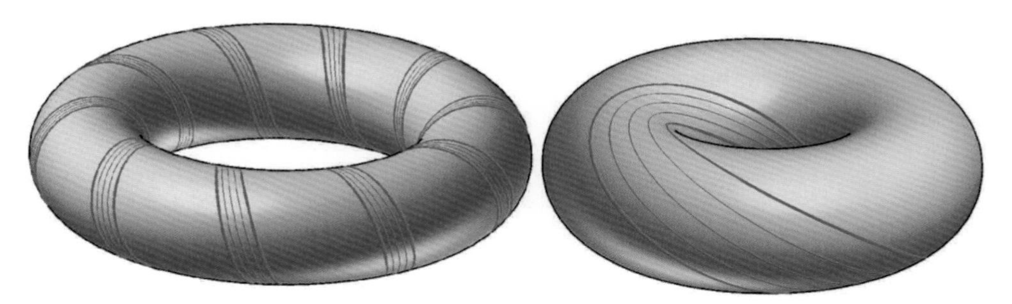

Fig. 6.68 Rhumb line on ring surfaces

Villarceau discovered even more: The circles named after him intersect all meridian circles – and thus all parallel circles – at the same angle. Such a curve is called a surface's *rhumb line* or loxodrome.
The *course angle* of this rhumb line is, of course, extremely special. If one changes the angle only slightly, then the curve no longer stays in a plane. In most cases, it doesn't even stay a closed curve but will wind itself around the torus "for all infinity" without ever returning to the starting point. For special angles, however, these curves are closed (Fig. 6.68) – and also algebraic, which means that they can be described by algebraic equations.
The sphere, if you like, is also a special torus where the spine circle is shrunk to a single point. Interestingly enough, rhumb lines on spheres are never algebraic

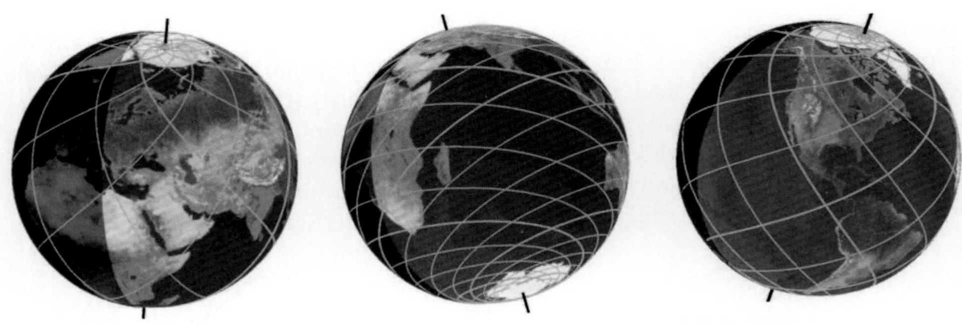

Fig. 6.69 Sphere rhumb lines (loxodromes) with different course angles

(except for course angles 0° and 90°). Sphere rhumb lines are spiral-shaped curves that wrap around the poles. A ship traveling at a constant course angle relative to the north direction travels along such a rhumb line! This is actually a productive strategy in tropical regions for traveling from A to B without great navigational effort. Upon reaching polar regions, however, the captain would have to change his strategy.

6.4 The torus as a prototype for all other surfaces of revolution

The importance of the torus in geometry is, among other things, due to the fact that each surface of revolution can be approximated by a sequence of torus segments. This directly follows from the insight that each planar curve can be approximated by a sequence of osculating circles (curves that are produced by the chaining together of circles are called *basket handle arches*). It can, thus, be said that:

From a local perspective, each arbitrary surface of revolution is curved like its "osculating torus".

Some surfaces of revolution found in architecture are parts of a torus even if this is not immediately visible. The dome of the Ibn Toulun mosque in Cairo (built in the 8th century AD) is part of a so-called *spindle torus* because the meridian circle intersects the axis of rotation (Fig. 6.70 on the left, Fig. 6.71). The inner parts of a torus are often found in nature (Fig. 6.70 on the right) though we have to be generous enough to speak only of the "basic form". Seen this way, even the pressurized stream of water escaping from a valve in Fig. 6.72 could be likened in its starting phase to a spindle torus.

- **On contact lenses and astigmatism**

Wearers of contact lenses (or glasses) often hear the words "spherical" or "toric" from their ophthalmologist or optician. The surfaces of spherical lenses (spectacle lenses) are simply sphere caps with a slightly different radius. This causes the focal point of the lens system to be offset (when "focused on infinity", the focal point

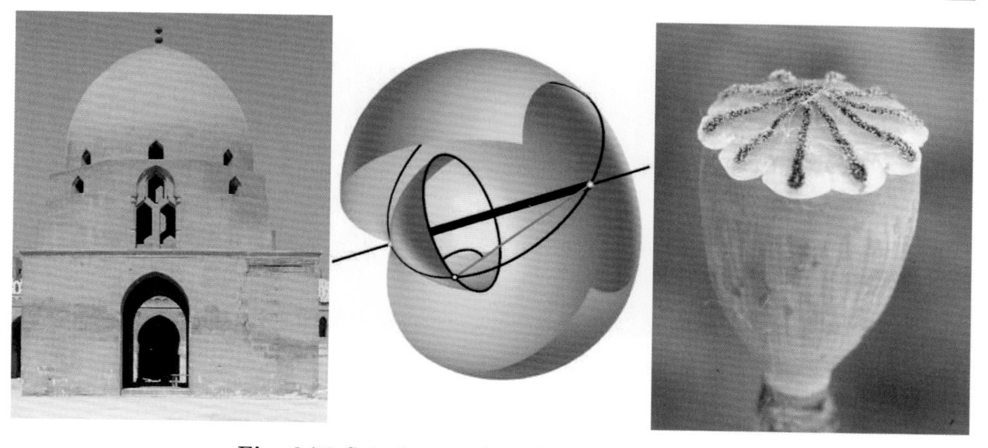

Fig. 6.70 Spindle torus in architecture and nature

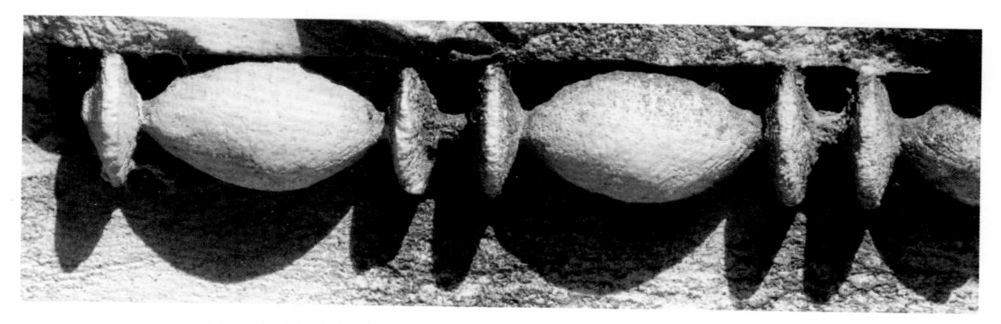

Fig. 6.71 Spindle torus as an ornament on an ancient wall

should lie precisely on the retina). This does not matter in spherical lenses, which attach to the eye by adhesion: The sphere has the same curvature in any tangential direction.

Firstly, the exact way in which the optical system of the eye works is relatively complex, and secondly, it is different from what most people believe it to be. The actual refraction happens on the cornea while the lens inside the eye is only responsible for fine tuning. More information can be found in application p. 329.

The issue becomes more complicated if one additionally suffers from astigmatism. In such cases, the principal curvatures of the cornea – and sometimes also of the lens inside the eye – are different. This causes points on the retina to be rendered in a "rod-like" manner. The corrective lens then has surfaces in the form of torus caps. As if that wasn't enough, the axis of the torus is often twisted so that a great multitude of different torical lenses exist while the number of spherical lenses (at granularities of quarter-dioptres) is relatively limited and often completely in store at the optician.

Hard spherical contact lenses are often capable of correcting astigmatism by the mere fact that they allow smaller cavities between the lens and the eye to be filled with lacrimal fluid, which possesses nearly the same refractive index as the cornea itself. The liquid, thus, complements the cornea into a sphere (Fig. 6.73 on the left). Soft lenses, on the other hand, adapt to the toric shape of the cornea. ♠

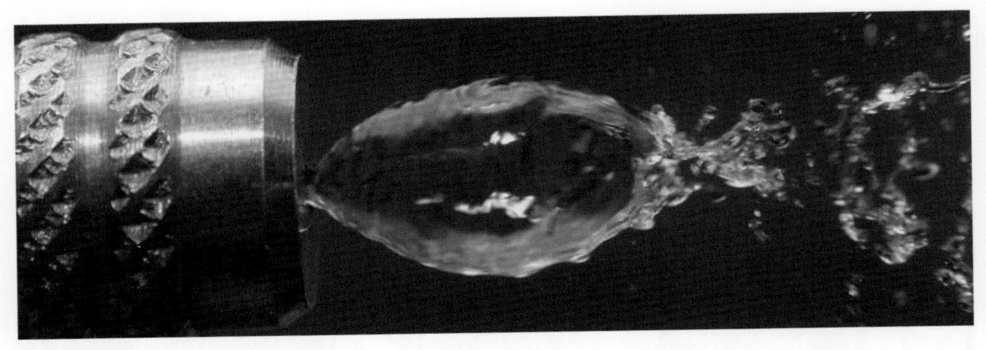

Fig. 6.72 A stream of water escaping from a valve

Fig. 6.73 Contact lenses

Cusps of contours

Torus parts – and by extension, many surfaces of revolution – tend to develop contour cusps during projection. Fig. 6.74 shows different perspectives while cusps occur in most of them. During a normal projection orthogonal to the axis, the contour degenerates into two circles and the lowest and uppermost circle that appear as a segment (Fig. 6.74 top left, also to some extent in the picture of the anemone 6.75). The equator circles form the contour in axis direction (Fig. 6.74 on the right).

The respective contour in space is an arbitrary space curve k (composed of two branches, as can be seen in Fig. 6.76) that does not have any cusp. In each point

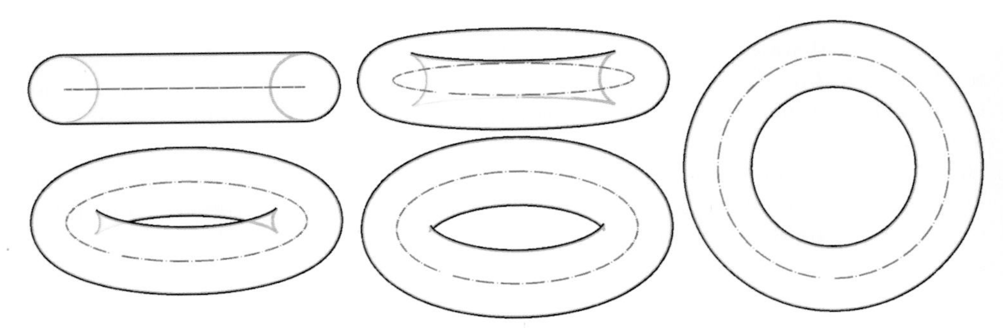

Fig. 6.74 Cusps of the contour of a torus

Fig. 6.75 Quarter-torus of an anemone

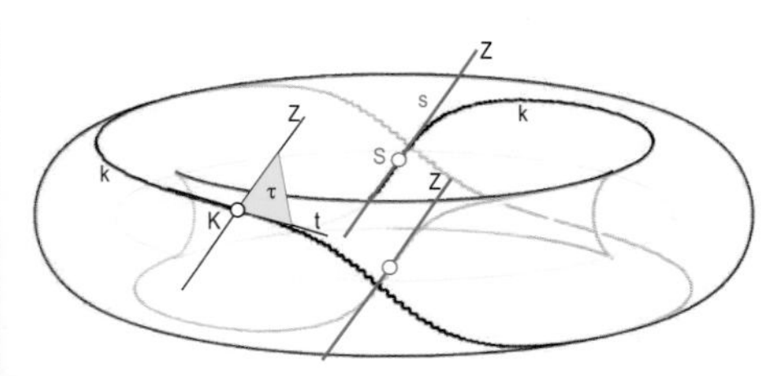

Fig. 6.76 How the contour cusps are formed: In each point K of the contour line k, the associated tangent plane τ passes through the center Z. A cusp S occurs when the tangent s of the contour k passes through Z.

K of k, the associated tangent plane τ – spanned by the tangent t of k and the respective projection ray – passes through the center of projection.

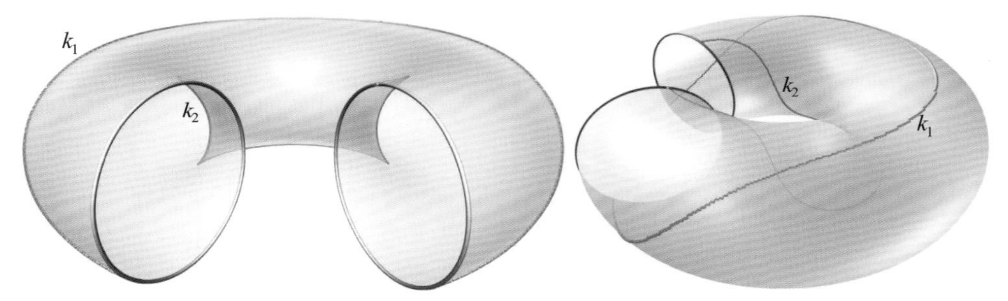

Fig. 6.77 Left: The contour of a torus in central projection (consisting of two curves k_1 and k_2) has four cusps in the projection. Right: Viewed from another projection center, k_1 and k_2 appear as space curves without cusps.

The probability that a tangent of k passes directly through the center Z of the projection at some point S is not even that small. In those exact cases, the curve appears, locally speaking, as a point and its *orthogonal view* possesses a cusp. The contour lines k in Fig. 6.76 and Fig. 6.77 are "jagged" in some parts, which is due to the fact that the torus is triangulated which causes the contour edges of the approximate polyhedron to become visible. In the projection from Z. These jagged edges cannot be seen.

Fig. 6.78 illustrates how the normal views of three visually similar surfaces of revolution may or may not have cusps despite the fact that the direction of the projection is the same. The catenoid is formed by the rotation of a catenary (Fig. 5.25). The contour of a hyperboloid is always a hyperbola (Fig. 6.78 center), which is why cusps on the contour never occur with this type of surface.

The question of which surface is shown in Fig. 6.79 can, thus, be approached from this point of view. It is actually a hyperboloid which, towards the roof, changes its shape into a torus.

Fig. 6.80 shows the cusps on the contour of general surfaces of revolution. The meridian is easily discernible on the right – no cusps occur there. A cusp is visible in both other pictures (see also the appendix about free-hand drawing).

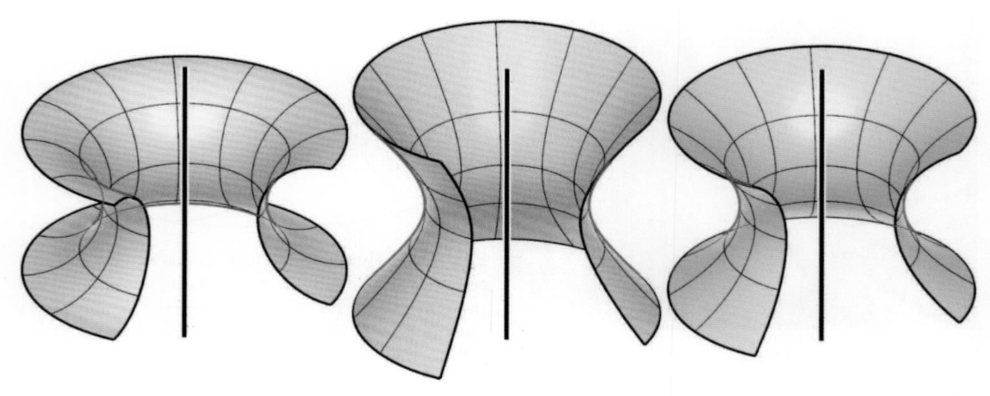

Fig. 6.78 Contour of a torus, hyperboloid, and catenoid. Cusps on the contour never occur with hyperboloids and very often with the interior parts of tori.

Fig. 6.79 Coop Himmelb(l)au: Great Egyptian Museum in Cairo

Intersection of surfaces of revolution

In the case of the bellflower in Fig. 6.80, boundary curves occur at the flower petals, which, when projected in axis direction, resemble circular arcs with great accuracy. It can, thus, be said that the boundary curves of the petals are created as the intersection curves of the surface of revolution with cylinders of revolution. This way of phrasing it is in no way exaggerated: A designer at a computer is easily and quickly capable of modeling the curve in this way!
In mechanical engineering, intersection curves of surfaces of revolution happen particularly frequently. The axes of these surfaces often intersect, and for this special case, Gaspard *Monge* has already specified a quick and efficient procedure.
If an auxiliary sphere is put around the intersection of the axes, it will intersect both surfaces along circles. The points of the intersection curve of both surfaces of revolution are then the section points of the circles (Fig. 6.81).
If the surfaces are projected orthogonally to the plane spanned by both axes of rotation, then any common circles with Monge's auxiliary sphere appear as line segments. This, however, immediately provides their section points. If

Fig. 6.80 Cusps on the contour of surfaces of revolution (left and center)

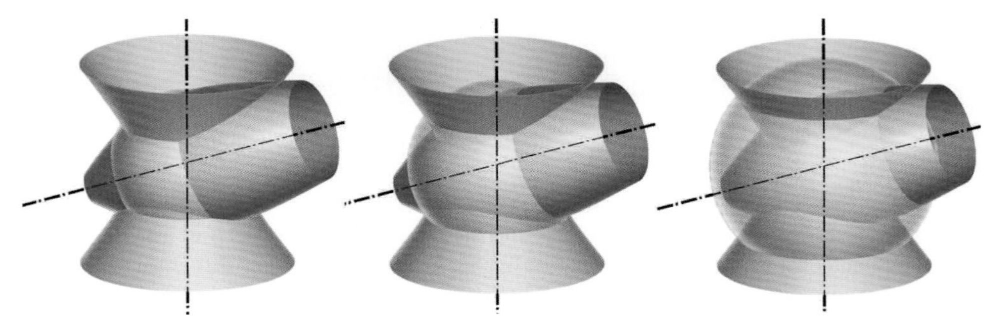

Fig. 6.81 Principle: Variation of Monge's auxiliary sphere ...

the size of the auxiliary sphere is now varied, an arbitrary number of points of the intersection can be found.

A note for advanced readers: Even if the circles on Monge's auxiliary sphere no longer intersect in real space, they still share complex conjugate intersection points. The *real* straight line connecting these points remains the straight intersection of both carrier planes of the circles on the sphere. It can be shown that the *normal projection* of the intersection curve, if "extended", passes through the orthogonal view of each such straight intersection. These points are called "parasitic points".

• **Reverse engineering**

Let us, once again, consider the bellflower from before: The difficulty seems to lie in finding the surface parts of seemingly arbitrary surfaces that belong to a known

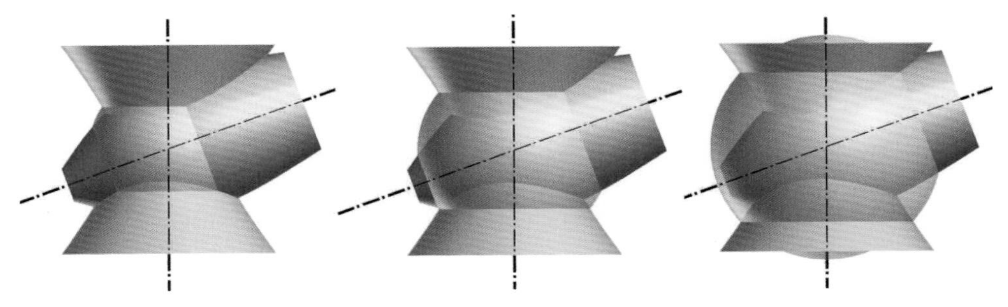

Fig. 6.82 ... during normal projection on the connecting plane of the axes

surface class, such as the surfaces of revolution. In computer geometry, the term for this procedure is *reverse engineering*. Special programs are capable of analyzing objects digitized by 3D scanners and trying to subdivide them into parts which are, for instance, surfaces of revolution or developable surfaces.

Fig. 6.83 Surface of revolution (an ellipsoid of revolution in its upper part)

This can be greatly advantageous in practice. Consider a brilliant architectural model with many curved surfaces. Once this project is due to be put into practice, the construction costs can be greatly reduced if, without substantial changes to the design, the surfaces involved can be reduced to certain standard surfaces.

Fig. 6.84 Inner part of the surface of revolution above (part of an affinely distorted torus)

Fig. 6.83 shows the limestone shell of a sea urchin – a general surface of revolution. Upon closer inspection, the entire upper part appears as an ellipsoid of revolution, which is further borne out by the shadow. The lower part appears to be an affinely distorted torus, which, after a mere glance at Fig. 6.84, may not be immediately obvious to the layman.
Given some practice, it is possible to recognize planar intersections of a torus in Fig. 6.85 on the left, and torus-shaped grooves on the right.
Reverse engineering attempts to develop methods which enable the "dumb but fast" computer to detect and prove circumstances which might not be easily seen with the naked eye. Much geometric theory is conscripted for this purpose:

• **Gauss map**

Points on a general surface can be associated with points on a sphere. We translate the associated surface normal to the center of the unit sphere, and intersect it with

Fig. 6.85 Planar torus intersections and torus-shaped grooves in a half-sphere

the sphere. In general, the associated points will distribute over the entire surface of the unit sphere.
If the initial surface is developable, it will be touched along each generatrix by one and the same tangent plane with a fixed normal direction. The Gaussian image of a curve on the sphere is thus reduced – namely the spherical binormal image of the surface's edge of regression. Fig. 6.86 shows the normal image of an oloid (blue). The curve could be called a "tennis ball curve". It divides the sphere into two congruent halves and possesses a congruent top and front view in relation to its symmetry plane.

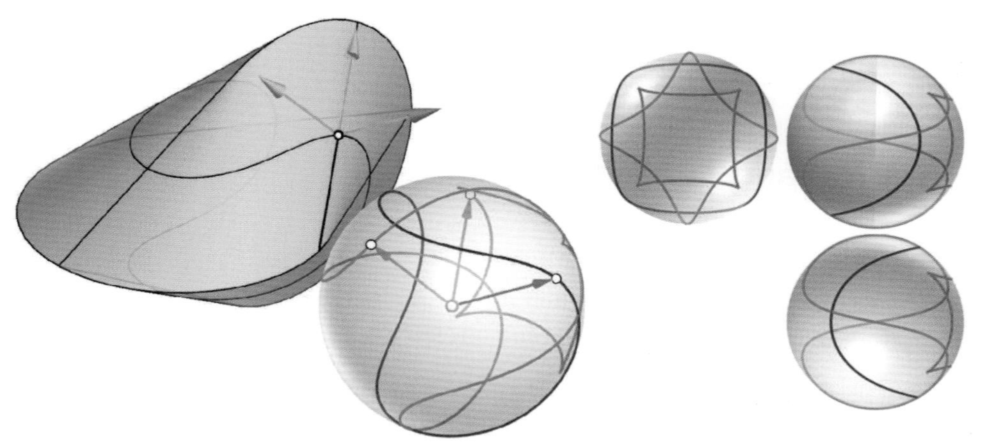

Fig. 6.86 Spherical tangent-, normal-, and binormal image of the oloid

If all surface generatrices are translated through the sphere midpoint, we get the *spherical generatrix image*. In the case of the oloid, this happens to be a curve with four cusps (in red) which also possesses a high degree of symmetry. This curve is the spherical tangent image of the surface's edge of regression.
Finally, the spherical primary normal image of the arris may be considered (which appears in green).
♠

• Automatic detection of cylinders, cones and developables

The spherical image of a surface is not only theoretically interesting but is actually utilized during *reverse engineering*: In order to detect whether or not a surface is developable, it can be scaled using a 3D scanner . This results in a "point cloud", which exhibits "noise" due to measuring inaccuracy. The cloud's surface normal is now interpolated and translated through the center of the test sphere. The more

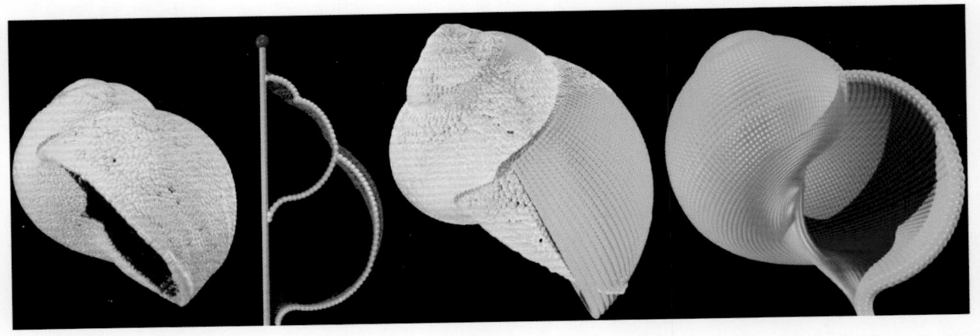

Fig. 6.87 Reconstruction of a snail shell according to Boris Odehnal et al: The shape is essentially a spiral surface (see the next chapter), which can be "idealised" by a great number of measured points. If one compares the perfect shape to the real object, one notices that the snail changes its building strategy in the last phase (`www.geometrie.tuwien.ac.at/odehnal`).

accurate this spherical image approximates the surface in the neighborhood of the curve, the more likely it is that the surface is developable.
This makes it very easy for the computer to detect general cylinders: Their spherical normal image is a great-circle of the sphere whose axis direction is the generatrix direction of the cylinder.
The spherical normal image of a cylinder of revolution is a small-circle of the test sphere. ♠

6.5 Pipe and canal surfaces

How to invent a new class of surfaces ...

A remarkable surface appears when a torus is inverted in a sphere which has its midpoint in the plane of the spine circle (Fig. 6.88).

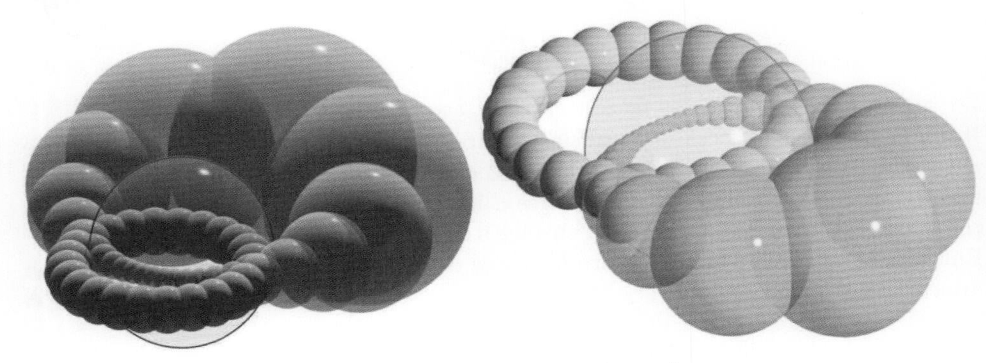

Fig. 6.88 Inversion of a torus in a sphere

The inversion in space preserves circles and angles like the planar version. The emerging surface, called the Dupin cyclide, was examined by the French geometrician Charles *Dupin* and inherits a multitude of beautiful properties

from the torus. Algebraically speaking, it is also, quite remarkably, of fourth order.
Dupin cyclides, like the torus, possess two planes that touch the surface along two entire circles and carry two families of circles which correspond to the circles of latitude and the meridian circles. Furthermore, in the case of the torus, two *Villarceau* circles lie in a double tangent plane. During inversion, this plane passes over into a sphere. For their part, the Villarceau circles pass over into circles on the cyclide which have to lie on this sphere and are, therefore, no longer coplanar (E. Hartmann).

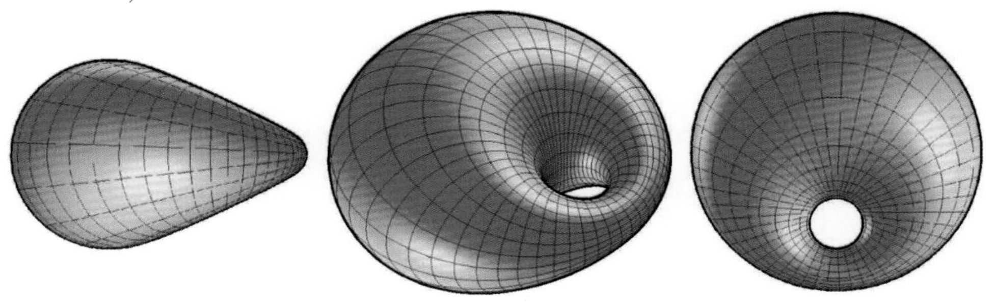

Fig. 6.89 Dupin cyclide (Ring cyclide)

Of course, such surfaces are no longer surfaces of revolution, with which they nevertheless share some properties. This is due to the fact that they can be generated in two ways as an envelope of spheres whose radius continuously varies. With surfaces of revolution, the spheres are centered on the axis of rotation (Fig. 6.90), and with the Dupin cyclides on the spine ellipse and spine hyperbola, or on both spine parabolas. Surfaces enveloped by spheres with varying radii are called *canal surfaces.*
The torus, in particular, can also be generated by spheres with a fixed radius. Such surfaces are called *pipe surfaces.*
Of course, every pipe surface is automatically a canal surface. Thus, the *torus is a canal surface in two different ways*: Firstly as a surface of revolution (sphere centered on the axis), and secondly as a pipe surface (centers of the spheres on the spine circle).

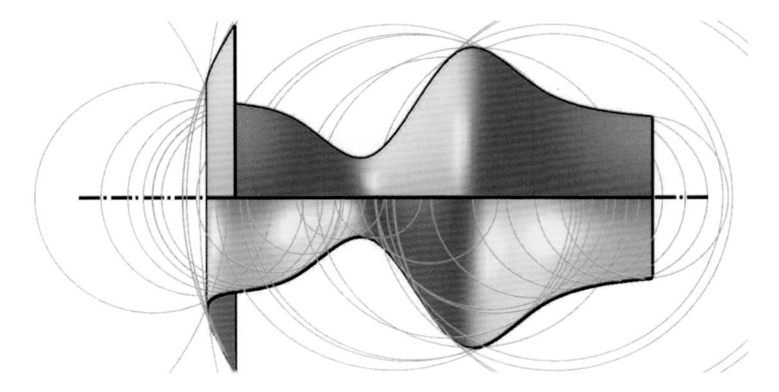

Fig. 6.90 Each surface of revolution is, in a trivial way, also a canal surface (see also Fig. 6.53).

By the way, we can immediately deduce why the spine curves of a Dupin cyclid are conic sections: An ellipse or hyperbola is the locus of all circle centers that touch two given circles (see page 156). It is, thus, possible to define each Dupin cyclid in two ways as the envelope of spheres: The centers of the spheres lie in a plane, and there, the spheres touch two circles.

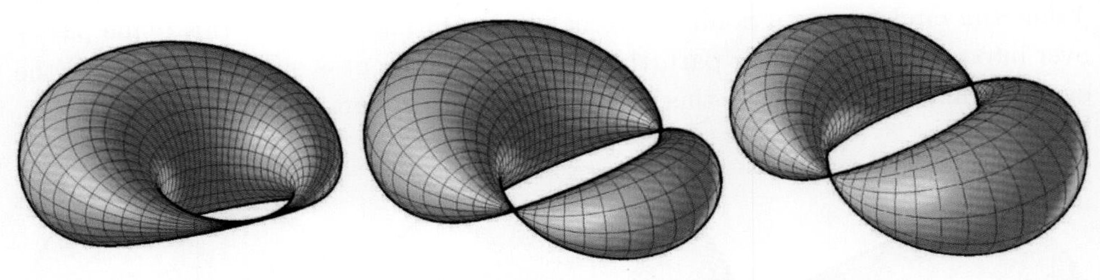

Fig. 6.91 Dupin cyclides ("single-horn" and two "double-horns")

If the situation is permitted that the given circles mutually touch or intersect, three types can be differentiated: The ring-shaped (Fig. 6.89), the "single-horn", and the "double-horn" type (Fig. 6.91).

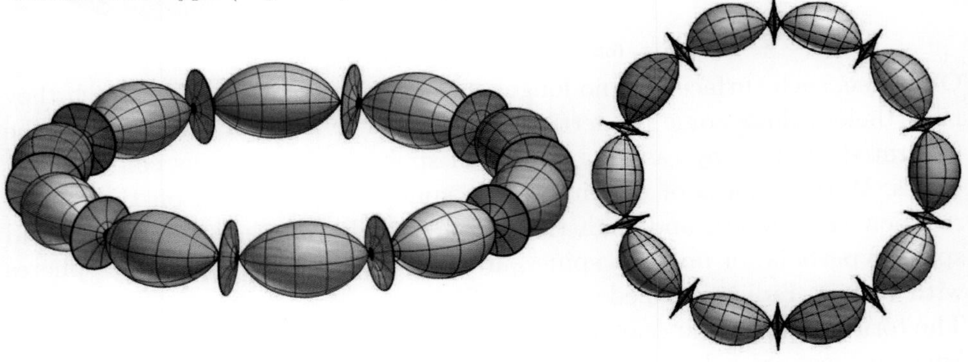

Fig. 6.92 Michael Schrott has strung together parts of such double-horns into a chain (Fig. 6.92) which is located inside a torus.

Pipe surfaces and blending surfaces

Pipe surfaces are created by the motion of a sphere along a *spine curve.* Fig. 6.93 shows in what way such surfaces depend on the form of the spine curve and the radius of the moved sphere. Two closely adjacent positions of the sphere share a circle – it lies in the symmetry plane of the two corresponding sphere centers. If we let both sphere positions move together, the straight line connecting the centers becomes the tangent of the spine curve, and the circle becomes a great-circle of the sphere.

If a circle is moved along a curve such that its axis is the tangent of the curve, it coats the same pipe surface as a concentric sphere of equal radius.

Pipe surfaces are often found in nature and technology (Fig. 6.94). They can be made flexible if the radius of the generating circle is kept small.

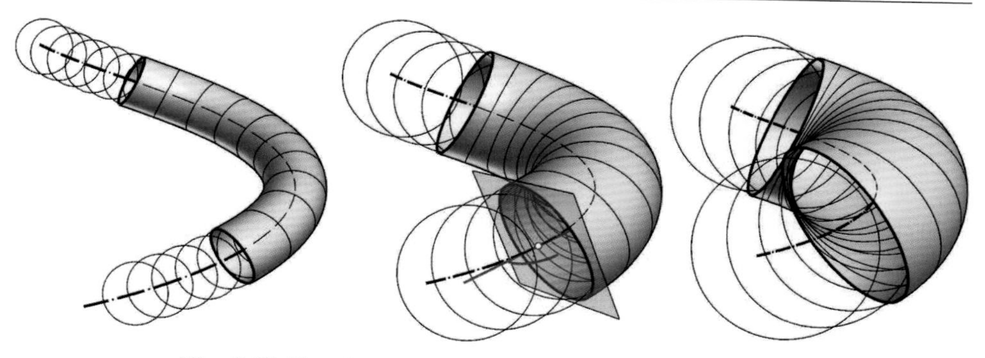

Fig. 6.93 The pipe surface as the envelope of a moved sphere

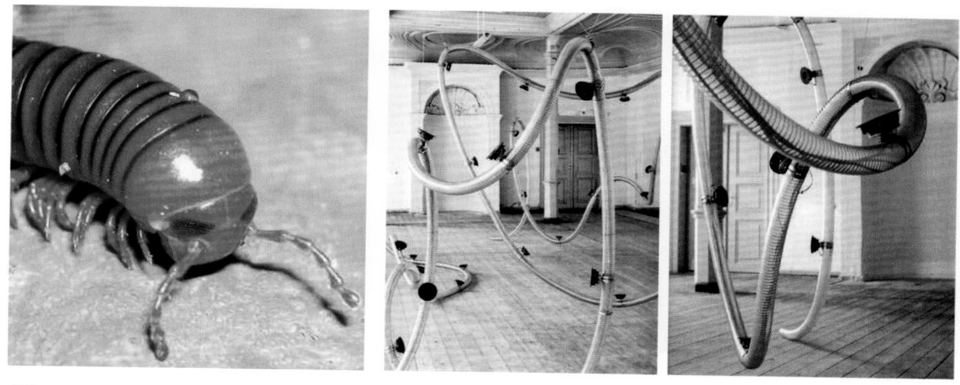

Fig. 6.94 Pipe surface in nature **Fig. 6.95** and art: Sound worlds by Bernhard Leitner

Fig. 6.95 shows an organically conducted tube through which the sound is directed. Openings, which serve as sound sources, were made in 40 places. Depending on programming, curvatures are emphasized, counter-punctually combined, or transformed into fugue-like sequences.

• **Blending surfaces**

Pipe surfaces are well suited for filleting (blending) planar or cylindrical side surfaces against a plane: A sphere of given diameter should be rolled along the plane so that it touches the work piece. The relevant part of the sphere's envelope is the sought-after blending surface.

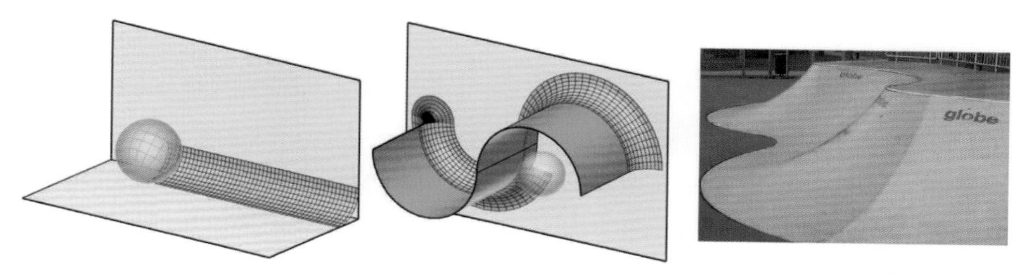

Fig. 6.96 Blending surfaces in theory and practice

Fig. 6.96 shows the case that is most common by far: A filleting towards a planar surface forms part of a cylinder of revolution whereas a filleting towards a cylinder of revolution forms part of a torus. In the right image many tori are glued together. In the case of the anemone in Fig. 6.75, the shaft shaped like a cylinder of revolution passes over into a quarter-torus – just as during classical blending.

• The altitudes of a tetrahedron lead to another aesthetic lamp

The designer lamp Fig. 6.97 shows a lot of geometric insight: Four cylinders of revolution with equal diameters and intersecting axes are arranged in a well-balanced way. This requires that they are the altitudes of a tetrahedron (Fig. 6.97 middle). The intersection without blendings, therefore, consists of parts of six ellipses in respective symmetry planes.

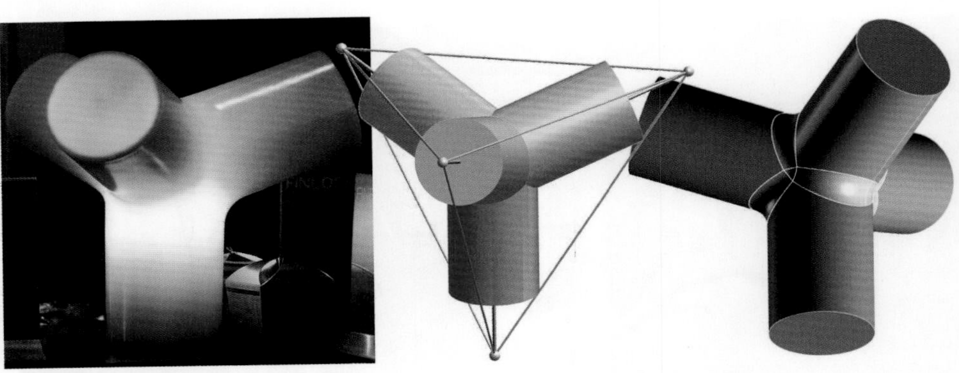

Fig. 6.97 The axes of the four cylinders of revolution are altitudes of a regular tetrahedron. The blendings are comparatively complicated.

When we try to blend the cylinders, we have to guide a sphere along a curve in the symmetry plane such that it touches those ellipses. Thus, the path of the sphere's center consists of parallel curves of ellipses. ♠

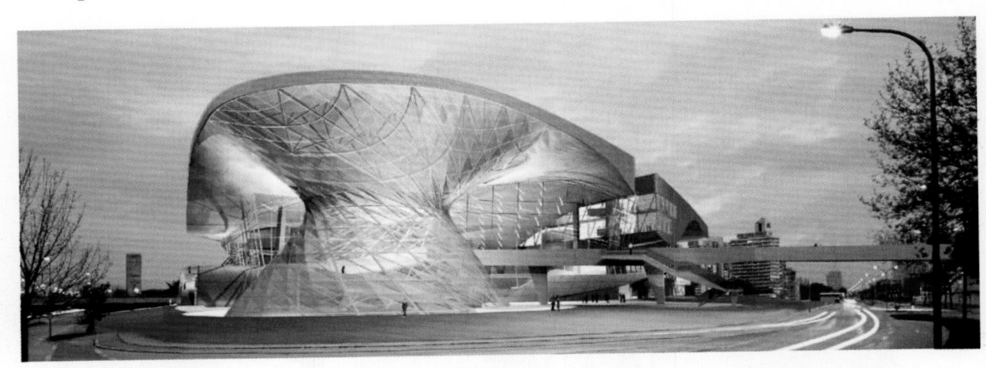

Fig. 6.98 Filleting surfaces in architecture: BMW world, Munich (Coop Himmel(b)lau)

In mechanical engineering, such blendings are important for increased stability, but are sometimes also employed so that sharp edges are avoided (Fig. 4.59), or simply for design purposes. Even "in the grand style", blending surfaces (Fig. 6.79, Fig. 6.98) are required to make the dreams of architects become reality. ♠

7 Further remarkable classes of surfaces

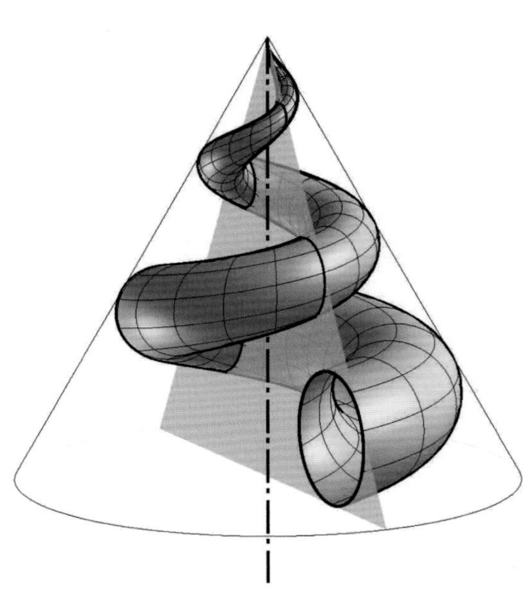

In the ideal case, surfaces can be defined by motion. Surfaces of revolution are produced by a rotation of a curve about an axis. In the case of helical surfaces, the generatrices additionally experience a translation along the axis of rotation. Surfaces of translation are generated by the translation of a curve along another curve. Ruled surfaces are generated by straight lines moving along arbitrary curves. Different surfaces (spiral surfaces, for instance) can be accomplished by moving spatial curves along other such curves and by changing the former according to certain rules as the motion.

All noted classes of surfaces play significant roles in nature and technology. Helical surfaces, for instance, may be employed to "change" rotation into translation, which makes them useful as propeller surfaces or drills. In nature, several types of spirals can be found, such as snail shells, animal horns, or even shapes of galaxies in outer space.

A special difficulty arises when a surface cannot be put into any of these categories. *Minimal surfaces*, for instance, are defined by a very characteristic curvature property and play a great role in physics, architecture, and nature.

Survey

7.1 Ruled surfaces

Ruled surfaces are, next to surfaces of revolution, particularly significant.

Ruled surfaces are generated by a moving straight line, and despite the fact that they carry only straight lines, they are not developable in the general case.

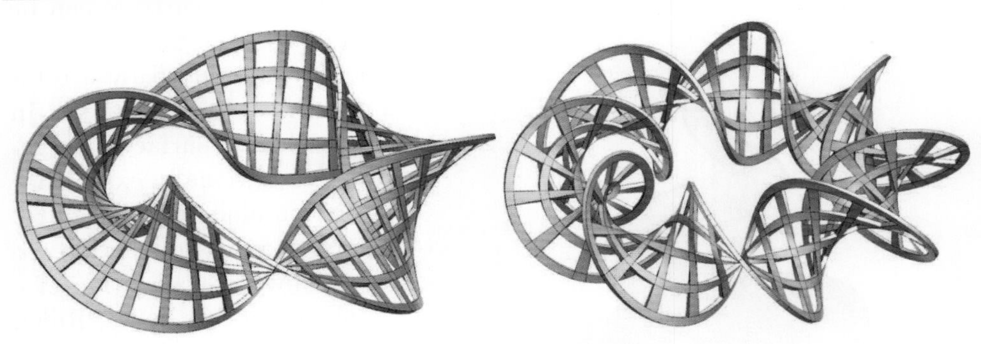

Fig. 7.1 Motion of a straight line

Fig. 7.1 is intended to illustrate the situation: A straight line is moved through space and generates a surface which is, in general, doubly curved (immediately to be seen from the curved contours).
Let us begin with the developable exceptions: Cylinders are formed by the motion of a straight line which is parallel to a fixed direction during the entire motion, and are, thus, ruled surfaces by definition. They are developable, and therefore, they always possess straight lines for their contours.
Cones are developable ruled surfaces too: They are generated by moving a straight line through a fixed apex in an arbitrary way.

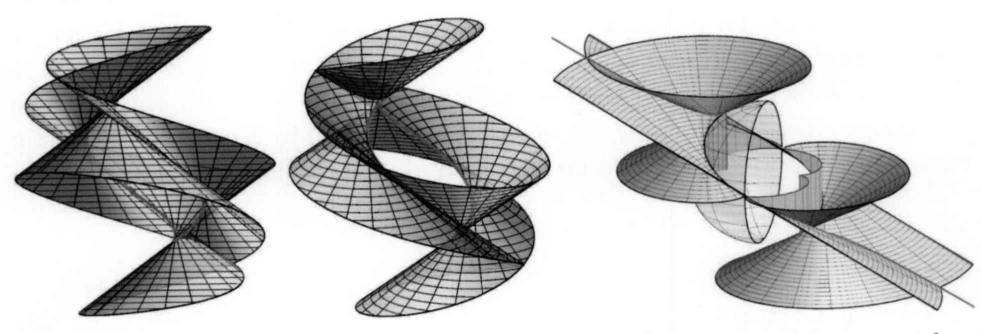

Fig. 7.2 Tangent surface of a spatial curve. On the right: a different section of the same surface; the edge of regression proceeds on an ellipsoid, and its top view is a "nephroid."

Finally, the developables (or torses) – the most general simply curved (and thus developable) surfaces – are also ruled surfaces because they carry the tangents of a curve. The associated motion of the straight line can, without proof, be interpreted as follows: A developable may be created by following a point K along a *directrix* k and considering the straight line as the tangent of k. In such cases, the directrix is called the *curve of regression* or *edge of*

regression, because it can be shown that along its trajectory, the generated tangent surface possesses a sharp ridge (Fig. 7.2).
Thus, we can say:

All developable surfaces (cylinders, cones, and developables) are ruled surfaces.

Now, on to the general ruled surfaces, whose shapes permit many (though not quite arbitrary) variations. In each case, an arbitrary tangential section through a surface point P will retain the direction of the *generatrix* which passes through it. Therefore, we can say:

All non-developable ruled surfaces are hyperbolically curved and, thus, constantly "saddle-shaped".

Conoids

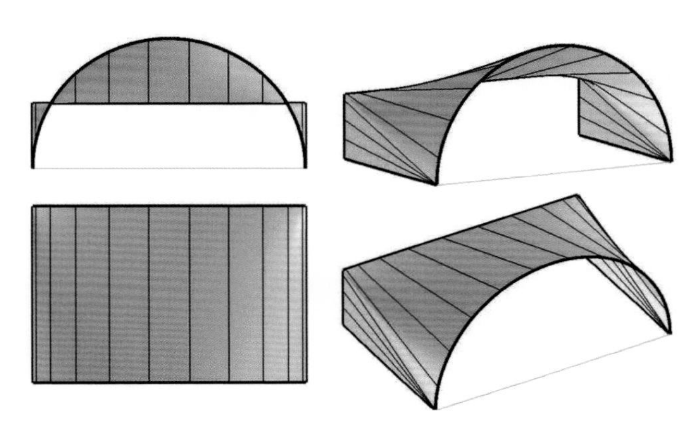

Fig. 7.3 Conoidal ruled surfaces as the shapes of roofs (circle conoid)

The so-called *conoids* represent an interesting category of non-developable ruled surfaces. The motion of their generating straight line is defined in a way that the straight line always intersects a straight *directrix* and another arbitrary given curve. As an additional condition, the surface must possess a *directing plane* which is parallel to all generatrices. Fig. 7.3 shows a *circle conoid* where the directrix is a half-circle.
Fig. 7.4 shows a conoidal roof shape, which is basically defined such that a horizontal directrix is connected to a wave-shaped curve by straight lines which appear parallel in top view (compare also Fig. 7.5). The question is whether or not the resulting ruled surface is developable, which would signify advantages for its practical construction.
This question cannot be answered by mere evaluation of the visual image since the recognizable contours are too short to discern any curvature.
On the other hand, it is theoretically possible to negate the question outright, considering that:

A ruled surface that contains a straight transversal of all generatrices is not developable. A conoid, in particular, is not developable.

Fig. 7.4 Arbitrary conoidal ruled surfaces as roof shapes. Pictured on the right is a model by Antoni Gaudi for practical construction.

Fig. 7.5 Model of a conoidal surface

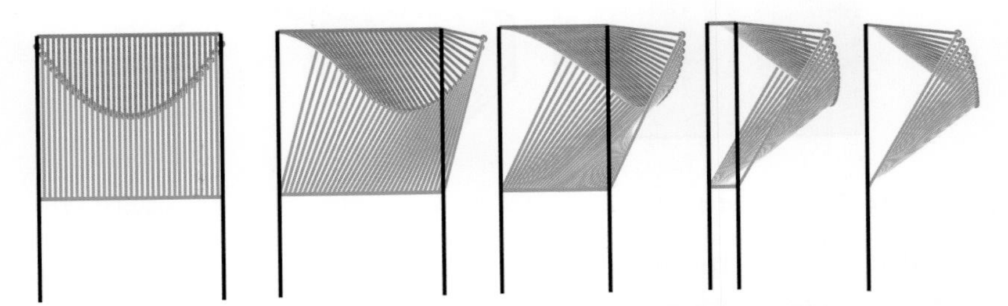

Fig. 7.6 Conoidal ruled surfaces being designed (Santiago *Calatrava*)

If P is a point on the directrix, then the tangential section in P is composed of the generating straight line and the directrix. Thus, the point is hyperbolic. Fig. 7.6 shows a similar problem. It again concerns two conoids with a horizontal directrix which share a spatial curve (from a certain perspective, it appears – as in Fig. 7.6 on the left – as a parabola). Due to a twisting of the surface, a slight curvature in the contours can be recognized, from which it can be deduced that the surfaces are "almost" but not "wholly" developable. In practice, this means that a slightly ductile surface – such as an elastic tent fabric – should be able to cope with the curvature.
Staying on topic: Both surfaces in Fig. 7.6 can be "animated" as in Fig. 7.7. Proceeding from a parabola, which determines the length of the rods in the initial surface, the lower "crossbar is positioned higher". Appropriate hinges, which prevent the rods from tilting sideways, have to be placed at the buckling spots.

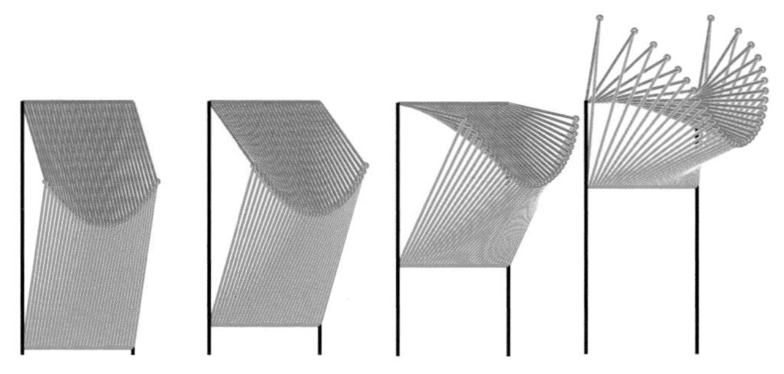

Fig. 7.7 ...or as a fancy sliding gate

Two important and two beautiful applications

There exist two conoids which carry a special name in academic literature.

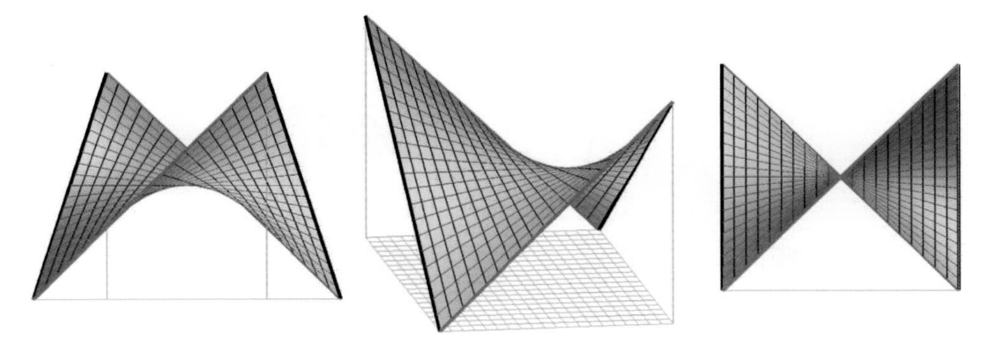

Fig. 7.8 The HP-surface is a conoid, a surface of translation, and a second-order surface.

Fig. 7.9 A cube intersected by a hyperbolic paraboloid seen from three different viewpoints (Palais de la Justice de Montreal)

The *HP-shell* (Fig. 7.8) is a particularly important conoid, and its second directrix is also a straight line. From this it follows, as we have already seen that the surface possesses an infinite number of directrices. Any two straight lines from one family form a *skew generatrix quadrangle* together with any two straight lines from the other family. From a special perspective, both pencils of generatrices appear as pencils of parallels, otherwise the contour is always a parabola.

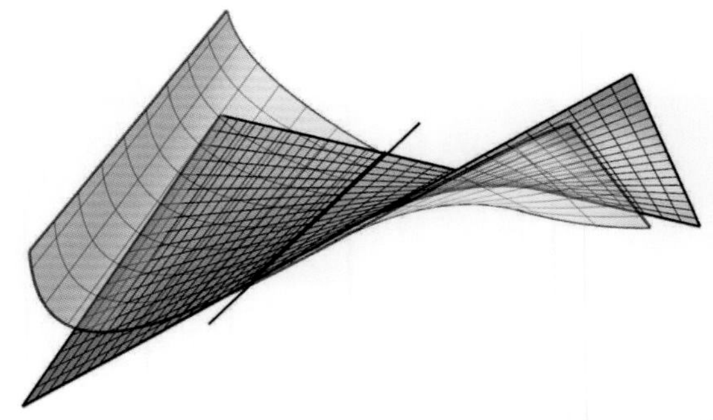

Fig. 7.10 Each ruled surface is being touched by an HP-shell along a whole straight line.

HP-shells are not just extremely interesting surfaces. They are also very practical, because locally, they can be considered as the prototype for any arbitrary ruled surface. Fig. 7.10 illustrates how a ruled surface is being touched by an HP-shell along a whole straight line.

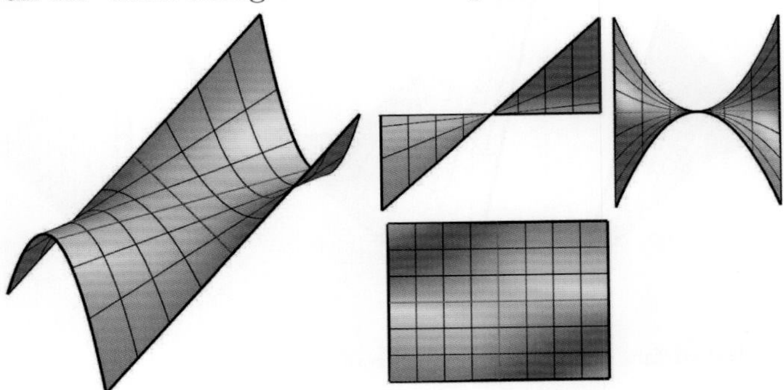

Fig. 7.11 Third-order conoid with a parabola as a directrix

A very visually appealing surface can be created by defining the *directrix as a parabola.*

Such a surface is of third order, and its prototype can be given by the simple equation $z = x^2y$. The prototype of an HP-surface can be given by the almost identical equation $z = xy$. Furthermore, arbitrary affinities can be applied to these prototypes. Due to the fact that affinities preserve parallelism, the surfaces stay conoidal, and their order remains likewise unaffected.

Beside the HP-shell, the *right helicoid* is also of great significance. It is created by rotating the generatrix about an orthogonally intersecting axis and translating it in proportion to the angle of rotation along this axis. The axis is the directrix, and all generatrices are parallel to the axis' normal plane. Each path curve of a point on the generatrix could serve as the second directrix. These path curves are *helices*. The right helicoid also belongs to the *helical surfaces*, and we shall examine it more closely in its respective chapter.

The right helicoid does not possess an "order" in the algebraic sense – it is instead called *transcendental*. It is, therefore, impossible to correlate the coordinates x, y, z of its points in an algebraic relationship. However, it is possible to formulate a

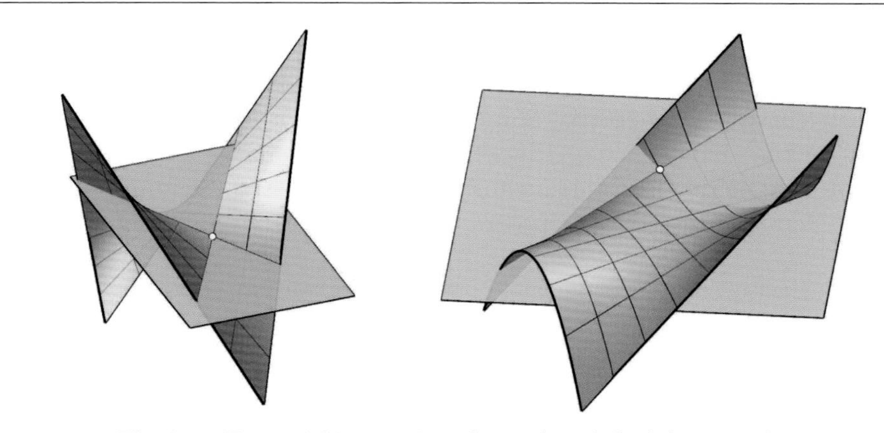

Fig. 7.12 Tangential intersection of second- and third-degree surfaces

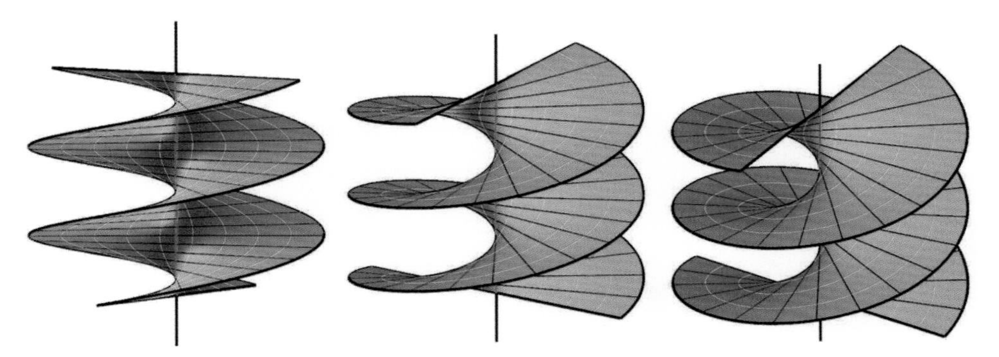

Fig. 7.13 The right helicoid is a conoid and a helical surface.

parameter representation or implicit representation as follows:

$$x = v \cdot \cos u,\ y = v \cdot \sin u,\ z = c \cdot u \quad (u \in \mathbb{R},\ \ v \in \mathbb{R}) \Rightarrow \frac{x}{y} = \tan \frac{z}{c},$$

where u is the angle of rotation about the z-axis and v the distance from this axis. Depending on the choice of parameter intervals, we get different parts of the surface.

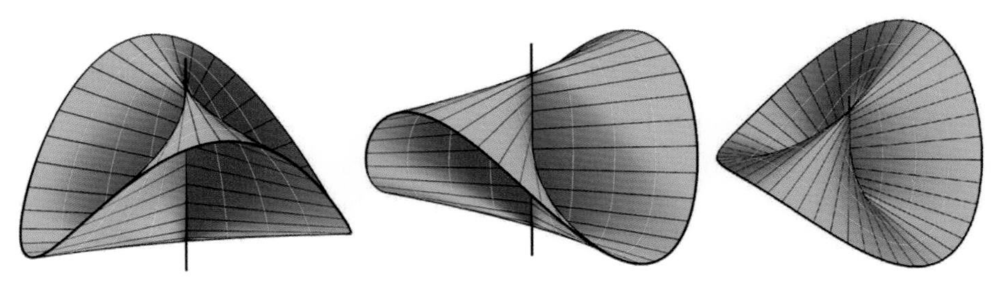

Fig. 7.14 Plücker's conoid

Our fourth and last conoidal example is the so-called *Plücker's conoid.* It can be generated in a similar way as the right helicoid – the only difference being that the translatory motion of the generatrix in the direction of the

directrix proceeds not in proportion to the angle of rotation, but to its sine. This produces the straight line's *harmonic oscillation.*
The path curves of generatrix points are *oscillation curves.* Despite its similarity to the right helicoid, Plücker's conoid is remarkably algebraic – of third order, to be precise. This means that each test straight line yields three exactly calculable intersection points, two of which may be complex conjugates.
Due to its generation, *Plücker*'s conoid possesses the parameter representation

$$x = v \cdot \cos u, \; y = v \cdot \sin u, \; z = c \cdot \sin 2u \quad (u \in \mathbb{R}, \; v \in \mathbb{R}).$$

It is now possible to eliminate the parameters due to $\sin 2u = 2 \sin u \cdot \cos u$ and $\sin^2 u + \cos^2 u = 1$, resulting in the implicit equation of the surface which reads

$$c \cdot x \cdot y = 2(x^2 + y^2) \cdot z,$$

from which it can be seen that Plücker's conoid is of third order.

Fig. 7.15 Ruled surfaces during the construction of bridges

Architecture often employs "wireframe models" of ruled surfaces. Fig. 7.15 shows a pedestrian bridge by *Calatrava* in Bilbao.

7.2 Helical surfaces

The fascination behind the helix

In planar geometry, there exists exactly one curve that can "cope" with an infinite number of reflections and rotations – the circle. This undoubtedly accounts in no small part for its beauty and practical applicability.
In space, there exists a class of curves that may allow self-motions – the helices.

A helical motion is defined as the combination of a rotation about a fixed axis and a simultaneous *proportional* translation along this axis. The *pitch* belongs to a full revolution as the distance of translation.

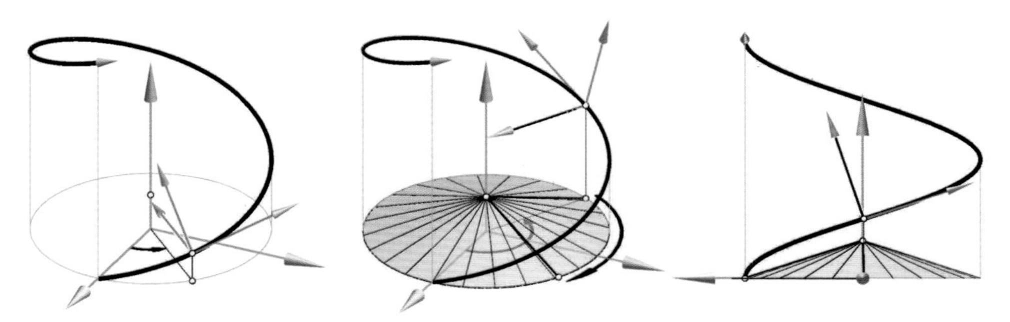

Fig. 7.16 Helices with an "accompanying tripod"

Helices are the path curves of points undergoing a helical motion. During a helical motion, they are being continuously translated into each other. They can additionally cope with an infinite number of point reflections (the center must lie on the axis) and translations in the direction of the axis if the distance of translation is an integer multiple of the pitch.
The factor of proportion is called the *screw parameter* and is henceforth specified as p. If a point is twisted by an angle φ about the screw axis, then it is simultaneously translated by an offset $p \cdot \varphi$ along this axis. If h is the pitch, then $p = \frac{h}{2\pi}$.
If the screw parameter is positive, one speaks of right helical motion. If it is negative, we speak of a left helical motion. The case $p = 0$ is special and yields a pure rotation, whereas $p = \infty$ yields a pure translation.
The rotation and the associated translation must not necessarily be regular, but what matters in any case is the proportionality. As a plant coils itself around a pole (Fig. 7.17), its speed of growth may vary, but the result is, nevertheless, the same. A mathematician would say that the helix need not be parametrised by time.

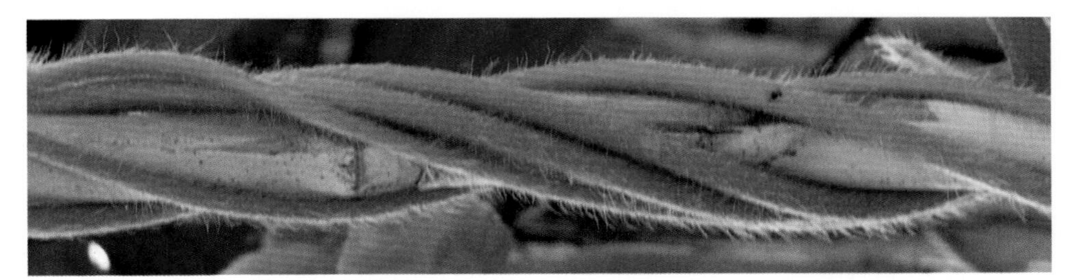

Fig. 7.17 Normal projections (orthogonal to the axis) of helices

The following properties are hard to improve in terms of elegance:

Helices always lie on cylinders of revolution about the screw axis a. They have constant slope with respect to a base plane $\pi \perp a$. In an orthogonal projection on π, a helix appears as a *circle*. In an orthogonal projection on a plane $\nu \perp \pi$ ($\nu \parallel a$), a helix appears as a *sine curve*. The development of the carrier cylinder shows a straight line as the development of the helix.

Proof:
Neither a translation in the direction of the axis nor a rotation about the axis change the distance from it. Thus, helices lie on cylinders of revolution, which appear as circles in a normal projection in the direction of the axis.
Owing to the definition of the helix, it intersects each generatrix of the carrier cylinder at the same angle. The ratio between the translation component and the rotation component is likewise constant – even if only an infinitesimally tiny segment (the tangent t) of the curve is being considered. The tangent is contained in the tangent plane τ of the cylinder, where the angles to the generatrix and the base plane π can be measured. This angle $\alpha = \angle t\pi$ has the value $\tan\alpha = p/r$, where r denotes the distance of P to the screw axis.
If t is projected orthogonally to a plane $\nu \perp \pi$, then the translation component remains undistorted while the rotation component is shortened depending on the angle $\psi = \angle\tau\nu$ by a factor of $\cos\psi$. The projection curve is, thus, a cosine – and, therefore, also a sine. Both curves do not differ in their shape but are merely phase-shifted. Since the helix intersects all generatrices of the carrier cylinder at the same angle, the development also intersects the parallel straight developments of the cylinder's generators at a constant angel and is, therefore, also a straight line.
Let us now use analytic methods in order to capture the path of a point P at a distance of r to the axis as a way of proving the above theorem.

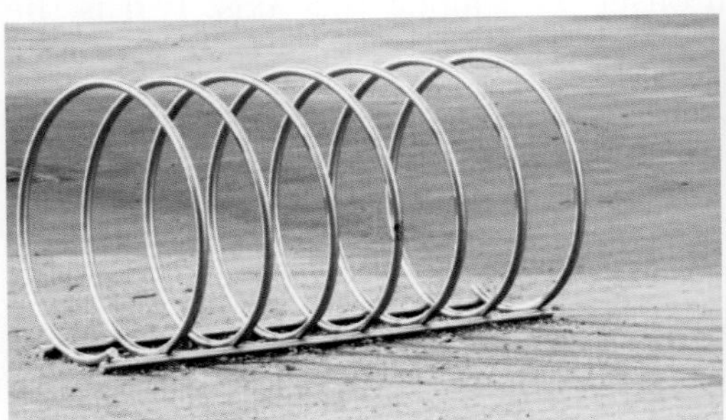

Fig. 7.18 Helix with a shadow

For this purpose, we shall adapt a Cartesian coordinate system to the situation as simple as possible: Let the axis be the z-axis and $P(r/0/0)$ lie on the x-axis. After the twisting through φ and the simultaneous translation about $p \cdot \varphi$, P has the coordinates

$$x = r \cdot \cos\varphi,\ y = r \cdot \sin\varphi,\ z = p \cdot \varphi.$$

The front view of the path curve can be represented by $y = r \cdot \sin\varphi,\ z = p \cdot \varphi$, and is, thus, a sine curve with the equation of $y = r \cdot \sin(z/p)$. Every other horizontal normal projection with viewing angle α also yields a congruent (and merely phase-shifted) curve in a (u, z) coordinate system:

$$u = x \cdot \cos\alpha + y \cdot \sin\alpha = r \cdot \cos\varphi \cdot \cos\alpha + r \cdot \sin\varphi \cdot \sin\alpha = r \cdot \cos(\varphi - \alpha),\ z = p \cdot \varphi.$$

The development of the curve is given in a (v, z)-system by the linear equation

$$v = r \cdot \varphi,\ z = p \cdot \varphi \Rightarrow z = p/r \cdot v,$$

which implies that the curve is a straight line. ◇

Arbitrary orthogonal views of helices (Fig. 7.18) stand – as we shall soon discover – in an affine relation to so-called *cycloids*. Central views as in Fig. 7.18 on the right have an appropriately complicated shape and pose a challenge to freehand drawing.

Fig. 7.19 On the left: The right helicoid alone may help to overcome the tedium of this car park. Center: Right helicoid in the Sagrada Familia (Antoni Gaudi). On the right: Intercity bank in Lima (Hans Hollein).

When straight lines or circles are undergoing a helical motion ...

All path curves during a helical motion are helices, except the paths of points on the axis. Which surfaces are produced if, instead of points, we let other objects – such as straight lines – undergo the same motions?

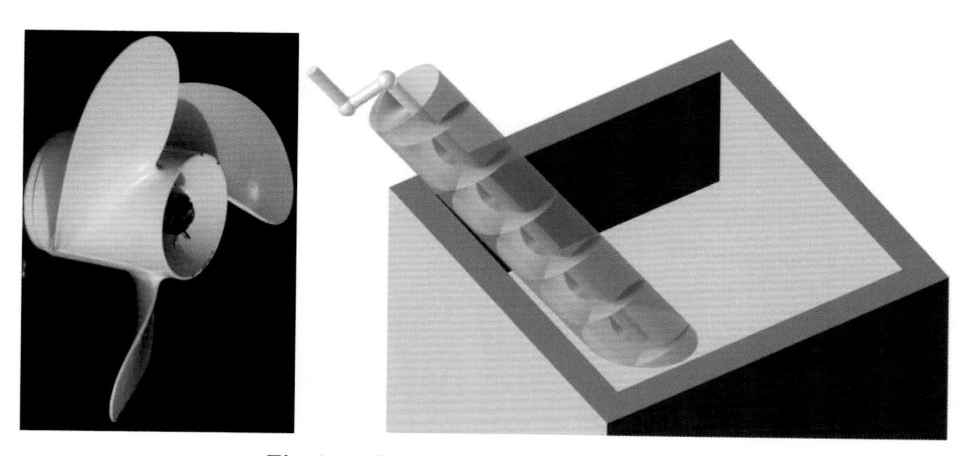

Fig. 7.20 Ship's propeller, Archimedean screw

When discussing the ruled surfaces, we have already touched upon an equally important helical surface – the right helicoid (Fig. 7.13). It is produced when a straight line which intersects the screw axis at a right angle undergoes a helical motion.

Right helicoids may, above all, be encountered in technology. For instance, the smoothly cleaned wall on the underside of right helicoids or the upwards leading driving path of the car park in Fig. 7.19 both belong to this class. Right helicoids are equally employed for the surfaces of propellers or cooling fans.

Using right helicoids, it is possible to induce propulsion by means of mere rotation – the screw motion is, thus, decomposed into its components of rotation and translation. Due to the inertia of the medium through which the motion should proceed (air or water), the surface "supports" itself on the medium and propels the airplane or ship forwards (or backwards, as is possible with a ship's propellers and pictured in Fig. 7.20). It was rightly recognized and utilized by *Archimedes* that right helicoids can be employed to lift liquids (see Fig. 7.20 center and on the right).

Fig. 7.21 Impeller of a snow blower

The *impeller* (Fig. 7.21) which is, for instance, built into excavator shovels (and turned electrically) or snow shovels is a further interesting application of a right helicoid. By mere rotation of the apparatus, the mixture of sand and cement in the left part of the left-handed screw is pressed into one direction, and in the right-handed screw into the opposite direction. Variations exist with a continuous outer screw and a counter-rotating inner screw.

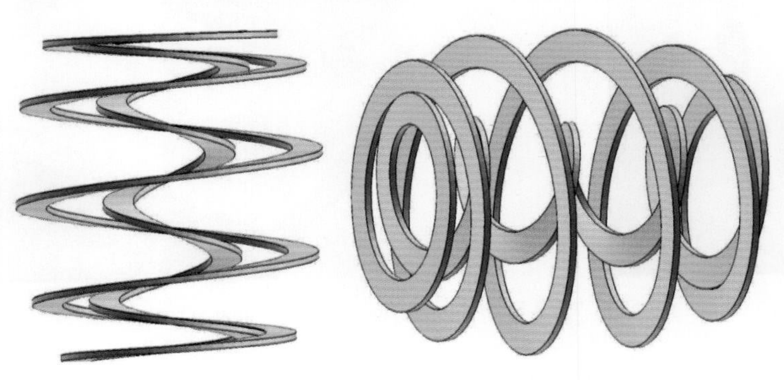

Fig. 7.22 Theoretical version of an impeller

Fig. 7.22 shows a "purely geometric" variation of a double screw, where the transition from one helix to the other proceeds "smoothly" (or differentiable).

• The DNA double helix

The *double helix* is something of a symbol of molecular biology. The hereditary material is preserved in the form of deoxyribonucleic acid (DNA), which persists as two strands (or a double helix) of molecules in the chromosomes of the cell nucleus. These DNA stands wrap themselves – similar to opposite helices on the right helicoid – about an imagined axis. More on this matter can be found in application p. 255. ♠

Fig. 7.23 Vices

The conversion of rotation into translation does not necessarily have to produce propulsion. Each olive press, each vice, and each corkscrew take advantage of the same principle. On a miniature scale, the conversion of rotation into translation is done with each ordinary screw!

Fig. 7.24 Bolts and nuts

The question arises: What kind of helical surfaces are employed in screw threads? They do not seem to be right helicoids – as, for instance, in vices. The surfaces are created by straight lines oblique to the screw axis. Let us consider the matter from a geometric standpoint.

Different types through different directions of the straight line

If the generating straight line of the right helicoid is translated in parallel so that it no longer intersects the screw axis, so-called *straight open ruled helical surfaces* can be obtained. Fig. 7.26 shows two such surfaces, where the left surface possesses a negative screw parameter and the surface at the center possesses a positive one. The helix with minimal axis distance can be seen in the related top view.

The general case is given if the generating straight line is oblique with respect to the screw axis. Both rightmost *oblique open* ruled helical surfaces

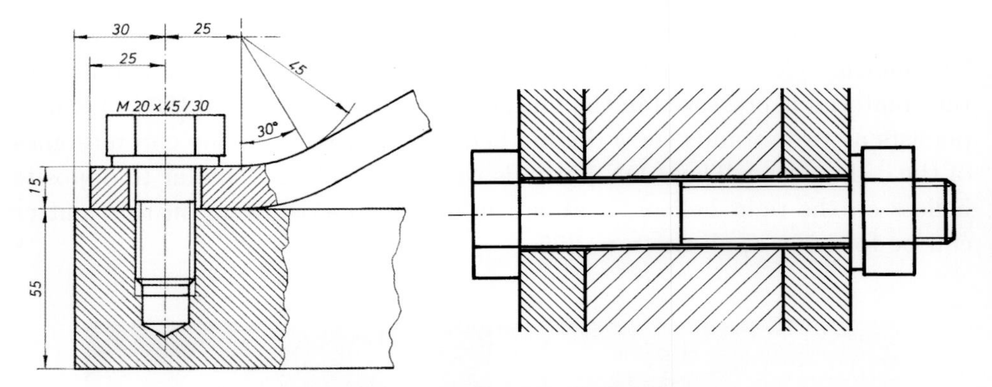

Fig. 7.25 Blind holes and clearance holes

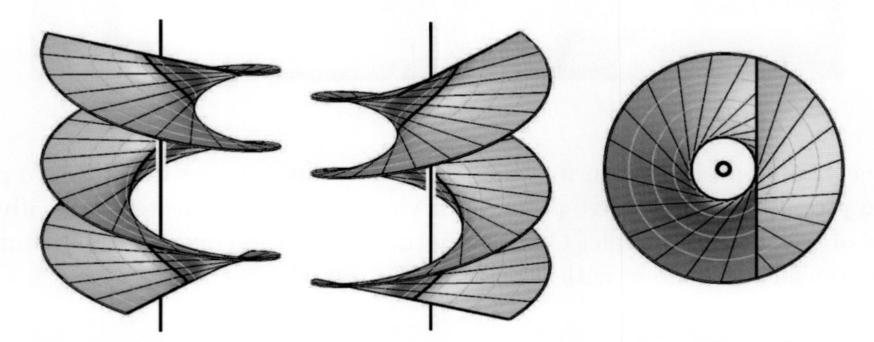

Fig. 7.26 Left-handed and right-handed straight ruled helical surfaces

in Fig. 7.27 serve as illustrations. If the oblique generatrix meets the screw axis, then we speak of an *oblique closed* ruled helical surface.
Oblique skew ruled surfaces possess a self-intersection, which is naturally also a helix.

There exists only one type of developable helical surface

None of the aforementioned ruled helical surfaces are developable in the general case, and this is recognizable from their curved contours. The surface in Fig. 7.27 on the right, however, seems to possess straight contour parts. It is, in fact, possible for an oblique open ruled helical surface to be developable under a single condition: The generatrices have to be tangents of a helix g. The whole surface is then a tangent surface (developable) of g, and g is a sharp arris (edge of regression) on this developable. Fig. 7.28 shows another helical developable – in the right image, another part of the surface is visible (every ruled surface is, in principle, unbounded).
Fig. 7.29 illustrates one further exception for ruled helical surfaces: If the twisted straight line is parallel to the screw axis, a cylinder of revolution is created. Following this interpretation, a cylinder of revolution can be seen as both a surface of revolution as well as a helical surface.
Helical developables are often used in engineering practice. They occur in mechanical engineering as gear flanks for so-called *involute gears* or mills

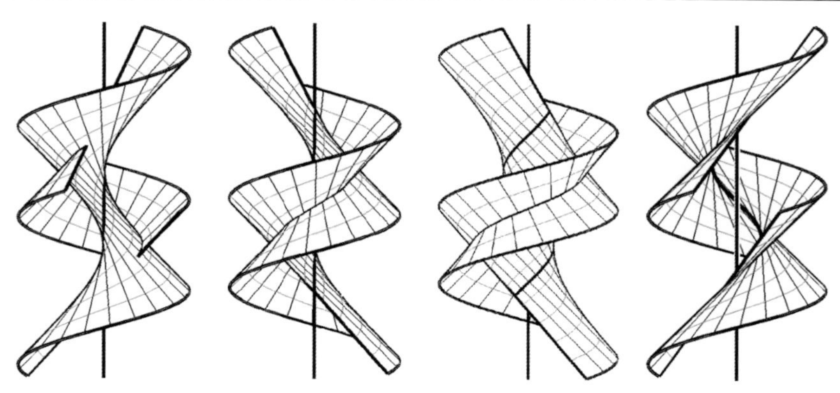

Fig. 7.27 Different oblique ruled helical surfaces

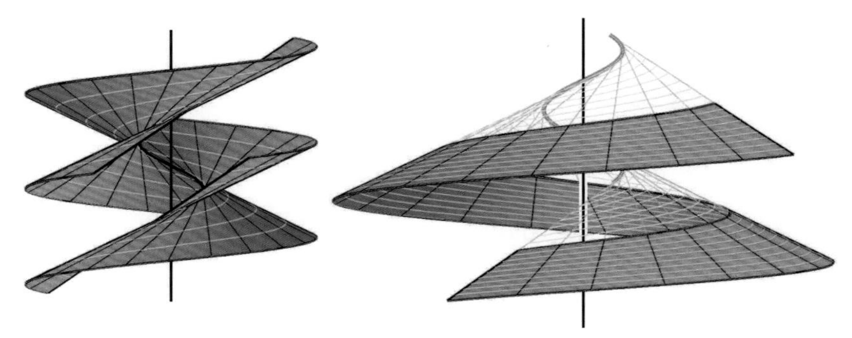

Fig. 7.28 Helical developable with a helix for its edge of regression

(Fig. 7.31), which serve to transfer rotations onto oblique axes. During road construction, they form the embankment for highway entrances (Fig. 7.32 and 7.33).

It is, of course, possible to apply screw motions to circles or spheres

By twisting curves other than straight lines, complicated surfaces may be produced. The most common case that occurs in nature is the application of screw motion to a circle. Even by restricting oneself to these *circular helical*

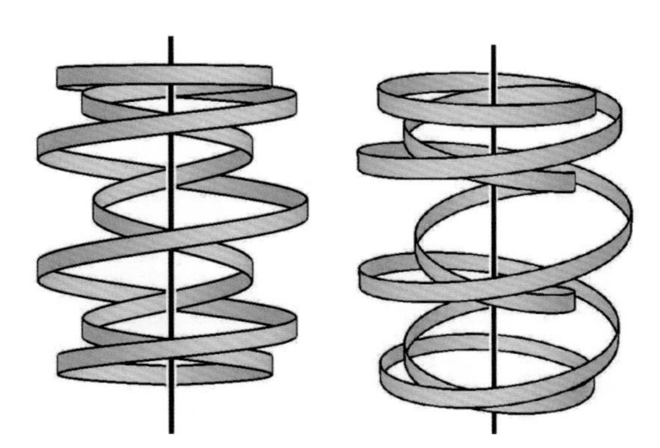

Fig. 7.29 A developable special case of a ruled helical surface (cylinder of revolution)

Fig. 7.30 Cone wheels with efficient gear flanks

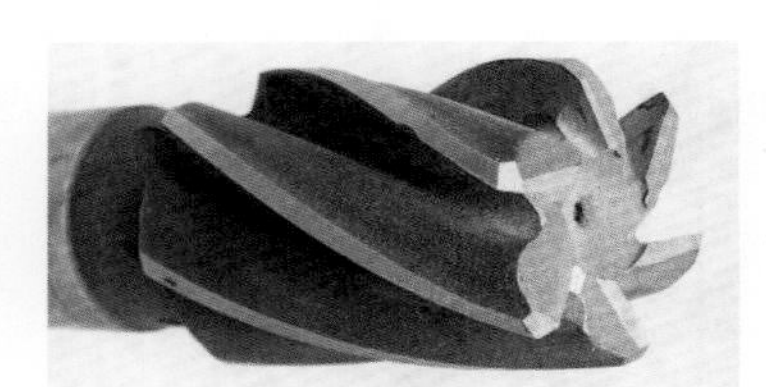

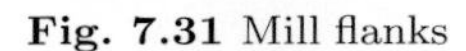

Fig. 7.31 Mill flanks

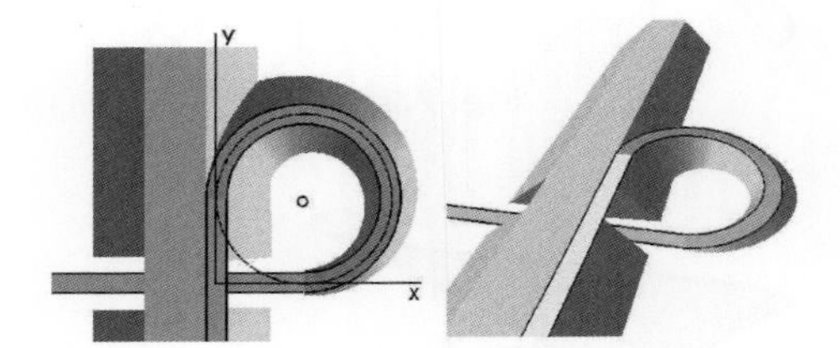

Fig. 7.32 Highway entrance in theory ...

surfaces, it is possible to distinguish between several types that differ in the initial position of the circle.

Fig. 7.35, for instance, shows a circular helical surface. The twisted circle is the "profile" of the surface in a cross section perpendicular to the screw axis.

By far the most frequent type of circular helical surface is the *helical pipe surface.*

A helical pipe surface is formed if a circle undergoes a helical motion such that its axis is tangent to the helix that is the path of the circle's center. It can also be created as the envelope of a sphere which counts the circle among the many possible great-circles.

Proof:

The circle undergoes a motion which, by definition, is orthogonal to its plane. Thus, the surface is touched by a cylinder of revolution along the circle, and a sphere fits into this cylinder which is in contact along the original circle. The generation of the helical surface by a sphere constitutes the significance of the surface. ◇

Fig. 7.33 ... and practice (Euler spirals occur in top view)

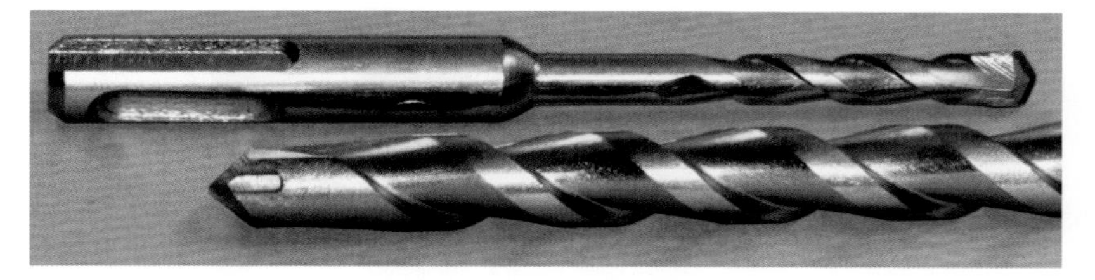

Fig. 7.34 Masonry drill with helical developables as flanks

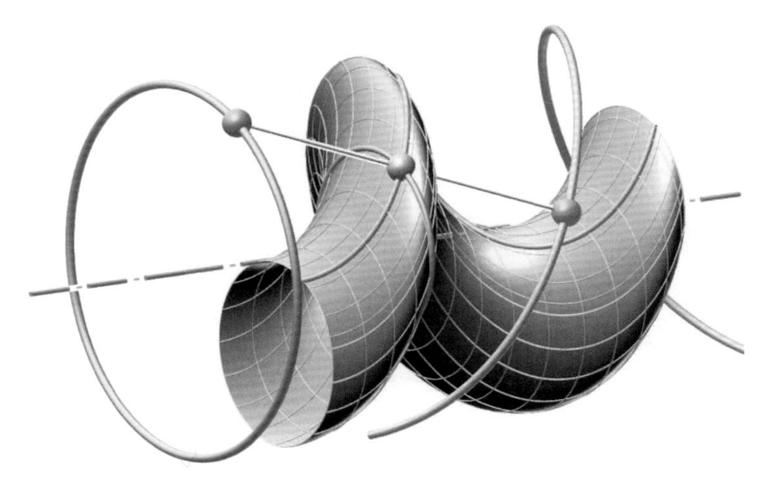

Fig. 7.35 Special circular helical surface which can also be generated as "midsurface" of a circle and a helix (compare section about translation surfaces)

Therefore, it is possible for a sphere to roll through the surface, while touching the surface along a great-circle in every moment. Many water slides in swimming pools are comprised of parts of helical pipe surfaces (Fig. 7.37 on the left). Another variation where the circle in a meridian plane is twisted can also be encountered (visible in Fig. 7.37 at the center). The thus generated surface is called the *meridian circular helical surface.*

Fig. 7.36 shows different normal views of the helical pipe surface (on the left, the projection is perpendicular to the axis, on the right we have a projection in the direction of the axis). It is clearly visible that the generating circle is not vertical,

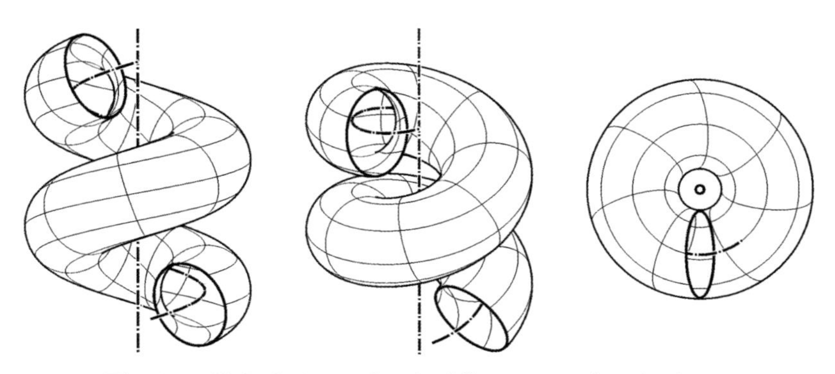

Fig. 7.36 Helical pipe surface in different normal projections

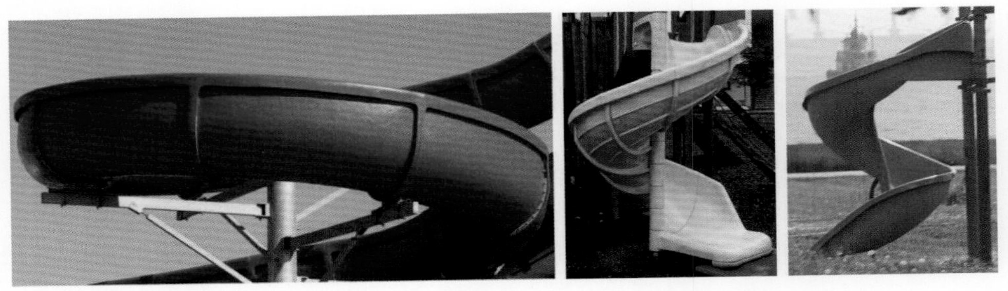

Fig. 7.37 Helical pipe surface and meridian circle helical surface as slides

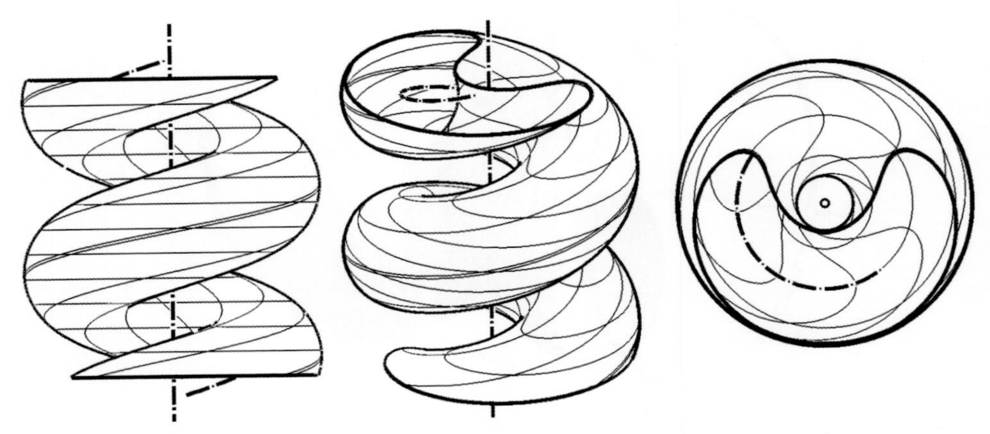

Fig. 7.38 Helical pipe surface, this time generated by the cross section

but instead has the respective tangent of the mid helix as its axis. The contours often exhibit cusps, depending on the radius of the twisted circle.

Fig. 7.38 shows the same helical pipe surface but generated by a kidney-shaped cross-section instead of a circle. Points of this cross-section can be created by proceeding from the originally twisted circle and applying the helical motion to its points P until they "reach" a cross-section plane π that is normal to the axis. The height difference $\Delta z = \overline{P\pi}$ corresponds to a proportional angle of rotation $\varphi = c{\cdot}\Delta z$.

Fig. 7.39 Steel cable (two helical pipe surfaces) under small workload

Helical pipe surfaces are among the most frequently encountered surfaces in nature and technology. The "helices" in Fig. 7.17 and Fig. 7.18 should, precisely speaking, also be counted among them.

- The secret of DNA (deoxyribonucleic acid)

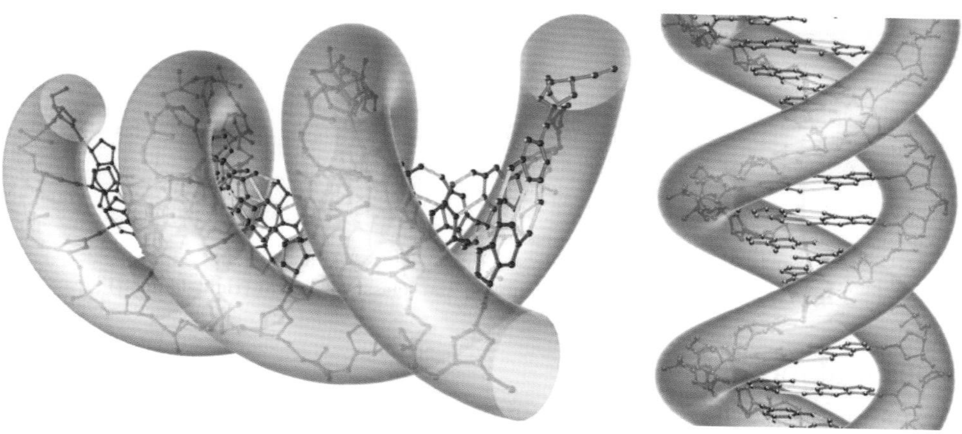

Fig. 7.40 Deoxyribonucleic acid encodes hereditary information and passes it on to the following generation via germ cells. The strands may be visualized from a geometric standpoint by two interlocked helical pipe surfaces.

It would be presumptuous even to attempt to describe the manifold secrets of DNA at this point, but as for geometry, this much can be said: As described in application p. 249, DNA molecules take the form of the famous double helix. Each strand represents a sequence of molecular building blocks (phosphoric acid, deoxyribose, and one of the four bases – adenine, guanine, cytosine, and thymine). The order of these bases determines the genetic code. ♠

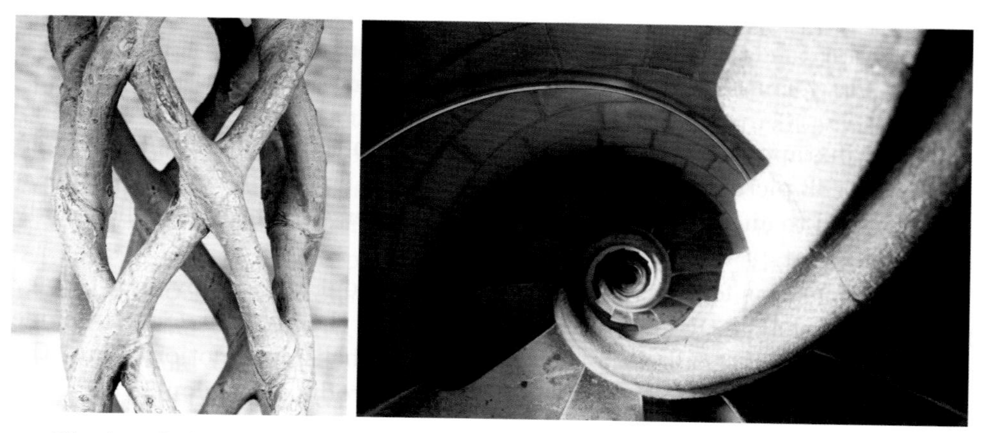

Fig. 7.41 Left: a strange kind of "double helix", right: helical stairs, looking downwards

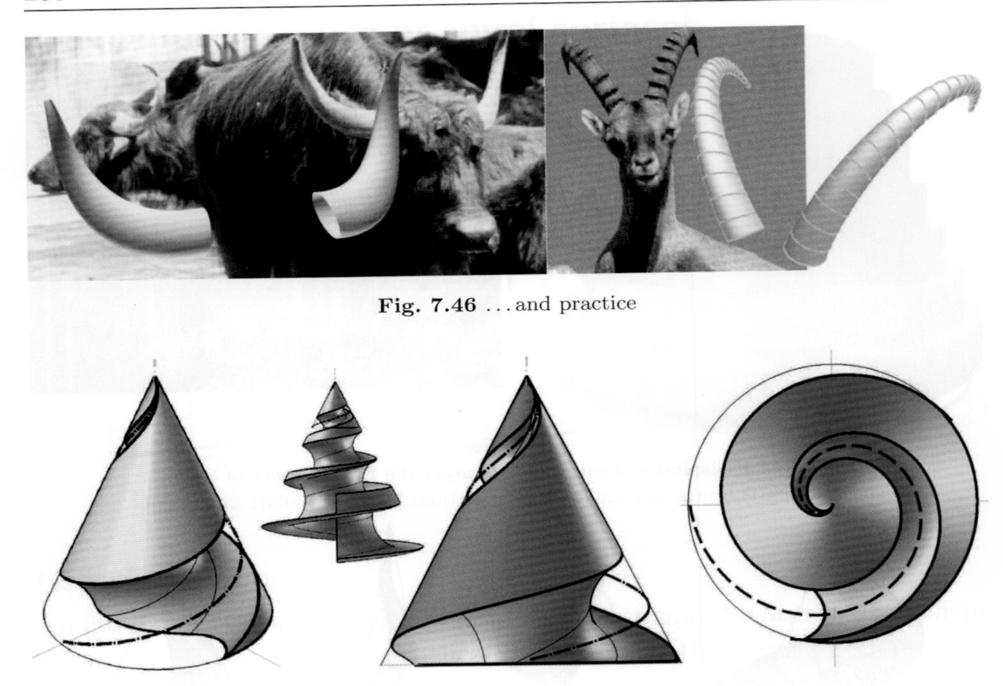

Fig. 7.46 . . . and practice

Fig. 7.47 Helispiral-shaped recess (Anita Aigner)

The classical spiral motion

Other than helispiral motion, there exist numerous other types of spatial spiral motions which all share the same principle: It always involves the composition of a rotation about an axis and a somehow proportional change of distance to a fixed axis point. The respective planar spiral motion proceeds by complete analogy.

The principle behind the classical (so-called *cylindroconical*) spiral motion is as follows: Perform a rotation about an axis by a very small angle φ and simultaneously increase the distance from the center by a value that is proportional to the size and rotation angle, and which also equals a multiplication by $(1 + p \cdot \varphi)$. The value p is called the parameter of the spiral motion. This

Fig. 7.48 The tangent surface of a helispiral (on the left and right) and a remarkable association (feather dust worm, Dalmatia).

Helical pipe surfaces are among the most frequently encountered surfaces in nature and technology. The "helices" in Fig. 7.17 and Fig. 7.18 should, precisely speaking, also be counted among them.

- The secret of DNA (deoxyribonucleic acid)

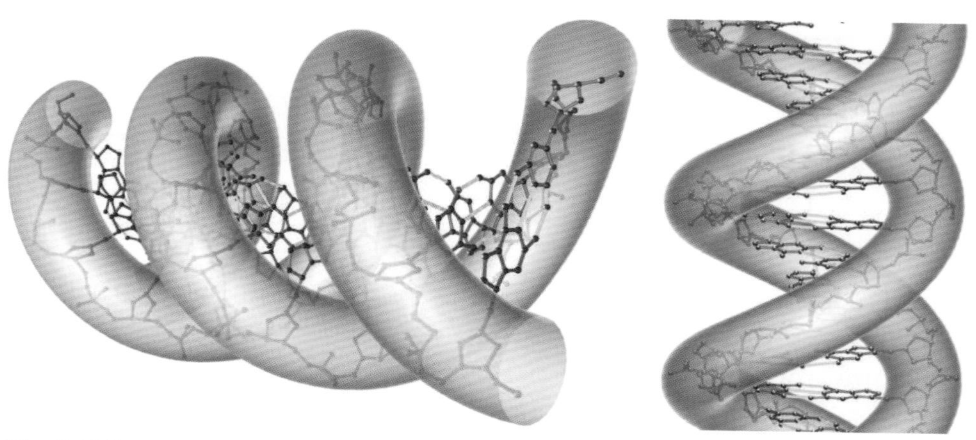

Fig. 7.40 Deoxyribonucleic acid encodes hereditary information and passes it on to the following generation via germ cells. The strands may be visualized from a geometric standpoint by two interlocked helical pipe surfaces.

It would be presumptuous even to attempt to describe the manifold secrets of DNA at this point, but as for geometry, this much can be said: As described in application p. 249, DNA molecules take the form of the famous double helix. Each strand represents a sequence of molecular building blocks (phosphoric acid, deoxyribose, and one of the four bases – adenine, guanine, cytosine, and thymine). The order of these bases determines the genetic code. ♠

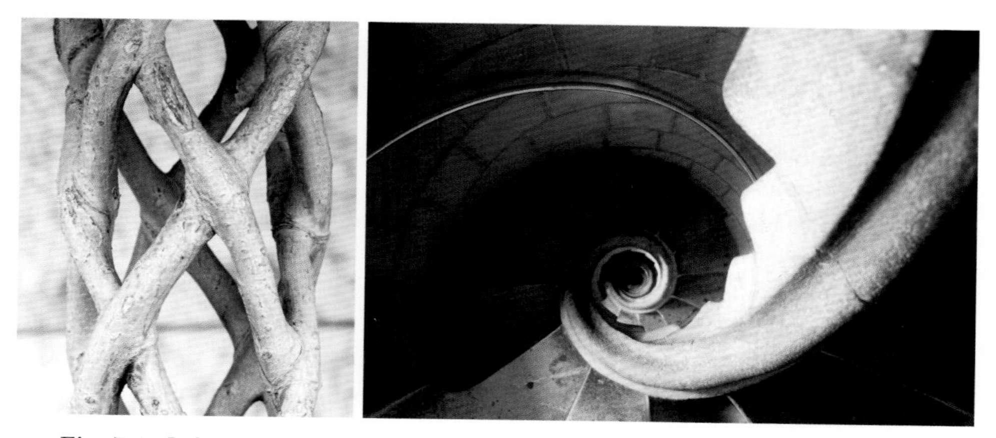

Fig. 7.41 Left: a strange kind of "double helix", right: helical stairs, looking downwards

7.3 Different types of spiral surfaces

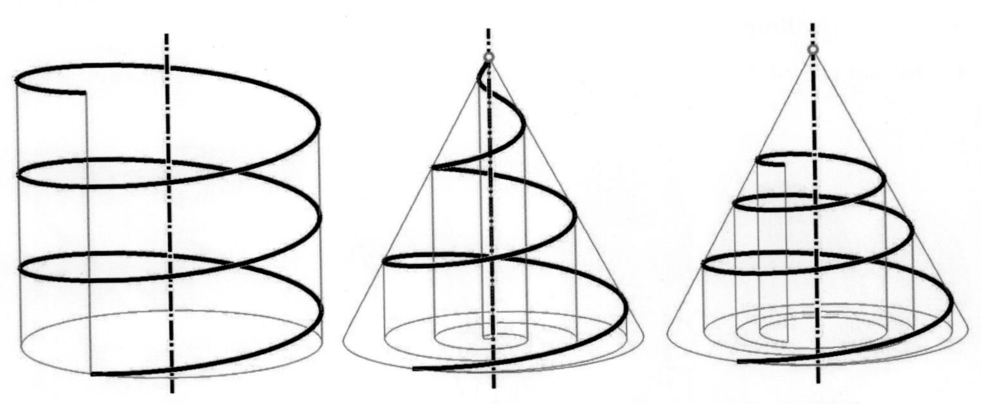

Fig. 7.42 Helix, helispiral motion, and the "classical" spiral (right-handed)

Helispiral motion

An interesting variant of helical motion is often found in nature – the helispiral motion. While the regular helical motion leaves the diameters unaffected, helispiral motion makes them smaller or larger *in linear proportion to the height.*

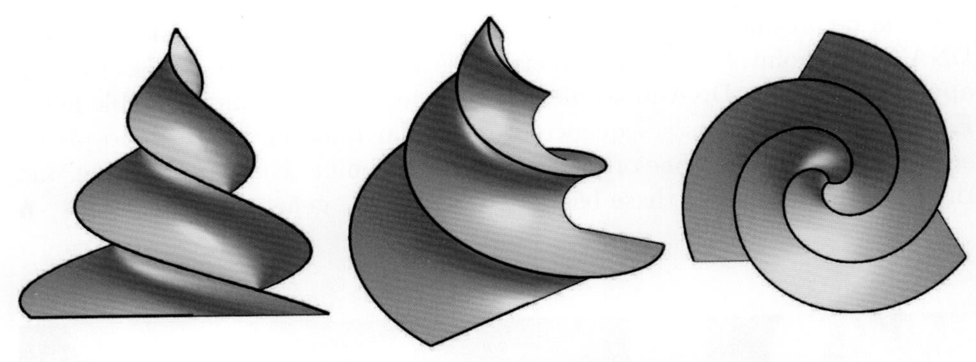

Fig. 7.43 Helispiral motion of a horizontal equilateral triangle (left-handed)

Fig. 7.42 juxtaposes the helical motion (left) with the helispiral motion (center). Fig. 7.43 pictures the surface that is created if this kind of transformation is applied to an equilateral triangle, which intersects the axis orthogonally at its midpoint.

The principle is as follows: Perform a rotation about the axis by an arbitrary angle φ and change the distance to the center by the factor $p \cdot \varphi$ at the same time. The value p is called the parameter of the helispiral motion. Thus, the growth in distance is linear, as the double angle produces the double distance.

A helispiral motion is given if the points in space are being rotated about an axis and simultaneously scaled proportionally in relation to a fixed point on the axis.

Fig. 7.44 Helispiral motion: Pieter *Bruegel* the Elder, Tower of Babel

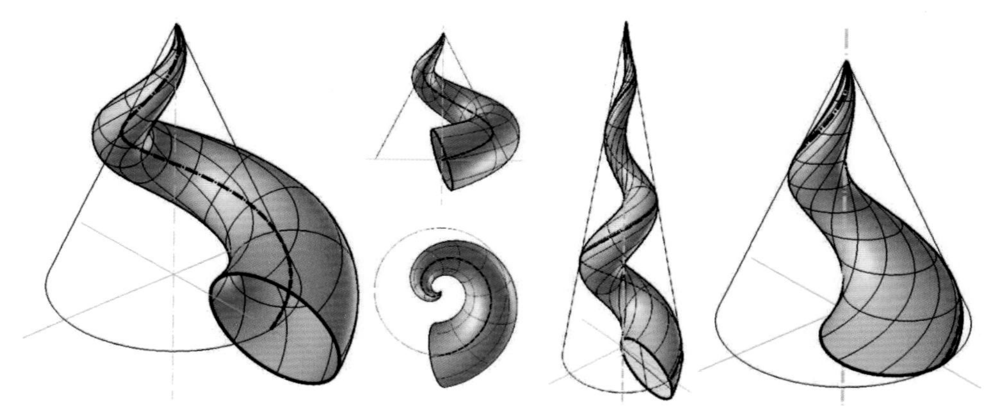

Fig. 7.45 Variation of circle helispiral surfaces in theory ...

This is analogous to the following "growth recipe" of nature: "Grow larger from a center at a linear speed and simultaneously rotate about an axis through the center."

This principle seems to be applied in the growth of animal horns, plant sprouts, and even certain trees.

Geometrically, this instruction indicates that, at every turn, all points change their distance to the axis by the same amount.

An exceptional point exists on the axis – the origin of growth – from which all path curves (called helispirals) make their way. Their top views are *Archimedean spirals.* These planar curves are created by letting a point move at equal velocity on a constantly rotating rod.

From a spatial perspective, a helispiral may also be created by letting a point move at equal velocity on a straight line, which itself is rotating about an obliquely intersecting axis. Another variation may be described as follows: If a point undergoing a helical motion is being moved at equal velocity on a normal to the screw axis, its path curve is a helispiral. During this type of motion, the point moves on a right helicoid.

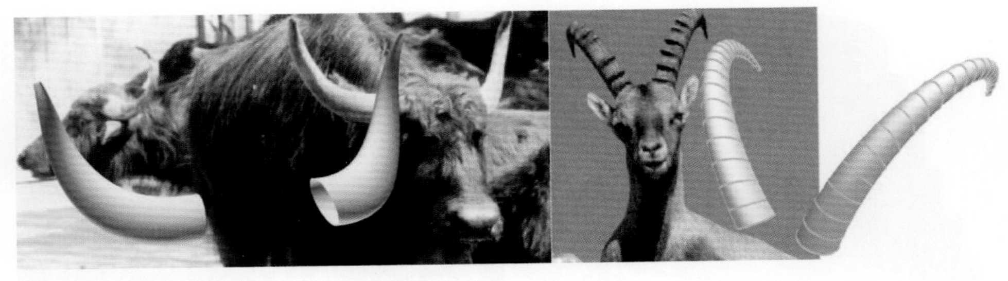

Fig. 7.46 . . . and practice

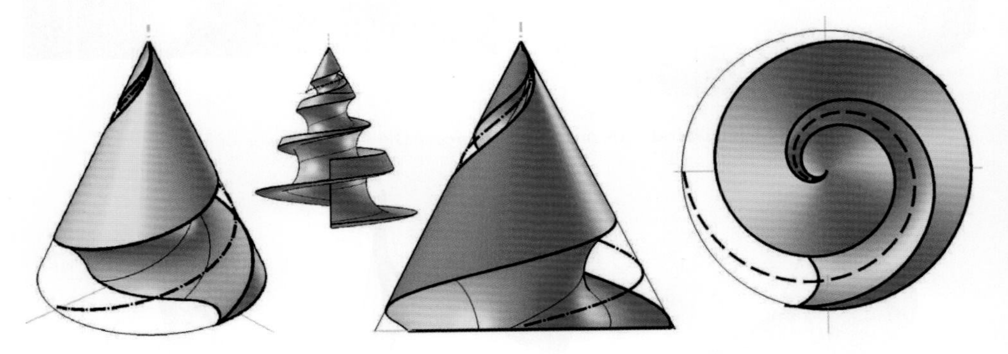

Fig. 7.47 Helispiral-shaped recess (Anita Aigner)

The classical spiral motion

Other than helispiral motion, there exist numerous other types of spatial spiral motions which all share the same principle: It always involves the composition of a rotation about an axis and a somehow proportional change of distance to a fixed axis point. The respective planar spiral motion proceeds by complete analogy.

The principle behind the classical (so-called *cylindroconical*) spiral motion is as follows: Perform a rotation about an axis by a very small angle φ and simultaneously increase the distance from the center by a value that is proportional to the size and rotation angle, and which also equals a multiplication by $(1 + p \cdot \varphi)$. The value p is called the parameter of the spiral motion. This

Fig. 7.48 The tangent surface of a helispiral (on the left and right) and a remarkable association (feather dust worm, Dalmatia).

Fig. 7.49 This spiral is "merely" a central projection of a helix.

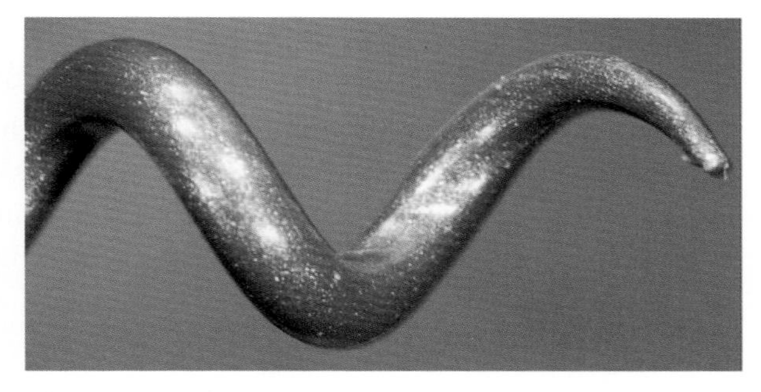

Fig. 7.50 In this corkscrew, the helical pipe surface transforms into a helispiral surface, which is no longer a pipe surface but a duct surface (S. 232).

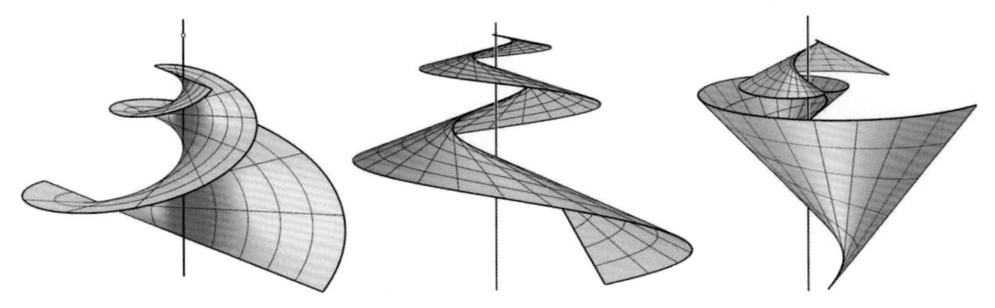

Fig. 7.51 Spiral right helicoid (on the left) and spiral developables

leads to a quasi "compound computation of interest" – for the "initial capital" grows bigger every time. Thus, the growth is exponential.
If we now allow for arbitrary rotation angles, we get the following definition:

A classical spiral motion is given if the points in space are rotated about an axis and simultaneously scaled from a fixed center on the axis in an exponential manner.

Fig. 7.52 Snail shells have a complicated interior.

This is equivalent to the following "growth recipe" in nature: "Grow by a certain percentage in relation to a center and simultaneously rotate in proportion to the percentage about an axis through this center." Such growth is obviously responsible for calcite shells of snails and shellfish, as these shells bear a remarkable resemblance to their computer simulations (Fig. 7.53). Studies by Boris Odehnal have shown that many such specimen are subject to irregularities (Fig. 6.87).

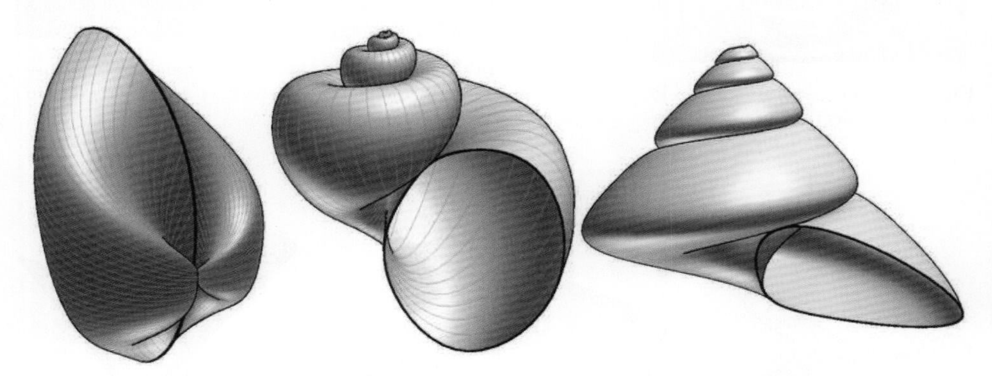

Fig. 7.53 Computer simulations of "real" shellfish and snail shells

• Reversely congruent "Christmas trees"

Christmas tree worms (*Spirobranchus giganteus*) are tiny creatures that live on corals. When disturbed, the worms quickly vanish into their holes. The

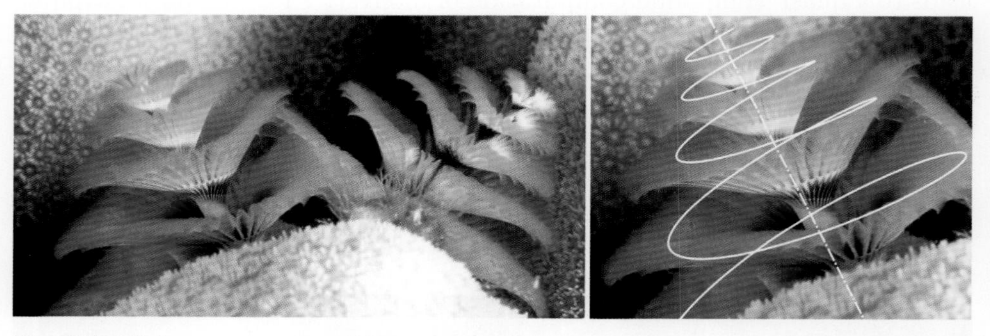

Fig. 7.54 Christmas tree worms: The multicolored spirals are merely the worm's highly derived respiratory structures.

spirals come close to "classic (exponential) spirals". It is interesting that the spiral parameters of two neighboring worms are often of different signs. ♠

The relation to the helix

Let us once again juxtapose the generation of the three aforementioned path curves (helix, helispiral, classical spiral), assuming the most equal initial conditions.

Let the axis a (e.g., the z-axis), a fixed angle of rotation $\Delta\varphi$, and a fixed distance of translation Δz (in axis direction) be given.

Let P be an arbitrary point to which we shall repeatedly apply the following instructions: First, P is to be rotated through $\Delta\varphi$ about a and then translated along a by Δz.

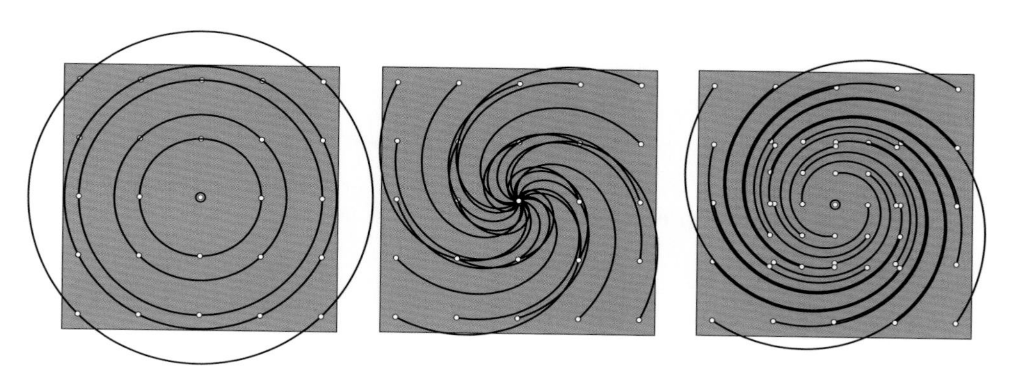

Fig. 7.55 Screw motion, helispiral motion, and classical spiral motion in top view

This already yields the points of a helix. For helispiral motion, P is additionally translated orthogonally to the axis direction by a distance proportional to the angle of rotation $s = p \cdot \Delta\varphi$, while for classical spiral motion it is changed in relation to the origin by a factor proportional to the angle of rotation $k = (1 + p \cdot \Delta\varphi)$. This, in turn, yields points on the helispiral and on the classical spiral.
If the objective is to calculate n times as many points, then the n-th part of Δz, $\Delta\varphi$, and the translation distance s are to be considered. The scaling factor of the classical spiral is to be replaced by its n-th root.
In contrast to helispirals, helices and classical spirals are *slope lines* – (lines with a constant angle to the axis or to the base plane) on cylinders with axis-parallel generatrices (in case of the helix, it is a cylinder of revolution. In case of the classical spiral it is a logarithmic cylinder).

7.4 Translation surfaces

Translation surfaces as locus of chord midpoints

Figure 7.56 illustrates the following theorem about a possible generation of translation surfaces:

If we connect all points P_1 of an arbitrary curve c_1 with all points P_2 of another arbitrary line c_2, the locus of all midpoints M is a surface of translation. The surface contains two families of congruent curves. The prototypes c_1^* and c_2^* of these parameter lines are similar to c_1 and c_2 (factor $1/2$).

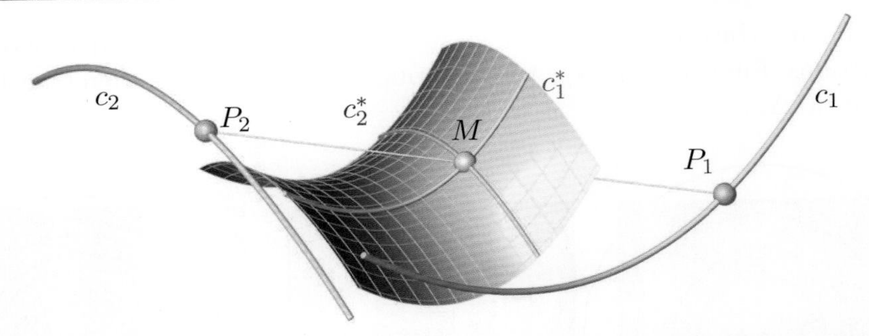

Fig. 7.56 Given two lines c_1 and c_2: We consider the locus of all midpoints of $P_1 \in c_1$ and $P_2 \in c_2$.

Proof:
Let

$$\vec{c}_1(u) = \big(x_1(u), y_1(u),\, z_1(u)\big),\ u \in [u_1,\, u_2]$$

and

$$\vec{c}_2(v) = \big(x_2(v), y_2(v),\, z_2(v)\big),\ v \in [v_1,\, v_2]$$

be parameterizations of c_1 and c_2. Then the the midpoint M is given by

$$\vec{x}(u,\, v) = \frac{1}{2}\big[\vec{c}_1(u) + \vec{c}_2(v)\big] = \frac{1}{2}\vec{c}_1(u) + \frac{1}{2}\vec{c}_2(v). \tag{7.1}$$

Therefore, $\vec{x}$ is always the sum of two vectors $\vec{c}_1^* = \frac{1}{2}\vec{c}_1$ and $\vec{c}_2^* = \frac{1}{2}\vec{c}_2$. ◇

Figure 7.56 shows already the most famous and easiest case: If c_1 and c_2 are parabolas with parallel axis direction and non-parallel carrier planes, the corresponding translation surface is a paraboloid (order two). When one of the generating lines is straight, the result is a parabolic cylinder. The trivial case that the other line is also straight results in a plane.
The above theorem also holds for identical non-planar curves $c_1 = c_2$, i.e., one single curve can define a translation surface. Henceforth, we will call such a surface *midsurface*.
Due to Equation 7.1, the following is obvious, although not trivial:

For $c_1 = c_2$, the midsurface contains the generating line. The parameter lines are congruent.

There is an additional geometrical interpretation for the two families of parameter lines that is true for any surface of translation (Figure 7.57, see also Figure 7.59):

The parameter lines of a surface of translation are the contour lines for all parallel projections of the surface in direction of any tangent of c_1 or c_2. Thus, the parameter lines form a net of conjugate lines.

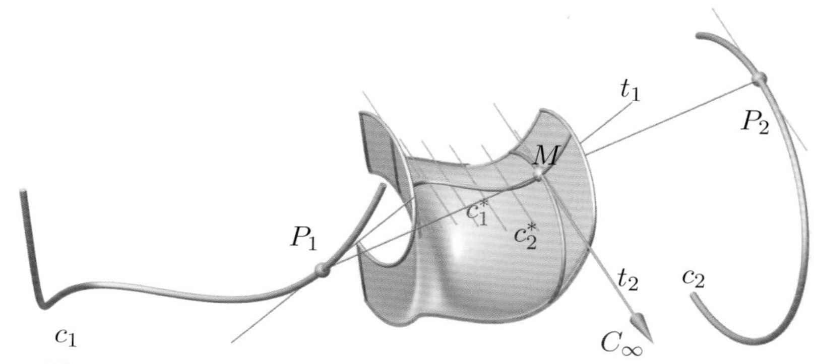

Fig. 7.57 The parameter lines can be interpreted as contour lines. The (magenta) contour of the surface is not a parameter line, since we did not project in direction of a tangent of c_1 or c_2.

Proof:
The tangent plane in a point M is spanned by the tangents t_1 and t_2 of the corresponding parameter lines c_1^* and c_2^* (Figure 7.57). We choose a projection center C_∞ in direction of t_2 at infinity. M is, by definition, contour point for this projection direction.

Now, we translate c_2^* along c_1^* (which generates the surface). The surface tangent t_2 remains parallel. The new tangent plane, therefore, contains the projection center during the whole motion, and M is contour point with respect to C_∞ at any position of c_1^*. Due to the fact that projection ray and tangent to the contour line are conjugate, we can, therefore, speak of conjugate parameter lines. ◇

The Helicoid as a surface of translation

Besides the classical example of a midsurface (the paraboloid), a non-trivial example is:

The midsurface of a helix is a helicoid. Therefore, a helicoid is not only a helical surface but also a surface of translation.

Proof:
First, c_1 and c_2 are identical (Figure 7.58). With constant radius r and screw parameter p, the equations can be written as

$$\vec{c}_1(u) = \big(r\cos u, r\sin u,\, p\,u\big), \quad \text{and} \quad \vec{c}_2(v) = \big(r\cos v, r\sin v,\, pv\big), \quad u, v \in \mathbb{R}.$$

The midsurface reads

$$\vec{x}(u,\, v) = \frac{1}{2}\begin{pmatrix} r(\cos u + \cos v) \\ r(\sin u + \sin v) \\ p(u+v) \end{pmatrix}$$

which can be modified by means of the addition theorems:

$$\vec{x}(u,\, v) = \begin{pmatrix} r\cos\frac{u+v}{2}\cos\frac{u-v}{2} \\ r\sin\frac{u+v}{2}\cos\frac{u-v}{2} \\ \frac{p}{2}(u+v) \end{pmatrix}. \tag{7.2}$$

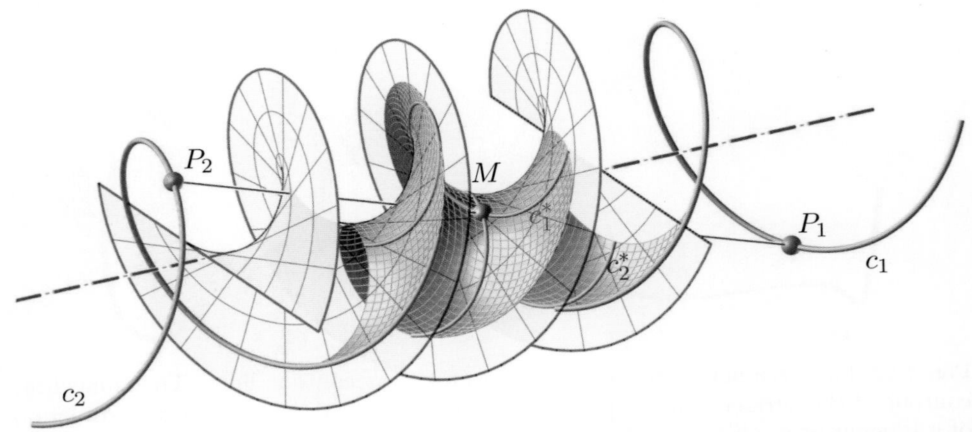

Fig. 7.58 If $c_1 = c_2$ is a helix, the midsurface is a helicoid which, therefore, is at the same time a surface of translation with two non-trivial families of helices on it.

With
$$s = r\cos\frac{u-v}{2},\quad t = \frac{u+v}{2} \tag{7.3}$$
we get a parametrization of a helicoid
$$\vec{x}(s,\,t) = (s\cos t,\ s\sin t,\ p\,t).$$
◇

Since $c_1 = c_2$, the prototypes c_1^* and c_2^* of the pencils of parameter lines are congruent. All parameter lines intersect the screw axis and are *not* helical orbits of the screw motion that generates the helicoid.
The generating helix $c = c_1 = c_2$ lies on the helicoid Φ which is, therefore, uniquely defined. If we vary the radius r of c, this will generate the same surface. Then, we get other two families of non-trivial helices on Φ with radius $\frac{r}{2}$ (Figure 7.59).

A helicoid (parameter p) carries a two-dimensional manifold of congruent helices (parameter $\frac{p}{2}$). Their projections in the direction of the axis yields circles through the image of the axis.

There is a well-known theorem about helicoids in context with the "non-trivial" helices[1]:

The contour of a helicoid for *any* parallel projection is a helix.

Proof:
As we know already, there is a two-dimensional manifold of non-trivial helices on the helicoid Φ that have constant parameter $\frac{p}{2}$ and lie on cylinders of revolution with variable radius ϱ through the axis a of Φ (Figure 7.59). Such a helix has a

[1]This theorem competes with the beauty of the theorem that every contour of a quadric for *any projection* is a conic section

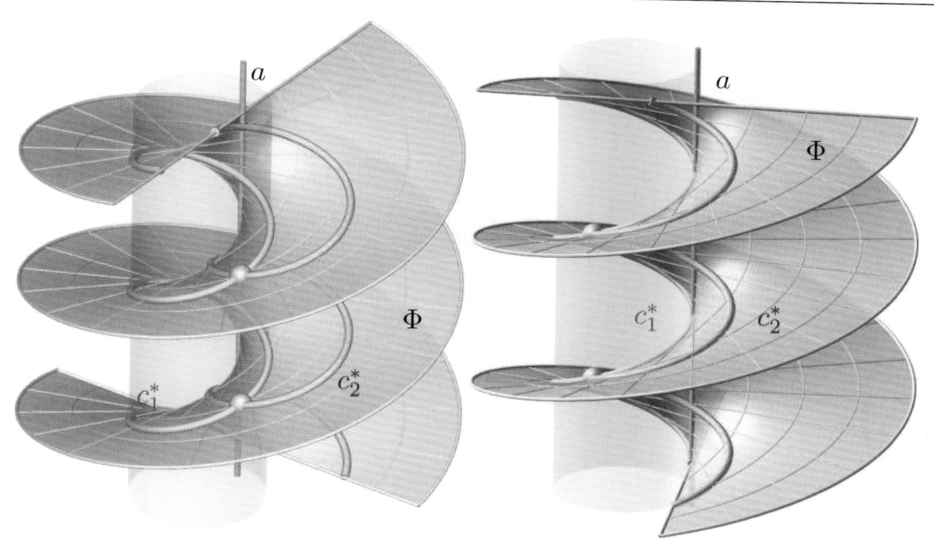

Fig. 7.59 Left: Two congruent conjugate non-trivial helices with arbitrarily chosen equal radius are displayed on a helicoid. Right: With an adequate change of the projection in direction of a tangent of the green helix, the red helix becomes a contour line.

constant slope and its tangents and the base plane enclose the angle $\alpha = \arctan \frac{p}{2\varrho}$. Now we project Φ from an arbitrary point C_∞ at infinity under elevation angle γ which is the inclination angle of a helix c_2^* with $\varrho = \frac{p}{2} \cot \gamma$. Therefore, a tangent of c_2^* will go through C_∞. According to the just proven theorem about conjugate lines, the conjugate parameter curve c_1^* is then the contour line with respect to C_∞. ◇

Another surface of revolution that is also a surface of translation

So far, we know two examples of surfaces of revolution that are at the same time surfaces of translation: the cylinder of revolution (as a trivial example), and the paraboloid of revolution (Fig. 6.16 on the left).
We now generalize the generation of the helical surface of Section 7 by allowing two different coaxial helices with parameters p and $\overline{p} = k \cdot p$. Equation 7.2 then modifies to

$$\vec{x}(u, v) = \begin{pmatrix} r \cos \frac{u+v}{2} \cos \frac{u-v}{2} \\ r \sin \frac{u+v}{2} \cos \frac{u-v}{2} \\ \frac{p}{2}(u + k\, v) \end{pmatrix}. \tag{7.4}$$

We obtain the helicoid if $k = 1$. We immediately see that $k = -1$ is another special case (Figures 7.60 and 7.61):

When we rotate a sine curve (amplitude r) about its axis, the generated surface of rotation can be interpreted as a surface of translation with two reversely congruent helices that lie on cylinders of revolution with diameter r that contain the screw axis.

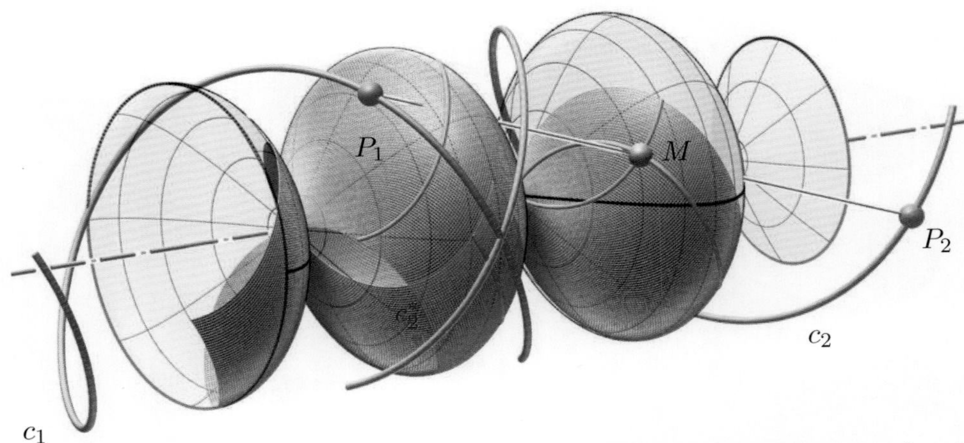

Fig. 7.60 If two coaxial helices c_1 and c_2 are reversely congruent, the midsurface is a surface of rotation.

Proof:
With the substitutions

$$s = \frac{u - v}{2}, \quad t = \frac{u + v}{2} \tag{7.5}$$

Equation 7.4 simplifies to

$$\vec{x}(s,\, t) = (r \cos s \cos t,\; r \cos s \sin t,\; p\, s) \tag{7.6}$$

which is the equation of a surface of revolution, generated by a sine curve with amplitude r rotating about the axis of the curve. ◇

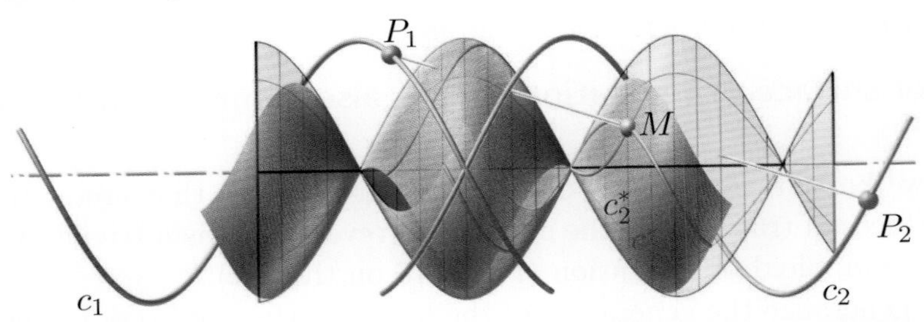

Fig. 7.61 The meridian of the surface is a sine curve. The non-trivial helices on the surface lie on cylinders that touch the surface and contain the axis of the rotation.

The general case ($k \neq \pm 1$) and other examples are discussed in a paper that can be downloaded from the website to the book.

- **About rhombuses and huge towers**

The aforementioned surface of revolution which is at the same time a surface of translation has its applications.
Fig. 7.62 on the left shows how polyhedra, entirely built up by rhombuses with constant side length but different interior angles, converge to the remarkable surface with a sine curve as meridian. The aligned adjacent sides of the rhombuses converge to helices. ♠

Fig. 7.62 Applications of the remarkable surface of revolution and translation. On the right the Gherkin tower in London (Norman *Foster*).

A remark for the next section: If the curves c_1 and c_2 are so-called minimal curves, the adjacent translation surfaces are minimal surfaces. Such minimal curves, however, have constant imaginary slope and do, thus, only exist in the complex extension of the real space. Therefore, we will define minimal surfaces in a different way.

7.5 Minimal surfaces

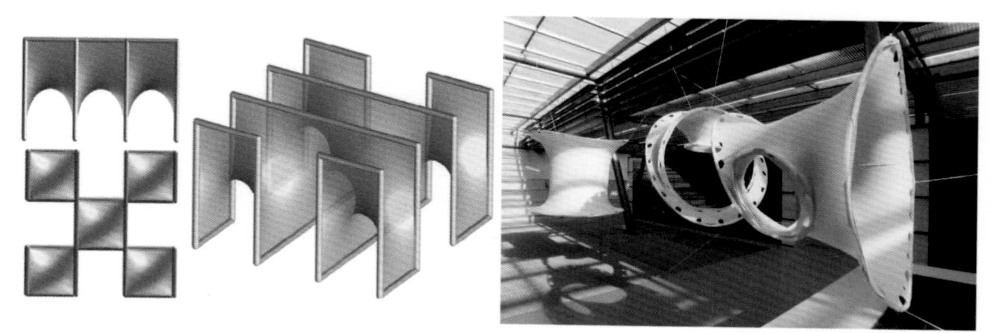

Fig. 7.63 Surfaces exhibiting tension in theory and art: Pictured on the left is the famous "chessboard minimal surface" by Scherk, on the right an installation by Maria Wambacher.

Due to surface tension, each liquid or elastic surface (such as a rubber membrane) strives to minimize its surface. Mercury, for instance, accumulates

into tiny round droplets if it is poured onto a planar glass plate. The spherical soap bubble likewise owes its shape to surface tension: A sphere simply represents the smallest possible surface for a given volume. Open umbrellas are likewise under pressure, as are tents and roofs. If it is bounded by a wire, soap film can arrange itself into many different forms.
Due to water's surface tension, the tiny metal needle stays afloat, even though its much larger density would otherwise make it sink.

• **Soap films with minimal area**
When we dip a wire frame into a soap solution, the soap film will, due to area minimization, form a minimal surface.

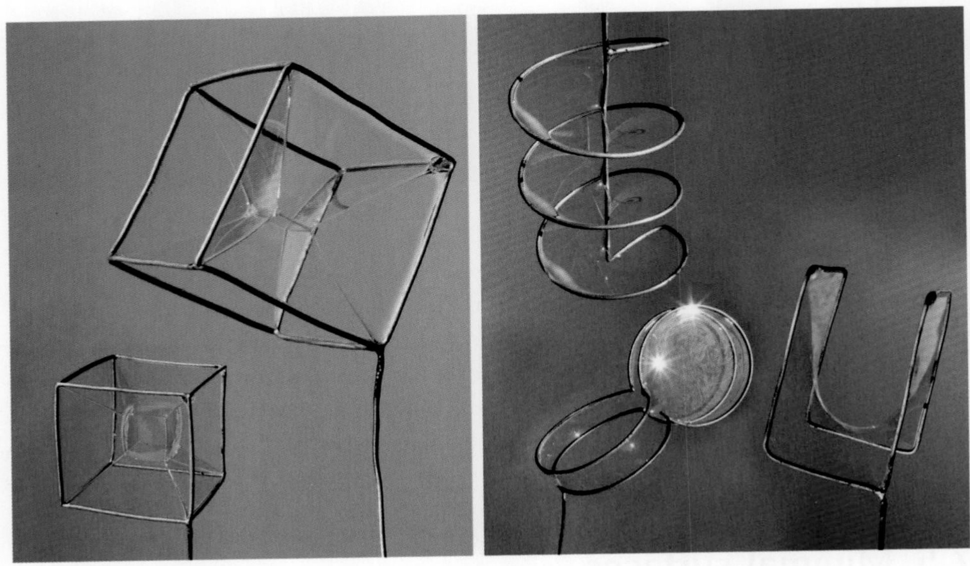

Fig. 7.64 Special minimal surfaces are formed when the wire frame looks like a cube, a helix, parallel circles or consists of only 8 edges of a cube (on the right).

Fig. 7.64 shows some examples of how the soap film may look like when the wire frame fulfils certain conditions. E.g., a helicoid can physically be generated by dipping a helix into the solution. ♠

Geometrically speaking, the membranes of minimal surfaces are distinguished by the fact that in every one of their points, the surface's principal curvature (see p. 209) is of equal magnitude. This is certainly the case for the three minimal surfaces that are already known to us: The sphere (the only elliptically curved minimal surface), the catenoid (Fig. 6.78 on the right) and the right helicoid (Fig. 7.13 and Fig. 7.64 on the right). Fig. 7.63 on the left shows a minimal surface by *Scherk* in top and front view, as well as from a general perspective.
A computer test will show that the HP-shell, indeed, comes close to a minimal surface at certain positions (in Fig. 7.65 on the left, these areas are marked red and

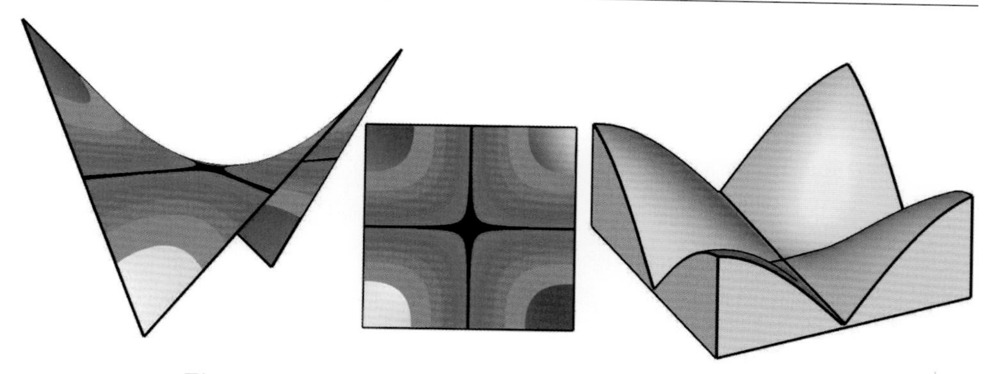

Fig. 7.65 HP-shell and its deviation from the "minimal state"

black – the top view is visible in the middle). How much of the area is marked by the computer depends on the choice of a tolerance interval. It is, nevertheless, possible to assume that the tent membrane in Fig. 7.63 is (at least visually) nearly identical to a HP-shell. Fig. 7.65 on the right shows the accompanying function graph where the mean curvature (the sum of both principal curvatures) is pictured by the height value.

• Roofs of minimal surfaces

Minimal surfaces can be characterized in two ways, physically (equilibrium of surface forces) and geometrically (principal curvature of equal magnitude throughout the surface). This is why they keep their useful properties when they are scaled to huge dimensions (Fig. 7.66 on the right shows the Olympic Stadium of Munich).

Fig. 7.66 Minimal surfaces can be scaled without losing their useful properties.

When working with other roof surfaces, one has to consider that the rapidly increasing weight may cause different problems. ♠

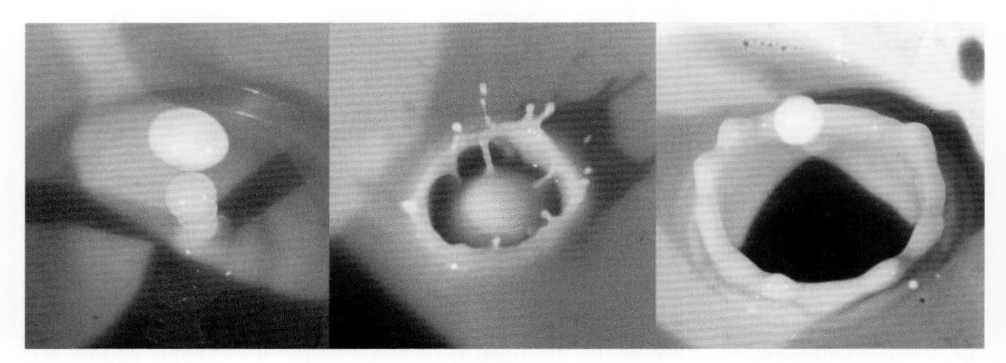

Fig. 7.67 The effect of surface tension ...

When the sphere-like droplets of a liquid "impact" on the surface of a liquid, much complexity may occur: In Fig. 7.67 it is possible to distinguish the white milk drops from the brown cocoa. In the left picture, the topmost droplet is still white – the column that piles up below is already a mixture which "rebounded" from a previous impact. The center picture shows "another" impact, and the picture on the right shows a "latecomer" which appears to come upon a massive crater ...

Fig. 7.68 Not all tense surfaces are necessarily minimal surfaces but some similarities can, nevertheless, be established. The skins of flying foxes are deformed by the same laws that govern an umbrella stressed by the blowing wind.

• Minimal surfaces with lightweight modules (Fig. 7.70)
Minimal surfaces can be quite useful for the creation of stable and lightweight sculptures, as the following experiment shows[2]: First, a minimal surface through four attractor points was created and subdivided by means of

[2] `www.springchallenge2012.wordpress.com` (Bence Pap, Irina Bogdan, Andrei Gheorghe, Trevor Patt, Clemens Preisinger, Moritz Heimrath).

Fig. 7.69 Wing flap according to Leonardo da Vinci: The computer simulation on the left (with the original sketch in the background) and the "fine-drawing" of the sketch by Harald A. Korvas on the right. When studying the wings of bats and flying foxes (Fig. 7.68), one may get the impression that *Leonardo da Vinci* based his famous wing sketches not primarily on birds, but on his observation of these remarkable mammals.

lightweight hollow modules. The geometry of the surface was then slightly changed according to circulation patterns (Fig. 7.70, left) and visual axes.

Fig. 7.70 Minimal surface "populated" with modules morphing from triangular to hexagonal shape

The advantage is stiffness that is needed at certain areas. The geometry was tested and improved by using the Grasshopper plugin "Karamba"[3]. ♠

• A special algebraic minimal surface and "something similar"

In Chapter 13, we will see that exact geometric shapes may not be seen in nature quite often. Sometimes, however, one finds rather good approximations to them, as Fig. 7.71 shall show. (computer drawing to the right) and "something similar" in nature ...

One of the simplest minimal surfaces is of ninth order, and serves as an algebraic example of such surfaces in mathematics. If we only consider the innermost part, there are no self-intersections (the surface is theoretically infinitely large). Compared with a brown alga from the northern Adriatic

[3]`http://www.karamba3d.com/`

sea, a certain similarity cannot be denied, indicating that the algae tend to minimize their surface.

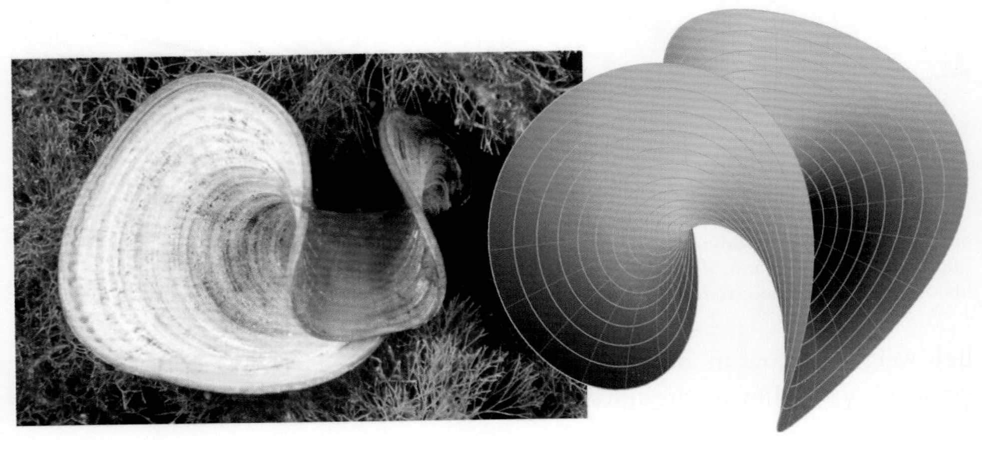

Fig. 7.71 The comparatively simple algebraic minimal Enneper surface ♠

• "Area minimizing surfaces"
Even though it seems to make no sense to speak of the surface of a snail (it is moving and changing shape all the time): In every moment it seems as if the surface tends to minimize its area with any new position.

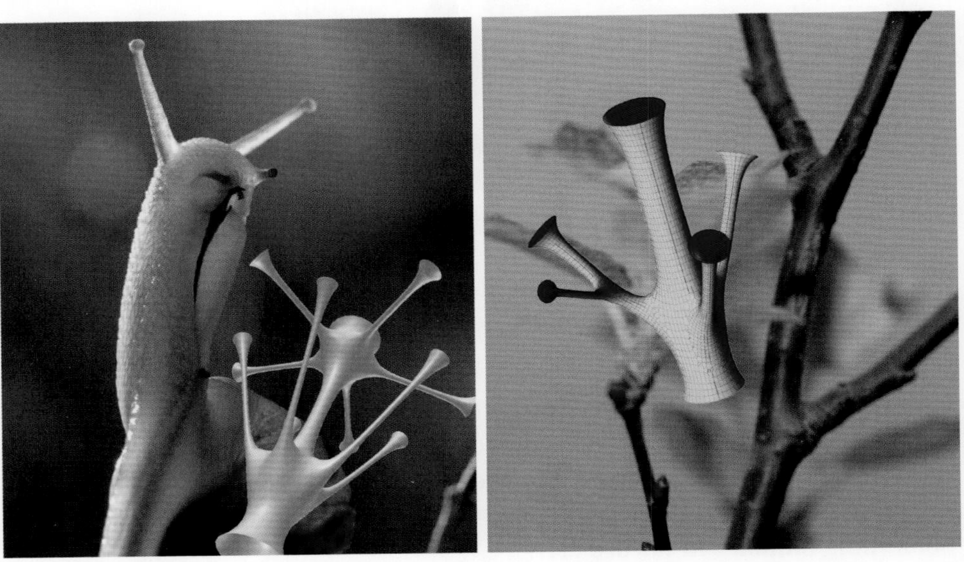

Fig. 7.72 Natural computer-generated surfaces by iteratively minimizing the surface

When we slightly change the shape of a given surface iteratively such that the area of the new surface is smaller than that of the previous one, the surface will eventually converge to a minimal surface. ♠

8 The endless variety of curved surfaces

So far, we have learned about several classes of surfaces. By a variation of the so-called parameter representation, however, many other classes of surfaces may be derived. In the age of creative (architectural) design at the computer, it is becoming increasingly relevant to create completely general surfaces by means of easy-to-control geometric instructions. Such *free-form surfaces* are the topic of this chapter.

The theory behind free-form surfaces is already several decades old, and the rapid nature of modern computers makes such surfaces interesting for everyone who designs from behind a computer screen.

The problems to be solved are as follows: How can surfaces be found, which pass through predefined points in space and additionally follow certain "boundary conditions"? How is it possible to work with such "control points" and retain the ability to change their individual positions without affecting the character of the entire surface?

The solution to these problems lies, as it often does, one dimension lower: One should define planar free-form curves that exhibit certain properties, and then transfer these properties into three-dimensional space. Thus, we are confronted with Bézier surfaces and B-spline surfaces by way of Bézier curves and B-spline curves.

Survey

8.1 Mathematical surfaces and free-form surfaces

Many well-known surfaces can be described by mathematical equations. This includes (but is not limited to) practically all surfaces that have caught our attention so far (Fig. 8.1 shows two surfaces of this kind that we haven't addressed yet).

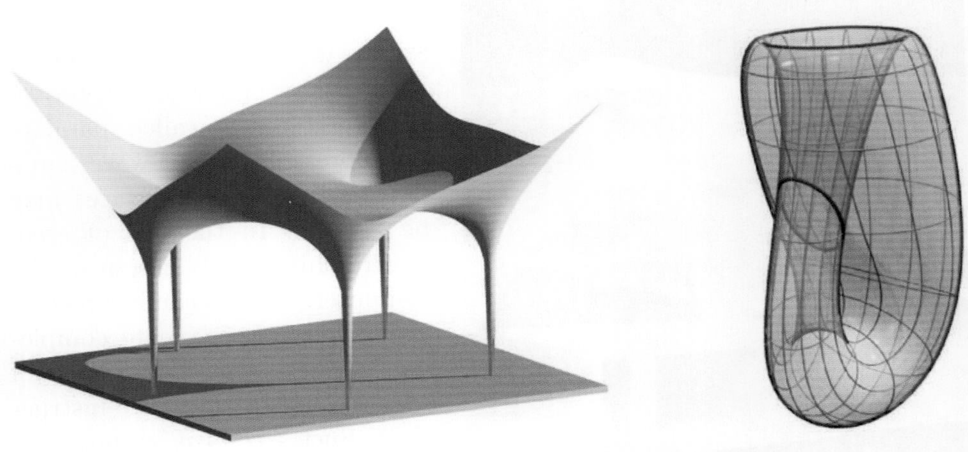

Fig. 8.1 Surfaces generated by mathematical equations

Such "mathematical surfaces" possess one or many formative parameters which permit a certain variation in their visual appearance. The employment of a certain mathematical dexterity allows for the generation of fairly complex surfaces.

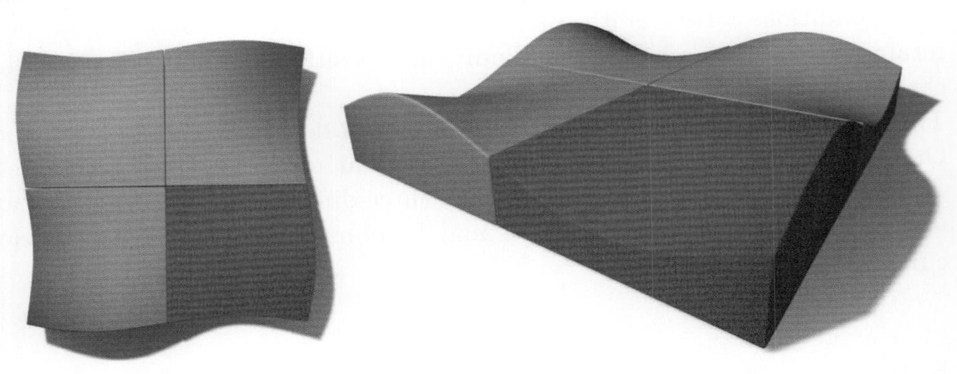

Fig. 8.2 Furniture as a mathematically defined function graph

A very elegant piece of furniture by Paolo Piva is pictured in Fig. 8.2. The idea behind it was to involve mathematics in the arts.
This surface is defined by a symmetrical equation, which may also be of interest to non-mathematicians:

$$z = c \cdot (y \sin x - x \sin y) \quad \text{where } c \approx 0{,}12,\ x, y \in [-\pi, \pi].$$

Parabolic points occur along the coordinate axes.

Different types of surfaces

Surfaces can generally be thought of as the "products" of space curves, which undergo arbitrary motions and deformations.
Surfaces of revolution, for instance, are created by the rotation of an unchanging generatrix. Scaling this curve in relation to a fixed point on the axis of rotation and proportional to the angle of rotation yields a spiral surface.

The addition and subtraction of surfaces

Many objects of highly alleged complexity are the result of so-called *Boolean operations*: Solids may be "added" to each other, but also "subtracted" from each other. To the geometrician, this does not necessarily indicate that a new surface was created: Each boundary segment, no matter how small, belongs to precisely one of the initial surfaces.

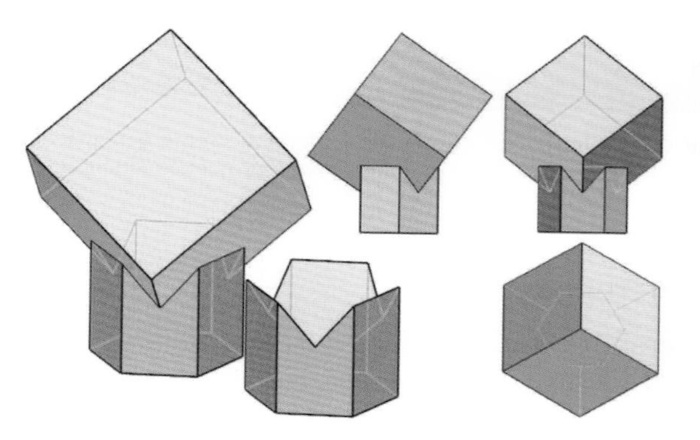

Fig. 8.3 Boolean operations in architecture and the arts: "Tree house" (cube apartment in Rotterdam by Piet Blom), where a cube was placed on its spatial diagonal and superimposed with a hexagonal support prism. At the center: Addition and subtraction. The right side pictures a tree house of a different kind: Erwin Wurm placed a house on top of the Vienna Museum of Modern Art – a classical Boolean subtraction is behind this effect ...

Fig. 8.4 Boolean operations produced, in projection, the initials of three great masters: Gödel, Escher, Bach. The *B* is put on its head, the *E* is rotated by 90° – both are translated along the coordinate axes, which yields the intersection of the desired residual solid ([16]).

Fig. 8.5 Booleean operations in art objects (Eduardo *Chilliada*)

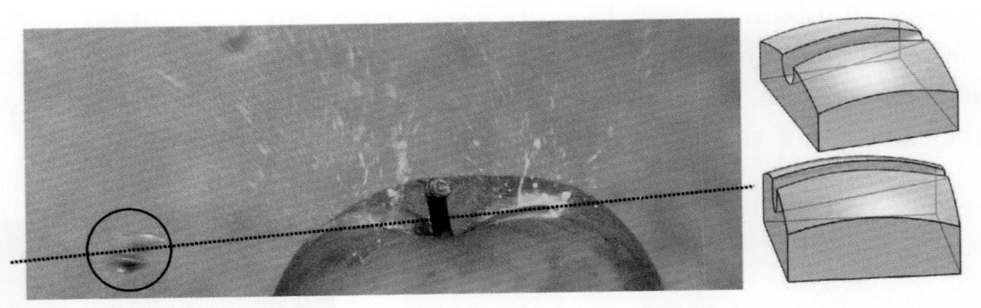

Fig. 8.6 Extreme Boolean "subtraction": The high-speed photograph shows an apple being penetrated from the right by the bullet of an air rifle. The double-cone-shaped projectile is encircled and fairly visible. On the right: Computer graphic.

Milled surfaces are also "difference surfaces"

During the milling process, all possible positions of the cutter envelop a manifold – the sum of all milling positions. The surface parts that are cut away are the difference surfaces of the original solid and this enveloped manifold. The more complicated the shape and motion of the cutter, the more complicated this enveloped surface. Fig. 8.10 and Fig. 8.11 show a surface designed by Florian Gypser. By looking at only the intermediate positions of the object in motion (the professional calls this a *discrete family*), one cannot fail to notice the aesthetic quality of these "imprints" – their possible applications in design are likewise obvious.

Fig. 8.7 Simple cutters (cylindrical and conic)

In practise, curved shapes are often produced by milling. Each milling pass creates a new and occasionally complicated shape (Fig. 8.8).

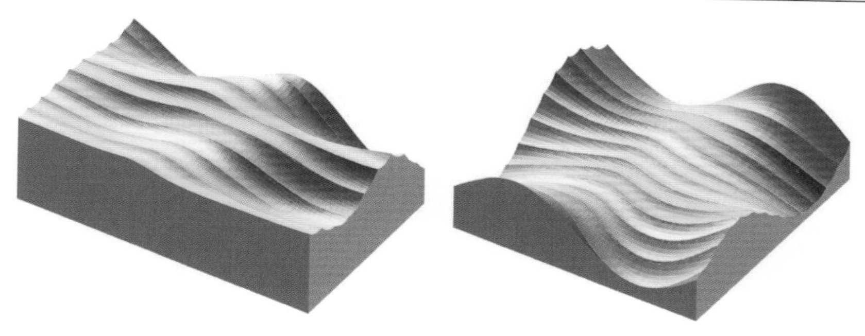

Fig. 8.8 Surfaces as the results of a milling process ...

Fig. 8.9 ... and an application of an altogether different type

How can we solve the problem?

In general, surfaces of arbitrary shape can, to some degree, be approximated by parts of standard surfaces. Due to the limited number of formative parameters, however, the result will rarely be satisfactory.
Let us from this point on define a *free-form surface* such that its appearance may be changed in an almost arbitrary manner through the variation of parameters and generating geometric elements. Thus, it should be possible – at least visually – to approximate mathematically defined surfaces very closely. Free-form surfaces play an important part in industrial design (auto industry, ship construction, and airplane construction) and in the modeling and construction of interesting architecture.

Fig. 8.10 Envelope or "imprint" of an object in motion

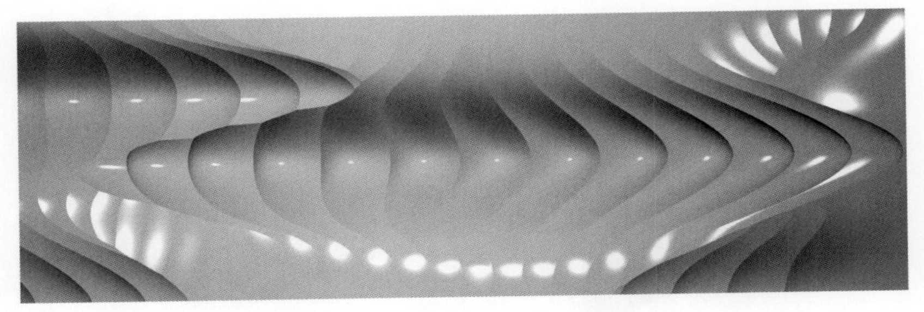

Fig. 8.11 Imprint of a "discrete" family

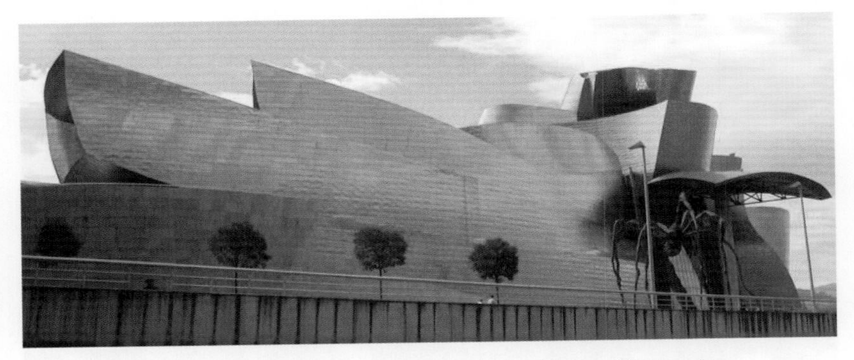

Fig. 8.12 Partly developable surfaces (Bilbao Museum)

Surface theory is helpful in the creation of categories

We know the following from the previous chapter: *Locally* speaking, there exist only three types of surfaces – no matter if they are free-form or mathematically defined: Each surface, regardless of its complexity, is curved like an elliptic or hyperbolic paraboloid, or like a parabolic cylinder, or a plane, but only in a small neighborhood of each of its points.
It makes no difference for our prior considerations whether the surface is mathematically defined or "free-form". As we shall soon discover, free-form surfaces also have to be calculated in some way – by relatively elaborate, "recursive" equations, to be exact. Due to their complicated definition, a greater quantity of formative parameters present themselves.

8.2 Interpolating surfaces

One particular technique of definition seems to be obvious: It should, for instance, be possible to fix single surface points as coordinates and interpolate the points in between. If a surface is to be given in the form of a computer model, it may be digitised (using keyers, laser scanners, or photographs) into a wireframe model on the surface, from which the remaining points in space may be computed.

In order to connect $n+1$ coplanar points by a smooth curve, the idea may present itself to position an *n-th order parabola* p through the points.

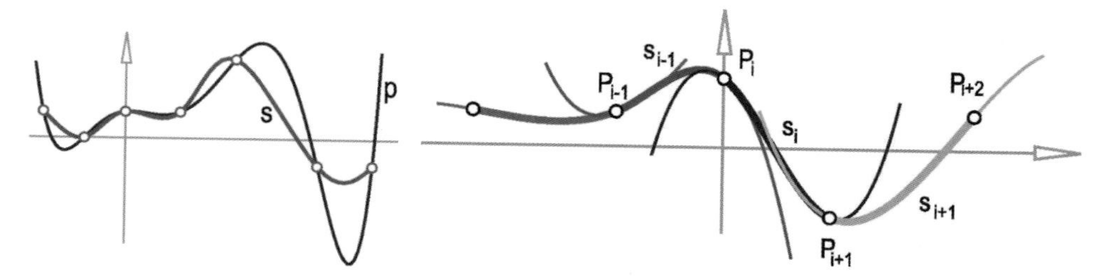

Fig. 8.13 Polynomial function vs. cubic spline

This approach may not be of much use to designers: Such curves tend to oscillate – especially at the boundaries (Fig. 8.13 on the left, curve p). Another technique assembles the curve piece-by-piece from third-order parabolas, while the separate curves are joined at the respective endpoints with "continuous curvature" (curve s in Fig. 8.13). These curves are called "cubic splines".

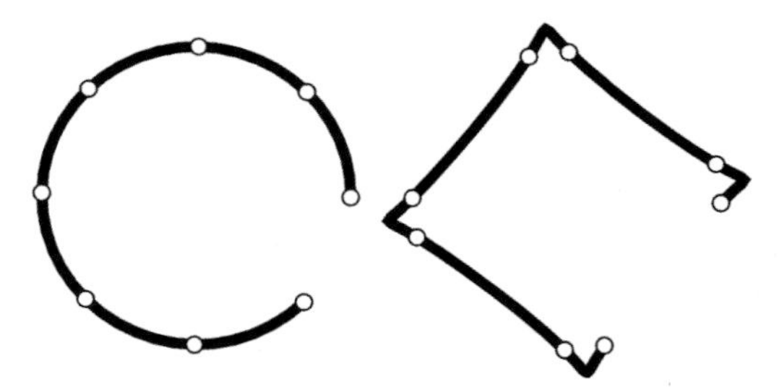

Fig. 8.14 The choice of control points may be significant.

In order to achieve a smooth curve, the control points of a cubic spline should be evenly distributed – unexpected results may otherwise be produced (Fig. 8.14). Curves through predefined points may also be produced by so-called NURBS (which will soon be described).

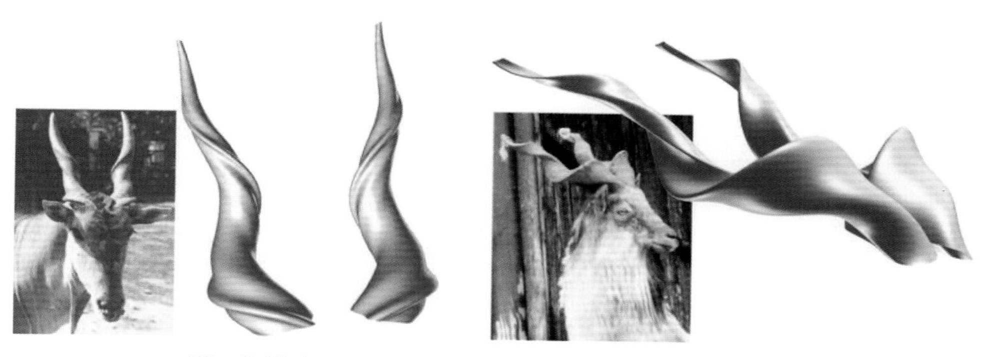

Fig. 8.15 Approximation of animal horns using cubic splines

By using interpolated splines, it is possible to describe surfaces whose mathematical definition is no longer feasible. In Fig. 8.15, for example, the cross-sections of animal horns are described by using interpolated splines before undergoing mathematical transformations (a *helispiral motion* is pictured).

Direct modeling at the computer

Designers are, of course, not interested in specifying formulas and would rather do their ingenious design work directly at the computer. The Bézier- and B-spline-surfaces, which allow for the easy definition of free-form shapes, were developed for this precise reason.
In order to comprehend the underlying principle, we shall first discuss Bézier- and B-spline-curves in the plane, before proceeding to the Bézier- and B-spline-surfaces in three-dimensional space.

8.3 Bézier- and B-spline-curves

Bézier curves

Next to B-spline-curves, Bézier curves are the most important mathematical representation of curves in computer graphics and computer aided design (CAD). They were introduced in 1962 by P. Bézier at Renault and are defined by a so-called *control polygon*. The number of polygon sides denotes the degree of the curve.

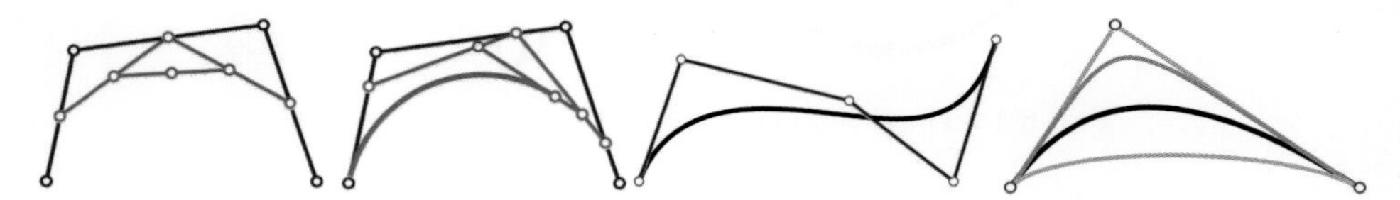

Fig. 8.16 The *de Casteljau* algorithm, inflection point, conic sections

The construction procedure for a curve point corresponding to the real parameter t was already defined by P. *de Casteljau* (Citroen, 1959) and goes as follows: Subdivide all n sides of the polygon following the same ratio $t:(1-t)$ (in Fig. 8.16, $t = 0{,}5$ was chosen on the left, and $t = 0{,}75$ on the right). The connecting segments of the new points form another polygon – with one side less than the original polygon. This procedure should be repeated for the new polygon, which yields another polygon with $n-2$ sides, and so on. After n steps, the polygon is only composed of a single point – the curve point. The related split segment is the curve tangent. In the special case of $n = 2$, this procedure always yields a parabola. Fig. 8.16 on the center right shows a fourth-order Bézier curve with an inflection point defined by a four-sided polygon. On the far right, a second-order Bézier curve is pictured in black – a regular parabola.

Due to the nature of their construction, Bézier curves possess many helpful geometric properties: They always lie within the convex hull of the control polygon and pass through (and therefore touch) the boundary points. The curve does not oscillate needlessly, which means that it does not intersect an arbitrary straight line more often than the control polygon. Furthermore, in the cases of rotations, reflections, translations, scaling, and parallel projections, it is sufficient to subject the control polygon to the same transformation. The construction procedure for curve points and tangents "copes" well with affine transformations. The control points are transformed into the control points of the image curve.
The algorithm by *de Casteljau* serves, beside the calculation or construction of points and tangents, another useful purpose: The decomposition of a Bézier curve at a point $t \in [0,1]$ into two parts (Bézier curves of the same order, Fig. 8.17) and also provides the key for the approximated calculation of all intersection points of two Bézier curves, or of a Bézier curve and a straight line.

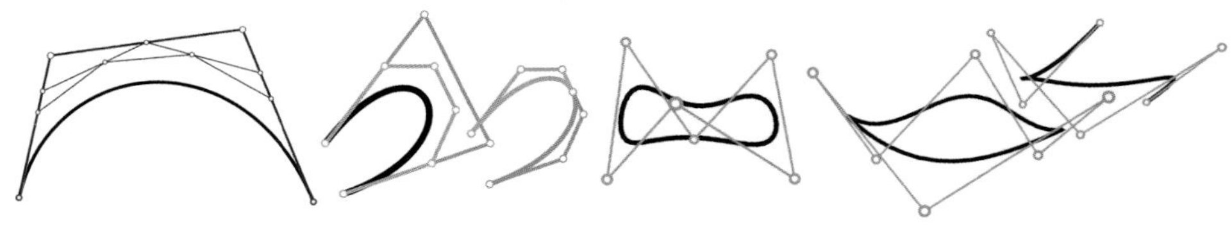

Fig. 8.17 Augmentation of the order, decomposition of a Bézier curve

If a single control point of a Bézier curve is moved, then the shape of the entire curve is affected (global influence of the control points)! This is a significant disadvantage of Bézier curves compared to the upcoming B-spline-curves. The same comparison can be done with Bézier- and B-spline-surfaces. It is further problematic that certain frequent types of curves, like ellipses or hyperbolas, cannot be precisely defined as Bézier curves (parabolas, however, are perfectly possible). In order to solve these problems, Bézier curves were generalized to so-called *rational Bézier curves.*

Rational Bézier curves

Weights are assigned to each control point. If all weights are the same, then a classical ("ordinary") Bézier curve is given. The geometric definition of a Bézier curve is independent of dimensions (Fig. 8.18):
Let us imagine the planar curve in a horizontal plane at height 1 and stretch the points of the planar control polygon from the origin, with the respective weight as the scaling factor. In the general case, we will thus get a non-planar control polygon in three-dimensional space that will, if exposed to the *de Casteljau* algorithm, yield a Bézier space curve. The central view of this curve from the origin into the original curve's horizontal carrier plane produces the rational Bézier curve.

This approach allows us to define parabolas, parts of ellipses, (including parts of circles), and parts of hyperbolas among other curves. (Fig. 8.16 on the right), which occur as central views of parabolas.

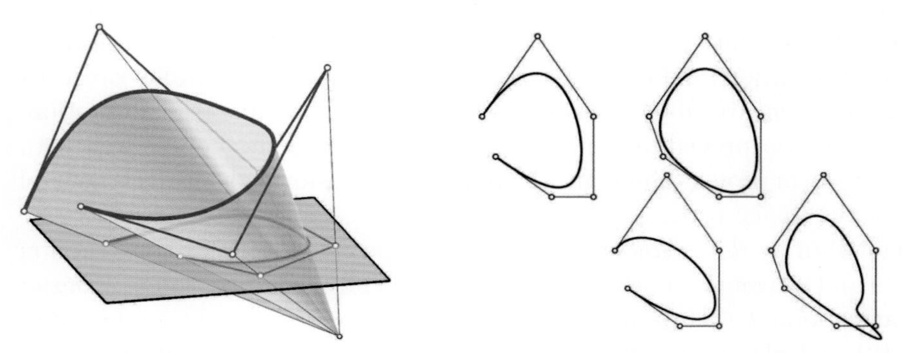

Fig. 8.18 Spatial interpretation **Fig. 8.19** Above: ordinary, below: rational Bézier curves

For rational Bézier curves, the properties of regular Bézier curves essentially persist. It may, however, be added that even projective transformations will not change the nature of the curve: It is sufficient to apply the transformation to the control polygon: The rational Bézier curve produced by the transformed polygon corresponds to the projective transformation of the original Bézier curve.

B-spline-curves

By piecing on rational Bézier curves (the transition can occur at constant curvature, if the degree is at least 3), so-called *Bézier splines* may be produced. They have the tremendous advantage that changing a single control point no longer affects the entire shape on a global level. On the other hand, the amount of data for the control points may grow to be very large, as Bézier splines are "hard" to adapt to arbitrary – and especially convex – shapes. In the general case, they, thus, require many control points for fine-tuning. This disadvantage was overcome in 1964 by J. *Ferguson* (at Boeing) and the invention of B-spline-curves, which otherwise possess properties very similar to Bézier splines.
For B-spline-curves, the algorithm analogous to *de Casteljau*'s is the so-called *de Boor* algorithm. The kinship between Bézier curves and B-spline-curves is such that each B-spline curve segment can be converted into a Bézier curve. The degree of the curve indicates how many control points around a curve point are responsible for its position.
In order to quantify this degree of influence, a "knot vector" is given to specify the strength of the influence that a control point's neighbors have. This vector must possess more components than the overall amount of control points. If a curve is to pass through the boundary points of the control polygon, the beginning and end components of the knot vector must be repeated accordingly (see the following example).

Rational B-spline-curves (NURBS)

If weights are provided for all points of the control polygon (as was the case with rational Bézier curves), a rational B-spline-curve is given, which is also called a NURBS (Non-Uniform Rational B-Spline) curve – "non-uniform" simply means that the knots do not have to be distributed at equal distances. The unit circle, for instance, can be specified by the control polygon (1,0), (1,1), (−1,1), (−1,0), (−1, −1), (1, −1), (1,0), the knot vector $(0,0,0,\frac{1}{4},\frac{1}{2},\frac{1}{2},\frac{1}{4},1,1,1)$, and the weights $(1,\frac{1}{2},\frac{1}{2},1,\frac{1}{2},\frac{1}{2},1)$. Based on this representation of a circle, it is possible to construct surfaces of revolution.
NURBS possess properties analogous to those of ordinary B-spline-curves: local control, the convex-hull-property, and continuity of curvature (if all weights are positive) as well as invariance under affine and projective transformations. There exists a de Boor algorithm for NURBS which can be used to calculate the points of the curve or decompose a NURBS into two parts.

8.4 Bézier- and B-spline-surfaces

Given a grid of quadrangles instead of a control polygon ($m \dots n$ sides), Bézier surfaces and B-spline-surfaces can be constructed in analogy to Bézier curves and B-spline-curves: If P_{ij} are the points of the grid ($i = 0, \cdots, m$; $j = 0, \dots, n$), then the $n+1$ points with a fixed i define the control points of a Bézier- or B-spline-curve. For each of these $m+1$ curves, there exist unambiguously defined points for each parameter t as calculated by the algorithms of *de Casteljau* or *de Boor*. These points may, in turn, be used as the control points of further Bézier- or B-spline-curves. We have thus defined a continuous family of curves on the surface. By analogy, an additional family of curves may be constructed by varying the index i while keeping j constant.

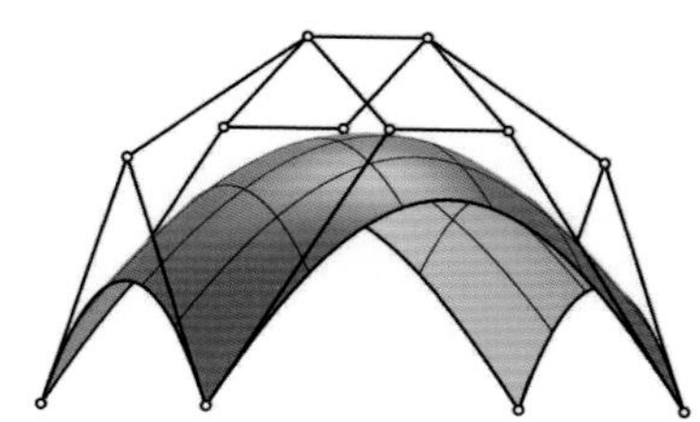

Fig. 8.20 Bézier surface

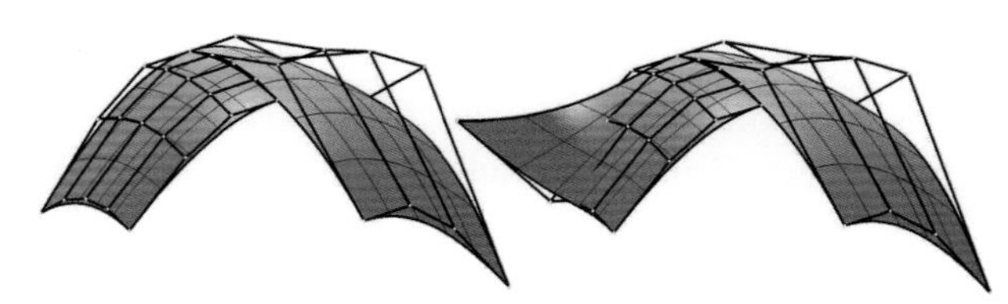

Fig. 8.21 Subdivision (Degree augmentation + local change)

When compared to their planar cousins – Bézier curves –, Bézier surfaces exhibit similarly practical properties for modellers: The surface passes through the outer edge points of the grid and is there touched by the planes spanned by the respective sides of the quadrangular net. The surface lies within the convex hull of the grid. In order to apply an affine transformation to the surface, it is sufficient to apply the transformation to the grid. Rational Bézier surfaces (which will be discussed shortly) are even projectively invariant. On-

ce again, a difficulty arises when the change of a single control point leads to the whole surface being affected. It may be helpful to increase the degree of the surface temporarily, decompose it into parts, and continue manipulating the separate parts only. If both surface degrees m and n are at least 3, then it is possible to configure the surfaces so that they meet at constant curvature. For a designer, constant curvature is of utmost importance: Shiny surfaces that do not have this property produce "light edges" .
Ruled surfaces and cylinder surfaces can easily be represented as Bézier surfaces by reducing one of the two degrees m or n to 1. In the case of the cylinder, the straight lines connecting corresponding control points must be parallel. Paraboloids occur for (2,2) grids (but only the surfaces of translation among the (2,2) grids are paraboloids). If the polygons of the quadrangle grid diverge as a result of translation, then a surface of translation is created. Mathematical surfaces can be exactly represented by Bézier surfaces if they possess a parameter representation composed of polynomial functions.

Generalized Bézier surfaces

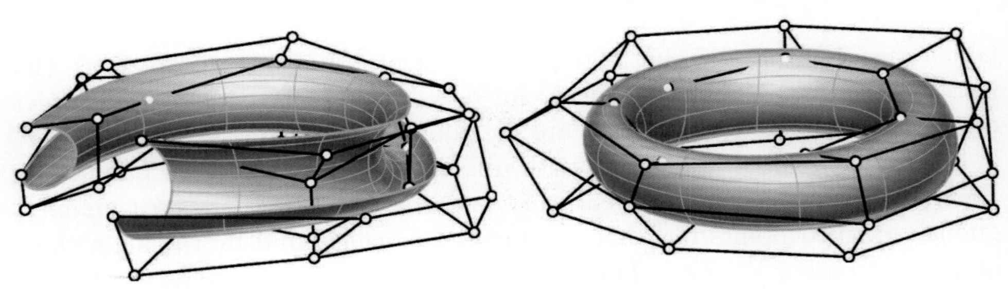

Fig. 8.22 Rational B-spline-surfaces, open and closed

If each grid point of a Bézier surface is specified with a weight, the result is a rational Bézier surface. Now it is possible to represent a greater class of surfaces in an exact manner – all surfaces, in fact, whose parameter representation can be specified as rational functions (a fraction of polynomial functions). But even such techniques can only encompass the tip of the iceberg of all possible surfaces, which is why it is recommended to work with sub-grids or sub-nets which can be edited separately.

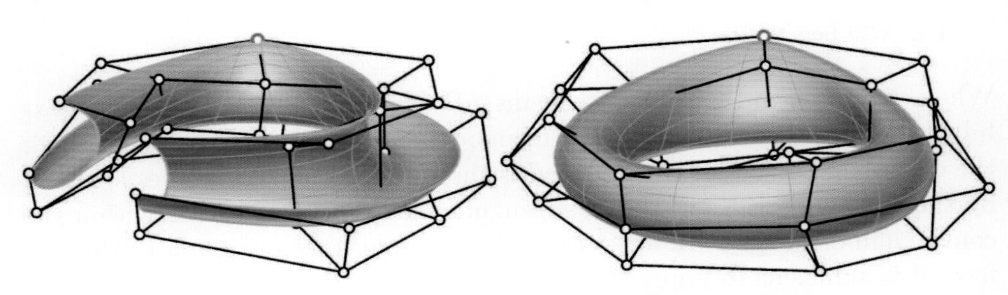

Fig. 8.23 Rational B-spline-surfaces (NURBS), open and closed

Like B-spline-curves, B-spline-surfaces can be manipulated locally. They also share useful properties analogous to those of B-spline-curves. Fig. 8.22 shows open and closed control grids and the corresponding surfaces.
Rational B-spline-surfaces (NURBS-surfaces) are the most flexible if, in addition to the control points, they can also be affected by two knot vectors for both directions. These surfaces are, thus, included in the design-modules of all relevant CAD systems.

Fig. 8.24 A nice application of NURBS (compared to Fig. 2.28)

• Practical construction

After a free-form surface has been designed on the computer, the problems generally tend to occur during the practical construction. *Rapid prototyping* permits the quick and easy generation of prototypes: Three-dimensional "printers" produce thin cross-section surfaces from synthetic material which can then be glued together.

Fig. 8.25 The practical construction of a NURBS blueprint

Large surfaces demand appropriately enormous means. A patent which specifies the computerized control of bending machines so that an exact wireframe model of the surface is produced from steel (Fig. 8.25) was submitted at the University of Applied Arts Vienna (by Sigrid Brell-Cokcan and Dumene Comploi), design by Mario Gasser, Cornelia Faißt, and Lis Ehses). ♠

• Nutshells made of plastic

An interesting idea by Marcus Bruckmann and Ursula Klein/schulteswien describes the fusing together of congruent, circular plastic foils which can then be inflated.

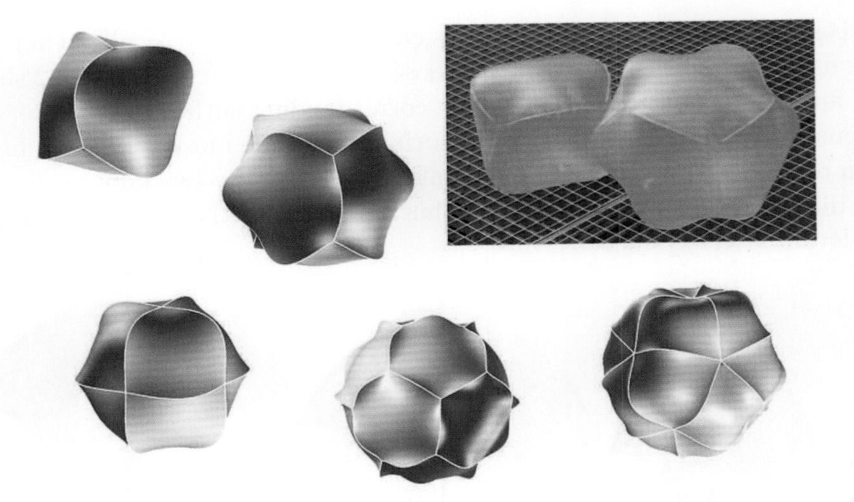

Fig. 8.26 Surfaces under pressure, derived from Platonic solids

Fig. 8.27 "Inflated" cube and dodecahedron using NURBS

The resulting formations have Platonic solids as their base frame (Fig. 8.26), and their shapes recall the shells of Brazilian nuts. They were modelled using NURBS in a particularly realistic way. ♠

8.5 Surface design in a different way

An interplay between art and geometry

An unorthodox question of the ceramicist Silvia Siegl and a surprising solution on the computer led, in turn, to a series of projects.
At first, the problem concerned merely the rigid connection of ceramic lenses using two orthogonal rods of equal length and the subsequent bending of the construction into certain shapes. This was to be accomplished by using spherical joints at the ends of the rods (Fig. 8.28). Geometric analysis was, at

Fig. 8.28 The initial question ...

first, aimed at fitting squares with their respective corners into the joints. This does not pose any problem in the plane, where it is simple to proceed along the square edges following the creasing of the "lens carpet". This creasing is only possible if it leads to the change of position of *multiple* lenses.

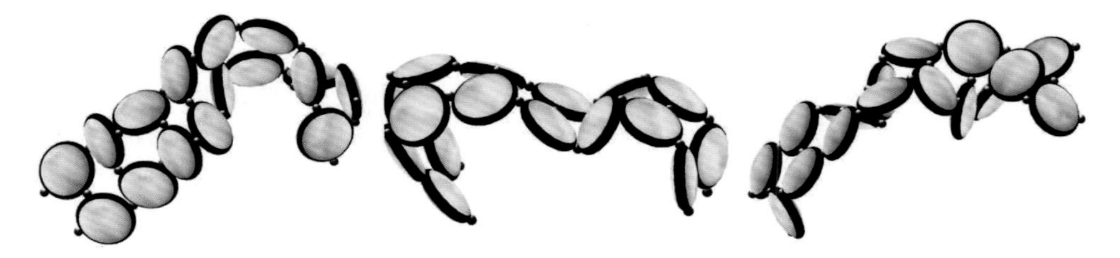

Fig. 8.29 ... was, at first, solved at the computer by Franz Gruber.

Let us now replace the lenses by small magnetic spheres, from which four attached magnetic rods of equal length protrude. This flexible magnetic net can, indeed, be fitted to the desired forms. In the illustration Fig. 8.30, the net was cast over two spheres.

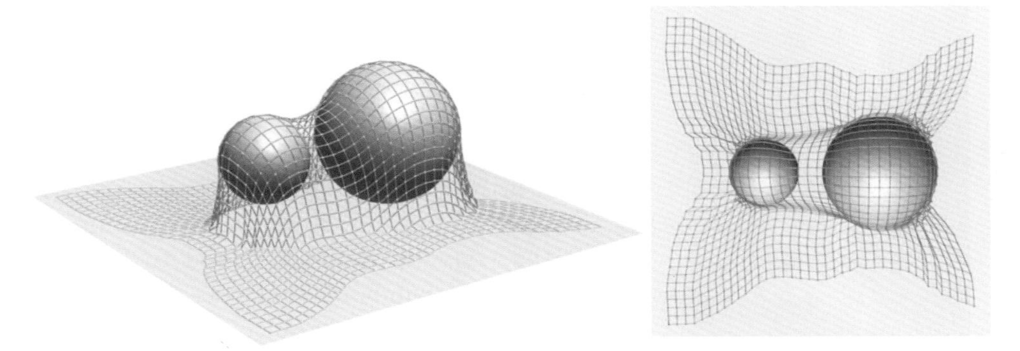

Fig. 8.30 Casting the net over a scene: A "quadrangulation" of a different kind ...

A possibility was thus discovered to approximate free-form surfaces through "magnetic nets". Groups of four rods are thereby arranged into "skew rhombi", which may be decomposed into two triangles along one of the two available diagonals. The resulting triangulation has the advantage that all triangles are isosceles and of "similar size".

Fig. 8.31 Load tests in theory and practice ...

As advantageous as the new method of triangulation may seem, the boundaries appear to be out of control. One may ponder what would happen if the four corners of a rectangular magnetic net are specified in their final position. The solution of this problem is more complicated and does not afford an exact solution in the general case. This seems to necessitate the stretching or compression of the rods – by an amount as small as possible. Fig. 8.29 on the left illustrates by use of yellow and red colors where a strong material strain may occur in a practical construction.

The project thus reached new dimensions and culminated in a simulation program for the manufacturer of a magnetic-geometric "toy". Several renderings of that program can be seen in this book (such as Fig. 3.35).

Fig. 8.32 "Triangulation" – detail

A reference to Fig. 8.32 seems appropriate: In her work "Triangulation of a dog", Oona Peyrer-Heimstätt investigates the philosophical and geometric question of whether or not everything may be triangulated ...

A comment from a geometric standpoint: The artist used exclusively *equilateral* triangles (!) – this only works if the attachment of the triangles is not required to be very precise, in a similar manner as the solution to the ceramic lens problem. The exact solution of a triangulation by using equilateral triangles theoretically leads to parts of much simpler base solids composed of equilateral triangles.

A different kind of architectural draft

We have often mentioned the fact that developable surfaces play an enormous role in architecture, for they can be constructed without resorting to the tedious technique of triangulation and lose none of their curved beauty. The examples seem limitless: Fig. 8.33 once again shows two lovely applications. A further question arises: How should architectural forms be designed so that they are guaranteed to be developable? Architects have their own methods to accomplish this task: They work, for instance, with bendable strips of varying materials which are then fitted into the desired arrangement (Fig. 8.34).

Fig. 8.33 Developable facades exhibit an attractive elegance. On the left: Santiago Calatrava's opera house in Tenerife (cone parts), on the right: Frank O. Gehry's New Zollhof in Düsseldorf

Once a model is supposed to enter an advanced phase, it needs to be transferred into the computer program. At this point, some small compromises have to be made: No program – so far – supports the construction by means of bendable strips. What was previously developable now becomes a general free-form surface. The difference may not be visually perceptible, but when the time arrives for the practical construction, the simple construction from developable parts is largely over, and the facades will have to be triangulated or manufactured in a complicated doubly curved way.

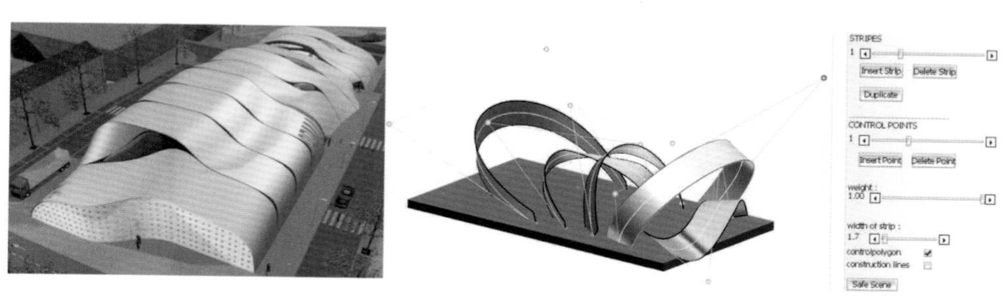

Fig. 8.34 Classic sketch using strips. On the left: Lars Spuybroek, Centre de Musique Pop, Nancy. On the right: The respective sketch using a custom-built computer program.

The University of Applied Arts Vienna attempted to develop such a software package and published the results ([14]). It permits the use of curves instead of surfaces during modeling (Fig. 8.34 on the right). These are then interpreted as the midlines of "paper strips", which are also the rectifying

developables of the curves. The curves themselves can be created and manipulated through NURBS.
During practical construction, some challenges necessarily arise, which can luckily be solved: The midline cannot possess any arbitrary form, because the developable that is derived from it must not self-intersect. Geometrically speaking, this can be avoided by calculating the edge of regression of the developable. If it approaches the midline "too closely", then the designer must not induce any further deformations in that area. This may be a restriction during design, but it also guarantees that the shape is realistic and does not need to be manipulated at a later stage.

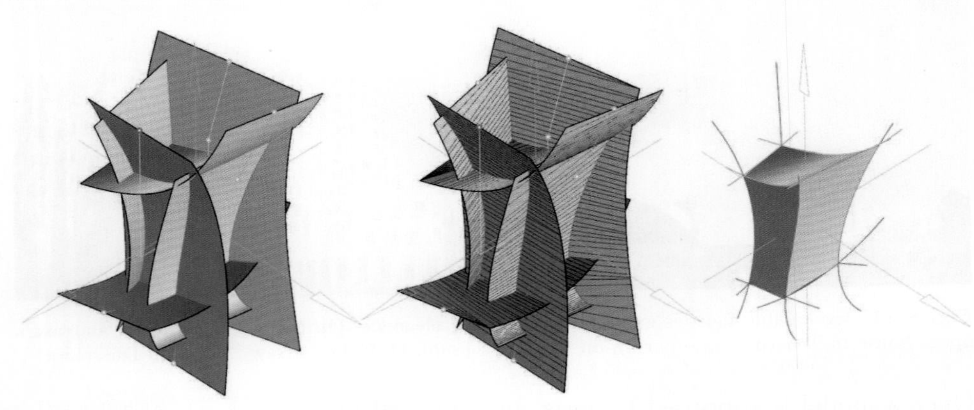

Fig. 8.35 Design using strips: In conjunction with modified Boolean operations, only small numbers of such elements are necessary to create interesting developable shapes.

9 Photographic image and individual perception

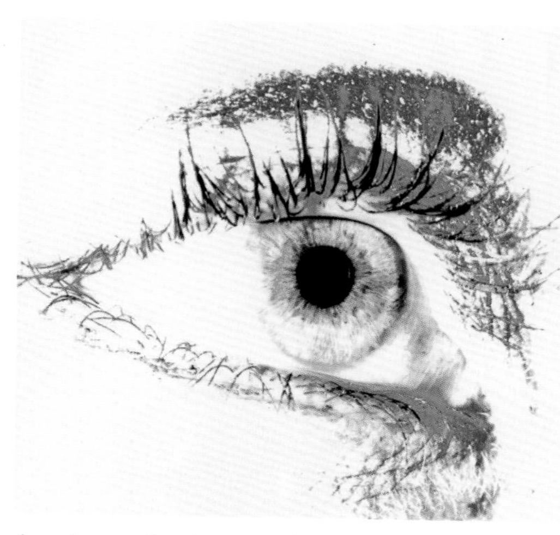

This chapter is dedicated to central projection, which plays its part in both single-eyed human vision and in photographic imaging. A simple pinhole camera is enough to realize this projection, but the way in which the human eye resolves the image is much more complicated due to the fact that the image plane is not flat. Ever since the painters of the Renaissance, "perspective" has been brought to a high standard. The knowledge about vanishing points and vanishing lines (such as the horizon) is especially important in this context.
We shall discuss several quick techniques for achieving good results in perspective drawing, even though the level of difficulty remains considerable. A complete understanding of the "intersection method" is essential for this purpose.
Although complex perspectives are not often drawn these days, a theoretic understanding is required in order to rectify and interpret such perspectives in a spatial manner (photographs, for instance).
Finally, we shall discuss alternate perspectives, where the projection falls onto cylinders or doubly curved surfaces – which is also true for natural vision. The decisive element of perspective is the position of a future spectator, because the observation of a perspective is another central projection. One should aspire to a reasonable combination of both projections in order to produce the right visual effect.

Survey

9.1 The human eye and the pinhole camera

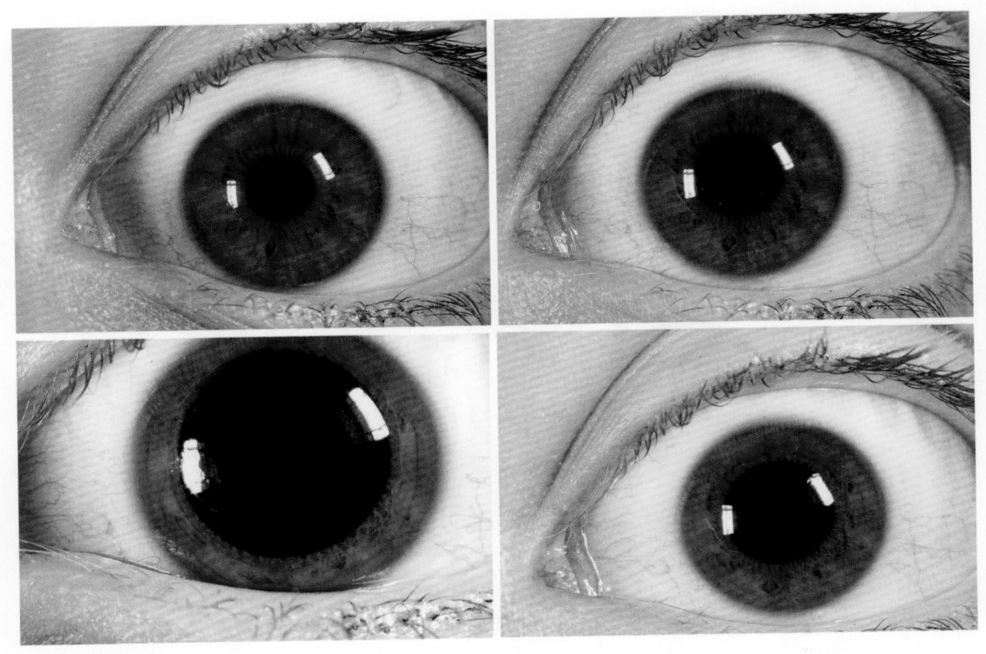

Fig. 9.1 Iris and different "apertures"

For a long time, it was believed that the perspective images created by the pinhole camera (and thus, essentially, any normal camera) are virtually identical to the visual image of the three-dimensional world that we perceive with one eye closed. The famous experiment by Filippo *Brunelleschi* was presented as "proof" of this claim in the year 1401(!):
A spectator looks through a tiny hole in the wall onto a spatial object (the experimental object was the Florence Baptistery). This hole is then covered, and a realistic perspective painting of the spatial object is inserted between the spectator and the object. When the hole is then uncovered, the spectator would fail to notice a difference and would still believe to see the original object (Fig. 9.2).
Brunelleschi was even more ingenious than that: He inserted a mirror instead of a painting, and the perspective was painted on the backside of the wall through which the spectator was looking – which amounted to the same effect.
In fact, if the eye is forced into a specified position with regard to the observed object, then the visual perception is identical to that of a central projection from that position. If this restriction is not given, then the spectator would, by a mere change of position, quickly come to the conclusion – even in a darkroom under otherwise perfect conditions for "illusions" – that he or she

is looking at a two-dimensional picture which appears unrealistic from other points of view.

Fig. 9.2 The ingenious idea of *Brunelleschi* – interpreted by Otmar Öhlinger

Spheres require a special sensitivity. In classical perspective (photography), the contours of a sphere most often appear to be an ellipse (the remaining conic sections rarely occur in practice). The contour ellipse is at its most eccentric if the center of the sphere is farthest away from the optical axis (the *principal ray*). If, however, a spectator is positioned in the right place, then the sphere contours will without a doubt be interpreted as circular, and no wonder: The projection cone of the ellipse through the eye is a cone of revolution. If the observer leaves the ideal position, then the projection cone becomes a general quadratic cone, and the observer will notice that "something appears to be wrong with the sphere".

Fig. 9.3 Almost circular contours of a sphere near the center of the image!

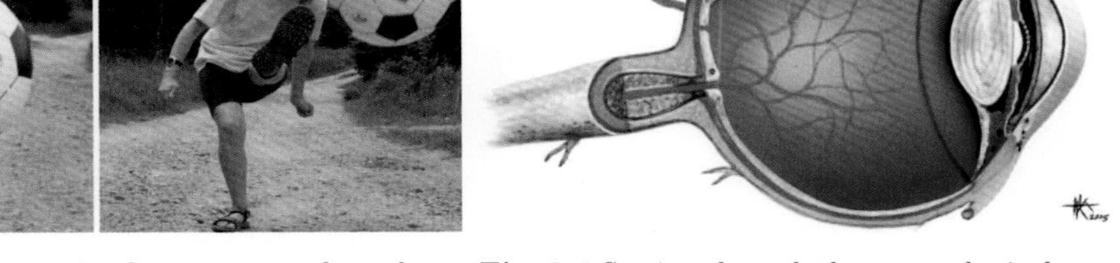

Fig. 9.4 Section through the near-spherical eyeball

By looking at a sphere, we unconsciously move it into the center of our field of view. The optical axis becomes the connection between the eye and the center of the sphere. This may occur either through a rotation of the head or through the largely equivalent rolling of the eyes. The contour of the sphere now appears as a circle. Yet, how is this circle transferred, via perspective, onto our retina? Due to the laws of optics (and our vitreous body with the built-in optical lens is exactly such an optical device), a cone of revolution, whose axis corresponds to the optical axis, is to be intersected with a sphere (the near-spherical retina). This, indeed, produces a circle on the curved retina. The optical nerves need only to transmit the image into the brain, where – due to visual experience – the curved impression is imagined as a spatial object. A second image, generated by the second eye, is simultaneously processed.

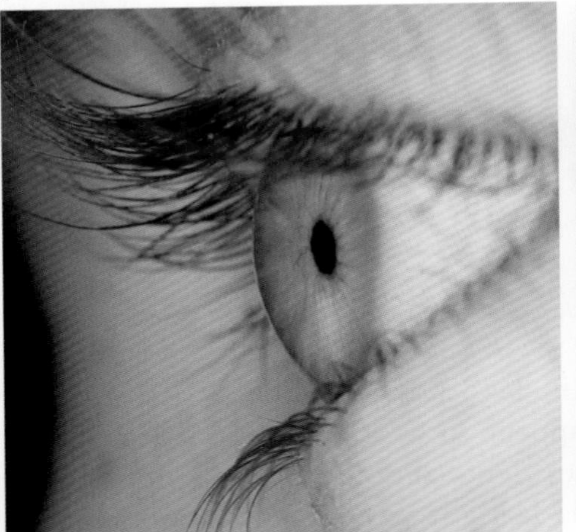
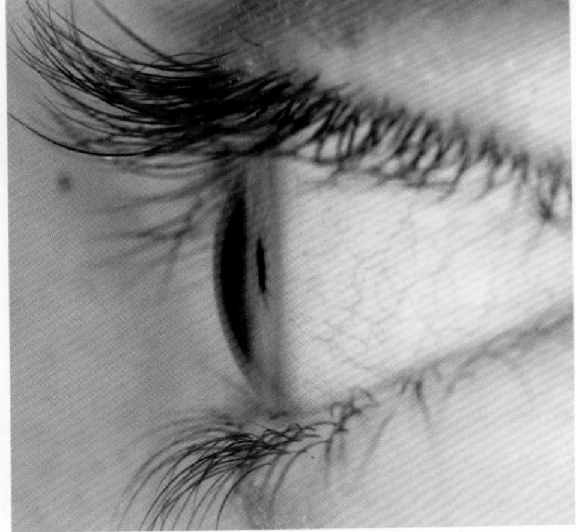

Fig. 9.5 Iris and pupil, curvature of the cornea; right: strange effects due to "image rising"

This whole procedure is anything but simple. However we may perceive three-dimensional space, the following model is very fruitful for its understanding: We do not, in fact, measure lengths but visual angles. For us, a circle is not a curve whose points are equidistant from a given center. Instead, a circle is a curve where the lines of sight through their points enclose constant angles with the optical axis, producing a cone of revolution. Curves that run across the same cone of revolution, indeed, tend to be intuitively interpreted as circles – until a change of eye position convinces us of the false impression.

Fig. 9.6 In perspective with vertical image plane, the images of all points at eye level lie on the horizon. The associated lines of sight are – as the name implies – "horizontal".

9.2 Different techniques of perspective

This chapter concerns perspective constructions which, in the days of old, provided considerable assistance during the creation of perspectives. In the age of computer graphics, where the art of manual perspective drawing is dying out, there still exist substantial reasons for understanding the procedures behind construction. Firstly, sketches with a correct perspective may still be in demand in the future, though this may only concern the artistically inclined. Secondly, knowledge about correct construction may help in the approach to highly relevant contemporary problems – especially those which fall into the purview of "reconstruction" – right up to the understanding of 3D-scanners!

Fig. 9.7 In both cases, the principal projection ray (the optical axis) is horizontal. On the left side, it is even perpendicular to the back wall (frontal perspective).

Fig. 9.8 In both cases, the principal projection ray is directed upwards – on the left in a perpendicular manner. In both cases, the goal was to preserve the symmetry of the main directions (Cathedral of Trier).

The intersection method

A perspective image is the result of a central projection. Points in space are depicted by projection rays that are intersected with the image plane. Technical objects tend to have pairwise orthogonal "principal directions". We shall first assume the image plane to be parallel to the upright direction.

This allows us to gain a better understanding of certain procedures – a later generalization should not cause any problems.
In Fig. 9.9, the object is positioned oblique to the image plane π (see the top view). The desired perspective is the intersection of the *viewing pyramid* emanating from the eye point E with the image plane π. Due to the special assumption ($\pi \parallel \pi_2$), the front view of the figure that is cut off from the pyramid is identical to the desired perspective image. The front view of E is the *principal point* H of the perspective. The orthogonal distance $\overline{E\pi} = \overline{EH}$ is the *distance* d. The circle centered at H with radius d is called the *distance circle*.

- *Horizon* and *principal vanishing points*

  The base plane γ is horizontal. Each point at infinity in γ possesses an image which is called the vanishing point (on the vanishing line through γ and the *horizon* h) and which in this constellation passes through H. In particular, the points at infinity of both horizontal principal directions possess the vanishing points F_1 and F_2, which can be generated by parallel translation of the directions through the eye point. This construction leads us to the following (compared to Fig. 1.38):

  The vanishing points F_1 and F_2 of orthogonal directions lie anti-inverse with respect to the distance circle.

  The distance circle should not be missing in any perspective image. All parts of the image that lie outside of this circle are highly distorted and would usually appear "unnatural" to a spectator – we shall later discuss this matter in more detail. "Classical perspectives" normally lie well-embedded in the "half distance circle". This means that the *viewing cone* around the optical axis, which is embedded into our scene, possesses an opening angle of more than 50°.

  If we change the image plane solely by parallel translation, the perspective image is merely enlarged or downscaled. If, however, the object is translated orthogonally to the image plane, its relative position to the eye point is affected – including, of course, the perspective impression. In computer programs which generate interactive perspectives on the monitor, a parallel translation of the image plane is equivalent to "zooming". The attempt to accomplish a usual enlargement of the image by moving closer to the object leads to new (and increasingly extreme) perspectives.

- Special and general points

  It is particularily easy to calculate perspectives of points which lie in π. These are simply the front views of the points. The drawing technique required for this situation is the reason why the front edge of the house (through the point A) was assumed to lie in the image plane. Let us now

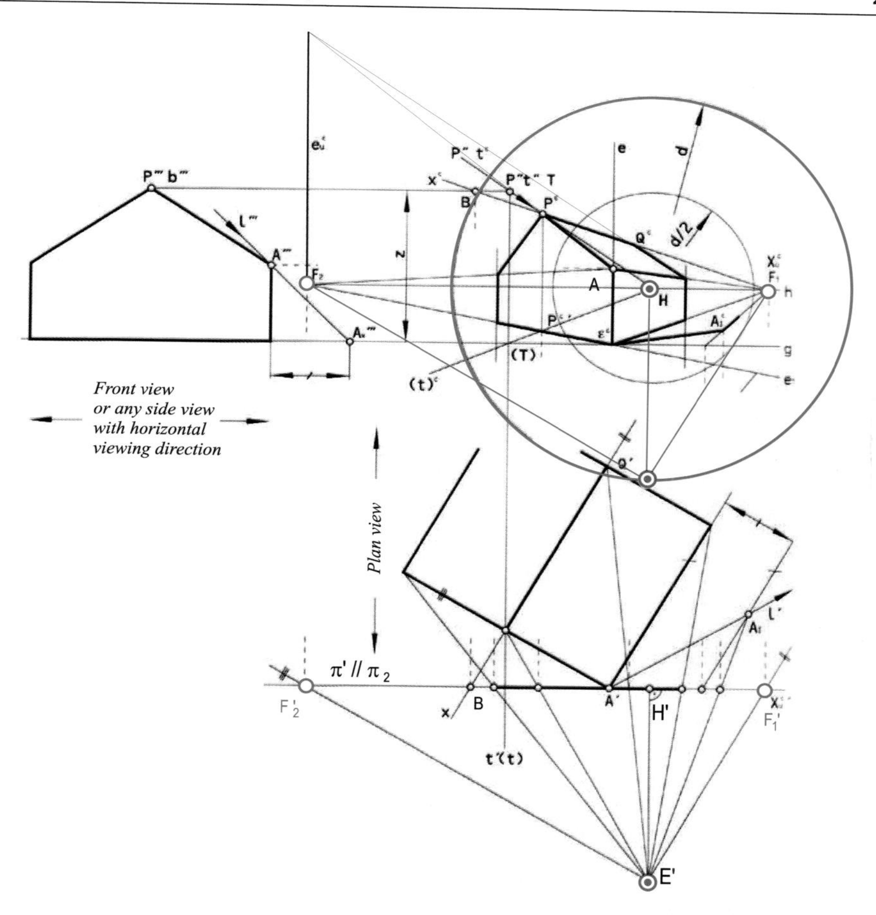

Fig. 9.9 Intersection method

try to depict a general point P at the crown of the roof. In top view, its projection $P^{c\prime}$ can immediately be found. In order to find the front view $P^{c\prime\prime}$ (and thus the perspective image P^c), we have to enlarge the crown up to the point B in the image plane. Its point of intersection B with the image plane lies at the undistorted height of the crown. Thus, the image of the crown can be drawn (BF_1). P^c lies on the order line. It is evident that in the best case, points should be "reconstructed" by special straight lines, where:

Straight lines are defined by trace point and vanishing point.

It would also be possible to reconstruct P on a so-called *depth line* perpendicular to the image plane. Depth lines vanish in the principal point H. Thus, P^c also lies on $P''H$.

- *Trace lines* and *vanishing lines*

Since our approach worked so well with straight lines, let us attempt an analogous one with planes. The horizontal base plane γ possesses a trace g in π, which is called the *base line*. The vanishing line of γ is the horizon h. Let us now consider the vertical plane ε defined by the left wall of the house. It possesses a perpendicular trace e and a parallel vanishing line e_u^c through F_2. All vanishing points (images of points at infinity) of ε lie on this vanishing line. In the image, the pairwise parallel oblique roof edges, thus, have vanishing points on e_u^c (compared with Fig. 9.15).

Fig. 9.10 Sunrise and moonset

• Shadow constructions in perspective

In perspective, a layman may not notice a substantial difference between parallel lighting and central lighting. Both sun and moon are points (or rather, small circular discs) on the firmament from which light rays are emitted. By use of vertical "light planes", it is possible to determine the shadow of a point P given its top view.

Fig. 9.11 The "sun point" and the not-so-parallel-looking sunlight

The additional condition that knowledge about the top view of a point must be given is usually met in architecture. The principle remains the same: Two planar views of a point in space must be known, otherwise its position is not uniquely determined.

A light source, in particular, is only "fixed" by knowing its top view. If the light source is the sun, then its top view – a point at infinity – lies on a perpendicular line through the image L_u^c of the sun and on the horizon. $L_u'^c$ is called the foot of

Fig. 9.12 Back light: parallel lighting on the left, central lighting (flashlight) on the right

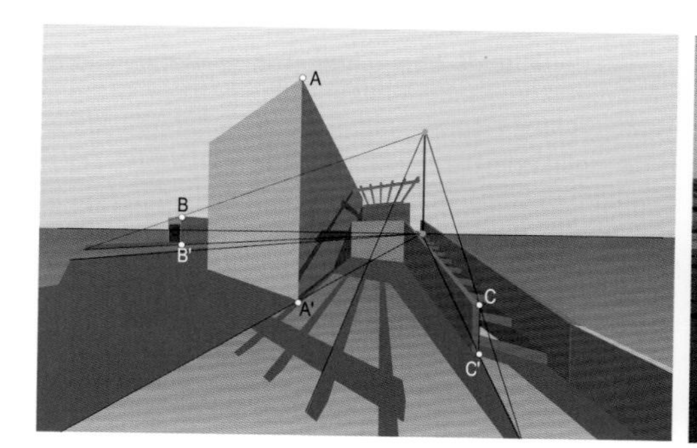

Fig. 9.13 Sun point and sun foot in theory and practice

the sun. If the sun is visible above the horizon, then its foot lies below L_u^c – and back light is given.

In cases of back light, the sun is not visible in the image. It can, however, be determined geometrically (like any other point), and its (theoretical) image lies below the horizon – although this simple fact may appear very unusual to the amateur. In both Fig. 9.14 and Fig. 9.15, the sun is blocked by the back of the spectator.

In the special case of sidelight, the light inclines parallel to the image plane, and both sun point and sun foot are points at infinity. All subsequent constructions are also possible by means of continuous parallel translation.

In Fig. 9.14, the shadow of P on the base plane is determined by connecting the point to the sun and intersecting the resulting light ray with its top view – the connection of the sun foot with P'. The horizontal straight line through P is parallel to the base plane, and its shadow is, thus, equally parallel and possesses the same vanishing point as the straight line.

The trace of the light plane λ through P in the horizontal base plane is determined by P' and the sun foot $L_u^{\prime c}$. The vanishing line l_u^c of the light plane through a vertical straight line connects the sun point and the sun foot. In Fig. 9.15, the shadow of a rod PP' onto an oblique roof plane is determined. For this purpose, the light plane λ is intersected with the building. The oblique portion of the shadow is part of this intersection figure. In order to verify the result of the construction, it is also possible to intersect the

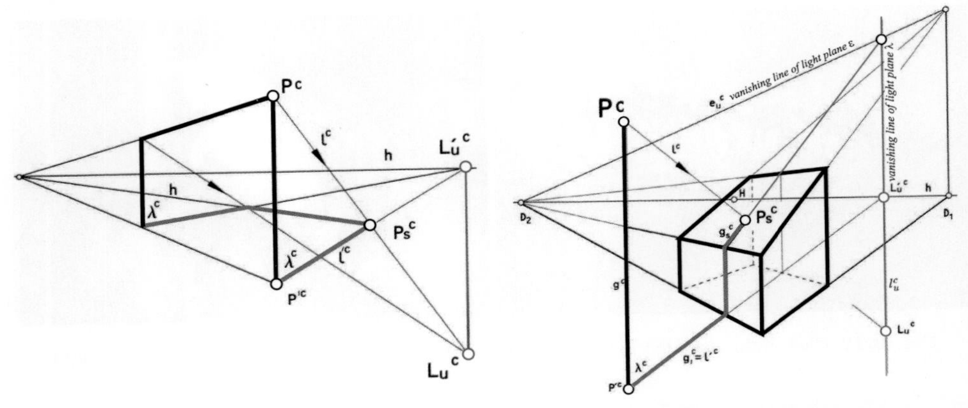

Fig. 9.14 Light planes ...

Fig. 9.15 ... and roof planes

vanishing line of the light plane with the vanishing line of the oblique roof plane.

If the sun were replaced by another (and much closer) light source, not much would change about the construction: The only difference would be that the foot of the light source would no longer lie on the horizon. ♠

• **Shadow of sidelight and general perspective** (Fig. 9.16)

General perspectives – central projections with non-vertical image planes – are usually much more difficult to construct. In the image, perpendicular edges pass through a third principal vanishing point F_3, and it is no longer possible to inscribe heights by simply transmitting affine ratios. In such cases, it is helpful to continue assuming that the image plane remains perpendicular when we apply the intersection method. Important points of the object, however, have to be drawn in a *tilted* manner in the top and front view, which requires one further side view (more detailed instructions can be found on the website to the book).

This example, however, should primarily concern the shadow construction with sidelight. Let P be a point in space with the top view P'. This produces a perpendicular line $P^cP'^c$ in the perspective image which possesses the vanishing point F_3. We shall define sidelight as follows: Light should incline such that *the shadows of vertical lines on the horizontal base plane appear parallel in the perspective image* (Fig. 9.16 on the left). Thus, the sun foot ${L'_u}^c$ becomes the vanishing point of the horizontal direction. The light rays can no longer be parallel to the image plane (as it occurred in the special case of the vertical image plane). The sun point L_u^c lies "perpendicular" above the sun foot, which explains why the order line ${L'_u}^c {L_u}^c$ passes through F_3 and is, thus, horizontal. Any point on this horizontal line could serve as the sun point. This reduces the rest of the construction to a mere providing with lines: The image of the shadow P_s on the base plane can be found by intersecting the straight line $P^cL_u^c$ and $P'^c{L'_u}^c$. ♠

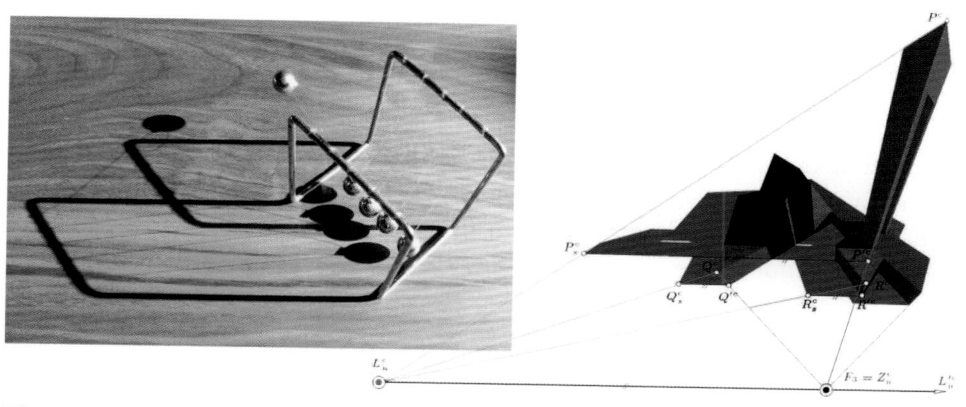

Fig. 9.16 The shadows, given sidelight and a tilted image plane, are defined in the following practical way: The shadows of perpendicular edges on the base plane should be parallel to the image plane and thus appear parallel in the image. The light rays themselves, as they appear in the image, are *not* parallel: They have the vanishing point L_u^c.

In principle, the intersection method enables us to construct arbitrary scenes, and by adding the aforementioned technique of shadow construction to our toolkit, we could reasonably consider ourselves to be satisfied. Computers calculate perspective images in precisely this way. In practice, however, it may be too cumbersome to "derive" an object's top- and front views every single time, especially as many cases permit a quicker procedure. More instructions can be found in the section concerning perspective sketching from page 460 onwards.

Measuring lengths

The following method is necessary to rectify photographs. Let a photo of a building facade or a photo of a planar landscape be given. The measurements of the various objects are to be determined.

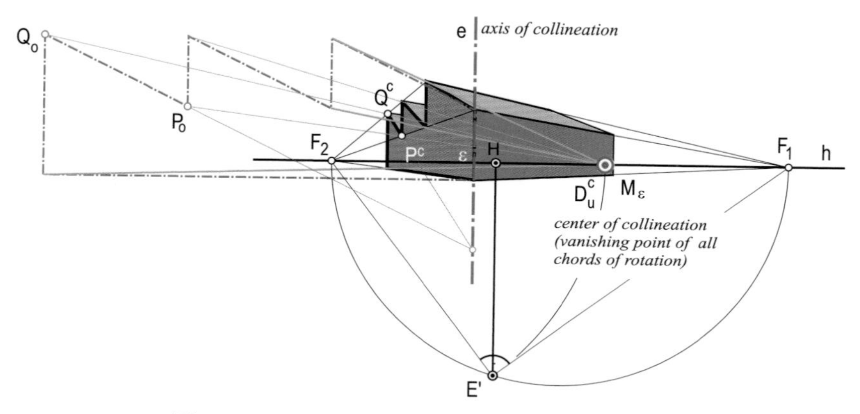

Fig. 9.17 Parallel rotation becomes "recalibration"

Let us first tackle the issue back-to-front. How would we inscribe a planar polygon in a vertical plane ε into the perspective (Fig. 9.17)? If, by mere coincidence, the plane does not happen to be parallel to the image plane, then the polygon will appear distorted in the image. Considered in space, the image polygon can be constructed by intersecting the "viewing pyramid" through the polygon with the image plane. Both planar sections of the pyramid are collinear images of each other.

In the drawing, it is possible to employ a trick which also boils down to the completion of a collineation – though in a different way. If e is the trace of the carrier plane ε, then we can imagine ε rotated about e into the image plane. This causes all points of the plane to become circles of revolution. The rotation, however, can also be replaced by a parallel projection into the direction of the rotation chords. The center of the projection is a point at infinity D_u. In three-dimensional space, the polygon in ε and the rotated polygon in the image plane are related via an affine transformation (axis e, direction D_u of rays).

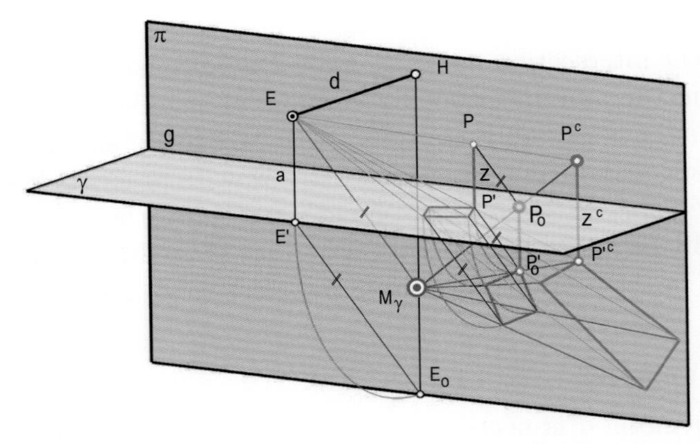

Fig. 9.18 Image of a perpendicular line segment and the corresponding collineation

This relation can be understood by looking at a sketch of the situation in space, where it is evident that the point at infinity D_u is associated with a vanishing point $D_u^c = M_\varepsilon$ at the horizon and at a distance of $\overline{F_2E}$. The measurement point M_ε can be constructed by rotating E about F_2 in a side view. All chords of the rotations vanish at this point in the image, and a planar perspective collineation derives from the relation between the rotated polygon and the polygon in space. The determination of a single pair of associated points allows us to construct images of all other points via a small number of lines.

Fig. 9.18 illustrates the most common case, where the carrier plane of the polygon corresponds to the horizontal base plane γ. The measurement point M_γ on this plane derives planimetrically through the inscription of the distance $d = \overline{EH}$ from the principal point. The axis of rotation (= axis of collineation) is the base line g.

Both the rotated and the regular top view are related via a parallel perspectivity by way of the parallel rotation chords. This makes the central view of

a figure in the base plane perspectively collinear to the "parallel-rotated" top view (center of collineation M_γ, collineation axis g).
The central view P^c of a point P at height z which does *not* lie in the base plane is located above the central view P'^c of its top view. Let us now consider the point P_0 which lies above the rotated top view P'_0 at a distance of z (Fig. 9.18). This point can equally be derived by intersecting the rotation chord through P with the image plane π. Thus, P_0 also lies on a collineation ray through M_γ – a by no means trivial conclusion! Planimetrically, P_0 can also be interpreted as the horizontal view of P (given a distortion ratio of $1:1$). The horizontal and perspective views are, thus, in a simple relation.

- Reconstruction from a photo

If the image of a rectangle is visible on the original photo, then the actual shape of any figure in the carrier plane of the rectangle can be reconstructed.

Fig. 9.19 Planar figures with varying levels of distortion (original photos)

The goal is to find the measurement point of the carrier plane, through which the shape of any planar figure can be derived – even without a ruler.
When dealing with original photos (that is, photos that have not been cropped as in Fig. 9.19), the assumption can be made that the principal point is located at the intersection of the diagonals. As soon as the image of a rectangle is provided, the two principal vanishing points F_1 and F_2 have also been found. The distance circle, on which the measurement point of the plane is located, can be reconstructed using the Thales circle above F_1F_2. ♠

- Spatial reconstruction, when the image of a cuboid is given

This useful practical case usually concerns a photograph where the horizon h does not pass through the principal point, i.e., when we have three principal vanishing points F_1, F_2, and F_3 with $h = F_1F_2$ (Fig. 9.20). The principal point (the orthogonal projection of the center C) is then the orthocenter of the triangle $F_1F_2F_3$. C itself is the intersection point of the three Thales spheres determined by each pair of the principal vanishing points. Again, one can find the measurement point of the base plane on the Thales circle above F_1F_2 on the altitude through F_3. ♠

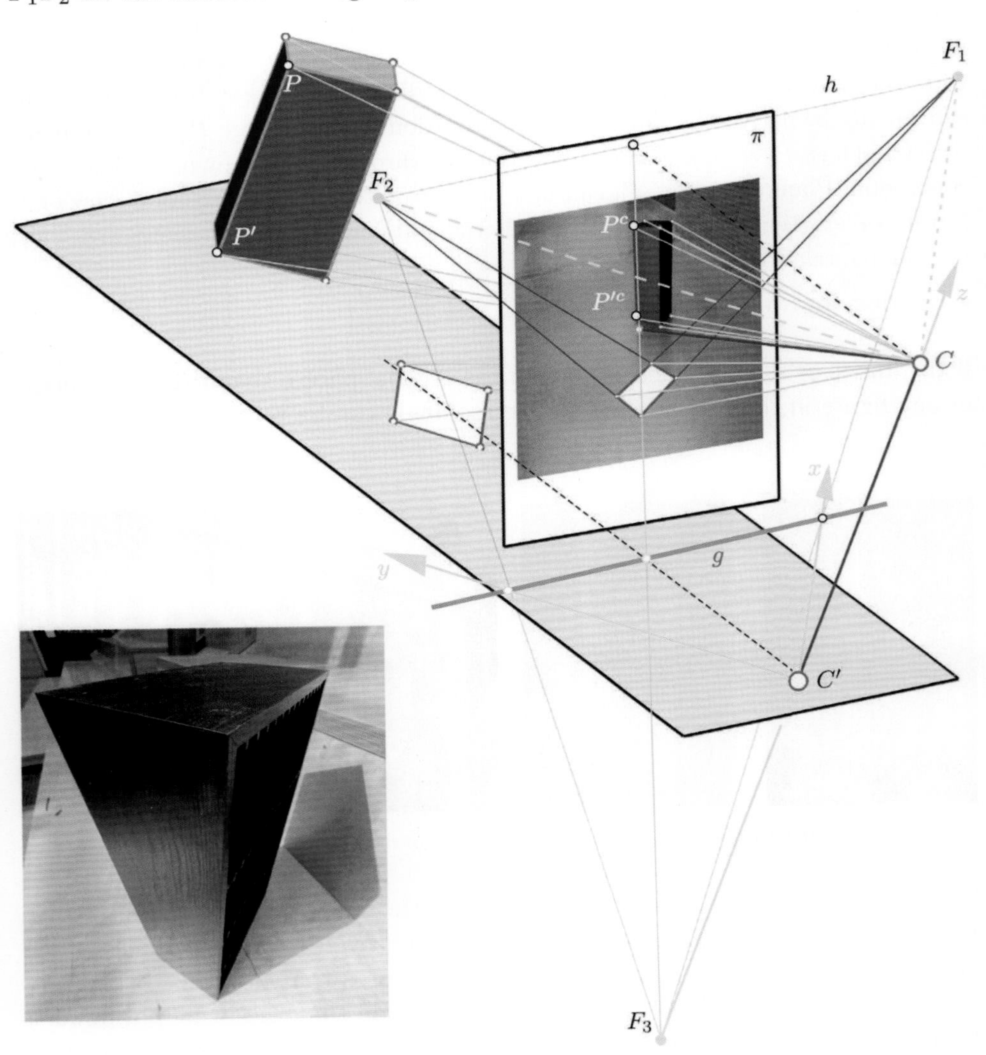

Fig. 9.20 Spatial reconstruction is, in general, possible when we know the image of a cuboid.

• Anamorphoses

In the 16th century, Hans *Holbein the Younger* painted his famous picture "The Ambassadors". An alleged shroud on the floor turns out to be a skull – but only if observed from an extreme angle. A figure that is only visible from extreme perspectives (such as curved mirrors as in application p. 184) is called an *anamorphosis*.

It is possible to "implant" such an anamorphosis into a perspective through a modified measurement point. In the painting, the skull is well concealed, since it only becomes apparent from a low perspective and after a 45° rotation

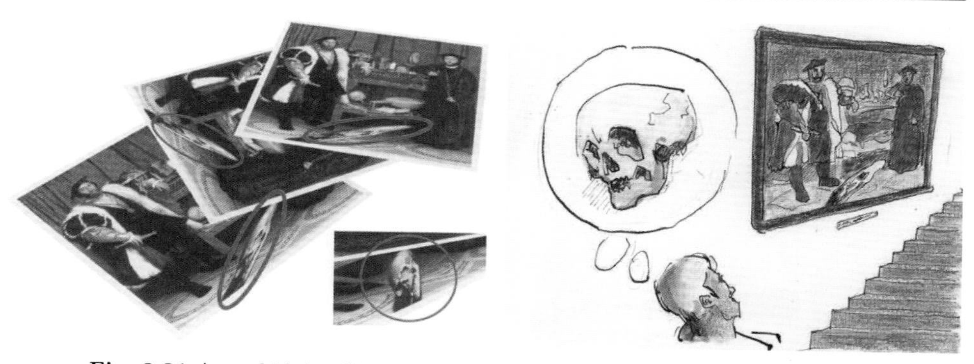

Fig. 9.21 A morbid detail in the famous picture by Hans *Holbein the Younger*

of the image. The picture is brought into position at the London National Gallery such that the skull becomes visible from the stairway (Fig. 9.21 on the right).

♠

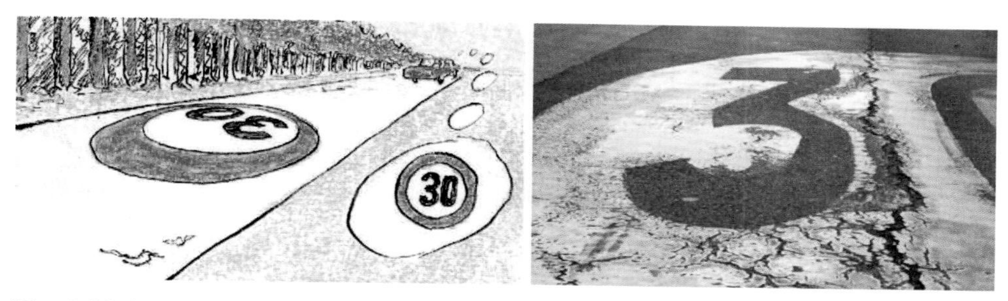

Fig. 9.22 Anamorphoses: "30-limit". The street inscription on the right is not actually a "pure" anamorphosis because the impression of a perspective distortion remains.

Fig. 9.22 (Stefan Wirnsperger) illustrates that traffic signs are often anamorphoses. Due to the extremely low angle of observation, the letters, numbers, and other symbols must often be printed with a certain perspective distortion. Similar challenges arise in the design of billboards on skyscrapers, which people usually observe from far below.

• Deliberate distortions in computer graphics

Perspectives are very frequently employed in computer graphics. One of the elemental tasks of such graphics programs is the so-called *clipping* procedure: Parts of the scene which lie behind the observer or too close to the front clipping plane (parallel to the image plane through the eye) have to be removed. This may be done by intersection with the *viewing pyramid.*

A fast and thus often employed method consists in applying a spatial collineation to all points in the scene – which turns the eye point into the point at infinity of the principal projection ray. This makes the viewing pyramid a coordinate cuboid and thus greatly enhances the efficiency of clipping algorithms.

Fig. 9.23 on the left shows a collinear-transformed object in top and front view. The view from the left is now equivalent to the perspective. It might

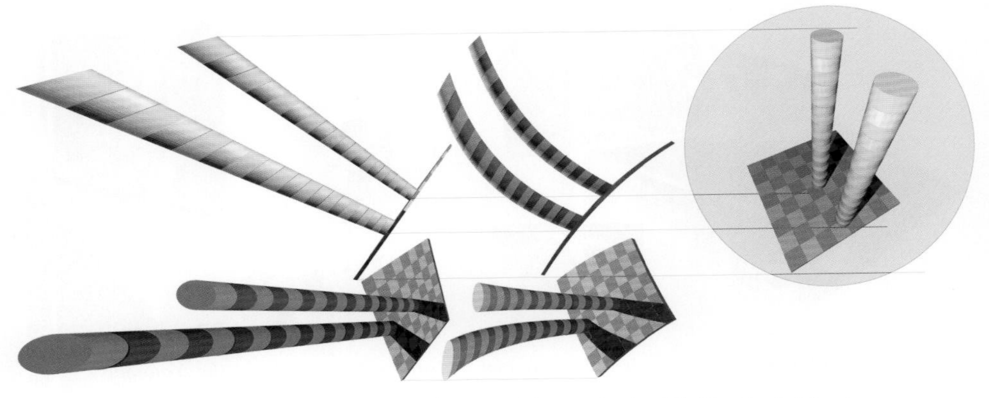

Fig. 9.23 Two different objects (top- and front view) with identical side views

now be maintained that it would be sufficient to inscribe the distance from the image plane above the image points (as illustrated in Fig. 9.23 at the center). Unlike a collineation, however, this transformation would not be linear. The objects are, thus, "distorted" which requires needlessly complex algorithms. ♠

The image of a circle

If the planar figure is a circle, then the corresponding image is a conic section, because the projection of the circle from an eye point E generates an oblique quadratic cone. An intersection with the image plane thus produces a conic section (see also Fig. 3.56). It follows from the definition on p. 49 that:

Fig. 9.24 Ellipses and hyperbolas as images of circles

The perspective image of a circle is an ellipse, a hyperbola, or a parabola, depending on whether the circle does not intersect, intersects or merely touches the vanishing plane.

Let us consider an elliptical image of a circle together with a circumscribed square of tangents, with two sides parallel to the image plane. The center of the circle is located precisely at the square's diagonal intersection. The

Fig. 9.25 Conic sections as images of circles

square's image is a trapezoid (straight lines that lie parallel to the image plane possess equally parallel images). The intersection of the trapezoid diagonals thus yields a point equivalent to the center of the circle – or, the intersection of the images of all diameters of the circle. However, it is not equal to the center of the ellipse appearing as the circle's image.

If the circle lies in a plane parallel to the image plane, then its image is also a circle. The 10-euro-bill contains a perspective image of a bridge which carries circular arcs without any apparent perspective distortion.

Circle depictions in ancient times

Fig. 9.26 Two more than 3200-year-old Egyptian reliefs (Abu Simbel). In principle, we deal with front views and "integrated side views".

Usually, it is believed that realistic and correct perspective drawings with geometrical context started with the painters of the Renaissance. For example, the ancient Egyptians often used "only" front views, and integrated side views into the pictures. (Note the 6-spoke wheel of the chariot in Fig. 9.26 (on the left) which appears undistorted as a circle). However, we have several examples illustrating that this conjecture is not quite true. One of the most impressive counterexamples is a tomb painting in the grave of Philipp III (the successor of Alexander the Great), drawn approximately 320 B.C.

(Fig. 9.27[1]). Not only that the entire painting is impressive, the wheels of the chariot are drawn almost as if their images were constructed. On the right you can see a detail with computer-drawn reconstructed wheels layered upon the image, including the 4-spoke wheels. Especially the wheel in front is amazingly accurate – the other wheel is too little and flat for a penile critic, but that does not diminish the accomplishment of the painter. The four spokes can even be interpreted as "conjugate diameters" of the image ellipse (in space, they are perpendicular to each other and the tangents in the end points of one diameter are therefore parallel to the other).

Fig. 9.27 A 2300 years old Greek tomb painting (Vergina, central Makedonia). A realistic perspective drawing: the closer wheel is depicted perfectly!

9.3 Other perspectives images

Optical illusions

Fig. 9.28 Painting on walls in Pompeii: Perspectives create the illusion of additional space.

[1] `http://commons.wikimedia.org/wiki/File:Painting_vergina.jpg`

Optical illusions seem to be very old. Fig. 9.28 shows two walls in houses of Pompeii (the ancient Roman city was destroyed and buried under ash and pumice following the eruption of Mount Vesuvius in 79 A.D.). One can still see that artists painted columns and other architectural objects on the walls in order to create the illusion of a spatial scenery behind the wall. This provides another example that certain knowledge about perspective drawing existed long before the Renaissance.

Fig. 9.29 Pseudo-perspective I

Set decoration employs the term *relief perspective*, which describes the illusion of a larger space that would never fit on the actual stage. Fig. 9.29 shows a hall of columns that appears realistic, but only if observed from the front. It would indeed be a fitting backdrop to a dramatic event such as the assassination of Julius Caesar. A spectator in a cheaper seat will, however, recognize that the supposed right angles are no right angles at all.

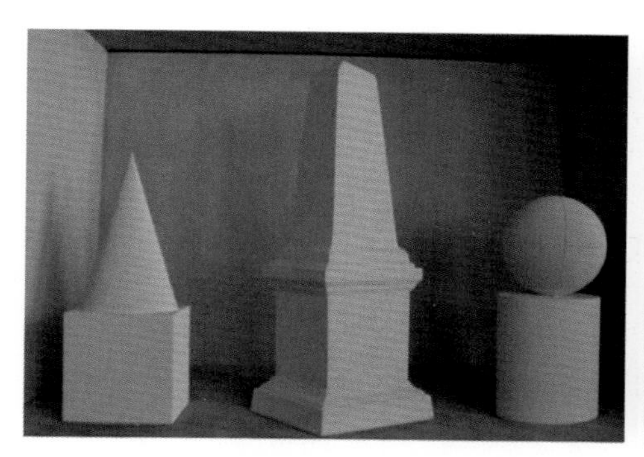
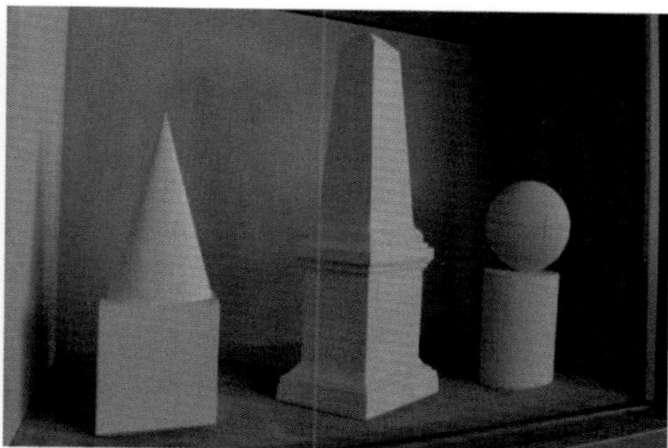

Fig. 9.30 Pseudo-perspective II

Fig. 9.30 demonstrates by using simple geometrical objects (square prisms, pyramids, spheres, and cones) to what extent our imagination still accepts an oval as being spherical before noticing a definitive geometric deviation.

• The Gaussian collineation: Virtual 3D-images

Similar to the already mentioned relief perspective, we have a collineation in optics (Fig. 9.31): When we apply a refraction at a convex lens, there are two rules according to the laws of reflection: Light rays h that go through the center C of the lens are not refracted ($h = h^*$) and light rays s parallel to the optical axis are bundled in the focal point F^* behind the lens. The refraction $s \to s^*$ occurs in the principle plane $\gamma \ni C$ perpendicular to the axis. A space point $P = s \cap h$ corresponds to a (virtual) space point $P^* = s^* \cap h^*$.

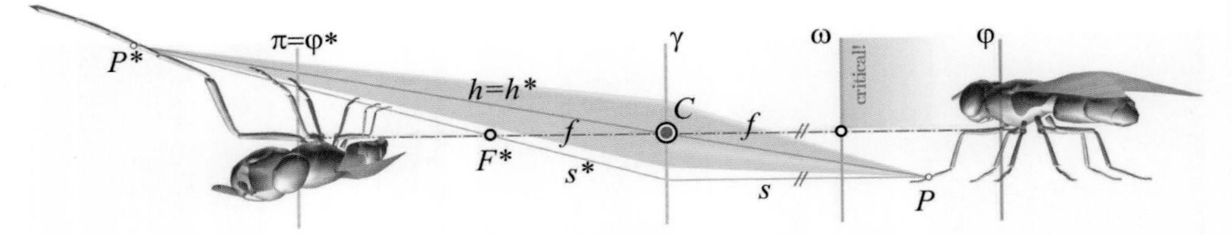

Fig. 9.31 A spatial collineation of space

After Gauss, this (reversible) relationship $P \leftrightarrow P^*$ deserves the name *collineation*, i.e., straight lines g are transformed into straight lines g^*.

Proof:
We consider g to be the intersection of a plane ε_1 parallel to the axis and the connection plane $\varepsilon_2 = gC$ with the center. The first plane is transformed into a plane ε_1^* through F^* and the intersection line $\varepsilon_1 \cap \gamma$, whereas $\varepsilon_2 = \varepsilon_2^*$ stays unchanged. Thus, g^* is the straight intersection of two planes. ◇

Parallel planes φ and γ are transformed into planes φ^* that are also parallel to γ. It is useful to know this when it comes to photography. Therefore, we will come back to the Gaussian collineation and its consequences in our geometry-based photography course (Appendix B). ♠

- **Two different sequences**

Fig. 9.32 Real and pseudo-perspectives

Fig. 9.32 shows two arrays of Buddha statues. On the left, all figures are of equal size and it is possible to move freely in their vicinity. The statues on the right, however, were stacked on top of each other in a cave. They can only be seen from a single position near the cave entrance. The figures are gradually smaller and smaller, which is barely noticeable due to the lighting effects, and this makes the cave appear deeper than it is. ♠

• **Geoglyphs in the desert sand**

The pre-Columbian Nazca culture (300 B.C. to 600 A.D.) in the southern Peruvian coastal valleys used to decorate their pottery and embroideries with stylized animal motifs. They are most famous for scraping gargantuan animal depictions, often many hundreds of meters across, into the dry desert ground. These images can only be seen in their entirety from an airplane (Fig. 9.33 on the right) and even appear perfectly symmetric from certain angles – a masterly accomplishment, which also led to far-out speculations. Some, for instance, have claimed that the Nazca people must have employed hot-air balloons!

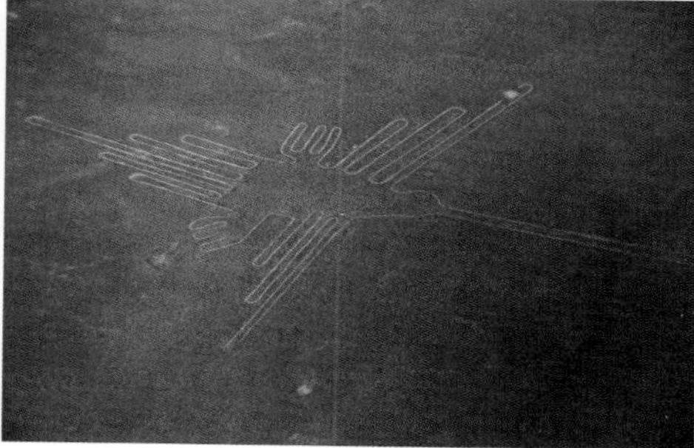

Fig. 9.33 In principle, these are planar figures.

It may be noticed that the giant animal figures *do not appear to be symmetric* from straight above, but rather seem to exhibit *perspective distortions* as in a skewed photograph from above (or a skewed photograph from below – see Fig. 9.33 on the left). Quite paradoxically, this leads to an appearance of symmetry, but only from certain positions, which some commercial airplanes luckily cross on their flight path.

Without any claim to truth, one may imagine the following story – inspired by projective geometry – of how these figures came about:

A high priest stands on a $h = 15$ meters tall wooden tower (apex Z, top view Z', Fig. 9.34 on the left) holding the *symmetric* drawing of an animal. He then directs his subjects to lay stone blocks in a way that would look symmetric and consistent with the image *from his point of view*. Of course, the subordinates have to work at quite some distance from the tower – otherwise, the priest would need to have a "panoramic view".

This would make the animal motif appear symmetric from the skewed, low perspective of the tower, while seeming distorted like a perspective view from far above. Now for the enlargement of the image: A rope is stretched from $Z_0 = Z'$ to one of the many contour rocks P and the rope length $\overline{PZ_0}$ is applied multiple times in the course of the enlargement process(e.g., $k = 3$

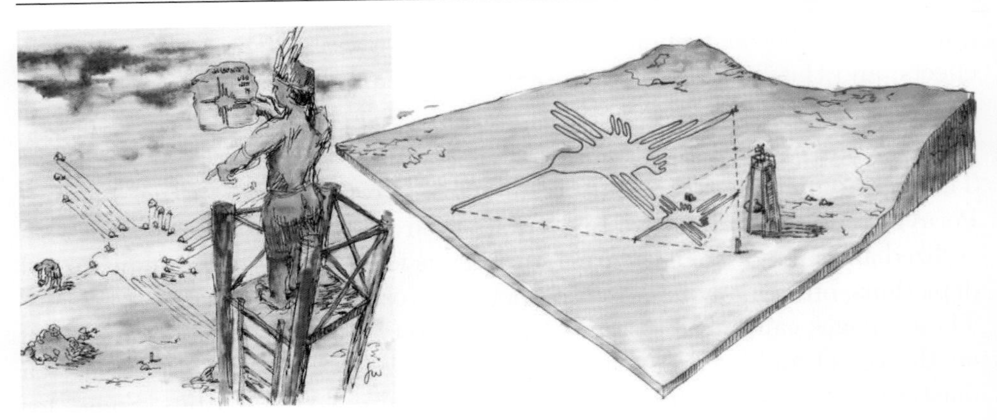

Fig. 9.34 It could have been like this ...

times as in Fig. 9.34, or at even larger scales). This must involve at least two people: One person at the center directs the other at the end of the rope, which ensures the straightness of the ray PZ_0. At the end of the enlarged distance, the point P^* is marked with another rock.
This technique makes it possible to construct an enormous contour in a relatively short period of time. As in the real Nazca lines, the image would appear perspective-distorted in top view and symmetric if seen from an enlarged tower (height $k \cdot h$ meters, in our case $3 \cdot 15 = 45$ meters)!
It can further be concluded that the scaling center Z_0 does not necessarily have to be the top view Z' of the tower apex. In the general case, the point from which the image appears to be symmetric lies in the elongation of the segment $\overline{Z_0Z}$, and it is scaled with factor k and not above the tower apex.
A further remark that might strengthen the plausibility of the above speculations: It would, of course, be possible to enlarge a smaller motif – perhaps only a few meters in diameter – using the "rope method". However, the scaling factor would have to increase, which would likewise lead to greater errors and imprecisions. What's more: Why would the Nazcas create perspective-distorted, non-symmetric images on a very large scale if their original images (hummingbirds, spiders, etc.) *are* perfectly symmetric? ♠

• Baroque illusions

In the 17th century, it used to be the fashion to paint ceilings in such a way that a spectator looking up from the center of the room would have the illusion of a much bigger space.
In producing a correct painting, a perspective image which, needs to be drawn on a plane (such as the ceiling). From a chosen point of view, it gives the illusion of a much larger space "with all the bells and whistles" (clouds, columns, etc.).
Andrea *Pozzo* was a true master of this genre. He was even capable of correctly painting barrel-shaped interiors, and used a special technique (Fig. 9.35) to accomplish this goal: He first constructed a planar "auxiliary ceiling" which

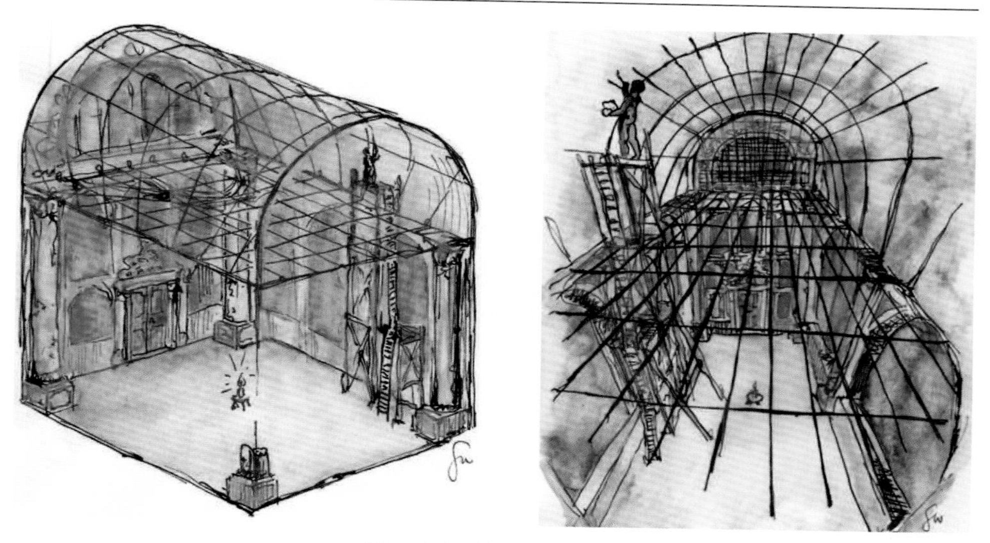

Fig. 9.35 *Pozzo* at work

consisted of a square net of tight strings. He then projected this net on the curved interior by placing a candle in the spectator's position and interpolating the painting on the "reference net" – a shadow of the net cast onto the interior by the candlelight. (The illustrations are by Stefan Wirnsperger.)

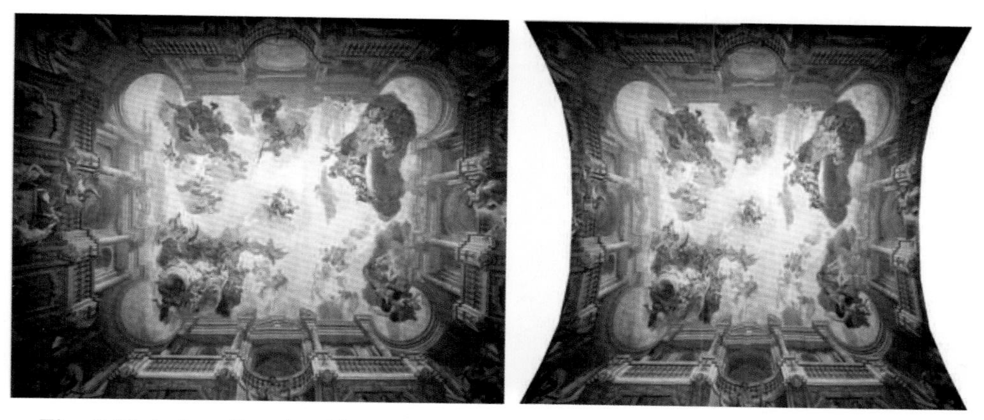

Fig. 9.36 Andrea *Pozzo*'s ceiling paintings and his projection to a horizontal ceiling plane

The inverse of the procedure would be the following: Consider the pattern of straight lines and the orthogonally intersecting half-circles on the cylindrical ceiling. If this curved net is then projected into the virtual ceiling plane, it would generate a net of parallel straight lines and hyperbolas. The latter result as section curves of the plane with the projection cones of the circles, which are oblique circular cones. In order to get the same impression of a *curved* painting from the projection center as in Fig. 9.36 on the left, one might choose to observe a distorted painting (as in Fig. 9.36 on the right) on a *planar* ceiling. ♠

• **We are only measuring angles!**
We follow an idea by Franz Gruber: Let us imagine ourselves sitting in a lecture hall, looking at images cast onto the screen by a projector. In such cases, it is advisable to avoid sitting at the sides. Under ideal conditions, we would want to sit near the projecting apparatus (Fig. 9.37), since this would enable us to see the pictures as they were taken.

Fig. 9.37 A dia projection seen from two different viewpoints ...

Let us now imagine that a prankster pulls on a rope, opening a trap door on the ceiling which causes paper snippets to rain from above – and straight into the light cone. Spectators on the sides would no longer be able to discern much of the projected image. Spectators near the projector, however, would barely notice a thing!

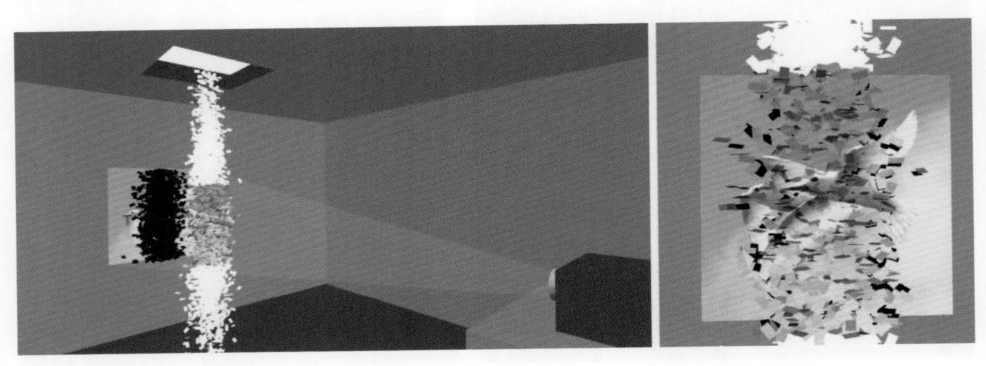

Fig. 9.38 ... suddenly disturbed by a rain of paper snippets

What actually happens is a perfect illusion: The paper snippets are, after all, illuminated by the projector as soon as they enter the light cone. Onto each snippet, a small part of the original image is being projected at varying levels of distortion. The view is further confused by shadows of the snippets appearing on the projection screen. Seen from the lens center, however, the projected image remains coherent, since only angles can be measured and the medium onto which the projection takes place is unimportant. ♠

• **Star signs ...** (Fig. 9.39)
To find the North Star, one must first discover *Ursa Major* (the Big Dipper)

in the night sky. The North Star is located approximately at the five-fold elongation of the first stars in the constellation.

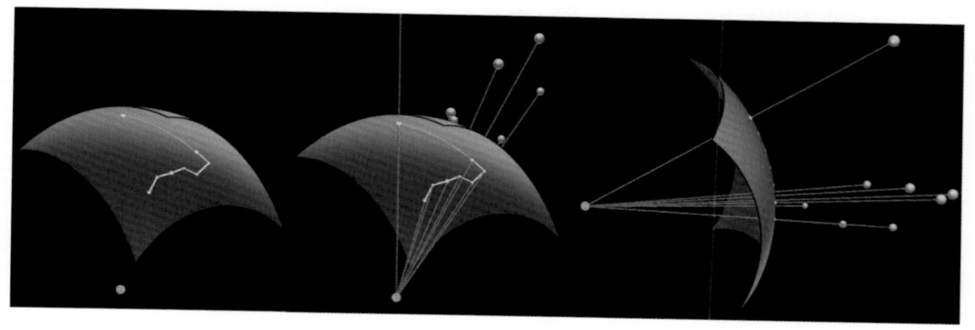

Fig. 9.39 The striking constellation of Ursa Major ...

The spatial constellation of the stars is indiscernible by the naked eye, since the stars may vary in luminosity and in their position on their projection rays. Due to the enormous distances involved, we also fail to notice a difference when observing the constellation over a half-year interval (and thus at a distance of 300 million kilometers). ♠

• "Smoothing" undesirably extreme perspectives

We already know that it does not matter onto which surface an image is projected, if only the spectator is forced into the "right" position. Certain circumstances may, therefore, necessitate the production of extreme perspectives.

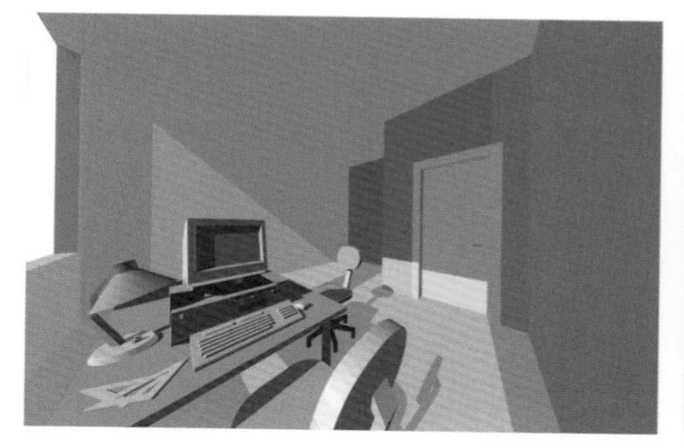

Fig. 9.40 Extreme perspectives

If a comprehensive picture of a small room is to be taken, an extreme wide-angle lens is usually employed. The result might look like in Fig. 9.40 on the left. If such a photograph is then projected on a wall, and if a spectator is forced into the same relative position as the photographer at the time the picture was taken, the optical impression will be realistic.
A similar problem arises during outdoor photography when a large object should be completely contained within a close distance photo (Fig. 9.40 on the right). This, too, calls for an extreme perspective. If such a photo would

then be projected, it would appear "normally distorted" from the perspective of a spectator sitting close to the projection screen. At larger distances, the image's unnatural appearance would become noticeable.

Image on a sphere

The question arises whether a compromise exists that would allow the entire object to be "contained within the frame" while still allowing the image to appear natural from greater distances. The answer, curiously, is both yes and no.

We would have to squeeze the "outliers" at the image edges into the picture by curving the image surface – a process that gives up linearity (straight line preservation)! In the case of the human eye, the image is being projected on a spherical image surface. It, thus, makes sense to project the image on a sphere. For the sake of simplicity, we shall project from the midpoint of the spherically curved surface, which may not exactly correspond to the projection in our eye but rather to the "stitching together" of many small images during the "scanning" of a room. The eyeball is constantly rotating about its center, and the many partial images are combined into a whole image inside our brain.

After the projection of the room from the center of the sphere to a spherical surface, we are faced with the next question: How do we save this image in order to show it to the spectators? Both of the following solutions may be technically feasible: The photo sensors on the electronic chip may certainly be arranged on a sphere, just as an approximately spherical projection screen may be produced. In both cases, however, the practical realization may be quite costly.

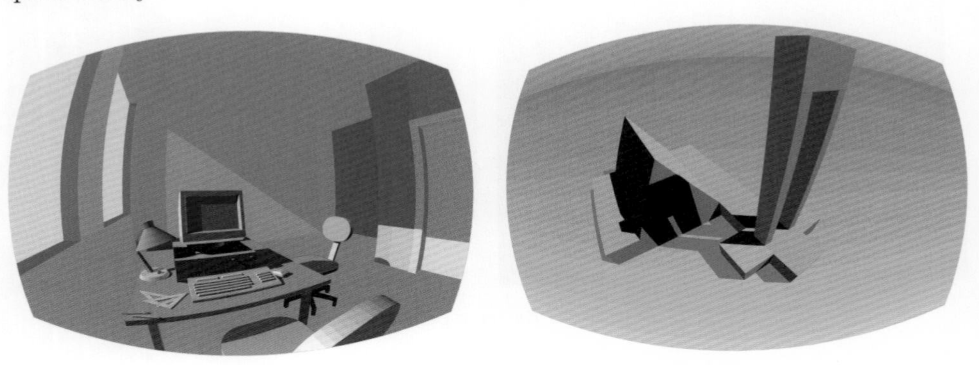

Fig. 9.41 Stereographic projections

As we have discovered numerous times before, the surface of a sphere cannot be developed into the plane without distortions. There are, however, two methods – among many others – which do this task particularly well for our purposes.

One of these methods is the *stereographic projection* of the sphere surface (with center M) from a point Z on the sphere to a plane orthogonal to the radius MZ. Such a projection has the great advantage that sphere circumferences remain spherical in the projection (Fig. 9.42 on the left). The projection cone around the sphere is a cone of revolution with the apex M, which intersects the sphere at a circle. Since the stereographic projection preserves circles, contours of the sphere remain circular after the projection. Straight edges in space are first represented by circular arcs on the sphere and then as circular arcs in the image plane. Fig. 9.41 illustrates how this technique can be used to "smooth" extreme perspectives.

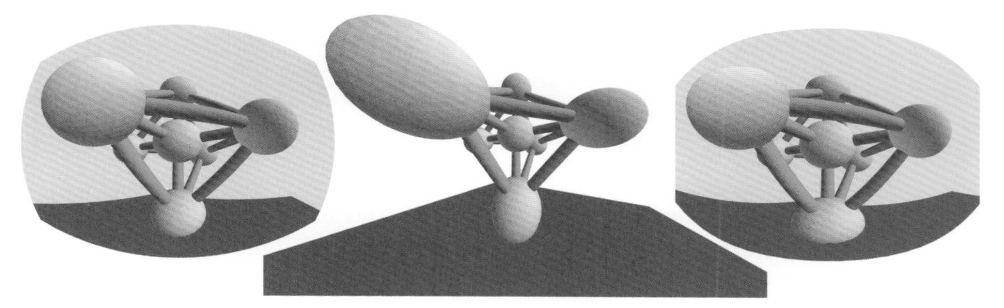

Fig. 9.42 Extreme views of spheres

The other method was propagated by Guido *Hauck* around 1900: It uses the already mentioned *cylindrical projection of the sphere*, where the spherical coordinates (latitude and longitude) are taken as planar Cartesian coordinates. Fig. 9.42 right shows that with this method the silhouettes of a sphere are irregular ovals.

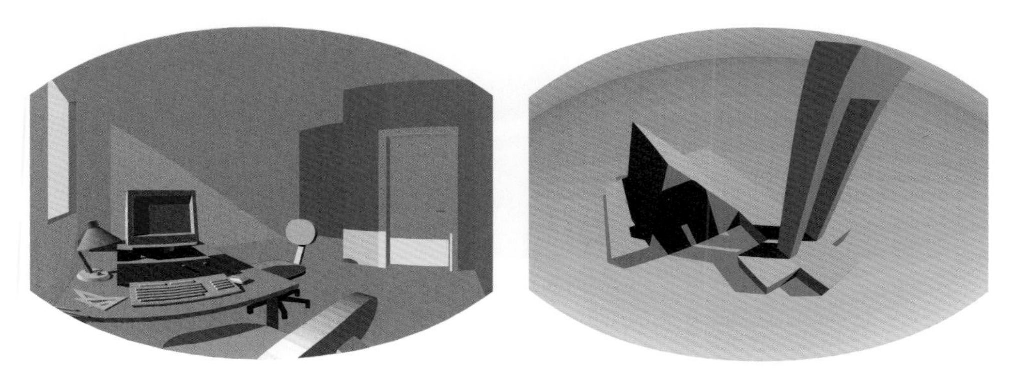

Fig. 9.43 Hauck's perspective

Given a horizontal optical axis (Fig. 9.43 on the left), the method of *Hauck* is well suited for the transformation of extreme perspectives: Vertical edges remain vertical and straight in the projected image. All other straight edges appear to be curved, though not in the form of circular arcs. In general views of objects, the edges may even be projected as curves with inflection points (Fig. 9.43 on the right). ♠

• **The geometry of the fish eye**

The aforementioned methods for the smoothing of extreme perspectives find most of their applications in computer graphics. There exists a further practical solution to this problem: The use of a fisheye lens.
This name makes sense when considering Fig. 9.46: By looking out of calm water (with a flat water surface), it is possible to see everything above the

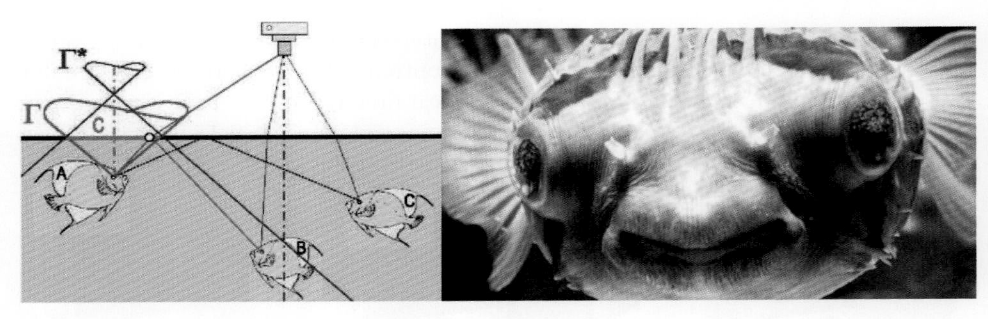

Fig. 9.44 What does a fish see?

surface. The image of the outside world is centered in a circle. Objects which are close to the water surface appear downscaled and flattened towards the edge of the circle.

Fig. 9.45 Another view from the water

The geometry on the water surface will be discussed in the next section. Fig. 9.46 shows an exception in the animal kingdom: the four-eyed fish. Its eyes are split in two, which enables it to see undistorted images above and below water.

Fisheye lenses are often employed in door spy holes. When one looks through them, the objects outside are being pressed inwards – an effect which in photography is done by using fisheye lenses. Standard photographic lenses are flat by comparison and symmetric with respect to a plane. The fisheye lens, on the other hand, is spherically curved to a large degree and is, therefore, capable of capturing "eccentric points" – the points whose angle between their connection to the lens center and the optical axis may be up to 90°. Compared to a flat lens, light rays have a smaller angle of incidence with respect to the surface normal, which leads to lower degrees of refraction. Furthermore, due to the inclination of the optical axis, the image points of eccentric points in space are pressed inwards into the image.

Fig. 9.46 A very strange fish eye

Fig. 9.47 On the left: circular ornament, on the right: concentric rainbow

Circles are usually pictured as non-elliptical ovals. Circles, whose axes coincide with the optical axes of the lens, remain circular, because both the circles and the lens are rotationally symmetric with respect to the optical axis.
Fig. 9.47 serves to illustrate this principle. In the left image, a circular ornament was photographed "at ankle level". In the right image, two (!) rainbows are visible upon closer inspection. When the sun is very low in the sky, the axis of any rainbow (or rather, of both rainbows) is almost horizontal. The primary and secondary rainbows [11], thus, appear to be nearly circular.
Straight lines are usually pictured in a very distorted manner while the angles that they enclose are spared the extreme distortion of traditional perspective (Fig. 9.40). ♠

Straight lines g which intersect the optical axis have their images in the projecting plane that is defined by g and the optical axis. With respect to the image, it can, therefore, be said that the closer a straight line comes to the lens center, the less distorted it appears. In practice, this becomes relevant when taking pictures of scenes with distinct horizons (such as Fig. 9.48 on the left). Due to the rotational symmetry of the lens, this principle also works for non-horizontal straight lines (Fig. 9.49).
The center of Fig. 9.48 shows that this knowledge can be employed to alter fish eye photos in very deliberate ways: A relatively small translation of the

principal point causes the boards of the patio to become distorted – one time in the frontal area, the other time in a ray-like formation.
What makes fisheye photos especially fascinating is the near-infinite depth of field: Both the sunflower and the mallow blossom nearly touch the lens in these shots. This property is also explicable in geometric terms: The smaller the focal length, the more distant the objects are positioned (measured in focal lengths). We know from far-away objects that their image points are very close to the light sensitive layer of the camera sensor.

Fig. 9.48 Diagonals remain straight (I)

Any regular photograph (points P^c) can be converted into a fisheye image (points P^f) – and the inverse, that any fisheye photo can be transformed into an image of a regular perspective, is also true. The associated points P^c and P^f lie on rays through the principal point H. The distances $\overline{HP^c}$ and $\overline{HP^f}$ are connected by a mathematical relation which does not depend on the polar angle of the ray. The law of refraction allows us to calculate several "support points" of the relation and derive further intermediate points by interpolation.

Fig. 9.49 Diagonals remain straight (II)

Finally, let us weigh the advantages and disadvantages of the fisheye lens. The one great disadvantage is the "distortion" of general straight lines. After half a millennium of "classical perspective", this visual feature is very unusual for our eyes. On the other hand, we have to remain fair: The fish eye would

have to be compared to an ultra-wide-angle-lens, and its images are equally unusual.
This disadvantage can be mitigated to some degree by a deliberate use of the lens: Straight lines which pass through the principal point remain straight, just as circles parallel to the image plane remain circular.
The advantage lies in the capability of photographing scenes from extreme perspectives which would otherwise be "impossible to capture". Fig. 9.50 on the right illustrates this very point: A cockpit can only be wholly captured from the inside by a fisheye lens. In this particular case, the image does not even seem to be very unnatural, because the original scene is not dominated by long straight lines. The human head is also largely undistorted in the photograph. The affine stretching of the circular image into an ellipse even enhances the "landscape effect" of the image.

Fig. 9.50 Everything is possible ...

Fig. 9.50 on the left demonstrates a further advantage: The toes in the image appear so large because they were only two centimeters away from the lens – the remainder of the scene (at the edges in particular) is downscaled, which neatly harmonizes with the extreme magnification. What is particularly noticeable, however, is the enormous depth of field offered by the lens. This, in turn, is an effect of the very small focal length[11].

9.4 Geometry at the water surface

In this section, we deal with the physical and optical peculiarities at the transitional surface between air and water (the latter being optically denser). In particular, we shall consider the reflections at the top and bottom sides of the division plane, as well as the refraction towards and from the perpendicular, depending on the medium that one inhabits. The refraction towards the perpendicular happens in the optically denser medium. By knowing a point of the light ray and another point of the refracted light ray, the calculation of the refraction point can be done by solving a fourth-order algebraic equation. In this particular context, its solution is always unambiguous. When looking through the water surface, a plane can never be seen in an edge view. Furthermore, a general object inside the water cannot be precisely reconstructed, no matter how many pictures of it are taken from outside the water.

Adapted eyes

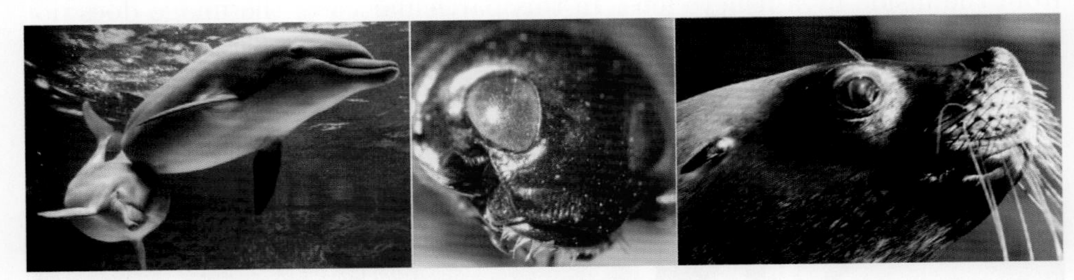

Fig. 9.51 Adapted eyes

Water is our elixir of life – it is almost universally beloved among our own species and among many of our mammalian relatives (Fig. 9.51 on the left and right). Animals who frequently alternate between land and aquatic life often possess specially adapted dichotomous eyes, such as the 5mm long whirligig beetle in Fig. 9.51 (at the center) or the four-eyed fish in Fig. 9.46.

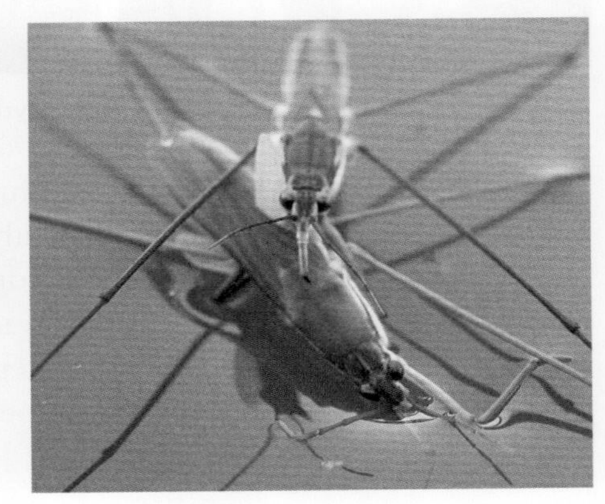

Fig. 9.52 Danger from above and below

The seal (Fig. 9.51 on the right) is capable of changing the focal length of its eye lens by muscle contraction, though it tends to be rather short-sighted on land (see also application p. 329). For the same purpose, some reptiles move a transparent eyelid in front of their eyes when descending below the water surface. Crustaceans possess telescopic compound eyes, which apparently enable them to see rather well above water.
Small animals can run across the water surface without sinking, such as the predacious water strider in Fig. 9.52 on the right or the wolf spider complete with its egg cocoon (not a water spider!) in Fig. 9.53 at the center. Other animals such as the

Nepomorpha in Fig. 9.52 on the left can even move at the bottom side of the water surface!

Fig. 9.53 Surface tension can carry many things but only when they're small.

Small creatures tend not to immediately sink in water because, like all small objects, they have a relatively large surface (in relation to their weight) by which they touch the water. If an object is k times as small as a similar one, then its surface is k times larger in relation to its weight (this relationship has very important consequences in nature and can easily be proven by the example of a cube). This is why a sufficiently small coin (Fig. 9.53 on the left) will float on the water surface under ideal conditions. In large creatures, the pressure exerted on the water surface is greater than the counteracting water pressure.

Reflection at the surface

Flatly incoming light rays (Fig. 9.54) are almost completely reflected by water, whereas total internal reflections occur under water (Fig. 9.56). As we shall soon see, this is only the case up to a certain well-defined threshold angle of incidence.

Fig. 9.54 Perfect mirror from above (flat viewing angle)

The total internal reflections in Fig. 9.58 show the tiles on a vertical wall or the floor of a basin. The right image shows the reflection of both types of tiles, as well as the abrupt ending of the total internal reflection at the threshold angle.

The Snell-Descartes law of refraction

The speed of light is 299 792 458 meters per second but only in vacuum. In the air, this speed is a little bit smaller, whereas in water, it is only about

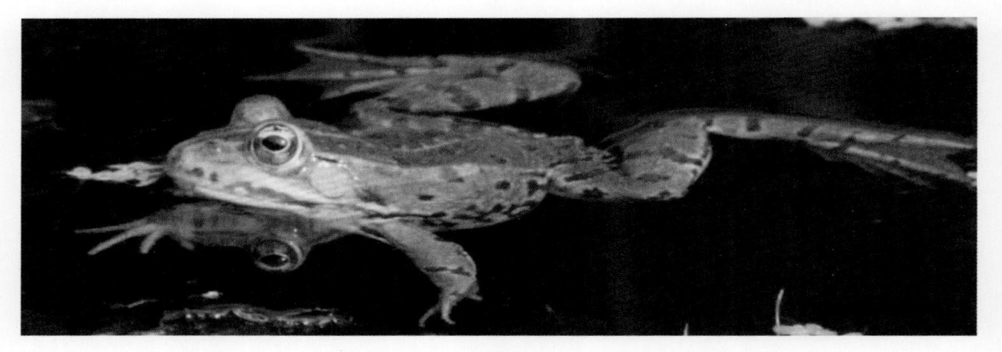

Fig. 9.55 Partial reflection (steeper viewing angle)

Fig. 9.56 Perfect mirror from below

75 percent of the original velocity. The ratio n of these speeds is called the refractive index. When usual (visible) light hits water, this index is about 4 : 3. Different light components (such as the rainbow colors) have slightly different indices.

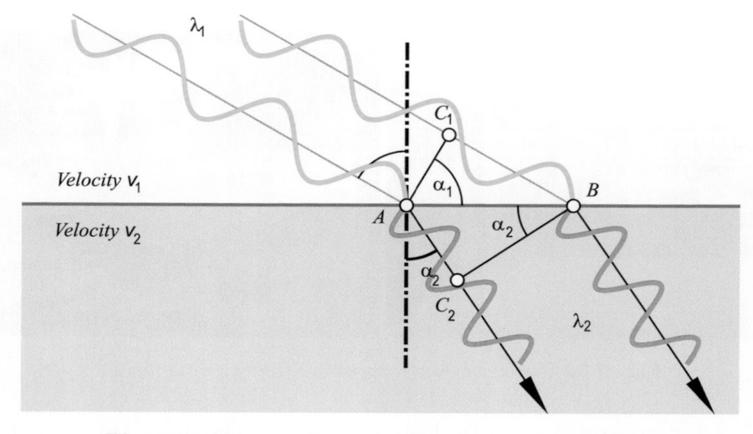

Fig. 9.57 The wavefront explains the law of refraction.

The wave structure of light plausibly explains the refraction of light rays (the wavefront tilts upon hitting the layer of separation). The law of refraction by *Snellius* can immediately be derived from Fig. 9.57 by

$$\frac{\overline{BC_1}}{\overline{AC_2}} = \frac{v_1}{v_2} = n \Rightarrow \frac{\overline{BC_1}}{\overline{AB}} = n \cdot \frac{\overline{AC_2}}{\overline{AB}} \Rightarrow \sin\alpha_1 : \sin\alpha_2 = n.$$

If we are now under water and wish to calculate the threshold angle for the total internal reflection, we only need to determine the inverse of the refractive index. Apparently, Snellius' equation does not always provide real solutions for the exit angle. A total internal reflection occurs starting at the limiting case

$$\sin\beta = 1/n = 3/4 \Rightarrow \beta \approx 48^\circ.$$

From then on, the water surface acts like a perfect mirror, and because the threshold angle is slightly different for different wavelengths, a rainbow effect can be seen at the transition (Fig. 9.45).

Fig. 9.58 Image raising from above and below

Who is looking towards the left in Fig. 9.44? This question is not at all trivial to answer. Fish A sees fish C both directly and as a reflection and fish B only directly, including its tail fin as a reflection outside of the cone Γ's base circle with the critical opening angle of $2 \cdot \beta$. The cone Γ may be called the *limiting cone* if seen from the perspective of A.

Image raising

The so-called effect of *image raising* produces confusing features when looking into the water. The flatter the viewing angle, the more pronounced these effects become. Fig. 9.58 shows an extreme example. The upper crocodile (which does not actually exist) is flatter than the lower one – which, in turn, is not exactly where it appears to be ...

Fig. 9.59 Image raising at the swimming pool

Fig. 9.59 shows a pool of constant depth, though one constantly has the impression of standing next to the deepest spot. The distortion effects appear at their most extreme if one moves sufficiently close to the water surface.

- **Polygons in an edge view?**

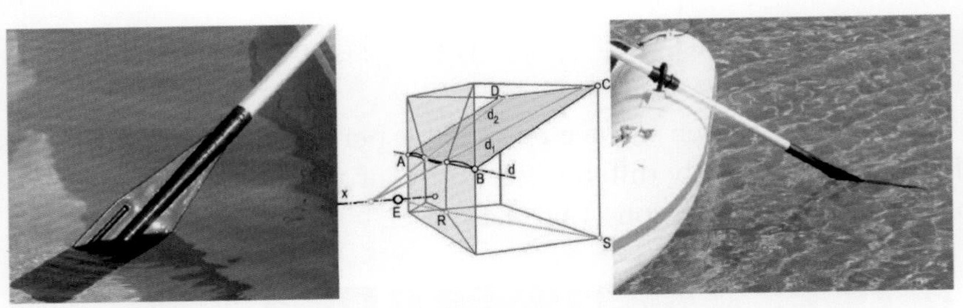

Fig. 9.60 Can a plane be made projecting through refraction?

In Fig. 9.60 on the left as well as in Fig. 9.60 on the right, a single question seems to arise: Is it possible to see planes through the water surface in an edge view? Fig. 9.60 at the center illustrates that the points of a surface are always distributed on a polygon (which, in the general case, is not convex). Only the planes which contain the normal to the water surface through the eye point appear as straight lines.

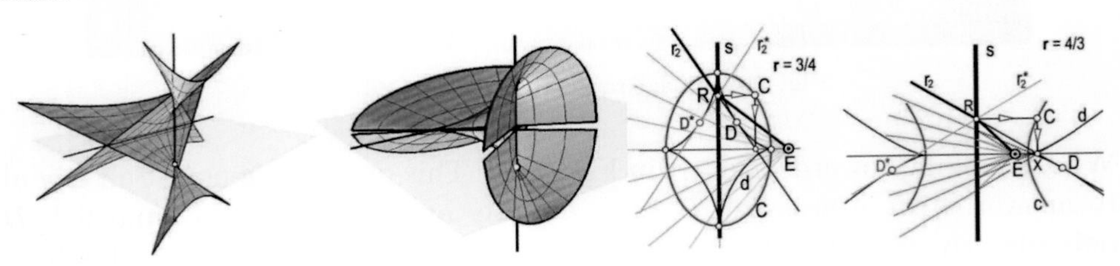

Fig. 9.61 "Folding" of a plane into a ruled surface

Fig. 9.61 on the left shows two ruled surfaces appear as straight lines from the position of the eye. They are generated by straight lines which correspond to the pencil of rays through the straight line in the division plane after the refraction. From an algebraic point of view, the symmetric parts of the surface also belong to it. The viewing rays emanating from the eye point E belong, after refraction, to the normal congruence of an ellipsoid of revolution or a hyperboloid of revolution (Fig. 9.61 on the right). The focal surface of the congruence is created by rotation of the respective focal lines (*diacaustics*).

- **Reconstruction with two eyes under water?**

Fig. 9.62 in the top left shows how a point S in space can be perceived at another, well defined position – at least in the case of planes which are normal to the surface and which also contain both eye points.

By moving both eye points closer together at the transition (Fig. 9.62 bottom left), the intersection point of the refracted rays moves towards a point on the focal line, which itself does not lie on a normal through S onto the water surface. If S moves at a constant water depth, then a locus curve whose

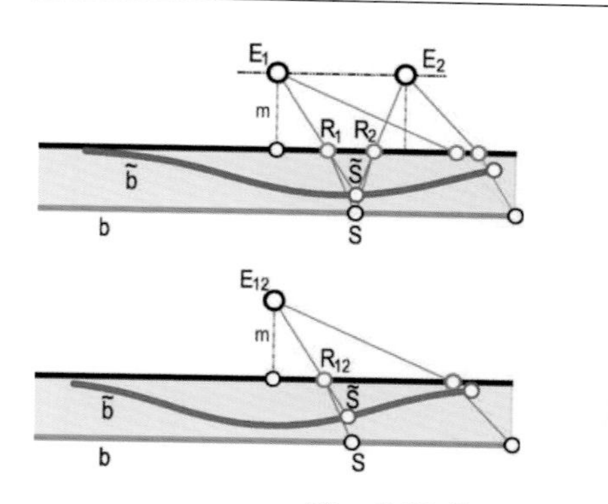

Fig. 9.62 Reconstruction of an object under water ...

deepest point is between the eyes (or, in the limiting case, below the eye) is produced. This explains the illusion of always standing above the deepest point of the pool (Fig. 9.59).
The point S in the water does not, of course, generally lie in the main normal plane through both eye points. According to the law of refraction, the refracted light rays lie in normal planes to the water surface and thus intersect on the normal m through S. On the other hand, the refracted rays coincide with m at various points (unless the eye points are equidistant from m). This means that the elongations of viewing rays cannot intersect in the points of refraction, and it is, therefore, not possible to reconstruct a point S in space precisely, no matter how many photographs are given. The error magnifies as the distance of S from the main normal plane becomes larger. ♠

Submerged

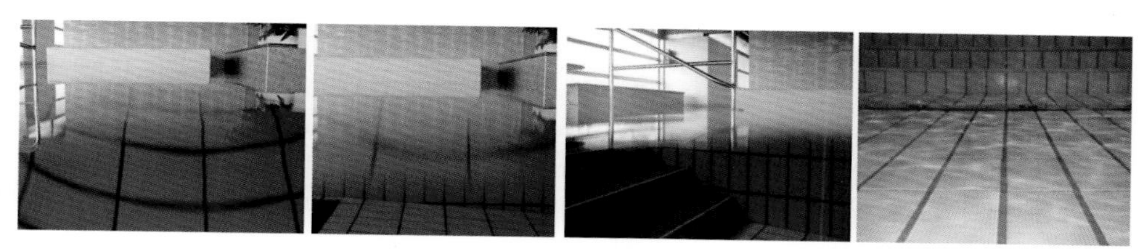

Fig. 9.63 The transitional layer and the world below

As a camera is lowered into water (like in Fig. 9.63), the photographer is faced with particularly strong refractions, which abruptly disappear as soon as the camera is wholly submerged – and the situation returns to regular perspectives, though the objects now appear enlarged by the factor 4/3. This leads us to overestimate the mass of underwater objects significantly which increases cubically if lengths increase linearly (see Fig. 9.64).

• **Refraction at curved (spherical) transitional layers**

We shall now discuss the creation of a rainbow as an example of refraction in a curved water surface (Fig. 9.65). The primary rainbow is produced by a refraction towards the perpendicular, a one-time total internal reflection,

Fig. 9.64 Simultaneous vision above and below water

and a final refraction away from the perpendicular (Fig. 9.65 center) – all due to the presence of water droplets in the atmosphere after a downpour. The total internal reflection at the back side of the droplets only occurs for light rays which strike the droplets at a flat angle. An exit angle of about 42° produces a maximum of scattering into spectral colors. Blue wavelengths are most strongly refracted, while red wavelengths experience the least refraction. This causes red colors to appear at the outer edges.

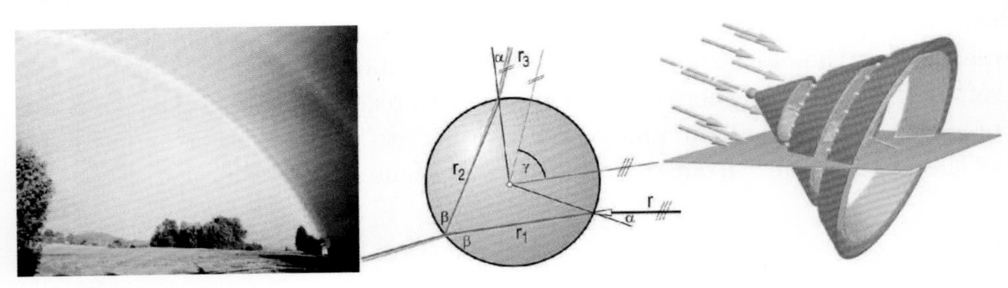

Fig. 9.65 Rise of a rainbow

A rainbow is not actually a circle but a cone of revolution which appears projecting to the observer and which possesses half an opening angle of about 42°. Driving into the foot of the rainbow is, thus, actually possible, but only at sunset, on a wet road producing a fine drizzle and at an angle of 42° enclosed with the shadow direction.

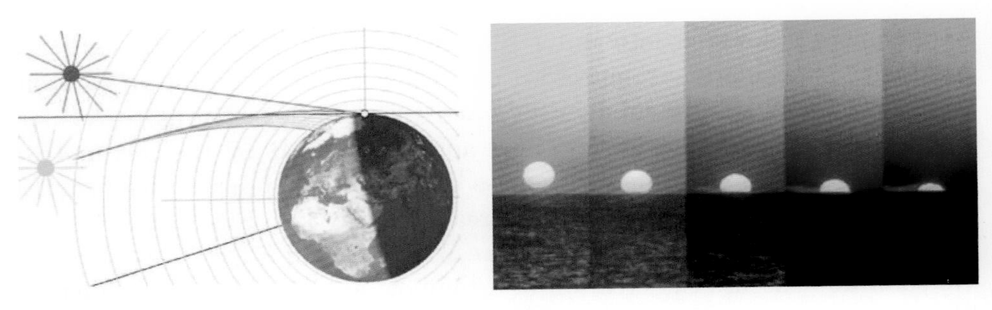

Fig. 9.66 Daily illusions

Part of the light which scurries about inside the droplets will only exit after a double reflection, and produce a very faintly visible secondary rainbow at 51°, which has its red strip on the inside (Fig. 9.65 on the left).
The spectral colors' varying indices of refraction cause a scattering of sunlight at flat angles of incidence. The more energy-rich components (blue, green) are so strongly scattered by the atmosphere that they do not even reach the observer. Due to the curvature of the light rays, the red sun remains visible on the horizon up to several minutes after it should have already disappeared (Fig. 9.66).

Fig. 9.67 Mirages

Mirages are also the effects of continuous light refraction (as in Fig. 9.67). The hot layers of air close to the desert surface or paved road are optically less dense. Light rays, which would otherwise reach the ground, are refracted away from the perpendicular and reach the spectator via a detour. The spectator thus notices an apparent reflection in the elongation of the curve tangent.

- The optical system of our eyes

The human eye is a highly interesting optical device. Light rays are first refracted at the cornea, proceed into the "area" of the intraocular fluids, and then only have to move through the inner lens before finally lighting the retina. First of all: Calibrating sharpness is only possible within a very small deviation from the optical axis, though a greater viewing cone is automatically captured by "scanning" (rolling of the eyes).

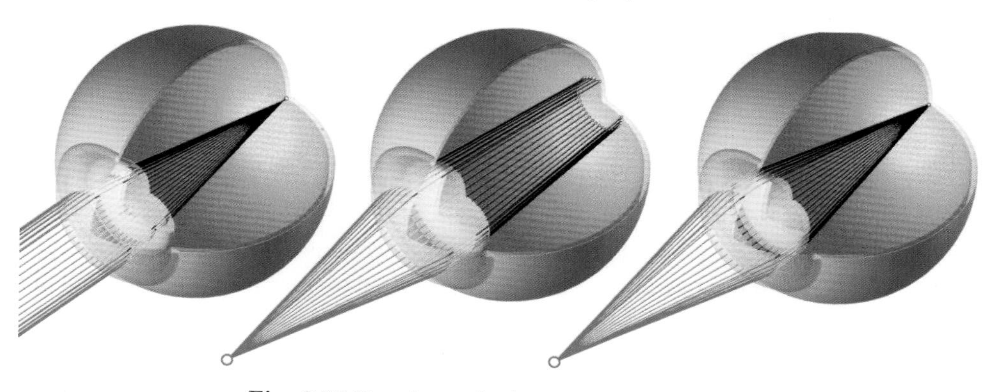

Fig. 9.68 Focusing and adaptation using the lens

In 1911, the Swedish ophthalmologist Allvar *Gullstrand* received the Nobel Prize for Medicine for his precise model of the optical rules that determine image construction in the eye – a model that is still being used today and

specifies curvature radii (including the indices of refraction) for all optical media that are involved. As mentioned in application p. 224, the light ray undergoes its greatest refraction at the cornea.
Fig. 9.68 shows a computer simulation based on standard values. Light rays which are parallel to the axis are precisely focused on the retina in several steps. A point on the axis near the cornea first appears in a very blurry manner (center image). *Adaptation* quickly changes the curvature radii of the lens via muscle contraction, though as close an object as the one pictured in the right simulation image can only be sharply seen by children and very short-sighted adults. ♠

• How does insect vision work?

According to the rules of geometry, three-dimensional vision – or rather the ability to estimate depth values – generally requires two separate images of the spatial situation.
It is sometimes stated that the simultaneous vision with two eyes and the subsequent reconstruction of a three-dimensional model from the images suggest an exceptional feat of the mind, of which an insect would simply not be capable. The logical conclusion would be that insects do not possess spatial vision.

Fig. 9.69 Fixed insect's eyes produce curved perspectives

Human vision can only vaguely be compared to the vision of a compound eye. The eyeball of "higher animals" is moving nearly spherically – it is rotating with two degrees of freedom. If we humans fixate upon a not-too-distant point, then we direct the optical axes of both eyes towards it, and "calculate" the distance of the point from the angle of the intersecting axes.
An insect eye is "fixed". A point in a shared field of vision is being observed by exactly one facet (*ommatidium*) of the left and the right eye (Fig. 9.71). If, however, a point is visible by the facet x of the left eye and the facet y of the right eye, then its spatial position can be determined – so to speak –

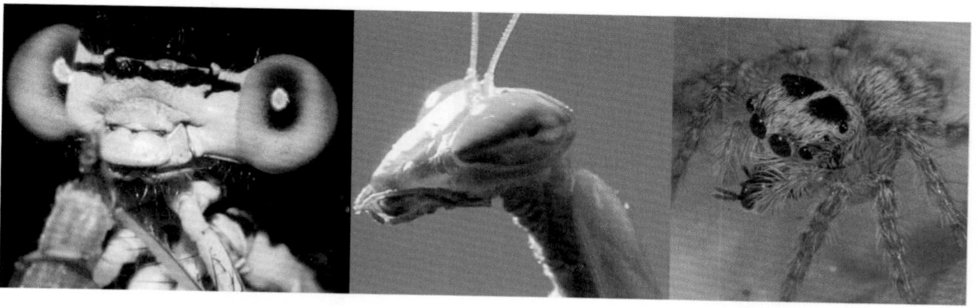

Fig. 9.70 An eye excellently adapted for spatial vision

"in hardware"! No further measurement of angles and reverse calculation of distances is required!

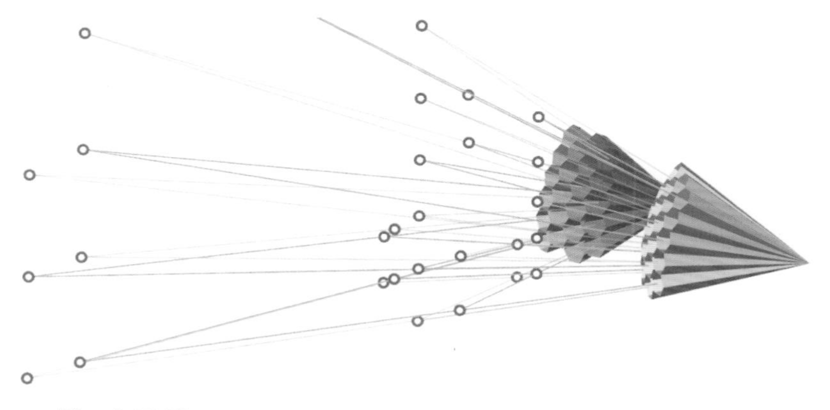

Fig. 9.71 The optical axes of the ommatidia intersect at fixed points

If, for instance, an insect has 1000 facets (ommatidia) in each compound eye, and 200 of them share the same field of vision in each respective eye, then there are thousands of such fixed positions that the insect may be able to interpret without "thinking about it" – just as the predefined joints and the rigid chitin exoskeleton allow for several thousand leg positions which the insect also probably does not "think about".

Jumping spiders (Fig. 9.70 on the right) possess multiple pairs of eyes which enable them to see excellently – and even in colors. Naturally, only a distance area of a few centimeters can be important to so small an animal. Fig. 9.72 on the left shows the way in which a reptile targets its prey. The insect may not even be aware of the immobile predator. The right image shows the superb eyes of a flying fox (not an insectivore, by the way), which are well suited for manoeuvring through dense treetops (Fig. 7.68 on the left). ♠

The eye has been independently invented hundreds of times in nature, and its enormous evolutionary benefit, thus, appears to be obvious. The mere fact that insects have been using compound eyes for many hundreds of millions of years proves what an important step they must have represented in the

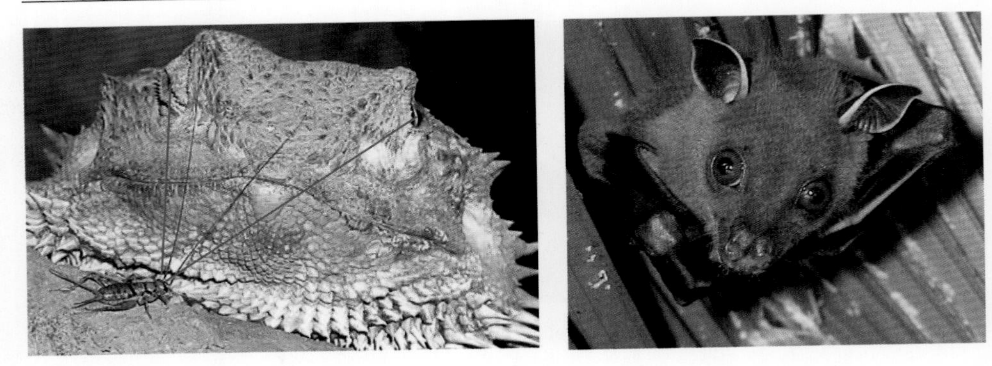

Fig. 9.72 Not only insects are good at judging distances.

evolutionary process. From the perspective of physics, it may be stated that insects are rather "wasteful" in their usage of light due to the absence of lenses: Light rays are barely focused, unlike in lens-based eyes. The following example shows, however, that an even simpler possibility exists.

- A pinhole camera is actually enough!

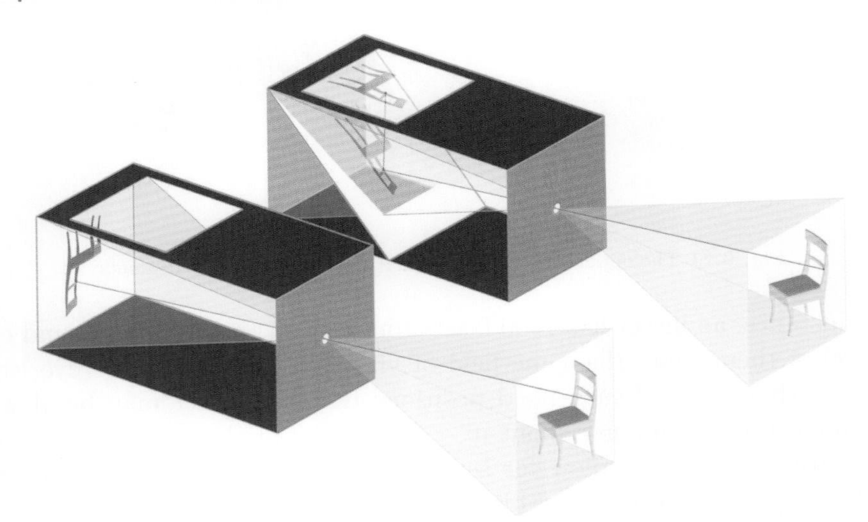

Fig. 9.73 The simple geometrical background of a camera obscura: The light rays enter a dark box and illuminate the back wall. Right box: When we insert a mirror under 45°, the image is deviated to the top wall.

Seeing, in its most basic form, is less complicated than one may imagine. The first attempts at photography show that a simple box impermeable to light is sufficient. A small hole is made at the center of one of its sides – opposite of a light-sensitive layer which captures the outside world (Fig. 9.73, left box). In these primitive conditions, a considerable amount of luminosity is required (like a flashlight) in order to illuminate the photo layer adequately.
When we insert a mirror into our dark box (Fig. 9.73, right box), we can deviate the created image and thus see it on a waxed paper (we just have to keep our head and the upper part of the box under a black scarf).

Fig. 9.74 Left: The principle of the "camera obscura". Right: A tent impermeable to light.

Fig. 9.75 Incredible: This is what the inside of the tent looks like once the eyes have had enough time to adapt. Almost razor-sharp images of the outside world are wrapped around the inside of the cuboid, and even around its edges!

An experiment, like the one by Marianne Kampel, Christoph Mandl, and Markus Jagersberger, may now be attempted: The "box" may be a cuboid tent – completely impermeable to light – with a very small hole. If an observer is put into the tent, he would be astonished by a projection of the outside world onto the walls and the floor (once his eyes have adapted to the darkness). This can also be "proven" fairly easily by taking a photograph from inside the tent with a very long exposure time (Fig. 9.75). ♠

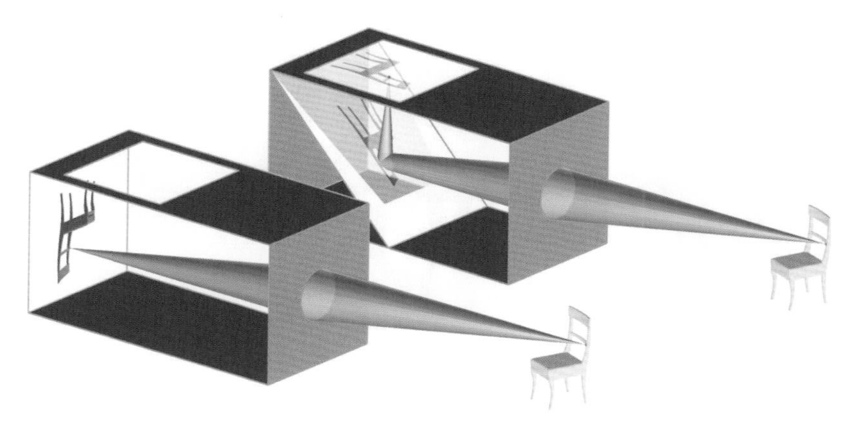

Fig. 9.76 When we enlarge the pinhole, we allow more light to enter, but we need a convex lens.

• The small step to a real camera

The pinhole camera has the disadvantage that only very little light can enter the system. When we enlarge the pinhole, the image gets immediately blurred. We have to insert a convex lens that collects the light on the back plane (Fig. 9.76 on the left). When the focal length of the lens is chosen such that a point of interest is "bundled" via a light cone on the back plane (according to Fig. 9.31), we get again a sharp (and bright) image. Remarkably, this cone can again be "deviated" by a 45°-mirror like in Fig. 9.73. The image of all sharp points is on our waxed paper (Fig. 9.76 on the right).

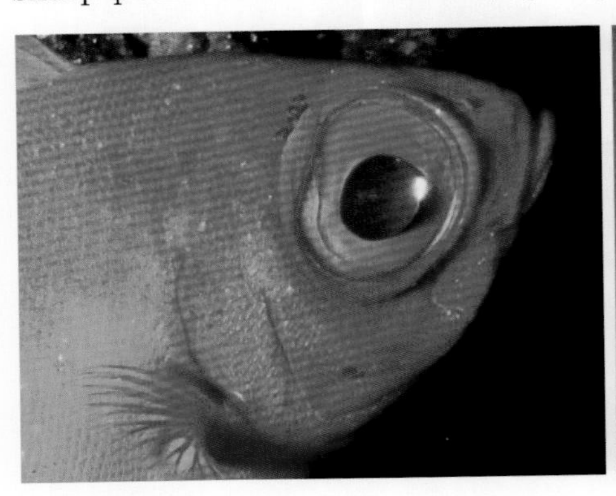

Fig. 9.77 Opening and closing the pinhole ...

This is how nature developed different eyes: A pinhole had to be covered by a refracting lens (Fig. 9.77). The "fine tuning" was to insert an additional lens to be able to slightly vary the focal length (Fig. 9.68). Adding more lenses for optical lenses is probably a human invention (Fig. 9.78).

Fig. 9.78 Placing several lenses in front of another makes the situation more complicated. ♠

10 Kinematics: Geometry in motion

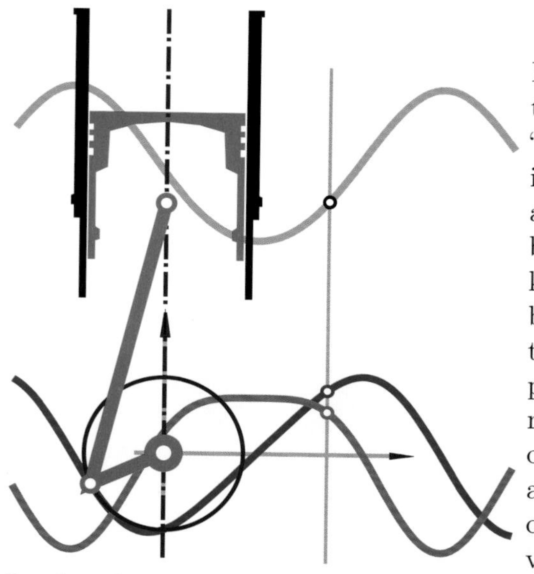

Kinematics – the geometry of motion – may very well be called the "paradise of geometers". Here, the insights and theorems of geometry are applied to solve difficult problems. Some of the basic laws of kinematics, then, appear remarkably simple, the basic point being the formulation of geometrical or physical propositions about moving mechanisms. The paths of points or the envelope curves (or surfaces) are of particular interest, as are the considerations about instantaneous velocities and acceleration.

In the plane, any system of arbitrary complexity is at any given moment rotating about a precisely determined (instantaneous) pole, which may also be a point at infinity. From this insight, conclusions may be drawn about the path tangents and the instantaneous velocities. Envelopes of moving curves can also be captured in this way as well as curvatures and instantaneous accelerations of path curves provided that certain additional information is given.

The most important types of transmissions will be categorized and studied, with particular attention being put on the central role of the mechanical linkages (such as, e.g., the hinge parallelogram or the antiparallelogram). Many technical challenges can be mastered by their use.

The elliptical motion, whose generalization is the trochoid motion, will frequently reoccur in our considerations. Trochoid motion plays a particular role in gear theory.

Survey

10.1 The pole, around which everything revolves

Constrained motion of two points

If an object can only be moved in a predetermined way, it is undergoing a so-called *constrained motion.* Let us imagine a triangle ABC in the plane (such as our set square). If we move A along a predefined curve a and simultaneously move B along another predefined curve b, then C (and any other point of the "triangle system") is left with no choice: It is constrained to move along a path curve c, which is already predetermined.

• **Two-point guidance with a bicycle**

The bicycle in Fig. 10.1 is driving on a wavy road. It is intuitively clear that points on the bicycle frame, such as the saddle's point of support, "have no choice".

Fig. 10.1 Two-point guidance

The bicycle would be moving exactly the same if its wheels would not spin but would simply be pushed along the road. The two contact points between the wheels and the road are precisely defined, and the path curve (including the motion) is thus determined.

As the picture shows, the cyclist is not rigidly connected to the bike. For him, the aforementioned rules only apply to a limited degree – small deviations from the theoretical path result, among other factors, from the pedaling motion and from the absorption of the impacts. ♠

Two path normals are sufficient

In the first chapter (Fig. 1.7), we have shown the following: Two pairs of (directly congruent) positions $A_1B_1C_1$ and $A_2B_2C_2$ of our triangle ABC can be transferred into each other by a uniquely defined rotation. The center of the rotation P_{12} is located at the intersection of the perpendicular bisectors of A_1A_2 and B_1B_2. The third bisector C_1C_2 automatically passes through P_{12}.

Let us now consider a kind of "stroboscopic pair of photographs" within a very short period of time. Both triangle positions will be nearly identical, but the construction of the bisector intersection point will be exactly the same. The bisectors of two points A_1, A_2 on a curve a and B_1, B_2 on a curve b are to be determined. A_1A_2 and B_1B_2 are, thus, the *chords* of the predefined curves a and b. We can use this insight to make the following limiting-case extrapolation: If the time interval for the stroboscopic photos is "infinitely small", then the chords of a and b will converge towards the *tangents* of the curves. The perpendicular bisectors, thus, become the *curve normals* at A and B (Fig. 10.2). Their intersection yields the precise instantaneous center of rotation – the *instantaneous pole* P. All other limiting positions of the perpendicular bisectors (i.e., the path normals) automatically pass through P.

No matter how complicated the motion of our set square becomes: If we mark the path curves a and b of two points A und B, we can then reconstruct the motion by moving A on a and B on b. This leads us to the following *always* pertaining, fundamental

Theorem of path normals: *At any instant* of a planar motion, all path normals pass through a single point – the instantaneous pole. During translations, the instantaneous pole is a point at infinity.

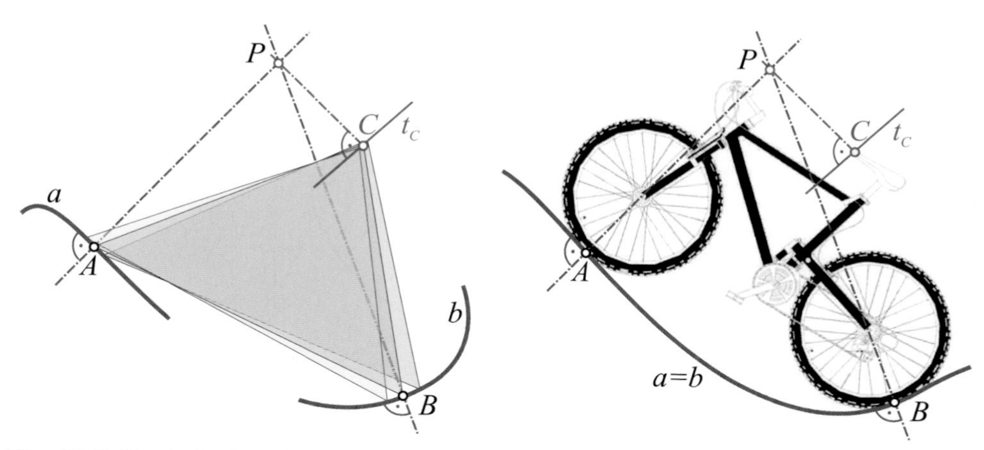

Fig. 10.2 Constrained motion: Two-point guidance ($A \in a$, $B \in b$)

Fig. 10.3 In which direction does the saddle move?

• Two-point guidance with the bicycle (continuation): Let us construct the path tangent of a saddle point C in Fig. 10.3.

Solution:

The path normals of the wheel contact points A and B are given by the shape of the road. All path normals pass through their intersection point, the instantaneous pole P. The path tangent t_C of C is, thus, orthogonal to PC. ♠

If the bicycle is moving on a planar road, it is undergoing a pure translation, where all path tangents are parallel. The instantaneous pole is the point at infinity of the normal direction.

Instantaneous velocity

During rotations, the distribution of velocities is very simple: The center of rotation does not move at all and has, thus, zero velocity, and all other points move the faster, the farther away they are from the center – the absolute value of the velocity increases linearly with the distance from the center of rotation.

Envelopes

By knowing two path normals of an arbitrary constrained motion, we can determine the path tangent for any point. This, however, is not enough, because it is further true that:

If a curve c is moved by a constrained motion, then the pedal points of the normals drawn from the instantaneous pole on c are points of the envelope of c.

Proof:
Points of the envelope can be calculated by intersecting two immediately succeeding positions of c. We shall now demonstrate that these intersection points are precisely the normal's pedal points T_i: Like any other points, they have path tangents which are orthogonal to the line connecting them with the instantaneous pole P. Following the definition of the normal's pedal point, however, this path tangent touches the curve c. This means that the points T_i are the ones that are currently moving along c. A "neighboring position" of c thus intersects c at the points T_i. ◇

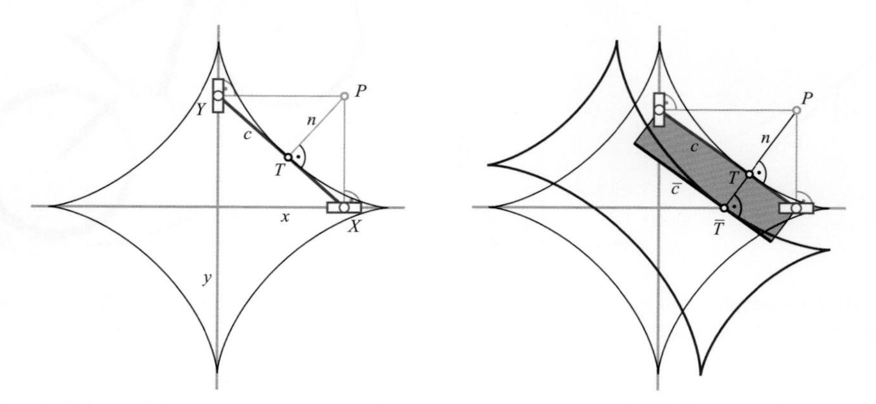

Fig. 10.4 Astroids and parallel curves as the envelopes of a straight line

In the most simple case, the curve c is a straight line. It, thus, contains exactly one pedal point of the normal through the pole. Fig. 10.4 (on the

left) illustrates that a curve is produced if the straight line c is moved such that a fixed point X is directed along the x-axis and another fixed point Y is directed along the y-axis. The resulting curve is also called the astroid, which we have already discussed in a prior chapter.
In Fig. 10.4 on the right side, a second straight line $\overline{c}$, which is parallel and therefore at a constant distance d to c, is carried along. It envelops a parallel curve to the aforementioned astroid, which can also be determined by the calculation of separate points and tangents.
Envelopes of straight lines often possess cusps. The cusps of the astroid's, considered as an envelope of a straight line, occur if the instantaneous pole lies on the straight line – a situation which may arise in the course of the motion. The cusps of the parallel curves occur at the same time.

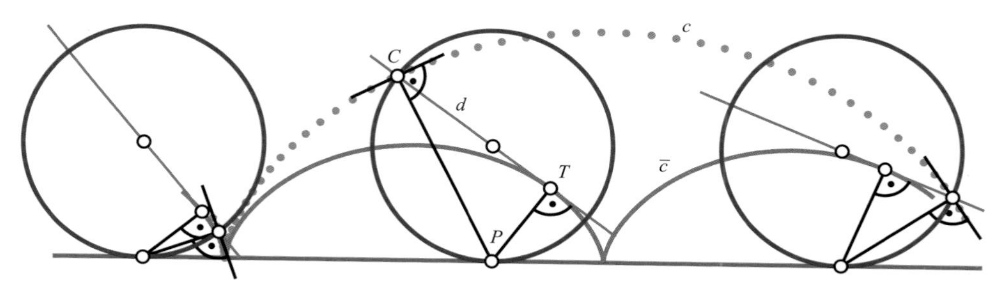

Fig. 10.5 Paths of points and envelopes of straight lines of a cycloid motion

A special situation occurs if a moving point happens to coincide with the instantaneous pole P. The rolling of a wheel along a straight line (Fig. 10.5) is an example of such an occurrence. This type of motion is called *cycloid motion*. In the course of the motion, each point C on the circumference of the wheel will coincide with the straight line, which produces a cusp on its path curve c. Interestingly enough, a wheel spoke (half of a diameter d of the circle) envelops a similar *cycloid* $\overline{c}$ of half the size.

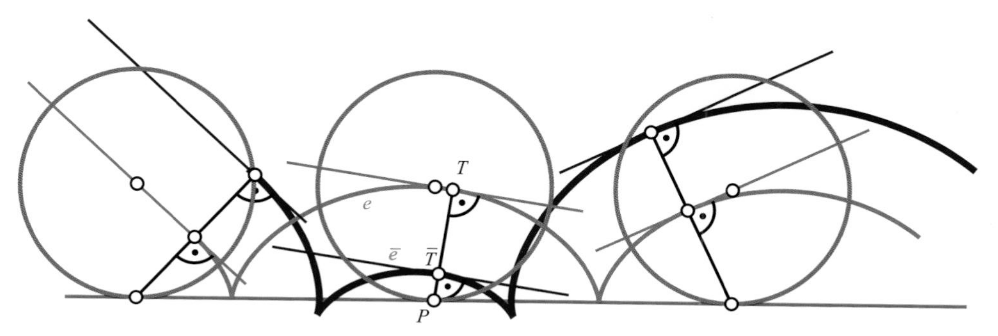

Fig. 10.6 Envelope of a generic straight line

The instantaneous pole of a rolling wheel is the contact point of the wheel with the fixed straight line. This point seems to have an instantaneous velocity zero, and it lies on the path normal of the center of the circle, whose path is a straight line parallel to the horizontal road.

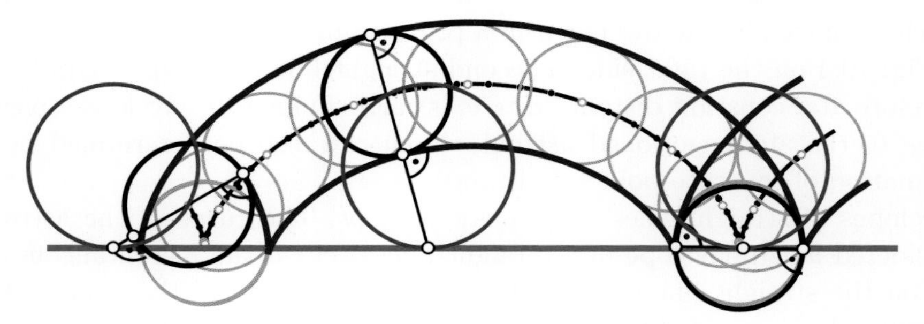

Fig. 10.7 Envelope of a special circle

The envelope $\overline{e}$ of a general straight line during a cycloid motion is a curve which is an *offset curve* of a cycloid e (Fig. 10.6): Here, it is sufficient to translate the straight line by a constant vector into the center of the circle. If we take a circle (instead of a straight line) into the moving system of the wheel, we get an envelope with two branches, which stem from two possible circle normals (respectively) through the instantaneous pole (Fig. 10.7). Let us first consider the special case where the center of the circle M lies at the boundary of the rolling circle. We are now faced with positions where M becomes the instantaneous pole. In this case, *every* point of the circle which was moved along is a pedal point of the normal through P. These circle positions, therefore, are parts of the envelope.

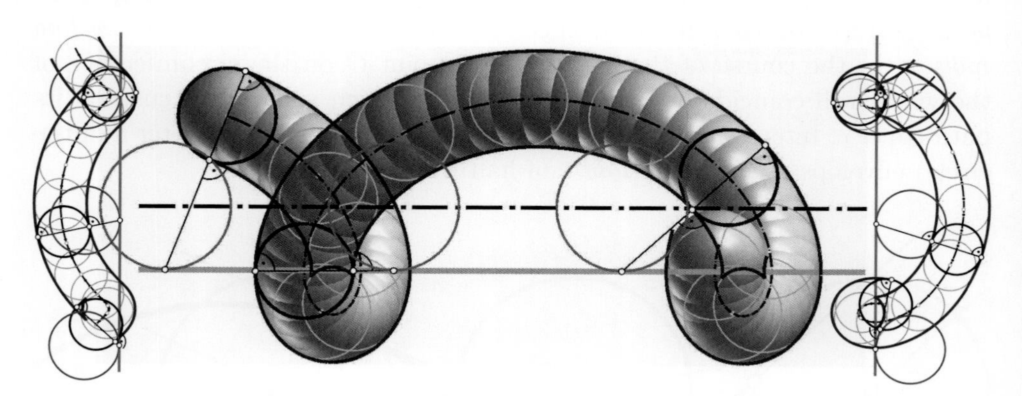

Fig. 10.8 Envelope of general circles and a spatial interpretation

In Fig. 10.8, the center of the circle is no longer a point at the edge of the rolling wheel (on the left it is inside the wheel, and on the right it is outside). The last two figures permit an interesting spatial interpretation: If we interpret the separate positions of the moved circle as normal views of spheres (Fig. 10.8 in the middle), then we can recognize the normal view of a helical pipe surface whose pitch is (at first) precisely defined. If we then

apply an orthogonal affinity in the direction of the straight line being the midpoint's path, we can vary the pitch arbitrarily.
The outline of a helical pipe surface in normal projection is, therefore, composed of parallel curves of an affinely distorted cycloid.

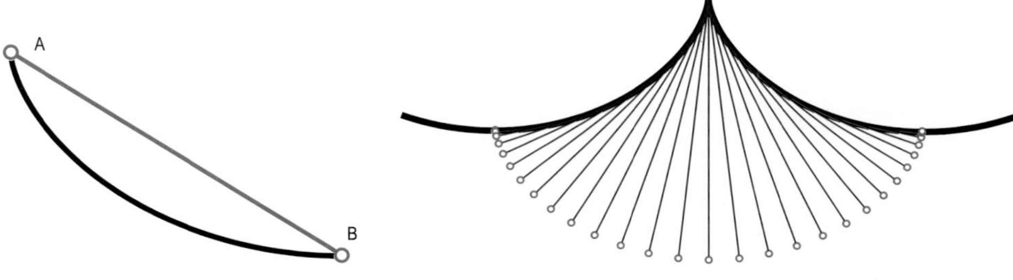

Fig. 10.9 Brachistochrone curve

Fig. 10.10 Cycloidal pendulum

The cycloid is an oft-reoccurring curve. As Johann *Bernoulli* discovered in 1696, it is the curve connecting two points A and B located at different heights, along which a particle under influence of gravity can get from A to B in minimal time (ignoring friction). In this particular context, it is called the *brachistochrone curve.*
A pendulum which swings on a cycloid in the manner pictured in Fig. 10.10 can be used as an exact harmonic pendulum. The frequency of oscillation is independent of the amplitude. In this context, the cycloid is called the *tautochrone curve.* The pendulum clock, which depends on this property of the cycloid pendulum, dates back to *Huygens*, who utilized the fact that the evolute of a cycloid is another cycloid. However, the error caused by friction was greater than the increase in accuracy due to the isochronicity of the oscillation.

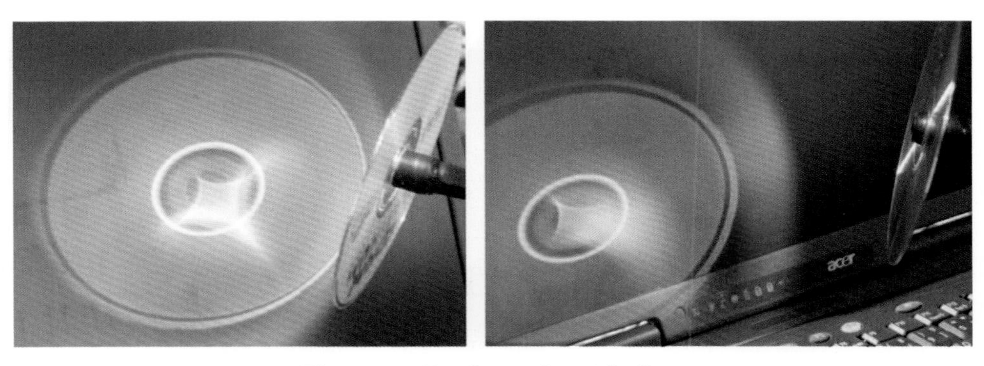

Fig. 10.11 Envelopes of straight lines

Envelopes often occur in reflections and tend to produce beautiful patters. Fig. 10.11 shows the reflection of a CD in the LCD-screen of a laptop. The ellipses around the center of the reflection come from the CD's circular edges. The pointed *focal line* at the center, however, is only indirectly caused by the reflection. As with the Newton rings in application p. 173, it is actually a result of interferences. The curve remains visible – if slightly less so – after the disk has been partially covered.
But what exactly do reflections have to do with "motion"? Well, it is a matter of interpretation. One might say: "'If a light ray moves along a curve ...". The scale

of the time during which this "motion" occurs is irrelevant. It may even happen at speed of light ...

Relative motion

Fig. 10.12 Relative motion: The excavator is being moved by its own scoop.

As a spectator, it is often practical to adapt to an existing motion. For example, in order to understand the path of the sun on the firmament, it is useful to switch from the "terrestrial system" into the "solar system". The solar system, however, is itself in motion: It is not only moving away from the position of the Big Bang at very high speed, it is also rotating about the center of the Milky Way.

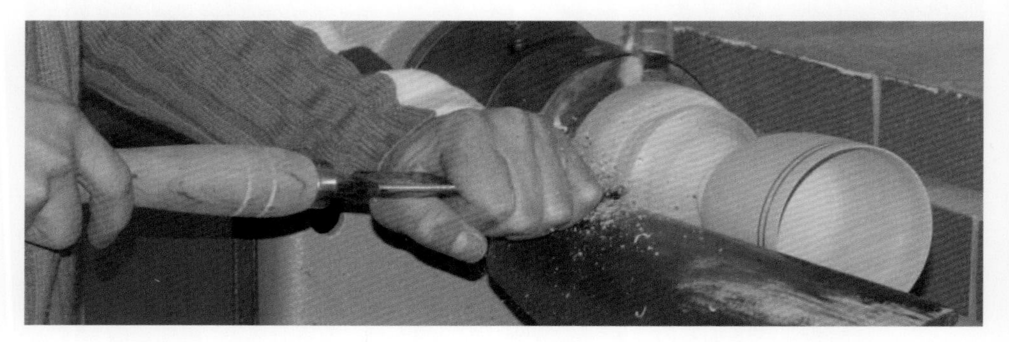

Fig. 10.13 Relative motion: Not the tool is moving, but the work piece!

If one follows this logic to its end, one must conclude that all motion is relative. We can hold onto a wooden piece and sculpt its outer layer (as sculptors do), or clamp it into a lathe and let it spin so as to change it by means of a lathe tool (as in Fig. 10.13).
We will soon deal with more examples of relative motion (such as application p. 358).

10.2 Different mechanisms

Let us shake hands ...

The act of shaking hands is a multi-layered process, and it mostly happens inside the brain, including the steering of the hand and the impressions of the "receiver". Being greeted by the wooden hand in Fig. 10.14, one is surprised by the realistic nature of the feeling (it is, however, a left hand, which relates to our android in the following example). The mechanism itself is as simple as it can be: A tug on a bundle of strings pulls the slightly curved fingers at their lower joints.

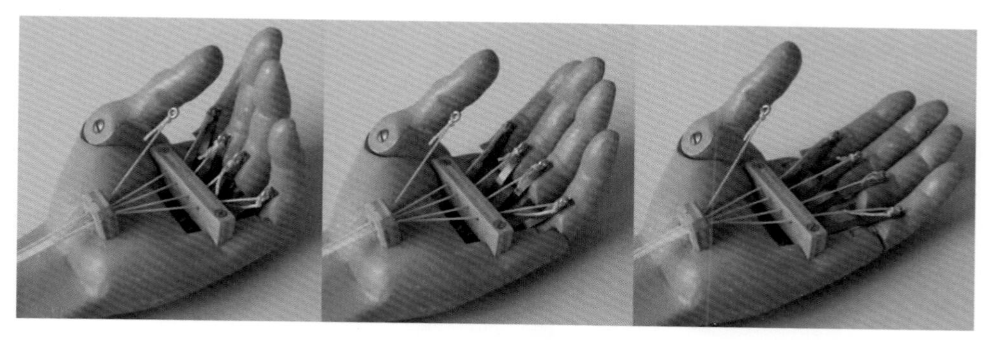

Fig. 10.14 Simulation of the opening and closing of a palm

The angles of rotation of the separate fingers are slightly different, which can be regulated by the distance of the strings' starting points. The thumb is rotating about its "own" axis, which causes the subjectively important embracing sensation. Such a wooden hand, as simple as it is, can already be used to grab and reposition certain small objects (like chess pieces).
One may insert an additional moving hinge joint to each finger, producing a so-called *two-bar linkage.*

The two-bar linkage

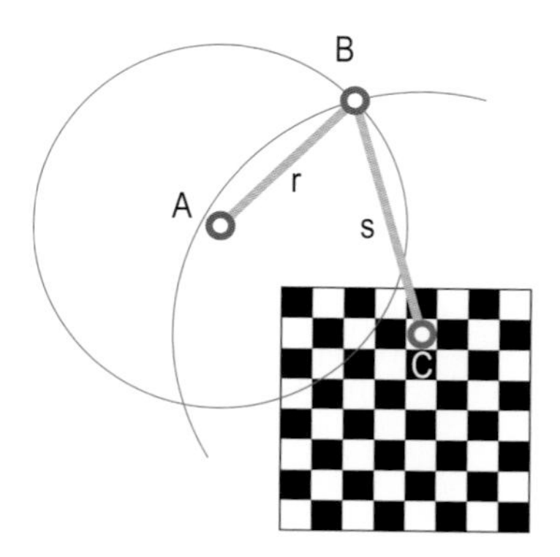

Fig. 10.15 Two-bar linkage: Any point C within a circle or an annulus about A can be reached. On the right: A pepper grinder the arms of which can be freely rotated in one direction.

A rod AB with length r is rotating about its fixed endpoint A while a second rod BC with length s is rotating about its end point B. This mechanism can be used to reach any point within a circle with the radius $r + s$ about A as long as s is greater than r (otherwise, it is an annulus with an inner radius of $r - s$): The position of B is derived from the intersection of the two circles – one about A with a radius of r, and another about C with a radius of s. *Two* solutions exist if A and C are assumed to be fixed.

- **The chess-playing android**

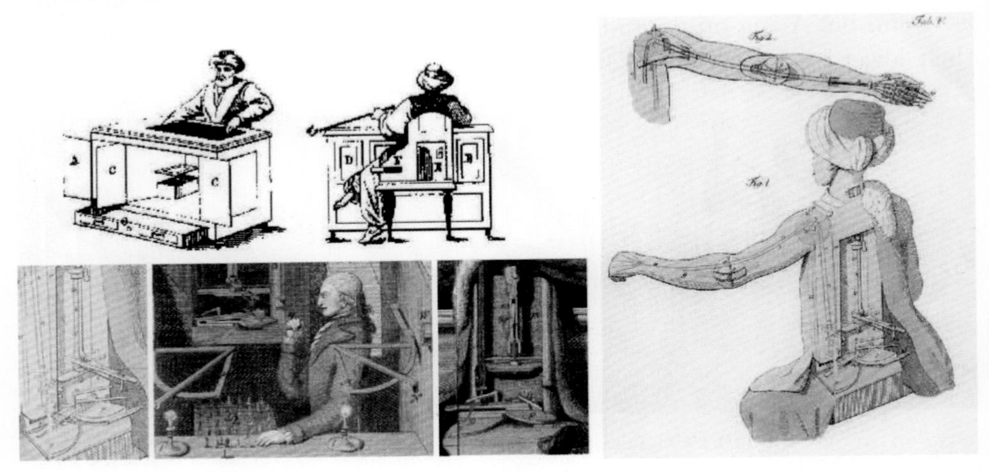

Fig. 10.16 Chess-playing Turk and historical explanatory attempts

The late 17th and the early 18th century were the ages of "androids" – humanlike automata which seemed to be capable of remarkable feats, all owing to their intricate mechanical interior. Wolfgang *von Kempelen* famously developed a "chess-playing Turk" (Fig. 10.16).

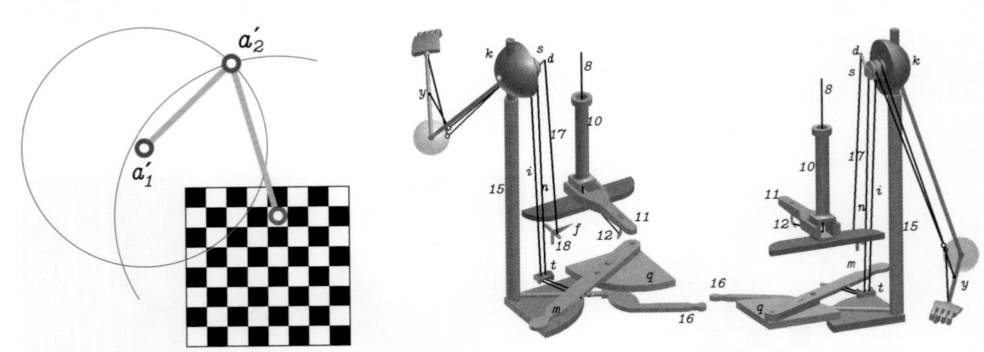

Fig. 10.17 On the left: The two-bar linkage relates to the top view of the upper and lower arm. On the right: Possible kinematic model of the chess-playing android (according to Joseph Friedrich *von Racknitz*).

Many contemporaries have racked their brains trying to explain the supposed artificial intelligence in mechanical terms, while the actual solution was something of a cheat: A little person who was hidden inside the box managed to receive information about what was happening on the chess board, and steered the left hand of the marionette so as to move the chess figures to the intended spot.

The real secret was, at the time, never lifted. An explanatory theory by *von Racknitz* was verified at the University of Applied Arts Vienna in a computer simulation and a subsequent mechanical realization (Fig. 10.17) was deemed to be successful. The hand in Fig. 10.14 was made of wood, like much of the rest of the automaton. Without further digression, it is sufficient to say that the core piece was a two-bar linkage formed by the upper and forearm. ♠

• The joints of articulate animals

In articulate animals (insects, spiders and crabs being among them), the mechanisms such as the two-bar linkage (or their generalizations) are implemented to perfection. The complex motions of the legs, claws, and pincers are not even directed by the brain but by "quasi-microprocessors" located directly in the joints!

Fig. 10.18 A multitude of two-bar linkages and their generalizations

No longer as trivial as in the case of the limbs – and only really visible in the photograph: The bee in Fig. 10.19 expands a sort of "auxiliary jaw", where the tongue is located, which then penetrates the farthest corners of the mallow blossom. ♠

Fig. 10.19 Something close to a two-bar linkage: One may observe the mandibles (including the tongue) of the black carpenter bee.

Fig. 10.20 Three sketches by *Leonardo da Vinci*

• **Leonardo's cranes**

Cranes have been "commonplace" for much of the last century, and it is, thus, easy to forget just how innovative the sketches of the genius from Tuscany actually were. It is a good practice of the spatial imagination to rotate the various wheels in one's own mind and to consider the directions into which the loads may be heaved. ♠

The hinge parallelogram

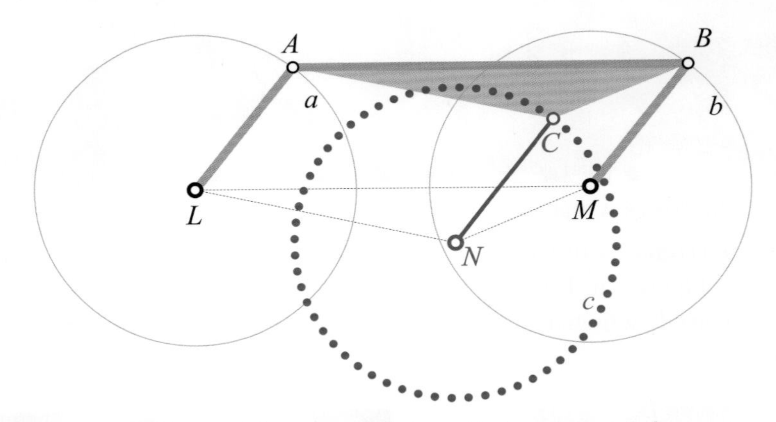

Fig. 10.21 A hinge parallelogram: Each path is a circle.

If a parallelogram $LABM$ is held at one side (for instance, LM), and if the parallel side (AB) is left to rotate, one deals with a hinge parallelogram (Fig. 10.21). Each point C that is connected to AB describes a path circle c congruent to the path circles a and b of A and B. Hinge parallelograms are often found in daily life, such as in desk lamps or in letter scales (Fig. 10.22). Parallel translations in all directions can be accomplished by a combination of multiple hinge parallelograms (old drawing machines in drawing offices employed this trick). A particularly clever variant is visible in Fig. 10.23, which shows a tripod that can be used to keep heavy cameras steadily parallel to the ground – even while walking.

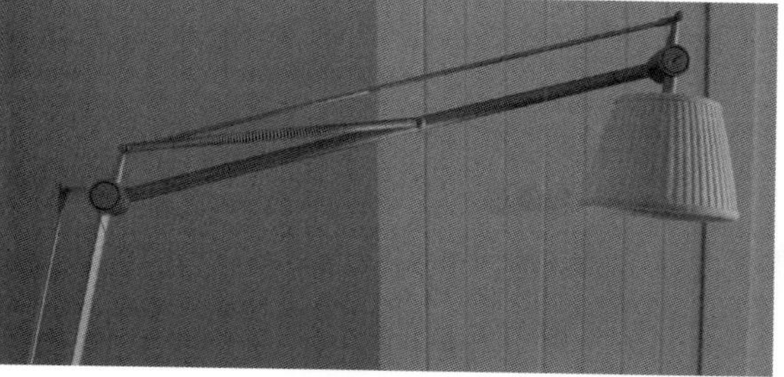

Fig. 10.22 The hinge parallelogram in a letter scale and designer lamp

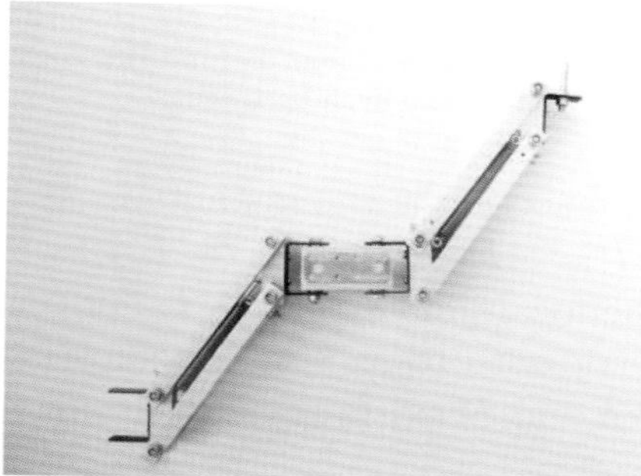

Fig. 10.23 Hinge parallelograms in a portable tripod (Robert Eder)

The pantograph, sometimes called "lazy tongs" (Fig. 10.24), consists of parallelograms strung together, and can often be encountered (for instance) as the adjustable mounting of lamps, sun blinds or even whole platforms, as in Fig. 10.25.

Antiparallelograms

If a hinge parallelogram is stretched into its longest form, its motion may continue in two different ways: either, as outlined before, as a parallelogram, or as an *antiparallelogram*, whose opposite sides are generally not parallel. In Fig. 10.27, such a hinge $LABM$ can be seen. A is moving on a circle about L and B on a circle about M. Both longer sides are, thus, path normals, and

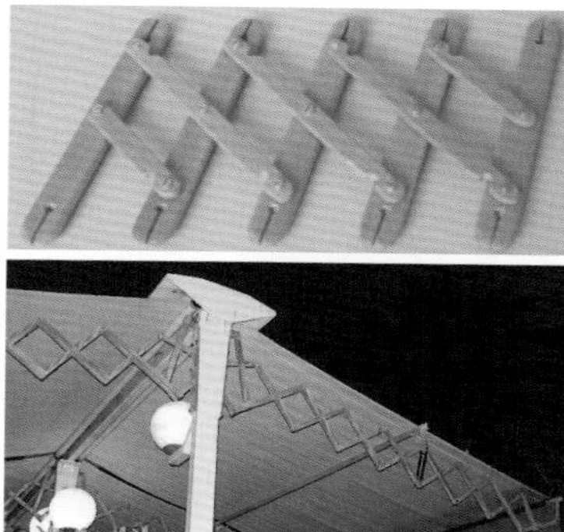

Fig. 10.24 Panthographs ...

Fig. 10.25 ... in theory and practice

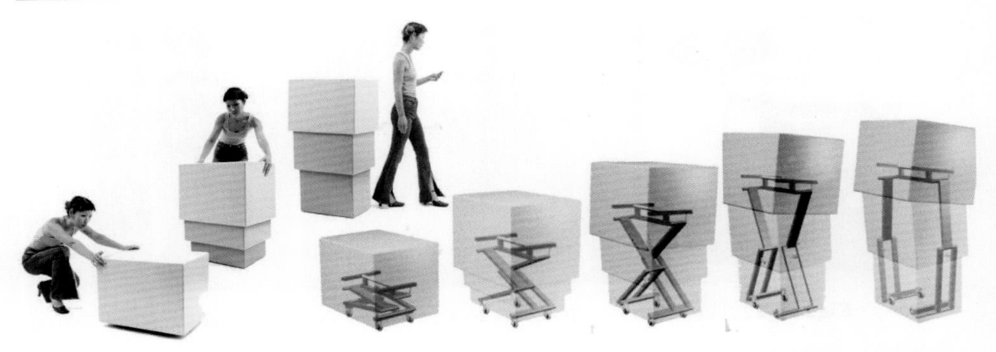

Fig. 10.26 As ingenious as it is simple: How to turn a cardboard box into a lectern (by Christian Bezdeka `www.bezdeka.com`): Behind this construction are the so-called lazy-tongs. In its final position, the lectern can even be tilted!

their intersection is the instantaneous pole P. The sum of the distances $\overline{LP}$ and $\overline{MP} = \overline{AP}$ is constant $(= \overline{LA})$, which causes P to move along an ellipse with the focal points L and M. For reasons of symmetry, an ellipse which is symmetric with respect to the ellipse's tangent in P and which has the focal points A and B exists in all positions and belongs to the moving system.

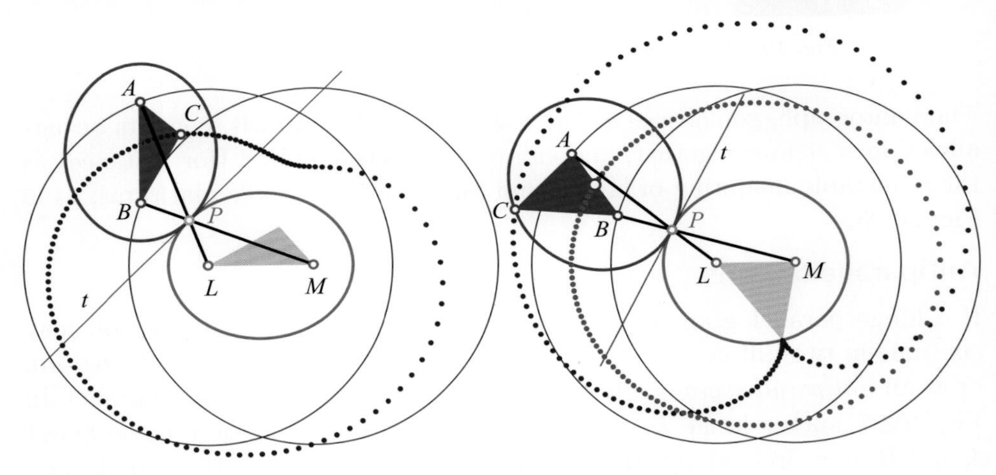

Fig. 10.27 The theory of the antiparallelogram

It does not always have to be a uniform rotation ...

The hinge antiparallelogram has an interesting application: The pole curves are congruent ellipses which roll onto each other. If these ellipses are furnished with gear flanks, then they can be made to roll as in Fig. 10.28.

If, instead of holding the principal axis of the ellipse at rest, the single focal points of both ellipses are fixed, then the uniform rotation about one focal point can be transformed into a periodically non-uniform rotation about the other focal point. By inversion, it is sometimes possible to diminish the effects of certain non-uniform rotations and even to equalize these rotations in ideal situations.

Fig. 10.28 Elliptical gears

Fig. 10.29 Elliptical gears move downwards

Fig. 10.29 shows a very original method of propelling elliptical gears: Both gears are placed within a cylindrical container – bounded by cylindrical discs – and are moved downwards by gravity.

Fig. 10.30 Ancient gears? Columns which have fallen apart!

Observing this slow motion may, to some readers, be more calming than an hourglass, though a certain similarity can be observed: Once the gears have reached the bottom, the entire cylinder can be turned upside down so that the motion proceeds anew. Notice the small pocket of air which has crept beneath the right gear.

The crankshaft

Fig. 10.31 Crank shaft propulsion

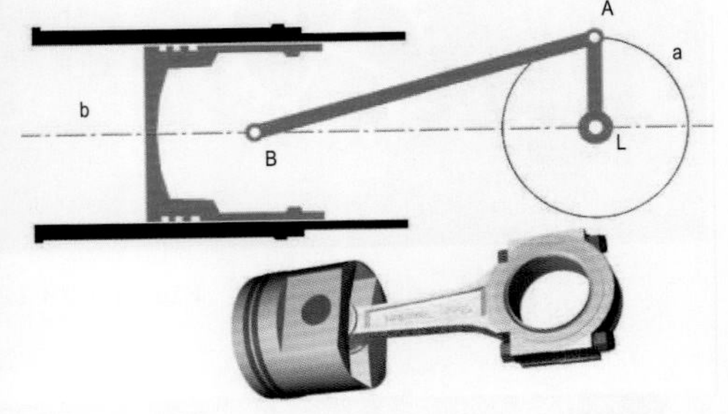

Fig. 10.32 Crank shaft of an Otto's engine

If a point A is moved on a circle a about a fixed point L, and a point B which is rigidly connected to A is also moved along a straight line b (Fig. 10.32), then one is dealing with a crank shaft. Fig. 10.31 shows the crank shaft propulsion of a workbench, and Fig. 10.32 the crank shaft of an Otto's engine (see also Fig. 10.33 in the top right corner).

• **Sawmill by Leonardo da Vinci**

Typical of Leonardo da Vinci was his ability to plan mechanical constructions in their smallest detail, without the apparent necessity of realizing these ideas in practice. His sawmill (Fig. 10.33 on the left) is still in use today and has undergone only slight changes. The crank shaft propulsion (Fig. 10.34), propelled by water, lies at its core.

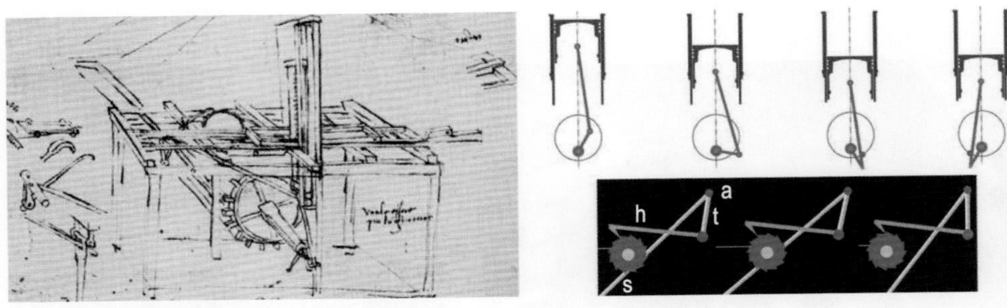

Fig. 10.33 Sawmill according to *Leonardo da Vinci*, with remarkably detailed solutions

The platform that pulls the pieces of wood towards the running saw blade is driven by a subtle mechanism:
A rod s, which is rigidly connected to an axis a, is being forced by a loop to resonate with the driving rod. Another rod t, to which the hook h is attached, is also rigidly connected to the axis a. This hook, when swinging back, attaches to a cog, and when swinging forwards, glides over the tips of the triangular teeth – thus, it can move the gear forward step by step, which in itself pulls the platform over a shaft. ♠

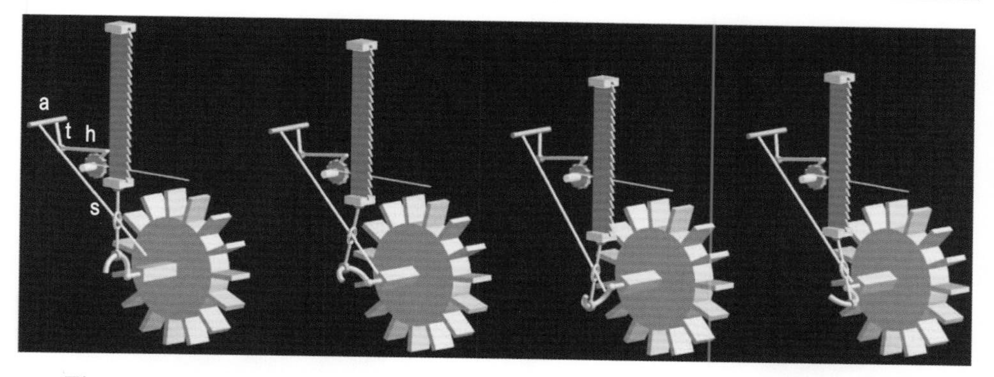

Fig. 10.34 The propulsion is comparable to the respective solution of a piston engine.

General four-bar linkages (hinge quadrangles)

A four-bar linkage (hinge quadrangle) is a mechanism made of four rods. Depending on the length of these rods, there exists a huge number of variations – and a corresponding multitude of applications.

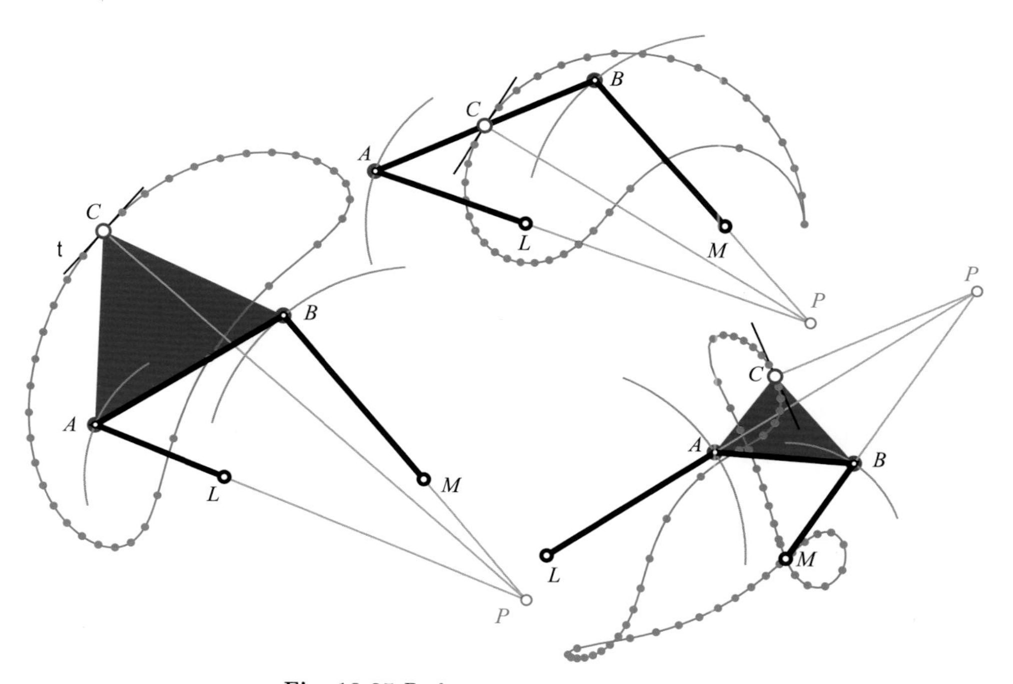

Fig. 10.35 Path curves of four-bar linkages

Let $LMAB$ be, once again, the hinges of the quadrangle (Fig. 10.35). If the points L and M are fixed, then the motion of the system connected to the rod AB is called a *linkage motion* and the path curves are called *linkage curves*. The paths of the points A and B are circles about L and M. The corresponding path normals are the linkages (or swings) LA and MB. The instantaneous pole P can be found at the intersection of the path normals.

Each point C of the linkage system is always moving orthogonally to the associated pole ray PC.

• **The front tires have to be rotated asynchronously!**

What happens if we turn the steering wheel of a car? First, the rotation is transferred by a Cardan joint onto a horizontal axis, where it is converted into a translation by a gear rack (Fig. 10.36). This propels a hinge trapezoid, which rotates the front wheels – and quite ingeniously at that!

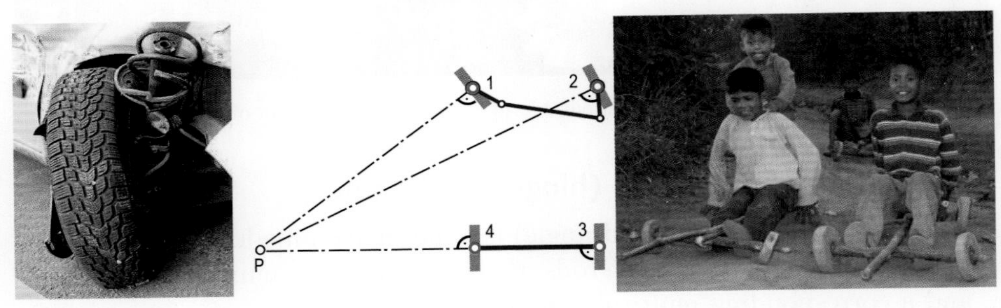

Fig. 10.36 Different rotations of the front wheels. Each tire has a different instantaneous speed (proportional to its distance from the instantaneous pole).

Let us now literally track back the process: In order to ensure optimal steering, the path normals of the wheel's four contact points must pass through a single point – the instantaneous pole. Pairs of wheels are mounted on an axis, and thus, all wheels must be perpendicular to the straight lines that connect their centers with P. This may be accomplished with a four-bar linkage. ♠

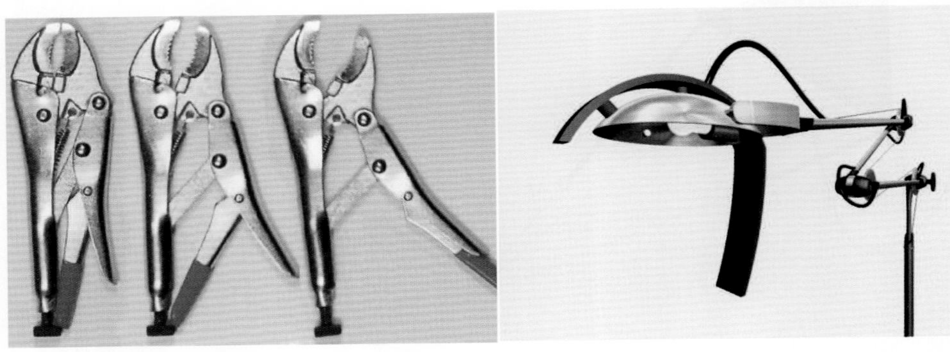

Fig. 10.37 Pliers with a four-bar linkage and a fixing mechanism

Fig. 10.38 Designer lamp with "hidden" four–bar linkages (Alexander Gufler)

In addition to the special four-bar linkages, the general forms are also widely in use (Fig. 10.37). The advantage always remains the same: The devices are easy to manufacture, to utilize, and beyond that, they are also very robust. The multiformity of the path curves is virtually unbounded, and the only question concerns the finding of the right rod lengths for a particular problem. Fig. 10.37 and Fig. 10.38 show examples of four-bar linkages.

Once an appropriate hinge quadrangle has been found, it is possible that the differing path velocities of a system point have to be compensated. Many critical points often move at very high speeds whereas other, less important areas are nearly fixed in dead-point positions. Here, one should attempt to configure the propulsion of the hinge in a flexible manner.

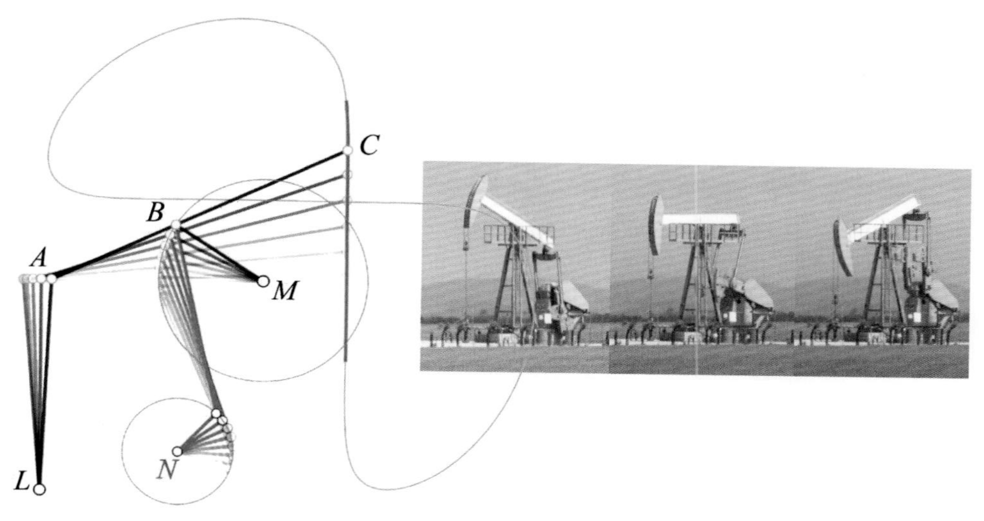

Fig. 10.39 Four-bar linkage with approximate straight line guidance (oil pump)

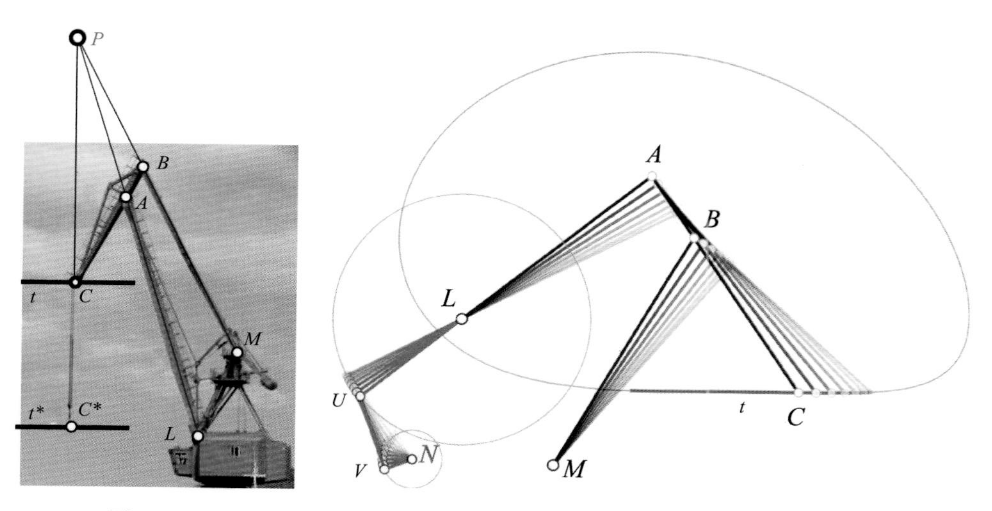

Fig. 10.40 Four-bar linkage with approximate straight line guidance (crane)

Fig. 10.39 and Fig. 10.40 show two special and very useful four-bar linkages. Drawn in red you can see an additional auxiliary four-bar linkage that can be driven by a rotation. Note that in Fig. 10.40 the point C^* is not directly connected to the linkage. Its path curve is congruent to the path curve of C, and the tangent $t^* \ni C^*$ is always parallel to the tangent $t \ni C$. The approximate straight line guidance is, in these cases, of utmost importance.

- **A sugar bowl and a four-bar linkage**

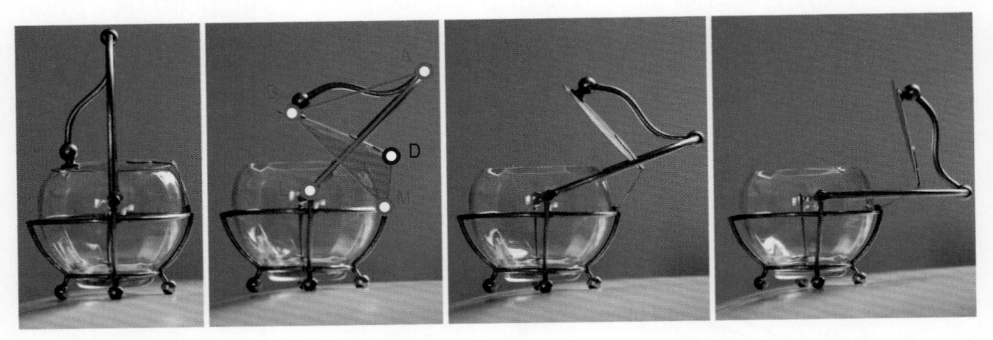

Fig. 10.41 The second image from the left shows which points remain fixed (L and M) and which have circular paths: A and B – and thus also D.

The sugar bowl pictured in Fig. 10.41 implements a classical, though barely recognizable planar four-bar linkage.
Let us consider the three points M, A, and B, which determine the vertical symmetry plane of the hinge. Now, let us further consider the fourth point L of this plane (it lies at the center of the imagined, horizontal axis of the mechanism, which appears almost projecting in the photo). A hinge quadrangle $LMAB$ is thus defined.
In order to fulfill the condition that the lid should be closed in the stretched position on the left, an imaginary, rigid triangle MBD is also moved. The segment BD materializes trough the circular lid, and the segment AB is likewise curved for aesthetic reasons. The designers of the Biedermeier period must have apparently had an interest in hinge quadrangles. ♠

If the arms could change their length

We have thus far discussed mechanisms with fixed arm lengths, which have obvious advantages of stability under heavy load.

- **A hinge quadrangle with a single variable**

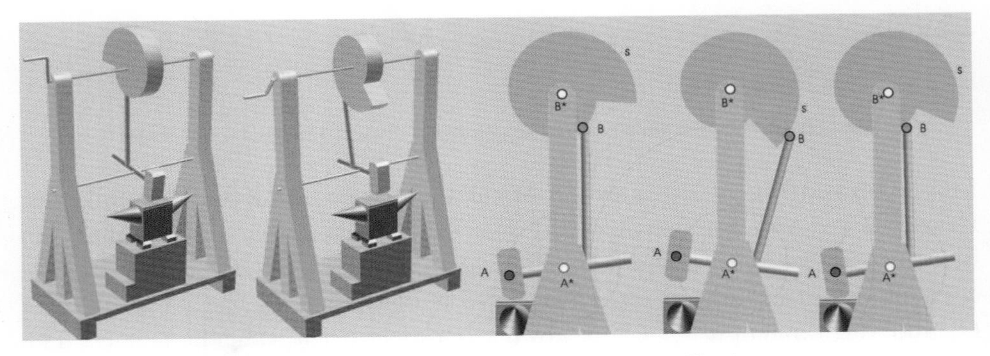

Fig. 10.42 Hammer mill by *Leonardo da Vinci*

After the example of the robot arms, the mechanism behind Leonardo da Vinci's hammer mill appears to be especially interesting. The simulation in Fig. 10.42 shows the propulsion of a hammer by the rotation of a spiral attached to a shaft.
In each position, we are dealing with a quadrangle (A^*, A, B, B^*) with three constant rod lengths and a variable one (BB^*). The angle $\angle AA^*B$ may be rigid. ♠

• The Stewart platform

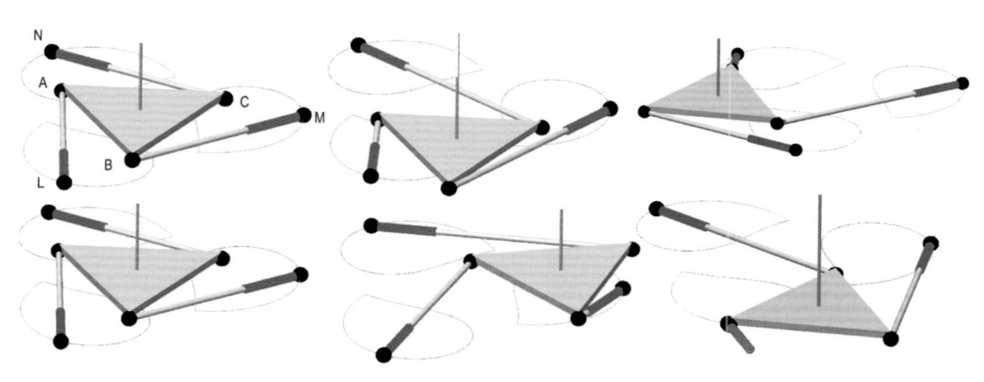

Fig. 10.43 A triangular platform ABC which is controlled by three variable-length robotic arms with the fixed points L, M, and N. In the top left corner: "Null position". On the right: Certain triangle positions are "dangerous" due to the indeterminacy of their robotic motion.

Contemporary robotics permits us to control any arbitrary motion of a plane by the use of three arms of varying length (Fig. 10.43). Despite the state-of-the-art technologies involved, there still exist "unsafe positions", such as the ones pictured on the right. ♠

10.3 Ellipse motion

If a rod of fixed length is directed such that its endpoints U and V follow the straight lines u and v, then it is undergoing an *ellipse motion*.
Such a motion is pictured in Fig. 10.44. Both guiding straight lines u and v are orthogonal in this case. We are concerned with the path curve of an arbitrary intermediate point C with $\overline{VC} = a$, $\overline{UC} = b$. Let $F = u \cap v$ be the origin of a Cartesian coordinate system and φ the rod's angle of inclination. C is, thus, located at $x = b \cos \varphi$, $y = a \sin \varphi$. This is the parameter representation of an ellipse with axis lengths $2a$ and $2b$. Remarkably, the midpoint R of the segment UV $(a = b)$ describes a circle.

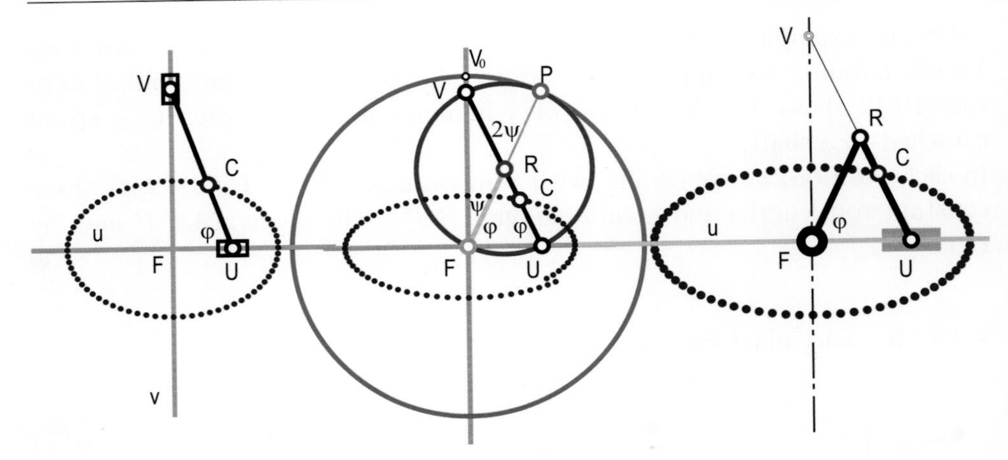

Fig. 10.44 Three related ellipse constructions

• Construction of paper strips

Ellipse motion, by its classical definition, can be accomplished by two orthogonally intersecting guidance rails even though the mechanism is only partially practicable: The motion is jerky, and the stretched positions are undefined, which is why the use of the "ellipse compass" has never really become widespread.

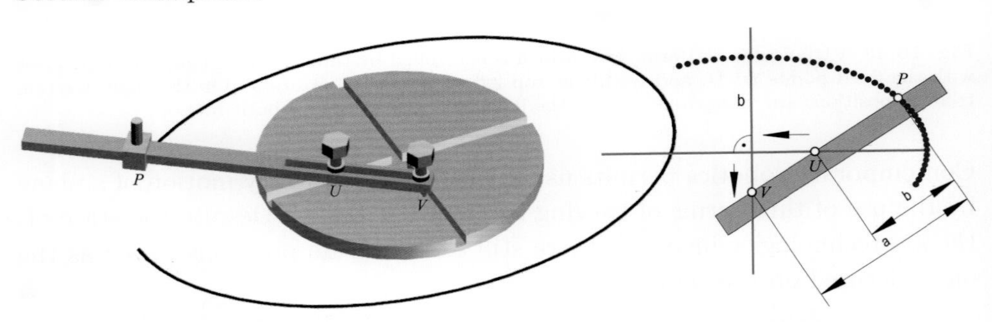

Fig. 10.45 Ellipse motion and reverse paper strip construction

The actual significance of this mechanism lies in its easy illustration of the so-called *reverse paper strip construction*: If an ellipse's principal apex (major axis $2a$) and a general point P are given, then half of the principal axis a needs to be inscribed between P onto the auxiliary axis (the perpendicular bisector of the principal apexes) in order to get V – though this is possible in two ways. The straight line connecting P and V yields U, and thus b. ♠

We can find the instantaneous pole P of the motion by intersecting the path normals at U and V that are parallel to the axis. Compared to R, P is always at twice the distance from the origin, and thus lies on a circle about F with a radius $\overline{UV} = a + b$ and, at the same time, on a circle about R with half the radius (Fig. 10.44 in the middle). The motion of the rod UV could also be forced by letting the Thales circle roll over UV within a circle of double

radius. Both circles are named *Cardan circles* after their discoverer Geronimo *Cardano* (1501–1576).
The Cardan circles roll on each other without gliding. Each arbitrary point on the smaller circle becomes, at some instant, the instantaneous pole – the point V, for example, does so in the point V_0 on the y-axis (Fig. 10.44 at the center) once R has rotated by an angle of ψ.
By adaptation of the Angle-at-Circumference-Theorem (p. 10), V has moved on a circle of half the size by an angle of $2 \cdot \psi$. We are, thus, always correlating circular arcs of equal length – $\widehat{PV}$ on the smaller circle and $\widehat{PV_0}$ on the larger circle – which requires a rolling motion.

- Ellipses as hypotrochoids (Fig. 10.46)

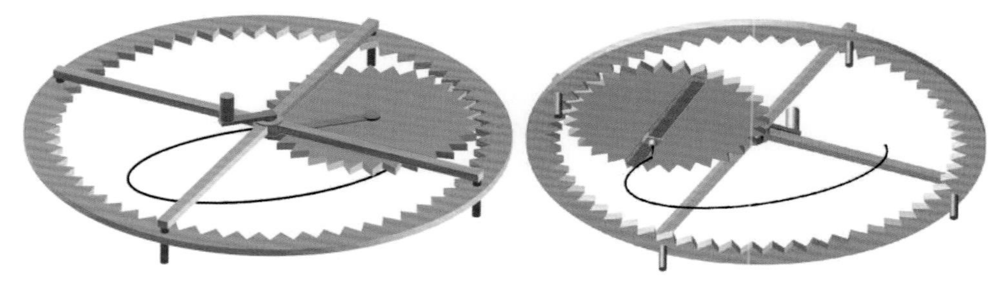

Fig. 10.46 Practical realization of the inner circle rolling

The rolling motion of the smaller Cardan circle within a second circle twice as large can be well simulated by a hollow wheel. ♠

- Milling machine for ellipses (Fig. 10.47, Fig. 10.48)

Fig. 10.47 Rolling of the outer circle instead of the inner circle, realization of the pole ray ...

The inner circle rolling with the given radial ratio (smaller wheel about R with a radius of 1 rolls within the larger wheel with a radius of 2) can be replaced by a much easier outer circle rolling by inserting an arbitrary auxiliary gear (midpoint H) which reverses the outer wheel's direction of rotation. The most important aspect remains the twice larger radius of the large wheel in relation to the smaller wheel about R. For the pole P, the relationship $\overline{FP} = 2\,\overline{RP}$ still remains.

Fig. 10.48 ...and a practical application

If we fix the centers of the wheels, then a point C on the diagonal of the wheel about R describes, of course, a circle about R – but that is only the case when referring to the fixed system. Relative to the rotating larger circle, however, C moves on an ellipse. The respective tangent to the ellipse is always orthogonal to the pole ray CP. It follows that an arbitrarily-sized circle through C about a point D on the pole ray always touches the ellipse. Wilhelm *Fuhs* has designed a mechanism (Fig. 10.47, 10.48) capable of realizing the pole ray on which the cylindrical mill is rotating about a vertical axis through D. If a sheet were to be mounted onto the large wheel, the milling process would turn it into an exact ellipse. ♠

- **Oval mill by Leonardo**

Leonardo da Vinci solved the problem of milling an ellipse in his own way and invented the *oval mill* in the process, which is also a beautiful example of relative motion (reverse motion):

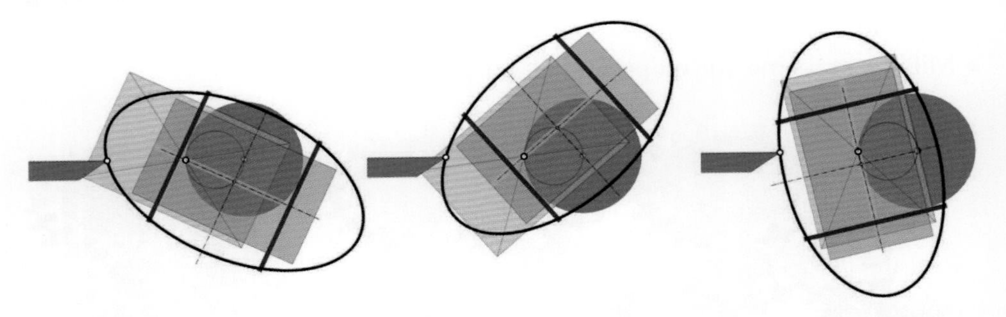

Fig. 10.49 Oval mill for the milling of ellipses

The lathe tool remains fixed. The "rotary table" consists of two plates (green and red). The red plate is rotating about its center P. The green plate rotates along with the red one, but is meanwhile translated so that two synchronously moved rails (blue in the picture) always touch a circle through P (also blue in the picture). This causes the center of the green plate to move along a Thales circle with half the radius of the blue circle. We shall now interpret the Thales circle and the doubly-sized circle as the pair of Cardan circles. The green plate, which is connected to the small Cardan circle, then performs an elliptical motion relative to the fixed blue circle: Each point of the green table, thus, has an ellipse as a path curve from the perspective of the fixed system. All that remains is to position a lathe tool at "some place", which would cut an ellipse out of the raw material attached to the green plate. ♠

• Ellipses produced by a crank shaft gear

Fig. 10.44 on the right shows a third possibility of producing an elliptical motion – by making use of a special crank shaft: If the shaft FR is rotating about F and propelling the crank RU along so that U is moved along the straight line u, then a point C on the crank describes an ellipse.

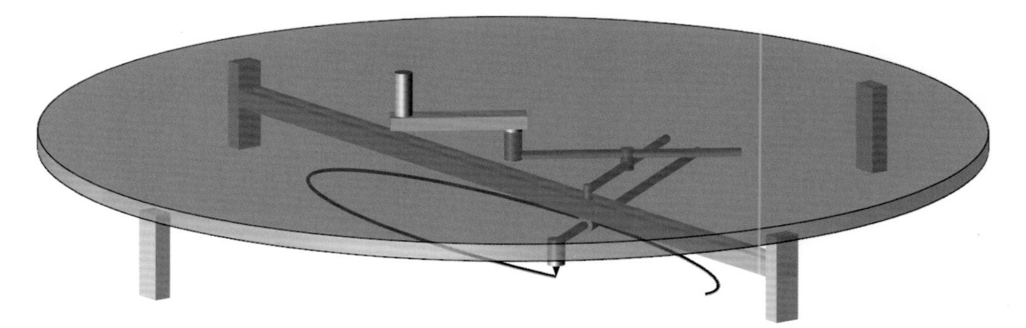

Fig. 10.50 Practical realization of the crank shaft gear

Fig. 10.50 shows a practical realization of the idea. The shaft is actuated over a table surface. The lengths of the shaft and the crank, as well as the distance from the gib, are adjustable. ♠

• Ellipse compass by *Hoecken* (Fig. 10.51)

In the pictured mechanism, two (yellow) rods are directed so that their end points move along circles with the adjustable radii a and b. After rotation about the angle φ, the intersection point has the coordinates $(a \cdot \cos\varphi,\ b \cdot \sin\varphi)$ and thus lies on an ellipse.

Fig. 10.51 Ellipse compass by *Hoecken*

In this case, the mathematical explanation is easier than the geometrical one, because the intersection point of the rods is, from a kinematic point of view, not so easy to capture. ♠

- **Another ellipse compass . . .**

Another beautiful ellipse compass (the last one in this context) can be seen in Fig. 10.52. It realizes the classical two-point guidance, though this is only apparent upon closer inspection.

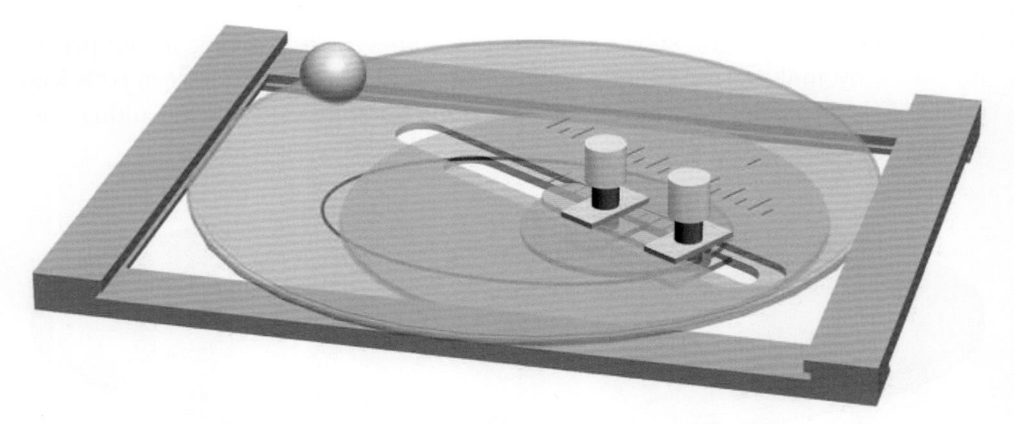

Fig. 10.52 Ellipse compass with "hidden" straight line guidance

A big (red) circular disk is being moved in parallel within a rectangular boundary. Its midpoint is, thus, moving on the perpendicular bisector of the shorter side of the boundary. A smaller (blue) circular disk is being moved analogously so that its midpoint remains on the perpendicular bisector of the longer side of the boundary. The distance of the centers of the circle is adjustable by slot guides, just as the distance of the drawing pen may be varied. ♠

General elliptical motion

Up to this point, the guiding straight lines of the elliptical motion were always mutually orthogonal. In the general case, however, it can be said that:

If a rod of fixed length is directed such that its end points U and V are moved along the arbitrary intersecting straight lines u and v, then each point which is rigidly connected to the rod describes an ellipse.

Proof:
The intersection $F = u \cap v$ is the midpoint of the path curve. The circle q through U, V, and F (midpoint R) can be interpreted as the peripheral circle over the segment UV due to the fact that the angle $\angle UFV$ is constant (see also Fig. 1.6). The radius of q is, thus, likewise constant. The instantaneous pole P is given as the intersection of the path normals in U and V and lies also on q according to *Thales*' theorem in relation to F. Thus, $\overline{FP}$ is constant, and P moves on a circle p about F. The radius proportion of the circles p and q is $2:1$, which means that the circles p and q are the Cardan circles of an elliptical motion.
Let us now consider the elliptical path of a point C which is rigidly connected to UV. The diameter CR of q intersects q at the points X, Y. These points descri-

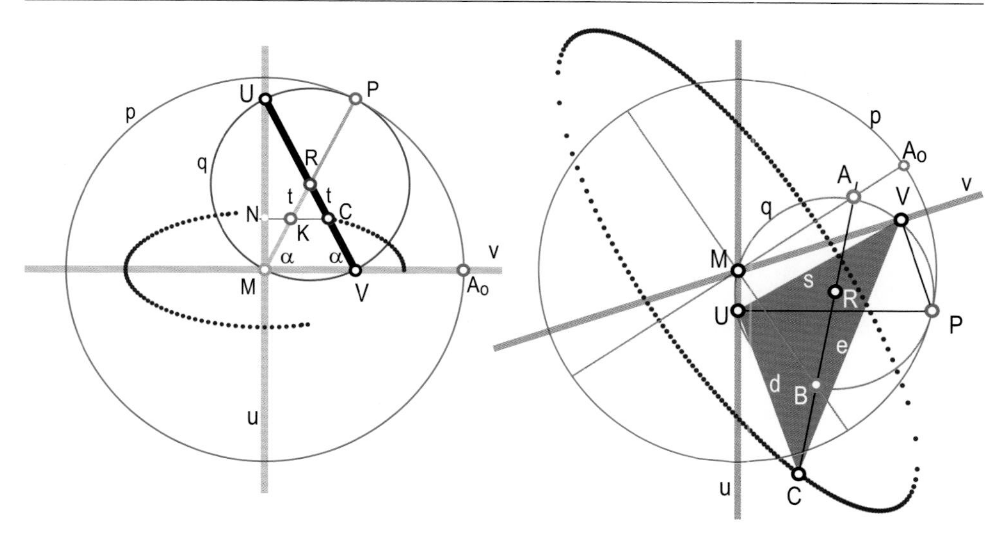

Fig. 10.53 Special and general elliptical motion

be "infinitely thin" ellipses through F, which are segments. According to *Thales'* theorem, the straight lines FX and FY are orthogonal. The given motion is, thus, also forced if the rod XY of fixed length is moved along two straight lines which intersect orthogonally. These are, therefore, the axes of the path ellipse of C. ◇

The point C might, for instance, be given through its distances to U and V. The triangle UVC can then be positioned onto UV in two ways, which, in turn, yields two different results (Fig. 10.53 left and right).

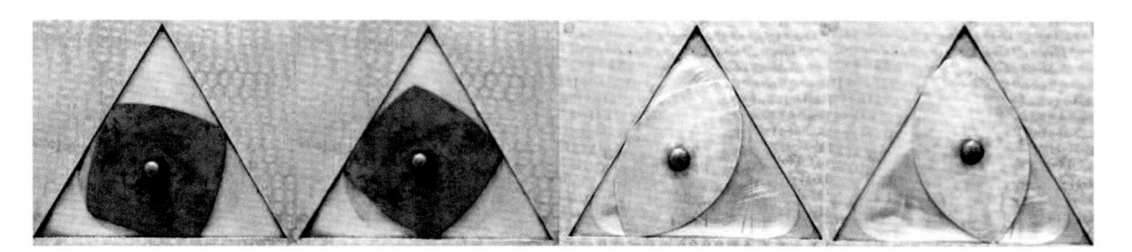

Fig. 10.54 Revolution within a triangle

• **Revolution within a triangle or a pentagram**

Fig. 10.54 shows the revolution of a square (or "digon") bounded by circular arcs within an equilateral triangle. The circular arcs have their centers on the opposite arc, producing a "curve of constant width".

During revolution, an alternation between three different elliptical motions occurs. The digon, in particular, produces traces of metal due to the frequent revolutions. They are the elliptical path curves of the vertices. Fig. 10.55 shows the revolution of a three-leaved curve of constant width within a pentagram. ♠

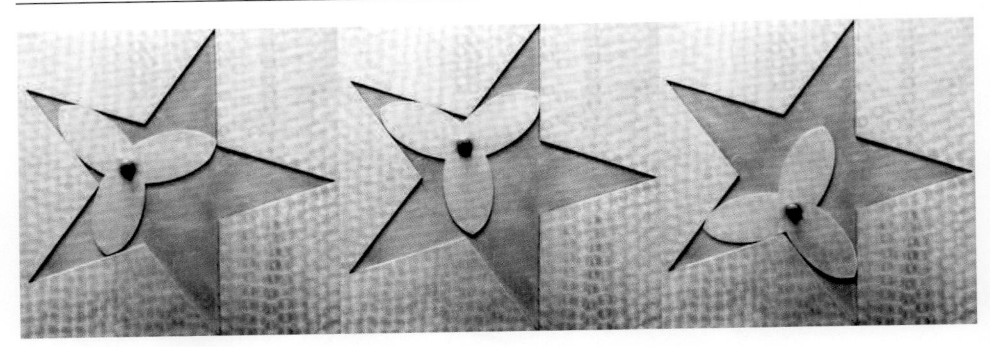

Fig. 10.55 Revolution within a pentagram

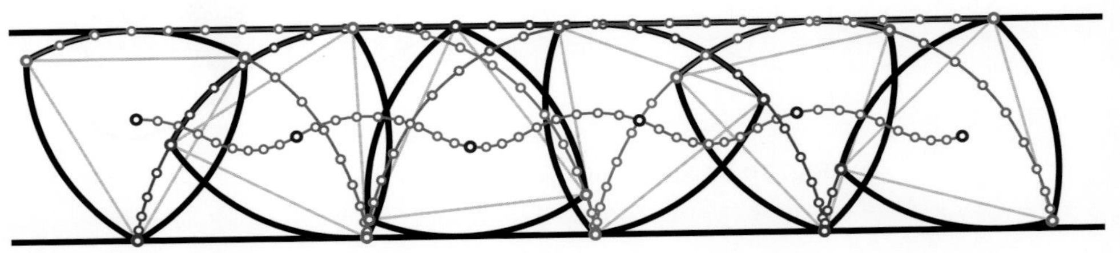

Fig. 10.56 Revolution of a curve of constant width within a parallel strip

Fig. 10.56 illustrates the revolution of a curve of constant width within a parallel strip. The path curves of the corners consists of pieces of straight lines and cycloids. The path of the midpoint resembles a sine curve.

10.4 Trochoid motion

We have seen how an elliptical motion can be forced by the revolution of a circle as long as the radii of both Cardan circles define the ratio 2 : 1. However, what happens if this ratio is changed?

Fig. 10.57 Gardeners often seem to exhibit an inclination towards geometry: The pictured rose patch in Schönbrunn is a good example (see also Fig. 10.58 on the right).

If a circle with an arbitrary radius rolls within a fixed circle, then each point connected to the rolling circle describes a (*hypotrochoid*).

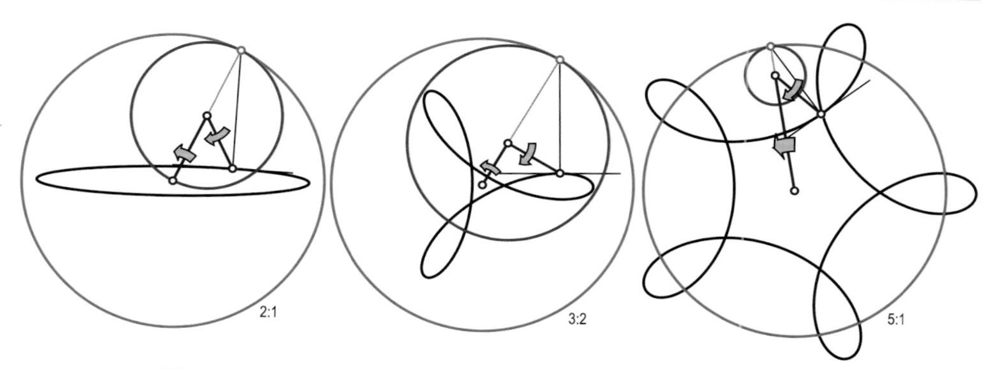

Fig. 10.58 Hypotrochoid motion as generalization of ellipse motion

Fig. 10.58 shows this generalization with the ratios $2:1$, $3:2$, and $5:1$ for the radii. The instantaneous pole is located at the point where the rolling circles meet and has an instantaneous speed zero. The corresponding path normal of a point's path curve passes through it.

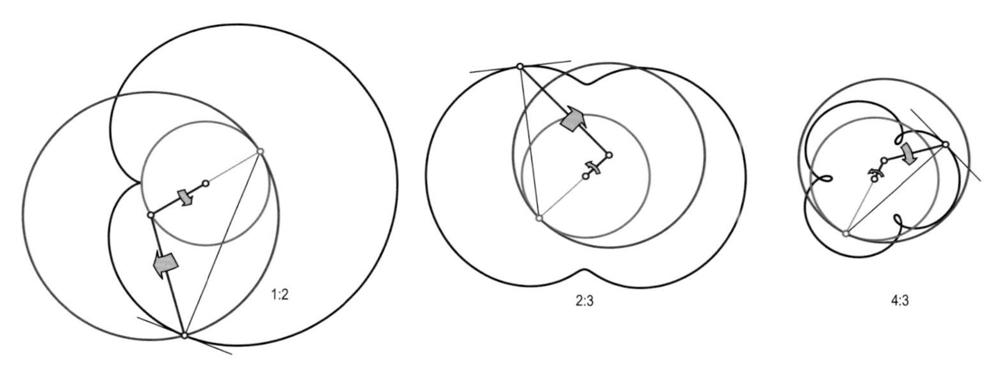

Fig. 10.59 Hypotrochoid or epitrochoid motion?

If the radius of the rolling circle exceeds the radius of the fixed circle, then the rolling circle encompasses the fixed circle, and the point of contact is located on the outside (this situation is sometimes referred to as a peritrochoid). Fig. 10.59 illustrates this case for three different ratios of radii. The path curve of points attached to the moving circle are stretched, have cusps, or are engulfed, depending on whether the point lies inside, on, or outside the rolling circle.
A further generalization is given if the moving circle is definitively rolled *on* the fixed circle. Fig. 10.60 shows the corresponding path curves for the same radius proportions as before.

If a circle with an arbitrary radius rolls on a fixed circle, then any point connected to the rolling circle describes an *epitrochoid*.

Trochoids are often found in geometry. If the ratio of radii is a rational number, then the curves are algebraic and closed. Examples include the ellipse

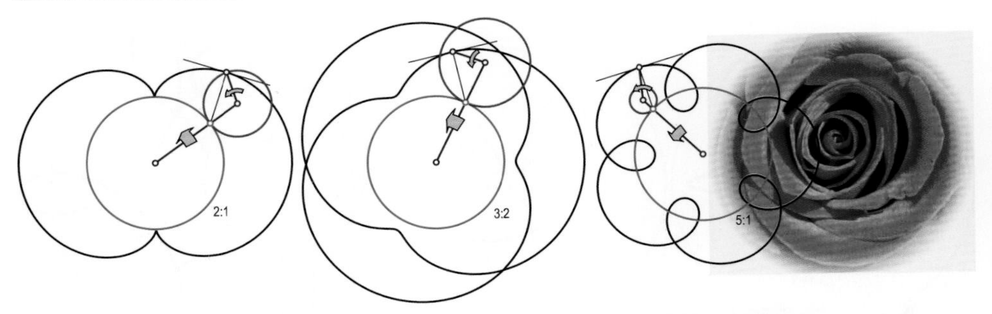

Fig. 10.60 Outside rolling for different radius proportions

(of second order) and *Pascal*'s limaçon (of fourth order), which often occurs as the locus (Fig. 1.21) or normal view of a spatial curve (Fig. 11.4 on the right).

• Wankel's engine

A classical technological application of trochoids is the *Wankel's engine.* The section of the piston chamber (Fig. 10.61) is an epitrochoid, which is thrice traced by the vertices of the synchronously moved equilateral triangle.

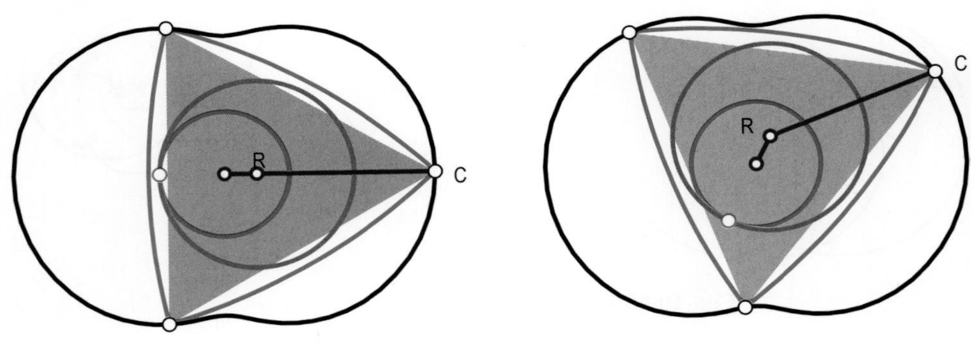

Fig. 10.61 Wankel's engine

The envelopes of the triangle sides fit easily into the piston chamber. In order to accomplish a greater compression, the central piece is thickened such that the envelopes are as large as possible while still fitting in the chamber. ♠

• Planetary motion

The trochoid motion was originally called *planetary motion.* An astronomer observing the path curves of our neighboring planets would eventually discover curves which are nearly trochoidal: Due to our own revolution around the sun, the rotation of both planets is superimposed.
Fig. 10.63 illustrates the rotation of the planets about the sun – in slightly deviating planes, but with highly varying angular velocities. If, at a specific time of day, we mark the position in the sky at which such a planet is seen, we will notice that it essentially moves on a great-circle – the equinoctial line. If the planet's path plane is additionally inclined relative to the path plane

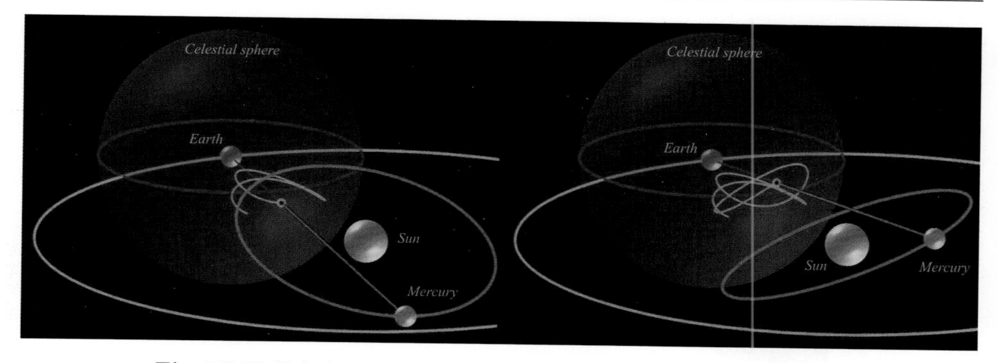

Fig. 10.62 Relative motion of an inner planet on the celestial sphere

of the earth (in Mercury's case, this inclination equals 7°), then the motion of the planet diverges from the equinoctial line.

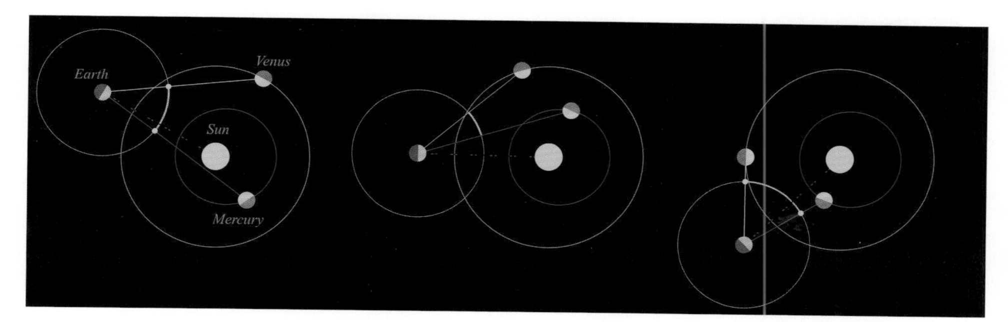

Fig. 10.63 Relative motion of the inner planets

In Fig. 10.62, the path of Mercury was simulated, but it was done with a threefold inclination of its path, in order to emphasize the effect. We can see that these cases may, indeed, have something to do with trochoid motions, but certainly not precisely so. ♠

• **An original folding chair**

We have often discussed "glide-free rolling" in cases of two circles rolling onto each other. As in Fig. 10.64, this motion can be forced by a ribbon of constant length, yielding (in this example) an original folding chair by Thomas Ehrenfried. ♠

Cycloids (trochoids) at a higher level

The trochoid motion can be generalized by letting a third rod rotate at a constant angular velocity about the endpoint of the second rotating rod. This approach produces a third-level cycloid with a much greater number of degrees of freedom, permitting the construction of many new curves. Fig. 10.65 shows three third-level cycloids from this infinite multitude – they were created by using rods of equal length. Even the angular velocity only differs for the center rod.

If k rods are rotating, then we speak of kth-level cycloids. The third-level cycloids in Fig. 10.65 are pictured in normal projection. It can be shown

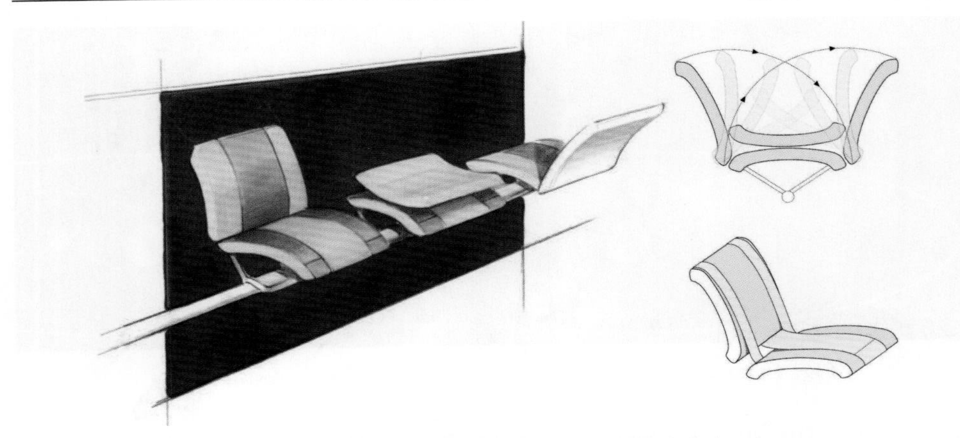

Fig. 10.64 Forced revolution of a folding chair

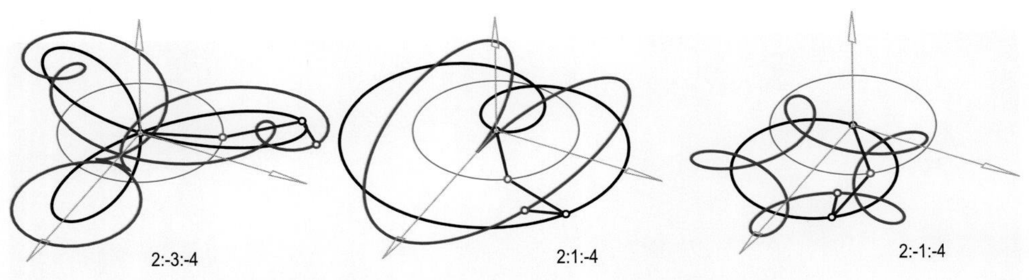

Fig. 10.65 Third-level cycloids with the rod lengths 2, 2, 1

that such a projection is itself a cycloid of a higher level. In general, it can be said that any curve can be approximated to any desired level by higher level cycloids.

Kinematics at "two-and-a-half dimensions": Ball bearing

Before proceeding to spatial kinematics, let us consider a ball bearing according to the master's sketches: The bearing was meant to be used for a precisely controlled rotation of heavy loads.

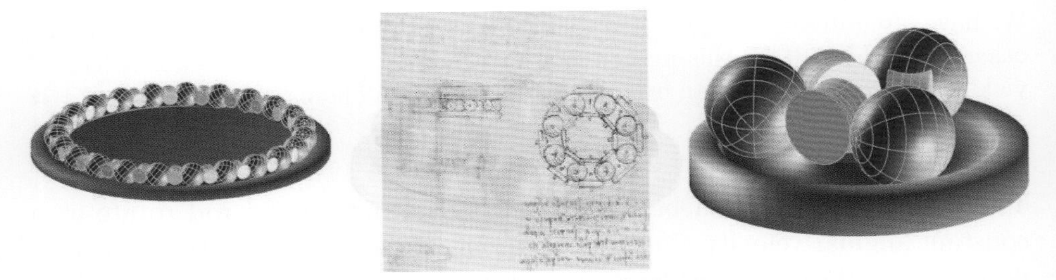

Fig. 10.66 Leonardo's ball bearing: Spheres alternate with ring surfaces

In this situation, spheres, between which torus parts (ring surfaces) have been inserted, propel each other. The number of elements may vary, as can the radii ...

11 Spatial motions

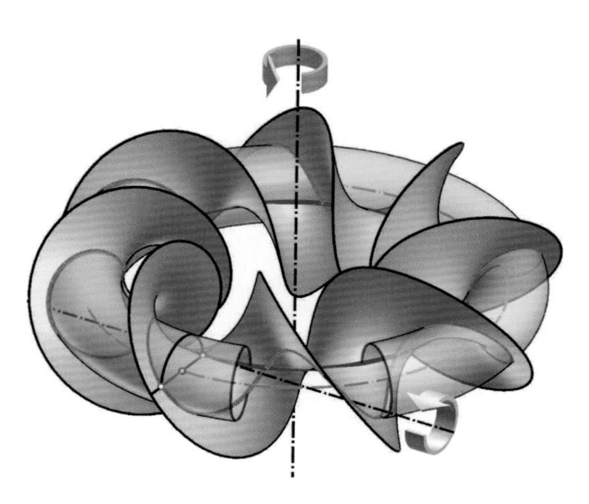

Planar kinematics is the basis for kinematics on the sphere. In a way, spherical kinematics can be conducted similarly to planar kinematics. In the spherical variant, everything rotates about straight lines through the center of the sphere. Purely spatial motions, however, make these matters much more difficult. In this chapter, both possibilities will be explored on the basis of certain examples.

Two general positions of a spatial Cartesian coordinate system can only be transferred into each other by a helical motion (including the special cases of a rotation or a translation).
Kinematics often deals with the transmission of motions from one axis to another. If these axes are parallel, then a two-dimensional problem is given. If these axes intersect, then we are dealing with geometry on the sphere. If these axes are oblique (as in the image above), then the problem is "truly three-dimensional".
We will try to shine a light on the motion of the earth around the sun, where two rotations with oblique axes are composed. To make matters even more complicated, the rotations are only approximately proportional. Important questions arise concerning the theory behind sun's position – such as when the sun is rising which angles of elevation it achieves, and at which position on the firmament it can be found at a given date and time. We will learn about the *equation of time* and become capable of analyzing sun dials at a precision of mere minutes.

Survey

11.1 Motions on the sphere

If a system is rotating about an axis r, and if this axis is moved so that a point M on it remains fixed, then a spherical motion is produced. (In Fig. 11.1, e.g., r is being rotated about the intersecting fixed axis f.) The path curve c of each point P within the moving system lies on a sphere.

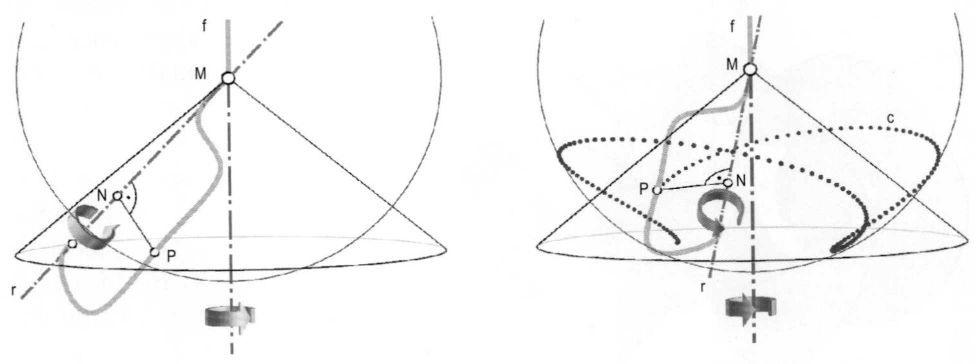

Fig. 11.1 The composition of two rotations about intersecting axes

In order to improve our understanding of this relation, let us consider the normal pedal point N of the normal from P to r. Both $\overline{PN}$ and $\overline{MN}$ remain constant in the course of the motion.

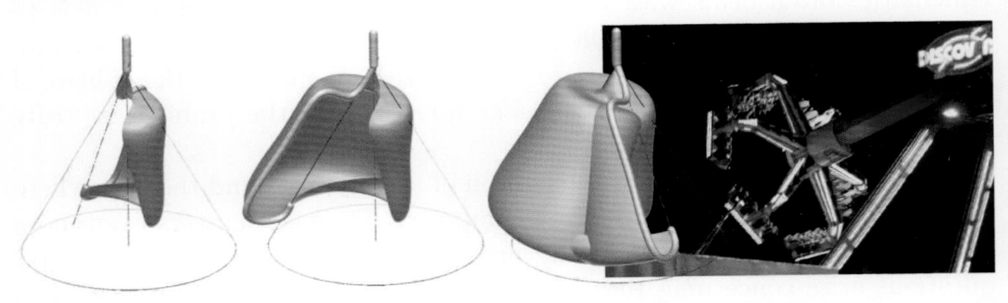

Fig. 11.2 Blender (including its enveloping surface): Composition of rotations about intersecting axes. On the right: Something rather comparable – this analogy, no doubt, contributes to the feeling of dizziness ...

Of the blender pictured in Fig. 11.2, the path curves are far less interesting than the enveloped surface that is being generated by the curved rod.

Let us focus our attention on spherical motions, where *two rotations about intersecting axes* r and f (intersecting at M) are composed. In that situation, we do not even have to prove the following assertion:

Two composed rotations can always be replaced by a single rotation.

The new *instantaneous axis* p lies in the plane spanned by r and f and passes through M (Fig. 11.4 at the center).

If the composed rotations are proportional, then the instantaneous axis forms a constant angle with the fixed axis, and thus, lies on a cone of revolution Δ_f with its apex at M. In the moving system, too, the angle enclosed by p and r remains fixed, and thus, p generates a cone of revolution Δ_r which is always touching Δ_f. It can be shown that both cones are performing a mutual glide-free rolling. The path curves of single points are called *spherical cycloids*.

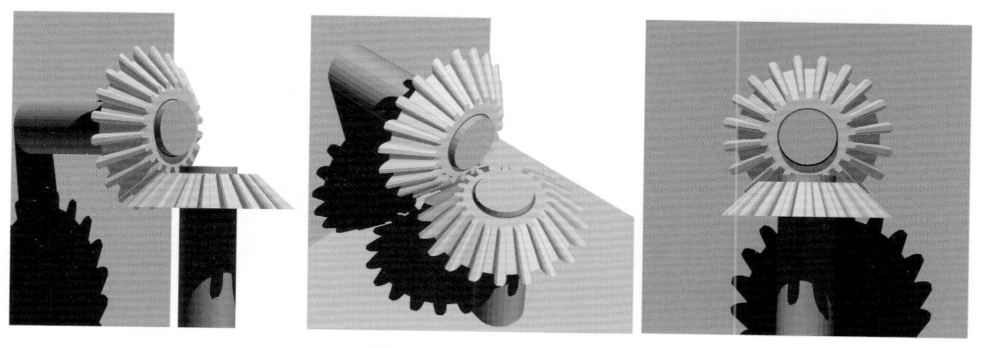

Fig. 11.3 Bevel wheels

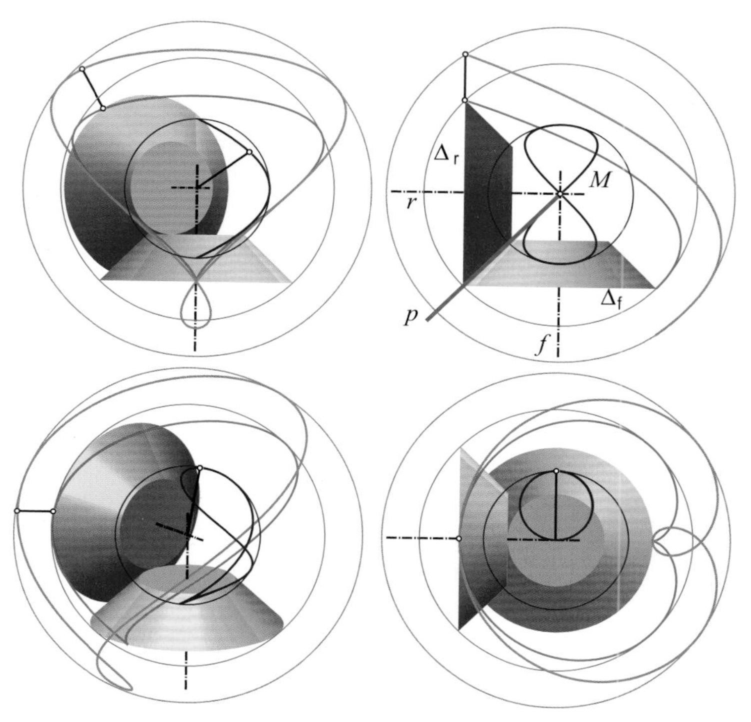

Fig. 11.4 Cone rolling, including path curves (principal views)

Fig. 11.4 shows a very special cone rolling situation: The axes r and f intersect orthogonally, and both cones have the same aperture.

• **The star handle – a mere rotation about a general axis**

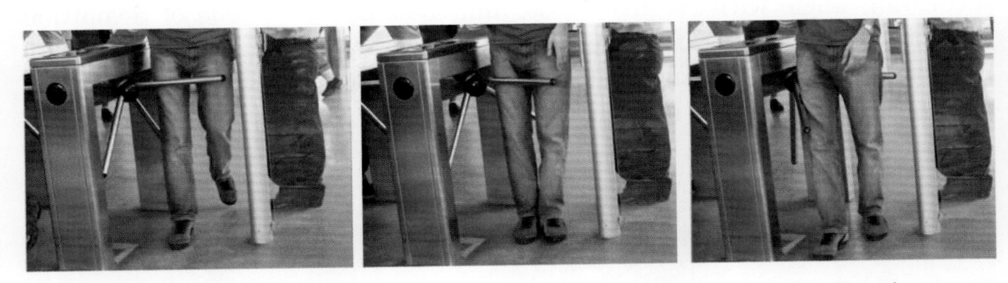

Fig. 11.5 Star handle in motion (Rotation about a generally positioned axis)

The star handle (Fig. 11.5) is an example of a very simple spherical motion. The end points of the three rods move on a sphere with a fixed radius – the rod length. The motion itself, however, is a fixed rotation about a likewise fixed (oblique) axis. The endpoints of the rods share their path circle! ♠

Spherical motion can also occur as relative motion – as the following example suggests.

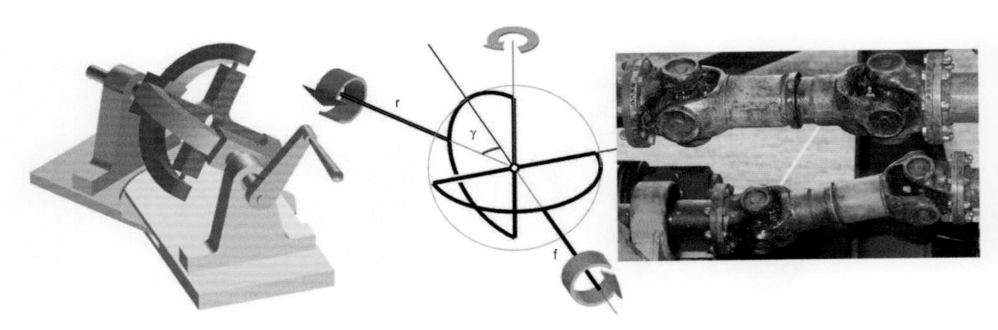

Fig. 11.6 Cardan shaft (non-uniform transmission)

• **The Cardan shaft**

By using a Cardan joint, it is possible to transfer rotations about an axis f onto a second, intersecting axis r. Fig. 11.6 demonstrates this technique in a star handle, which is connected to one axis via one fork (in the shape of a half-circle) at f and to another axis at r.
The transfer of the rotation, however, is non-uniform. In a uniform powertrain, this produces a "wobbling" of the driveshaft, which increases in magnitude in relation to an increasing axis angle γ. A combination of two Cardan shafts (Fig. 11.6 on the right) cancels out this problem. ♠

The uniform transmission of a rotation to a rotation about an orthogonal axis is a recurring problem in technology. In the simplest case, it can be solved by letting two rubber-coated cones of revolution with aperture angles of 90° roll on each other, as in Fig. 11.4. If greater forces are involved, then the use of cogwheels will be required. Fig. 11.3 shows a possible solution to this problem. In practice, *worm gears* are often employed, as in Fig. 7.30.

• A sophisticated ball joint

If the axes f and r are positioned at a variable angle, and, if the use of a Cardan joint is not desired (presumably because the transmission of rotation has to be proportional), then the ball joint pictured in Fig. 11.7 may be employed. This particular ball joint has been the subject of detailed research by Hellmuth Stachel[1].

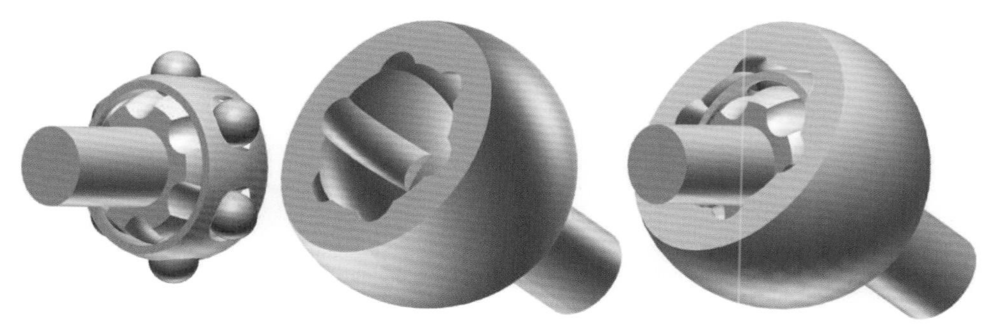

Fig. 11.7 An efficient ball joint, which may replace many bevel wheels.

♠

• "Madhouse" (Fig. 11.8, see also Fig. 5.73)

A (red) circle rotates about a vertical axis. A second (blue) circle is hung up horizontally within it, and rotates proportionally about this horizontal axis. Within the second circle, hung up vertically to the horizontal axis of rotation, there is a third (black) circle. It also rotates proportionally about this third axis. Which path curves are described by points on these circles?

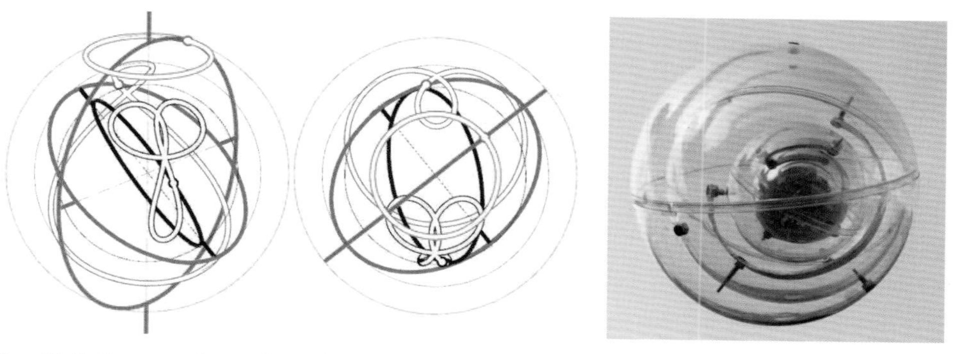

Fig. 11.8 Compound rotations about intersecting axes are always spherical. The shape on the right, however, is slightly eccentric and possesses no constant gear transmission ratios.

The first rotation naturally produces path curves, and compound rotations about axes which intersect at a single point produce spherical motions – the

[1] `http://www.geometrie.tuwien.ac.at/stachel/dres.pdf`

path curves, thus, move on concentric spheres. If the angles of rotation are proportional, then (as in the case of the cone rolling in Fig. 11.4) spherical cycloids are produced, whose top views are planar cycloids of second or third level (p. 365). Fig. 11.8 on the left illustrates this motion for three equal angular velocities. In Fig. 11.8 at the center, the respective top view is visible.

A variant by Maria Walcher is visible on the right. It is, however, slightly eccentric, and thus, does not resemble exact spherical motion despite all the spheres involved. Due to the absence of a propulsion (the outer sphere is hung up like a puppet and can be moved in much of the same way), the angular velocities are not proportional. Walcher was inspired by the motion of the eyeball. ♠

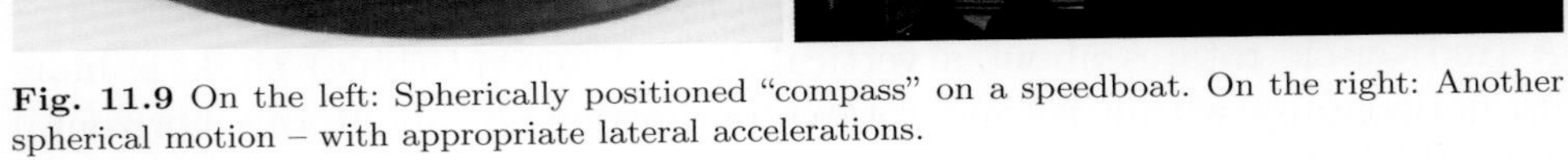

Fig. 11.9 On the left: Spherically positioned "compass" on a speedboat. On the right: Another spherical motion – with appropriate lateral accelerations.

Construction on the sphere

We live on a sphere, even though we may not recognize this at first sight: Therefore, we also construct on a sphere whenever we construct on the earth surface. Thus, it might be said that our planar kinematics are actually "slightly spherical". What happens if we actually let certain simple things happen on the sphere, such as the string construction of an ellipse (Fig. 1.27) or its construction according to *de la Hire* (Fig. 3.55)?
The result is a "spherical ellipse", or rather a "spherical conic section". It turns out that it makes no sense to differentiate between an ellipse and a hyperbola on a sphere (as seen in Fig. 11.11 on the left). The spherical parabola, defined as the locus of all points which are equidistant to a given "straight line" (=great-circle) and a fixed point, is not an exception (and looks rather like Fig. 11.11 at the center left where, however, a small circle is being touched): All spherical conic sections are intersection curves of the sphere with certain quadratic cones concentric with the sphere[2].

[2] `www.uni-ak.ac.at/geom/geom/harald-tranacher.pdf`

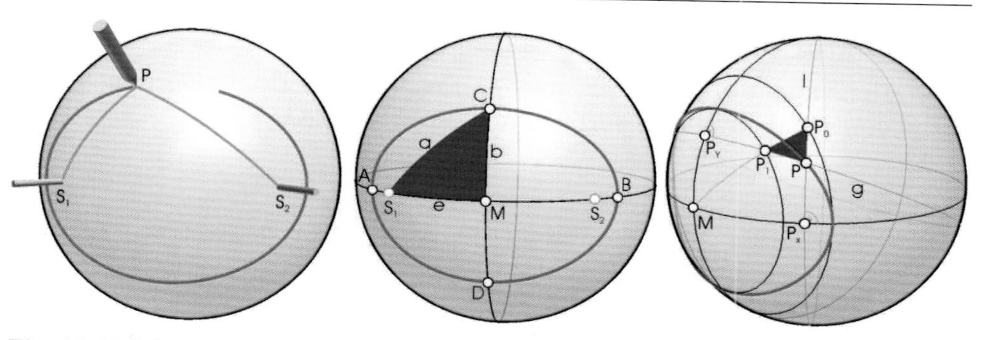

Fig. 11.10 Spherical conic section: Some planar constructions can be well comprehended on the sphere. On the left: String construction. On the right: Construction according to de la Hire.

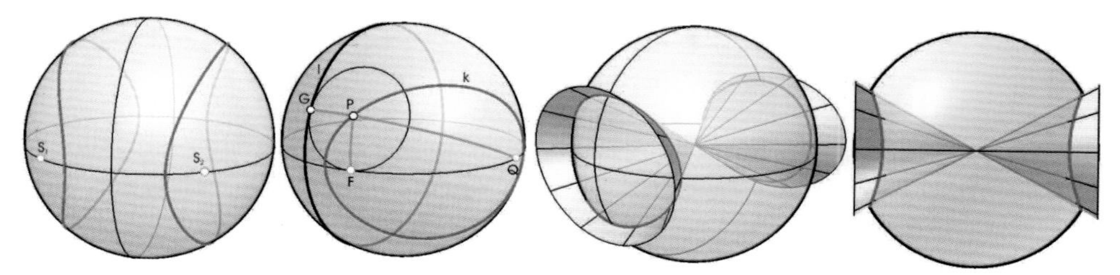

Fig. 11.11 Spherical conic sections and intersection curves of quadratic cones and spheres around the apex. It is impossible to make any further type differentiations.

11.2 General spatial motion

If a system is *somehow* moving in space, then it naturally becomes difficult to imagine. The *main theorem of spatial kinematics* is here applicable (a proof may be found in [11]). It declares that any spatial motion – no matter how complicated – can at each point be approximated by an infinitesimal helical motion about a well-defined *instantaneous axis*.

Fig. 11.12 A sphere with a variable radius

Fig. 11.12 depicts a very interesting and sophisticated "gimmick" – a "sphere with a variable radius". Due to an ingenious system of joints, the object is movable such that all points on the rays pass through the center of the circumsphere.

Two rotations about oblique axes

If rotations about oblique axes are composed, then certain assertions about path curves and enveloped surfaces may be attempted.

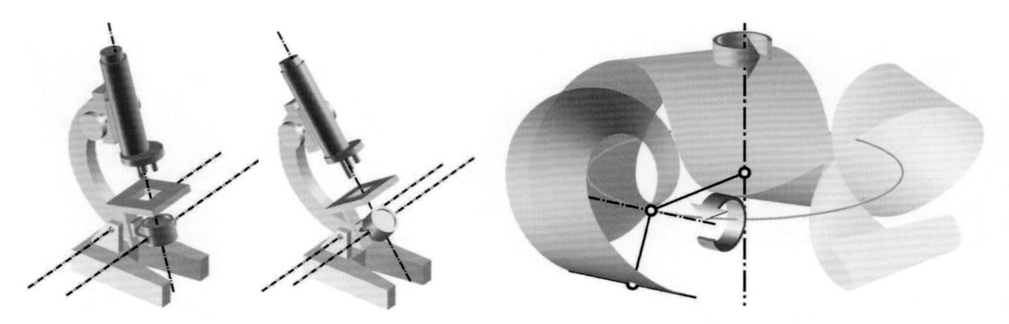

Fig. 11.13 Two composed rotations about oblique axes

If the rotations are proportional, then the problem is further simplified. We have already pointed towards the one-sheeted hyperboloid of revolution: Using two *hypoid gears*, a rotation about an axis can be converted into a proportional rotation about another axis (Fig. 6.30).

• **A loving effort at a tiny scale . . .**

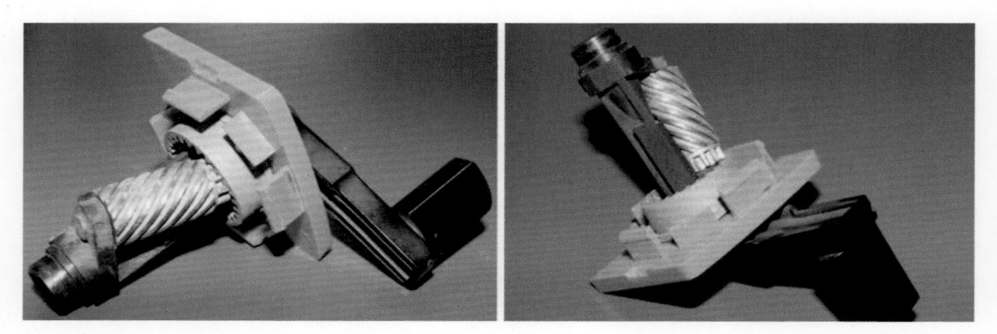

Fig. 11.14 The inside of a very efficient pencil sharpener: A cylindrical mill is rotating about an oblique axis.

Fig. 11.14 shows the complicated inner life of a very efficient pencil sharpener. The axis on which the pencil is centered stays fixed: In contrast to simpler pencil sharpeners, the pencil itself does not undergo any rotation! Instead, a cylindrical mill with an oblique axis is rotating and moving around the pencil. First, a crank is turning the cylindrical mill by moving along two points of the axis. On the top of the mill, there are gear flanks which force its relatively quick self-rotation by means of a fixed, hollow gear. ♠

• **"As if it was turned by gears"**

What happens in the above example at a tiny scale occurs in nature at tremendous extents: The Earth is rotating about an axis of fixed direction while it is also rotating about the sun. The axis of the latter rotation is oblique in relation to the Earth's axis. For a viewer in outer space, the motion of the

Earth is akin to the motion of the mill about the pencil. Spatial kinematics at its finest! (For the interested: The sun itself has a self-rotation, and its midpoint describes an elliptical path about the gravitational center of the entire solar system.)

Of course, the motion of the Earth is not caused by gears, but essentially by a constant reaction towards gravitational and centrifugal forces. Gravity is of particular significance in outer space – and this significance is perhaps best exemplified by the observation that due to self-gravity, virtually every solid object with a diameter greater than 500km is spherical. ♠

• The Stewart platform in space

In three-dimensional space, robotics allows us to control any motion of a plane by six arms of variable length (Fig. 11.15). The points of these arms lie on a regular three-sided prism.

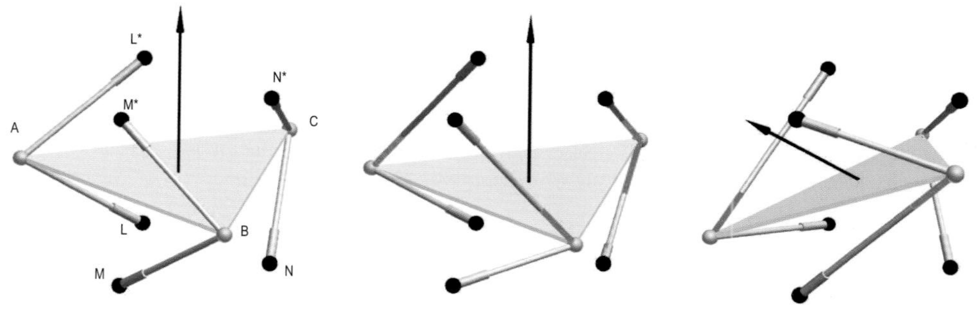

Fig. 11.15 Triangular platform ABC controlled by six variable-length robot arms with the fixed points L, M, N, and L^*, M^*, N^*. On the left: "Zero position". At the center: The triangle was translated horizontally. A rotation about the inscribed axis (about the center of gravity) can easily be done. On the right: This time, the triangle was tilted, and the instantaneous axis of rotation is inscribed (Images: Georg Nawratil).

Fig. 11.16 An amusement park "3D shaking machine" with fixed-length arms and robot arms of variable length for precise and safe control

Such platforms are, among other areas, used in flight simulators. Fig. 11.16 shows something similar in an amusement park, where the "platform" consists of a row of seats, in which the hapless victims have a chance to get sick for a small change. ♠

• Non-rigid connections

The construction in Fig. 11.17 on the right serves a similar purpose as the "3D shaking machine". During uniform rotations, an equilibrium between centrifugal and gravitational forces is quickly established. Phases of acceleration and deceleration are especially interesting from the perspective of physics due to the then pronounced effects of inertia.

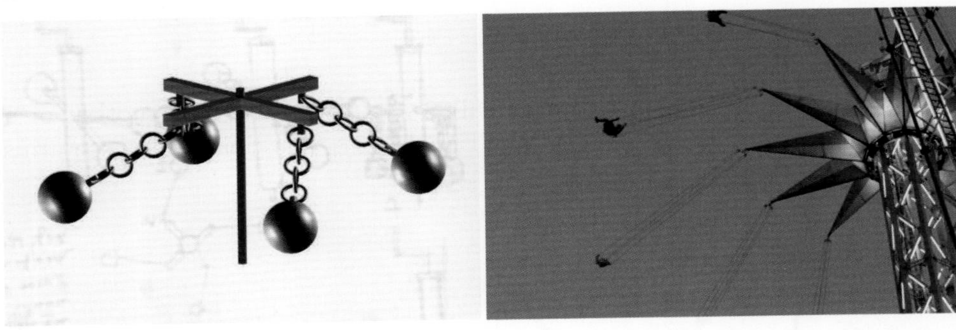

Fig. 11.17 Flywheel according to *Leonardo da Vinci*, and a large-scale modern application. During uniform rotation, the path curves of all points naturally describe circles. Acceleration or deceleration brings about spatial spirals which approximately lie on a torus.

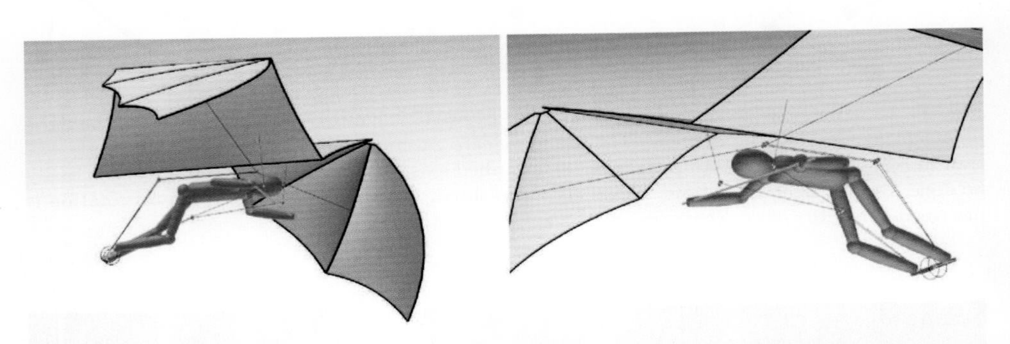

Fig. 11.18 Flying in the manner of *Leonardo da Vinci*: In theory, this model works perfectly fine. It can be animated and verified by methods of computer geometry. There is no doubt that its weakest link is the human passenger, who would have no chance of generating the necessary uplift by pulley-induced wingflaps (not to speak of maintaining it). Bats and birds are capable of flight due to their small size, which automatically increases their relative muscle power ([11]).

Once again, it was the proverbial Renaissance man who already investigated issues of a similar kind (in the computer simulation pictured in Fig. 11.17 on the left, the corresponding sketches are overlaid in a transparent manner). The blueprint for a flying machine (Fig. 11.18 on the left) provides one further example of Leonardo's tremendous imagination. From a purely geometric point of view, it is entirely correct and was probably derived from the observation of bat flight. ♠

11.3 What is the position of the sun?

In this section, we shall employ methods of geometry to illuminate the relative position of the sun in the course of the calendar year, as well as during a whole solar day. Certain simplifications allow us to get an easy grip on the issue and answer not only day-to-day questions, but also several important ones from the field of architecture:

- When and where does the sun rise and set?
- At a given day and time, whence does the sunlight come?
- How can a wristwatch be used to exactly determine the cardinal directions on a sunny day?
- When is the sun exactly in the west?

As it turns out, the first and the last questions are completely equivalent. For the mathematically inclined readers, the results will be described by approximative equations. The quality of these approximations can be checked by a comparison to astronomical tables.

Does the sun not rise in the east and set in the west?

The sun famously sets in the precise west on only two days of the year. In the winter half-year (and in northern latitudes), it disappears somewhere between the south west and the west – in the summer half-year, between the west and the northwest.
Let a living room be precisely directed towards north. At which time of year, and for a duration of how many minutes or hours, can one count on the morning or evening sun illuminating the room? This problem leads exactly to the same question. The naive assumption that the sun stands at exactly 18^h "of true apparent time" in the west – in Vienna at about 19^h of central European summer time – turns out to be false. Let us, therefore, get to the bottom of the matter using the methods of descriptive geometry.

First of all, the facts

The Earth takes a year (365,24 days) to revolve around the sun and is continuously rotating about an axis of fixed direction which currently points approximately to the North star. The time for a full revolution is a little less than 24 hours (about 23^h56^m). 4 further minutes are required to compensate for the Earth's rotation about the sun ("Revolution").
The path curve of the Earth is a near-circular ellipse. The relation $e \approx a/60$ holds for the great half-axis a and the ellipse's linear eccentricity e. The normal of the *ecliptic* (carrier plane of the ellipse) defines, along with the axis of the earth, a constant angle $\varepsilon = 23{,}44°$. Due to the *precession motion* (gyroscopic motion) of the Earth, this value slightly changes over the course of the decades. Thus, in older books, this value is sometimes given as $23{,}50°$.

The idealized Earth's surface is well approximated by a sphere. The flattening at the poles measures only about 1/300 of its radius. A point on the surface can be precisely given by its geographical longitude λ and latitude φ (Vienna, for instance, is located at $\lambda = 16{,}3°$, $\varphi = 48{,}2°$).

Let us now simplify several things ...

In order to lessen the difficulty of our problem, let us assume the following – wholly acceptable – simplifications:

1. Time designations generally refer to the *apparent time*. It is defined such that the sun reaches its culmination point (highest point on its apparent path on the celestial sphere) at exactly 12^h noon. The time difference between the culmination points is always varying and does not amount to exactly 24 hours. In order to see the sun culminate at 12^h on each day, we would need to adjust our clocks all the time. The daily errors accumulate, which leads to deviations of about ± 15 minutes over the course of weeks. This is caused by Kepler's second law, the *law of planetary motion*: The ray from the sun to the earth generates equal areas at equal times.

Fig. 11.19 Shadows become continuously sharper from left to right – the sun is not point-shaped.

2. Earth's path can be approximated by a circle (the maximum deviation of the path ellipse measures below 1,5% of the circle diameter).

3. The sun's rays are parallel due to the sun's great distance. In the approximate model, it is, therefore, enough to calculate by using the sun midpoint. However, it should be kept in mind that despite its distance of ≈ 150 million kilometers, it still appears on our firmament – due to its enormous diameter of $\approx 1{,}4$ million kilometers – at a viewing angle of $\approx 0{,}5°$.

4. If each month is assumed to be 30 days long (and thus each year to have 360 days), then earth rotates roughly 1° about the sun each day. This value, too, is subject to Kepler's second law and thus varies slightly. During its north winter, the Earth has a higher rotational velocity than during summer. On the northern hemisphere, this causes the winter half-year

to be about six days shorter than the summer half-year! For observations that only last for a single day, however, this relatively tiny change of position is disregarded.

When compared to the actual values (e.g., without the various simplifications), it can be shown that for not too extreme latitudes ($\varphi < 52°$), all values will be correct to a maximum deviation of 4 minutes (though usually not more than 2 minutes). Given an average length of day of $12^h = 720^m$, this amounts to a negligible maximal error of about 0,5% – and this even though we have applied several simplifications!

Two important auxiliary theorems

For a viewer on the Earth's surface, the following first auxiliary theorem can be applied:

The northern celestial pole ($\approx$ North star) can be found in northern direction at the elevation angle φ (Fig. 11.20).

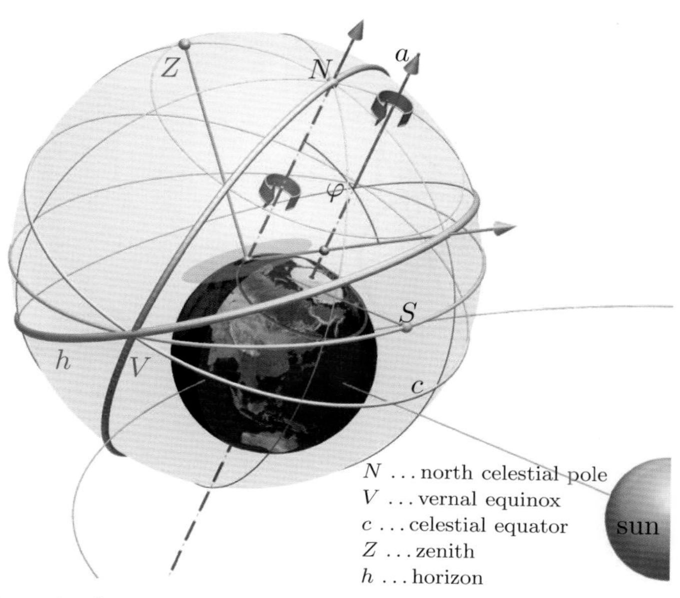

Fig. 11.20 The axis through the celestial poles is parallel to the Earth's axis a and inclined under the elevation angle φ.

From now on, when we – rather sloppily – refer to the "North Star", we shall mean the northern celestial pole – the point on the firmament that defines the direction of the Earth's axis. There may not be a prominently visible star at this precise position, but the literal North star is only offset by about 0,9°. This star, the α *Ursae Minoris* (main star of Ursa Minor), is the current *North Star*. Due to the precession motion of the Earth, the direction

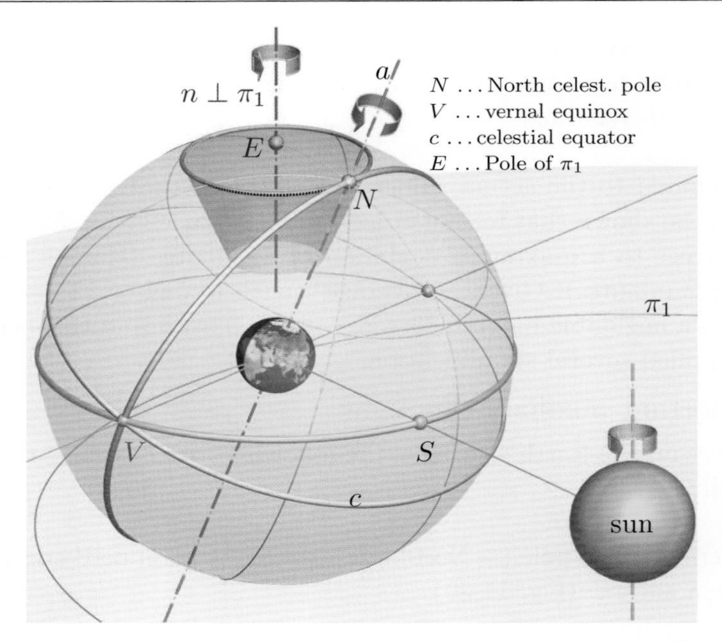

Fig. 11.21 For investigating Stone Age celestial geometry, the precision motion of the Earth has to be taken into account.

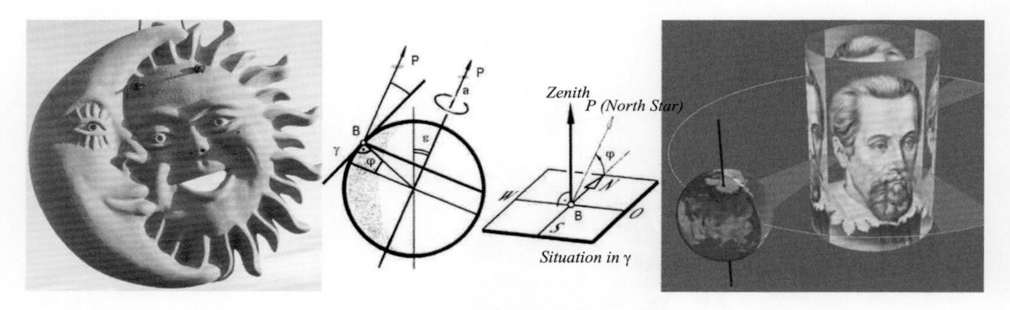

Fig. 11.22 As for the North Star, about which everything turns (on the right: *Kepler*)

of the Earth's axis continuously changes, and so does the celestial pole. In a few hundred years, another star will assume the function of today's North star. The ancient Egyptians even spoke of multiple *circumpolar stars*, which were visible at the time of the great pyramids (4500 years ago) in northern direction at an elevation angle of φ.

Proof:

The normal projection of the Earth's axis a on the horizontal base plane γ – the sphere's tangential plane at the vantage point B – points towards the geographic north pole. The inclination of the axis corresponds to the geographic latitude φ of the observer (Fig. 11.22). ◇

The auxiliary theorem may even be of some use in daily life. From our perspective, the whole of outer space rotates about the axis through the North star. All stars, including the sun, describe circles about this axis.

From a heliocentric perspective, due to the daily relative motion of the Earth about the sun (which is negligible due to its small impact), the direction of the sun's rays s during the course of a day stays constant. Geocentrically speaking, we can thus derive a second auxiliary theorem:

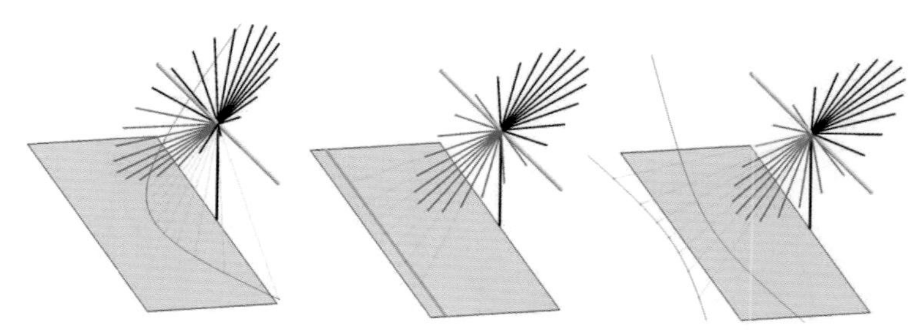

Fig. 11.23 The "sun-cone-of-revolution" at northern latitudes

The direction of the sun's rays s rotates almost uniformly about the Earth's axis a during the day and in due course generates a cone of revolution Σ with the axis a. The half angle of aperture σ of this cone of revolution is independent of the geographic position and varies according to the dates in the interval by $90° - \varepsilon \leq \sigma \leq 90° + \varepsilon$ ($\varepsilon = 23{,}44°$).

In astronomy, we mostly speak of the complementary angle of σ, which is called the "sun declination". The value of σ changes slightly during the course of 24 hours, but never more than by 0,4° (near the equinox).
Fig. 11.23 illustrates the change of the cone of revolution during the four seasons (summer is on the left, equinox at the center, and winter on the right). The aperture of the cone is, as already indicated, independent of the geographical latitude φ. However, the inclination of the cone's axis (geographical latitude φ) significantly affects the inclination angle of the sun's rays.
Within the tropics ($\varphi = \pm 23{,}44°$), the sun at noon alternates between north and south. In this area, if $\sigma = 90° - \varphi$, it is precisely vertical at noon (at the *zenith*). At the moment of the equinox ($\sigma = 90°$), the cone of revolution "degenerates" to a plane orthogonal to the axis a. On this day, the shadow of a point on the horizontal base plane moves on a straight line. On all other days, the shadow point moves on a cone of revolution. With the exception of the vicinity of the poles, this conic section is always a hyperbola.
This situation is especially transparent at the north pole and south pole, where, during the day, the sun simply moves about a perpendicular cone axis. For 24 hours, it is located at the same angle of elevation (a maximum of ε). During winter (north winter or south winter), the sun does not appear at all.

- **The locus of all possible sun positions**

The sun may occupy many positions in the sky during the course of the year, but the totality of all these positions covers only a part of the visible sky. If one imagines all cones of revolution with half angle of aperture σ intersected with a huge sphere centered at the viewer, one gets a set of circles on the

sphere which form a *sphere layer*. This layer is symmetric with respect to the normal plane of the direction of the North star.
All parts of the layer above the base plane (the sphere's tangent plane at the viewer's position) belong to positions above the horizon. From Fig. 11.24, it is, therefore, possible to derive a lot of information about the length of days (see also Fig. 11.44).

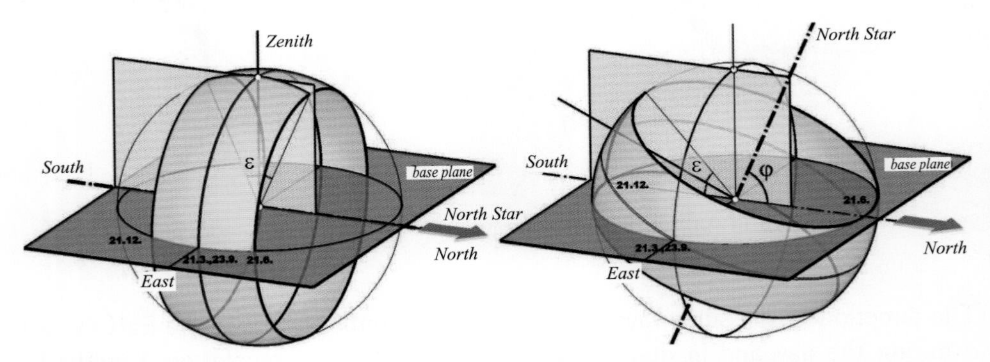

Fig. 11.24 Locus of all sun positions on the equator and on the northern polar circle

Fig. 11.24 shows the sphere layer on the equator and on the northern polar circle. It is immediately apparent that on the equator, all days have a duration of 12 hours. Even on the equator, the sun is only at the zenith twice per year. On the northern polar circle, the sphere layer apparently touches the base plane two times – on the 21st of June (the only 24-hour-day) and on the 21st of December (the only polar night). Fig. 11.28 shows a usual day-length-diagram. ♠

• **Determination of the angle between the Earth's axis and the sunrays**
As we have seen, the size of the date-dependent angle σ is important. We now want to determine it constructively, to an accuracy of half a degree.

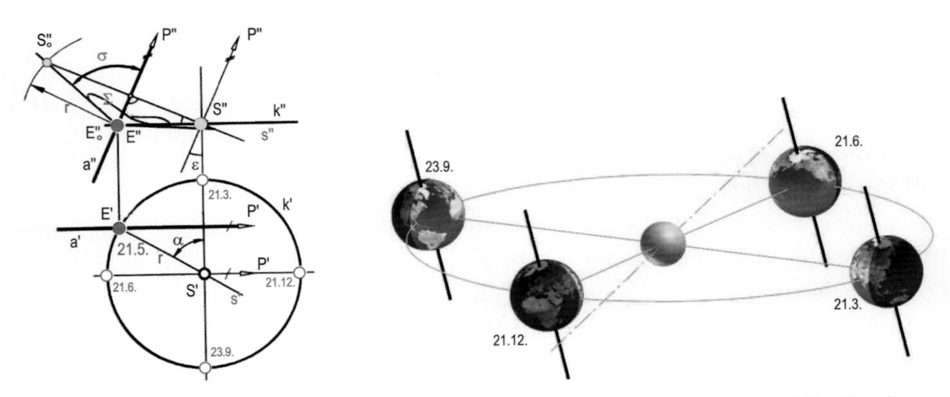

Fig. 11.25 Calculation of σ

Fig. 11.26 The main orientations of the Earth

Let us imagine the path circle k of the Earth E in the base plane π_1 and the Earth's axis a parallel to the front view plane π_2. The "zero position" of the

light ray direction may be reached at the beginning of spring (around the 21st of March) and is assumed to appear as a point in the front view.
At any given date – on the 21st of May, for instance – there is a well-defined angle of aperture for the cone of revolution Σ. Let this date be α days after the onset of spring (in this particular example, it is 60 days). The Earth E is then located about $\alpha°$ after the zero position.

For the sake of simplicity, let us now assume *all* months to be 30 days long and thereby kill two birds with one stone: Firstly, we get 360 days in a year so that each day corresponds to one degree – and secondly, the days are "weighted" in accordance with Kepler's law of equal areas. Our supposedly arbitrary calendar, where February only has 28 days and both July and August have 31, roughly compensates for the differences between the summer and winter half-years.

The half-opening-angle σ of the cone of revolution Σ now results, by the terms of descriptive geometry, through a parallel rotation of the light ray direction $s = ES$ about the Earth's axis a, until an orientation parallel to the front view has been reached: The axis of rotation is a frontal line, and from this it follows that the circle of rotation of S appears as one of its diameters. The length r of the segment ES appears undistorted in the top view ($\Rightarrow \overline{E_\circ''S_\circ''} = \overline{E_\circ'S_\circ'} = r$).

Compared to the exact value of σ on the 20th of May, the error measures only about one fourth of a degree – which we shall gladly accept in this case. During the course of the year, this error never exceeds half a degree, and is usually substantially lower than even that.

♠

• Sunrise and sunset

Let us now determine the length of day or the moment of sunrise and sunset for our given date at a location with a geographical latitude of φ (for instance $\varphi = 48{,}2°$). Fig. 11.27 shows how this calculation may proceed in the simplest of terms: The axis a of the cone of revolution Σ is assumed to be perpendicular. The parallel circles of Σ then appear undistorted in top view. The horizontal base plane γ, if set against a, comprises an angle φ. It cuts out two points S_1 and S_2 from an arbitrary parallel circle (such as the one with the unit length for its radius), and these points represent sunrise and sunset. The corresponding circular arcs can be interpreted as the lengths of day and night. By inscribing a 24-hour "clock-face" on the "unit circle", the points in time corresponding to S_1 and S_2 can be read off directly.

In reference tables, day lengths are usually described as being about 15 minutes longer than their theoretical values. This is because sunrise and sunset are defined such that they correspond to the subjective impression of an observer. Thus, the sun is already said to rise as soon as the first light rays – and not the center of the sun – reach the horizon plane. The same is also true for the setting of the sun.
During transition from the vacuum of outer space, the sunrays are always refracted into the gradually densifying atmosphere towards the perpendicular so that the sun (which appears to increase in redness as it approaches the horizon, due to the non-

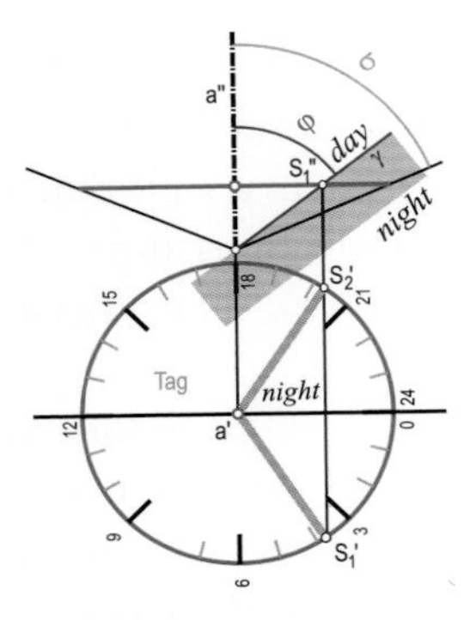

Fig. 11.27 Sunrise

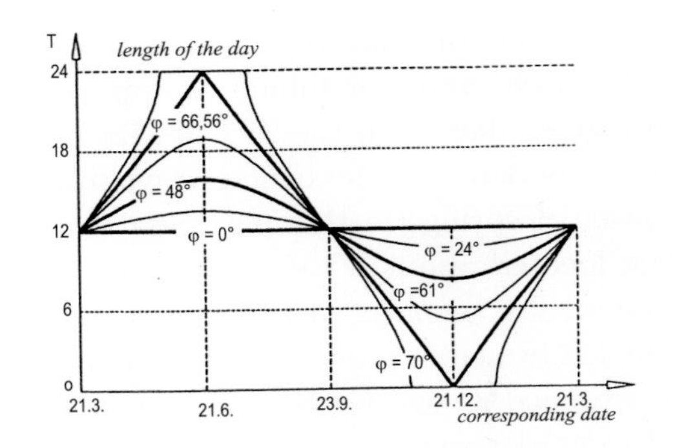

Fig. 11.28 Lengths of day on Earth

red color components being scattered more strongly) remains visible on the horizon, even though it should, in theory, have already set. ♠

• At which angle of elevation does the sun culminate?

Fig. 11.29 How high does the sun make it on the 21st of December? By the way, the angle measurement on the left is a relatively simple way to verify some of our conclusions quickly.

From Fig. 11.27, we can immediately conclude the following:

The sun reaches its maximum angle of elevation of $180° - \varphi - \sigma$ at noon.

This angle reaches its minimum on the 21st of December. On this day, the sun on the 50th degree of latitude reaches a mere angle of elevation of $180° - 50° - (90° + 23{,}44°) \approx 16{,}5°$ (Fig. 11.29). On the same day on the northern polar circle ($\varphi = 90° - 23{,}44° \approx 66{,}5°$), it merely touches the horizon.

On the northern polar circle, the sun rises during its summer solstice to an angle of elevation of $2\varepsilon \approx 47°$. If one considers that the sun never sets on that day, one can imagine the gargantuan quantities of solar energy which are induced into the Earth. Under ideal conditions (on a clear sky day), this amount is greater than the induced solar energy on the same day in equatorial areas! It stands to reason that an incredible amount of pack ice must thaw on that day – especially on the southern hemisphere. ♠

The incidence of light at a given moment

With all the aforementioned knowledge as part of our arsenal, it should be a small step to determine the direction of light incidence *for any given time of any given date.* This time, we draw the cone $\Sigma(a;\sigma)$ not in a special orientation ($a \perp \pi_1$), but adjusted to the direction of the North star (let the north direction, to a viewer, be the y-direction of our coordinate system). Fig. 11.30 shows how the situation in Fig. 11.27 can be restored by employing a side view. In this view, the given time (helping point H''' on the light ray s''') can be inscribed on the clock-face and also transferred into front and top view by the rules of descriptive geometry ($\to H'' \in s''$ and $H' \in s'$).

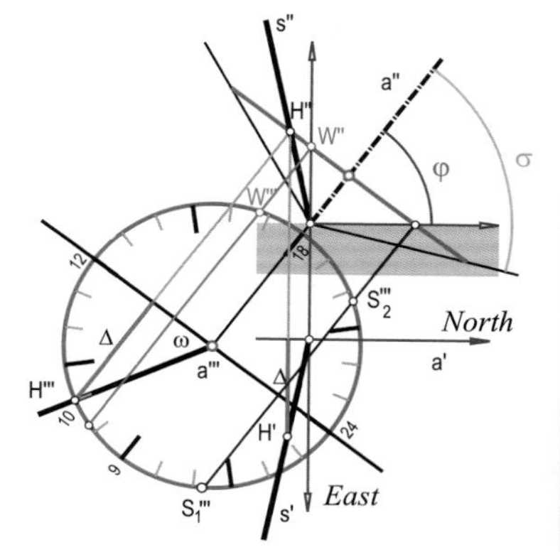

Fig. 11.30 Let the geographical latitude, the date, and the time be given: From which direction does sunlight incide?

Fig. 11.31 This picture was taken on the morning of December 21st on the northern tropic. How late was it, and what was the position of the sun at midnight?

The questions of Fig. 11.31 can already be answered: From the photo, it is apparent that it must have been a sunrise. The geographical latitude of the northern tropic is $\varphi = 23{,}5°$. On the 21st of December on the northern hemisphere, $\sigma = 90° + 23{,}5°$. A construction analogous to Fig. 11.27 yields about 7 in the morning for the sunrise. The construction in Fig. 11.30 for midnight shows that at the moment the picture was taken, the sun was located precisely at the *nadir* – the opposite of the zenith. On the same day, on the southern tropic, the sun at noon stands at the zenith ($\sigma = 90° - 23{,}5°$).

It is also apparent from the construction that the sun stands exactly in the west (or in the east) whenever the y-value of the helping point H disappears (in the drawing, it is labeled as W).

Let us now imagine the cone Σ rotated by $-90°$ about the x-axis so that it corresponds to the geographical latitude of $\varphi - 90°$. In such a case, the synchronously rotated point W for the new cone takes the role of the point S_2, which marks the moment of sundown. Thus:

The sun stands on a circle of latitude with the geographical latitude of φ in the precise east and west whenever it rises or sets on the "complementary circle of latitude" $\varphi - 90°$ on the other hemisphere (insofar as it does the latter at all).

• **Where is south?**

Fig. 11.30 can be read in an interesting manner. If we interpret the 24-hour-clock-face as a clock, then we notice that its hour hand (corresponding to the central angle ω) is the normal projection of the light incidence direction onto the clock-face, provided that the 12^h-marking (central angle $0°$) points towards the normal projection of the south direction.

If we imagine the usual 12-hour-clock-face on the clock, then the central angle $\overline{\omega}$ of the current time towards the 12-hour-marking is twice as large as on the 24-hour-clock-face in Fig. 11.30 ($\overline{\omega} = 2\omega$). The southern direction is, thus, marked by the moment which lies precisely between the current time and 12 o'clock.

Thus, by making use of any analogue watch, the southern direction can be derived from sunlight in the following manner:

For now, one proceeds according to the commonly known "boy scout rule": One imagines the moment S that lies between the current time and the culmination point of the sun (around 12 o'clock, or 13 o'clock during summer time). We also have to consider the geographical longitude: For each degree of longitude, the culmination point moves by 4 minutes so that on the Greenwich meridian in France or Spain, 60 minutes would have to be added! We must neither forget the deviation of the culmination point from 12 o'clock caused by Kepler's law of equal areas (*equation of time*), which affects the calculation by up to ± 15 minutes.

Now, we take off our watch and turn it around so that the hour hand points to the sun. The marked position then points pretty accurately towards the south.

For a more exact determination of the south direction, we would now have to tilt the watch: One should be oriented towards the currently determined south and position the hour hand at the time S so that it points into the same direction. Now, one must merely tilt the clock-face towards the body by $90° - \varphi$ (in our latitudes about $42°$) and repeat the procedure. The tilting action yields a slightly modified south direction. The error, without subsequent correction by tilting, is the greater, the earlier in the morning or the later in the evening we make use of the boy scout rule, or the farther south we are. By this rule alone, the sun would *always* have to be located in the precise west at 18 o'clock! ♠

• **A tiny and versatile gem on your necklace** (Fig. 11.32)

The gem mainly consists of a refracting semi-sphere placed on a 24-hour time scale. Let's say you know what time it is. Now try to get the focused sun rays onto the adjacent point on the time scale (usually. You have to tilt the "tool"). Then you already have a good direction towards south in direction of the letter "S".

If you do not know what time it is, but you know the south direction: Turn southwards and try to get the focus onto the time scale (again tilt the tool).

Fig. 11.32 A very small and versatile gem. It can be used similarly to the boy scout rule to either determine the south direction (if you know the time) or to find out, what time it is (if you know the south direction).

Then you read the (true) time on the time scale. In Fig. 11.32, the focus shows that it is 8 a.m. *true time*. With daylight saving time, you have to add an hour . . .

♠

A constant and a varying angular velocity

All our previous deliberations referred to the *apparent time*, which is defined as the time when the sun reaches its culmination point precisely at noon. We now have to consider that the path of the Earth about the sun is an ellipse, and its angular velocity on its path about the sun is, therefore, not constant. On the other hand, the angular velocity during its rotation about its own axis is not affected: The time for a whole revolution about the Earth's axis is about 23 hours and 56 minutes.

The equation of time

In the remaining four minutes (which would complete the full 24 hours), the Earth rotates about another degree, which causes the sun's culmination point to occur at a slightly different time than on the previous day – during these 24 hours, the Earth has moved by another degree about the sun, but only approximately due to its constantly changing angular velocity. Each day, several seconds of error are added or subtracted, culminating in a deviation on the order of ± 15 minutes.

A precise calculation of this phenomenon is very elaborate. There exists no precise equation for it – rather, one always has to fall back on *Kepler*'s non-algebraic *equation of time* and try to solve it approximately [11].

• Detective games

If we are faced with a picture of a well-known object (such as a building) on which the sun's shadows are visible, then we can determine the date and the time at which the picture was taken. The date, however, is dually ambiguous,

as we shall soon discover. We must, however, know the north direction and the geographical latitude of our scene.
Actually, we only have to "reverse engineer" the light ray construction at a given date and time. (Fig. 11.33):
Let s be the light ray that is given in top view and front view. Let us now consider the plane which appears as a line in the top view contains the Earth's axis a (it points towards the north). In a side view, the geographical latitude φ appears undistorted (auxiliary point $H \in s \to H''' \in s''' \to a''' \ni O'''$).
In a second step, let us rotate the light ray s about a such that it becomes undistorted ($H''' \to H_0 \in s_0$). Let us then rotate the projecting pat circle of H so that its axis a appears projecting in a fourth view. We have now killed two birds with one stone: In the third view, we can see the date-characteristic angle σ, from which we can derive the date by reversing the construction of σ.
The angle of rotation ω appears undistorted in the fourth view, which gives us the time (15° correspond to an hour). From the indication, we can see whether or not the sun is still largely in the east. The time difference from 12 o'clock is to be subtracted or added accordingly.

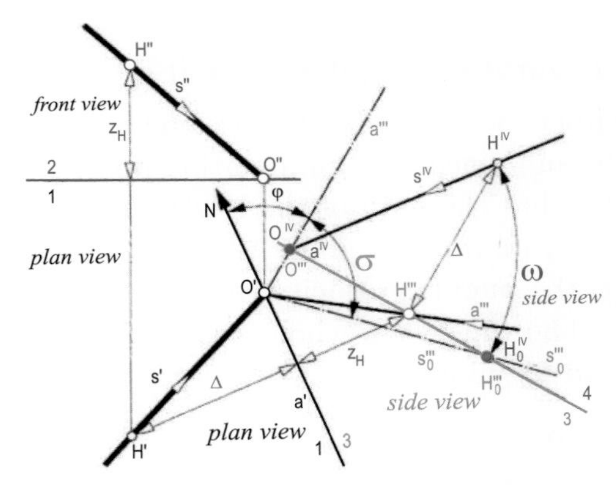

Fig. 11.33 The date σ and the time ω for a given light direction s

The 53th degree of latitude was chosen in this example (Berlin, for instance, lies on it). The construction produces $\sigma \approx 77°$, $\omega \approx 53°$.
The angle α is to be determined via Fig. 11.25: In the front view, we draw a cone of revolution with the axis direction towards the North star, half of the aperture σ, and the base circle of radius r. It is to be fitted into the scene so that its apex E lies on the path circle k and the base circle of the cone includes the sun S. The top view of E leads to α ($\alpha \approx 34°$).
The situation then unfolds either 34 days after the onset of spring (25th of April) or 34 days before the onset of autumn (19th of August). The angle ω corresponds to about 3,5 hours. This makes the actual time 15 : 30. Both possible dates occur in central Europe during summer time, which means that we need to add another hour. We must neither forget the deviation

from the 15th degree of longitude, to which central European time refers. Berlin lies about 1,5° westwards from the reference circle, which produces a further delay of 6 minutes. A public clock which may have accidentally been included in the photograph would indicate about 16 : 36. The clothing of the people in the picture would provide us with further clues about the time of year: If many people are pictured in summer clothes, it would probably have to be the 19th of August ... ♠

The path of the moon on the firmament

The interplay of sun and moon has been a constant object of fascination for all cultures. Like essentially all planets, the moon moves on the same path plane as the Earth, give or take periodic fluctuations of ±5,2° and a period duration of 18,2 years.
It may be tempting to conclude that the path of the moon – just as the paths of the planets – essentially overlaps with that of the sun. This may approximately be true during the equinox or new moon, but the full moon's much higher angle of elevation during winter is still noticeable.
Whenever one talks to an astronomer about the moon's path, one is inevitably met with a slight grin and a suggestion that the whole issue is not quite as simple and that it, therefore, has to be calculated to a much greater degree of accuracy. We shall, nevertheless, attempt to solve the problem qualitatively and geometrically.

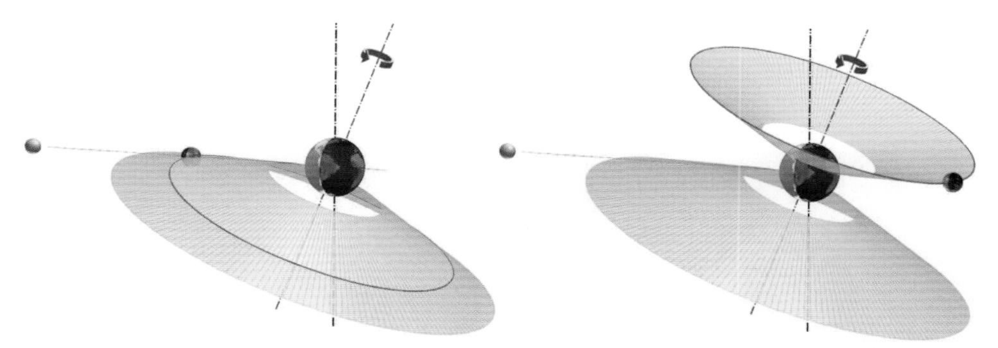

Fig. 11.34 Whenever the days or nights are at their longest, the path of the full moon deviates considerably from the path of the sun. The full moon stands very high at the onset of winter.

Let's have a look at Fig. 11.34 on the left: It is north winter – the southern hemisphere is much more strongly illuminated – and we have new moon. Only the dark side of the moon is visible from the Earth. In the following 24 hours, the Earth rotates about its axis while the moon always points its same side towards Earth. It takes 4 weeks to rotate about the Earth. When considered from the perspective of the Earth, the sun rays and their reflections from the moon overcoat the same half cone of revolution. It seems that the new moon, indeed, follows the path of the sun (see also Fig. 11.22 on the left).
The situation during full moon (Fig. 11.34 on the right) is completely different: The moon is now opposite to the sun, and the half cone generated by the reflecting rays is now practically the elongation of the sun cone. During the self-rotation of the Earth, the full moon rises to the same height that was essentially accomplished by

the sun at the start of summer. This explains why, during long and clear winter nights, the full moon is a considerable source of light from sunrise to sunset.

According to the theorem about the sun's culminating point on p. 384, the sun on the 50th circle of latitude is visible at a maximum angle of elevation of $90° - 50° + 23{,}45° \approx 63{,}5°$. If we now apply – as in the winter of 2005/06 – the maximal deviation of the moon's path, then we have almost reached 69°. Such an angle of elevation is subjectively almost interpreted as "perpendicular". During the day, however, the sun moves at a maximum of $90° - 50° - 23{,}45° \approx 16{,}5°$.

On the poles, the sun is not visible for the winter half of the year. Every few weeks, however, the moon shows up as the half moon and reaches its culmination point as the full moon at up to 23,45°. This happens whenever the full moon occurs precisely on the 21st of June or the 21st of December.

According to the previous considerations, it should be clear that the paths of Mercury and Venus on the firmament should always be somehow approximate to the path of the sun (see also [11]): Both planets are significantly closer to the sun than the Earth and are, thus, to be treated geometrically like the new moon. Mars and Jupiter, however, can "change sides" on the days surrounding the full moon, and their path on the firmament can, thus, significantly deviate from the sun's path.

- **What is different on the southern hemisphere?**

Actually it is unfair to talk always about the northern hemisphere, and readers from "down under" or elsewhere on the southern hemisphere please apologize that many statements in this book seem to prefer the northern hemisphere (this is due to the fact that the book first was published in German).

Fig. 11.35 The southern cross rotates clockwise about the southern celestial pole.

One of the first things a traveler should do when arriving on the southern hemisphere is to sit down in a park and watch the shadow of a tree going round "the other direction". The sun still rises in an easterly direction and sets in a westerly direction, but in general, it reaches its culmination point in the north. Due to the fact that our traveler from the northern hemisphere is – compared to his or her homeland – "standing on the head", the orientation of the revolution seems to be inverted. (In space, the orientation of a rotation about an axis can only be determined if the axis is oriented. If you are oriented southwards, the sign of the orientation swaps.)

On the northern hemisphere, the sun rotates clockwise if we naturally turn somehow southwards. The same is true for the moon which, to a certain extent, follows the path of the sun (delayed or in advance). On the contrary, the stars rotate counter-clockwise about the northern celestial pole – if we, again naturally, look northwards.

On the southern hemisphere, the directions are opposite. Fig. 11.35 shows how the night sky at Cape of Good Hope (South Africa) twists clockwise a little more than 90° within more than 6 hours (≈ 15° per hour). There is no significant star around the southern celestial pole. Therefore one tries to find the *Southern Cross* or *Crux* which is so popular that it is part of several flags of states on the southern hemisphere, including Brazil, Australia, New Zealand, or Papua New Guinea. Then, the southern celestial pole can be found as indicated by yellow lines in Fig. 11.35.

Although it does not quite tie in with the theory of sun positions, the following example still fits here (and is to some extent additionally an example of a spherical motion):

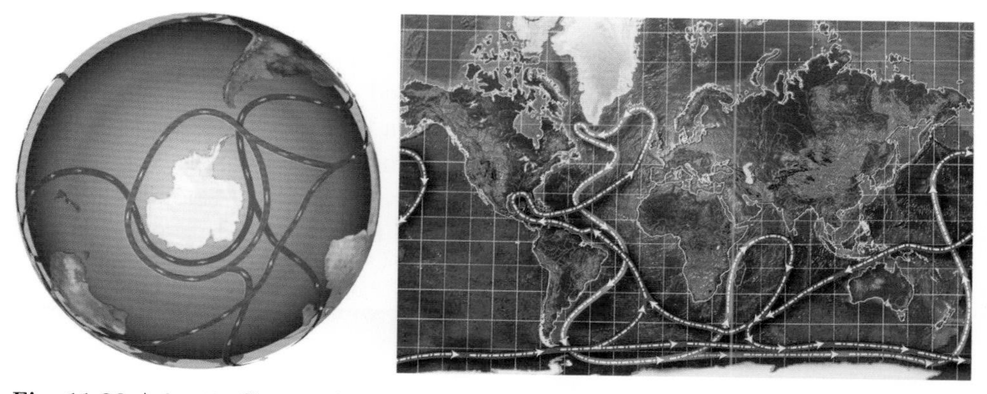

Fig. 11.36 Antarctic Circumpolar Current (ACC): On the left displayed to be understood, on the right only readable for people with extremely well developed spatial imagination.

The Antarctic continent appears completely distorted on maps like the one in Fig. 11.36 on the right. One either needs to apply a stereographic projection from the North Pole or display a "bottom view" (or "top view", if you are on the southern hemisphere) of the globe. ♠

11.4 About minute-precise sundials for the mean time

In this section, we shall introduce sundials for the determination of the mean time. Two types function with shadows of curved surfaces on the indicators. Another sundial does its job by an intricate mechanism – the sundial of *Oliver* and *Bernhardt*, a variation by *Riegler*, an idea for a sundial by *Hofmann* and another by *Pilkington* and *Gibbs*. All sundials make use of the equation of time and are precise up to the minute, given a careful enough construction and a great enough size. The calculations of the arrows and of the described mechanisms will now be discussed.

True and mean time

Sundials used to be the early instruments for the measurement of time, and today they still exert a great degree of fascination. Their history dates back to antiquity, when the duration between sunrise and sunset was divided with the help of sundials into twelve approximately equal parts. The position of the sun was projected on a surface where a grid of date and hour lines was drawn by means of a shadow cast by a rod or by a circular aperture (Fig. 11.37). An hour used to be a variable unit of measurement that was dependent on the time of year – a temporal hour.

In the modern division of hours, the duration between the sun's two consecutive meridian passes is divided into 24 equal parts. The meridian of a point of observation is (with exception of the poles) the great-circle on the celestial sphere that passes through the zenith, the north point, and the south point of the horizon. The midpoint of the celestial sphere can be assumed at the point of observation due to its "unimaginably great" radius. As will be shown, these lengths of days vary during the course of the year – this is the true local time. Today, a mean time is employed – a uniform unit of time. It is most precisely measured by atomic clocks. In harmony with the revolution of the Earth about the sun, a second of mean time is defined by means of the duration of 9,192,631,770 periods of the radiation corresponding to the transition between the two hyperfine levels of the ground state of the ^{133}Cs atom.

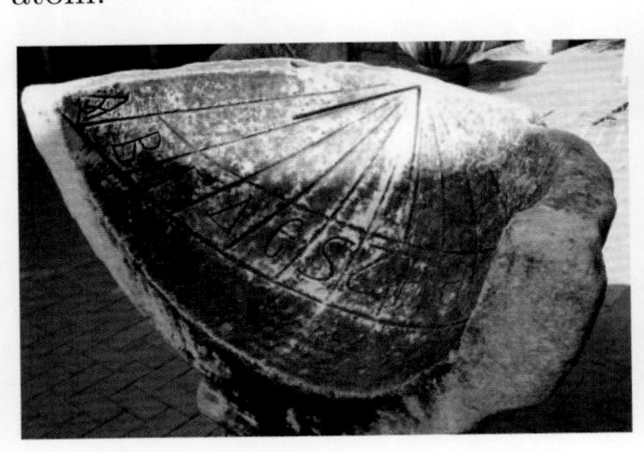

Fig. 11.37 Ancient clock

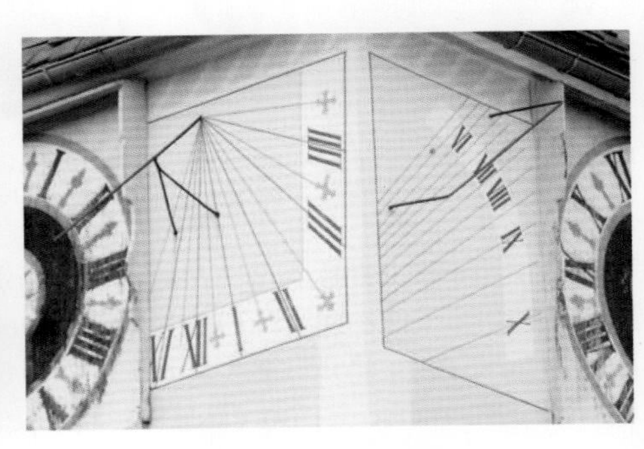

Fig. 11.38 Tower clock

Next to the true time of a location, the *mean location time* was first observed. The railroad industry required the introduction of timezones, and of mean times for greater areas. The central European time, for instance, is the mean location time for locations at 15° of eastern longitude. Our summer time corresponds to eastern European time (30° of eastern longitude).
Sundials – true analogue clocks – measure the true local time. We speak of true noon whenever the sun is precisely at the south. Mean time and true time correspond to each other on four days in a year (mid-April, beginning of June, beginning of September, and the end of December). The *mean noon* deviates from true noon during the course of the year by the equation of time, which we shall soon discuss – this can amount to up to 16 minutes.

Simple sundials

From the perspective of a location on the Earth's surface, the sky seems to rotate about a celestial pole. On the northern celestial sphere, the North Star is located at that point at a distance of under one degree. The axis of the firmament's seeming rotation points towards the celestial pole. It lies in the vertical north-south-plane and is tilted against the horizontal at the angle of the respective geographical latitude φ. The axes of many sundials point in that direction (Fig. 11.38).

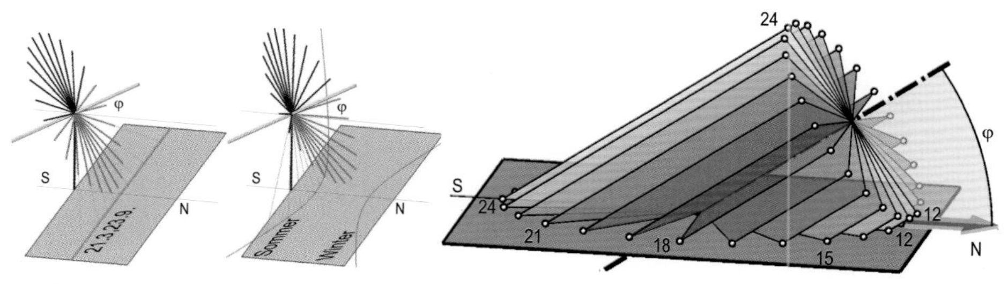

Fig. 11.39 Date lines ...

Fig. 11.40 ... and hour lines

The planar and horizontal trace of the cone (with half aperture σ) overcoated by the light rays through a fixed point is a conic section, a hyperbola in non-polar regions, and a straight line during equinox (Fig. 11.39). Path curves of this kind are called *date lines*.
If we wish to construct a simple sundial, we must – for the full hour moments – intersect a pencil of planes with planes of 15° to 15° through a straight line parallel to the celestial axis with the carrier surface of our clock-face (Fig. 11.40). The intersection lines are the *hour lines*: The Earth rotates by 360° during 24 hours.
Equatorial clocks are special sundials with planar, equator-parallel clock-faces or with clock-faces shaped like cylinders of revolution whose generatrices are parallel to the celestial axis. In both cases, hour lines are distributed evenly.

The equation of time

The motion of Earth about the sun is complicated. In order to accomplish a minute-precise sundial, we have to work with tables or, better yet, with a computer simulation. The Earth's precession motion (the rotation of its axis during the course of 26000 years, Fig. 11.21) and the nutation (small disturbances in the precession motion) are not significant for sundials.
Fig. 11.41 shows a "caricatured depiction" of the Earth's positions during the equinoxes and the solstices (see also Fig. 11.21). The precession motion is visualized on the great sphere (the Earth's axis generates a cone of revolution).

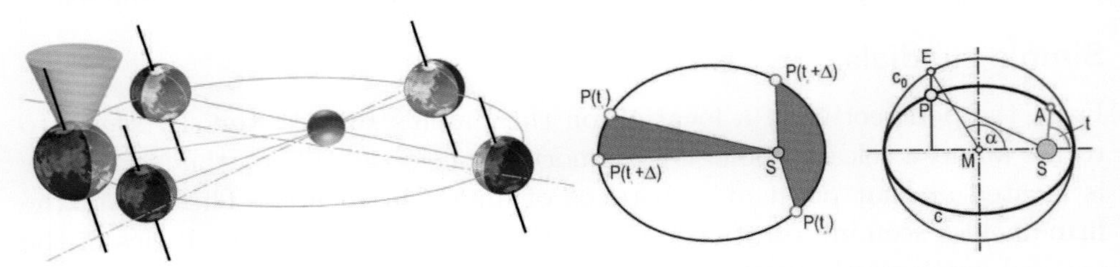

Fig. 11.41 Precession motion

Fig. 11.42 *Kepler* equation

In order now to achieve a computer simulation, we must explore the issue geometrically with a certain foreknowledge in physics. A mere application of Kepler's laws is possible, but not easy to implement: The path ellipse of the earth must be parametrized according to time, which requires the solution of the transcendental Kepler equation. By using the nomenclature from Fig. 11.42, this equation is given as

$$f(\alpha) = \alpha - \varepsilon \cdot \sin\alpha - t = 0 \text{ with } \varepsilon = \sqrt{1-(b/a)^2},$$

where this time ε is the numerical eccentricity of the ellipse. One would do better to employ the theorem about the conservation of angular momentum, from which Kepler's laws necessarily derive.
From the perspective of differential geometry, the path curve with its physical properties may be deduced from the knowledge of the initial velocity and the corresponding direction of motion. The angular momentum is calculated with respect to the Earth's axis through the sun which is orthogonal to the Earth's path plane.
The Earth is moving according to the theorem about the conservation of angular momentum on an elliptical orbit about the sun (Kepler's first law). The numerical eccentricity of the path ellipse is rather small (ε=0,017). By the same principle, the Earth must constantly be changing its orbital or angular velocity about the sun. A diminution of the distance from the sun leads to a square increase of the angular velocity with respect to the diminution factor (Kepler's second law: equal times – equal areas).

1. The "physical" component

While the Earth is moving non-uniformly about the sun, it is also rotating with a uniform angular velocity about its own axis. This leads to a small daily deviation, until the sun once again returns to the south. This deviation contributes to the equation of time as the daily difference "real time minus the mean location time".

2. The "geometrical" component:
The equation of time includes another substantial component produced by the tilt of the Earth's axis. The rotation about the sun and the self-rotation have mutually oblique axes of rotation. Thus, it is impossible to imagine both rotations in one plane! A circular clock-face whose plane is parallel to the equator plane with a uniform division of hours appears slightly distorted if projected orthogonally on the path plane of the Earth. Thus, in this normal projection, the clock-hand's shadow on the clock-face does not rotate at uniform angular velocity!
Let us assume, for the sake of simplicity, that the Earth rotates about the sun at a constant daily angular velocity of $1°$. Thus, every single day, an angle of $1°$ would have to be compensated on the elliptically distorted hour index in order to reach the sun's relative position. The revolution of the shadow on a circular clock would not be uniform!

3. "Equinox line" and path ellipse
At both equinoxes, the Earth's self-shadow boundary passes through both poles. Sunlight incides the Earth's axis orthogonally. The connecting straight line of both positions on the Earth's path – the equinox line – encloses an angle with the principal axis of the path ellipse. This angle changes slightly due to the precession motion and the perihelion precession (the rotation of the ellipse apexes) – though this change has been negligible during recorded history. It also affects the values of the equation of time, if not to a great degree.

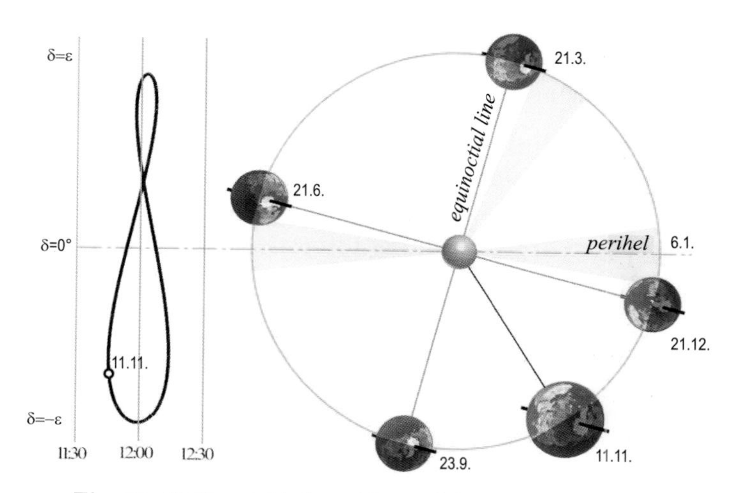

Fig. 11.43 The eight-shaped loop of the equation of time

In Fig. 11.43, the equation of time was solved for the Earth's data. Each position of the sun at midday corresponds to a point on the eight-shaped loop. If the angle between the equinox line and the principal axis of the path ellipse is 90°, then the eight-shaped loop is axially symmetric. Fig. 11.43 also shows the self-shadow boundaries of the Earth.

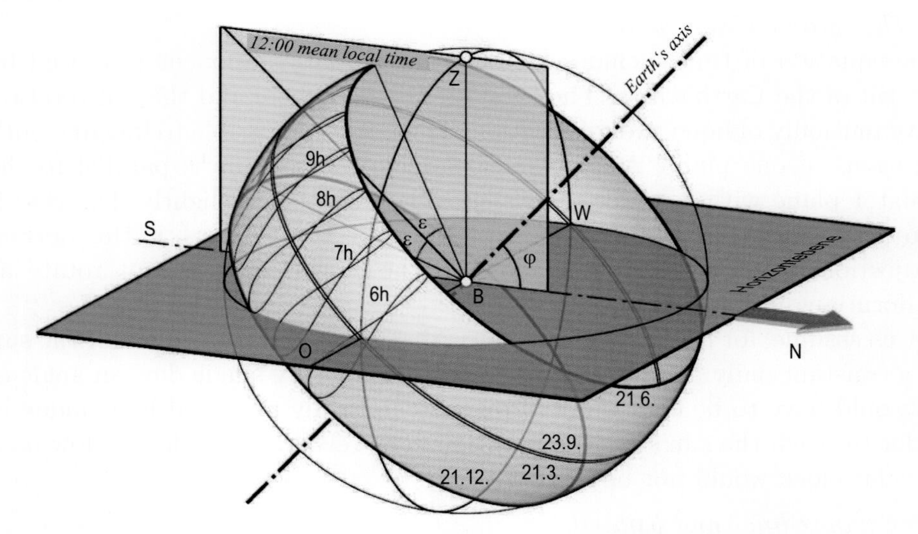

Fig. 11.44 Eight-shaped loops on the celestial sphere

Fig. 11.44 shows an image of the celestial sphere on which the stars were projected. The horizon plane for 48° of northern latitude (Vienna), the zenith Z, the celestial axis, the plane of the local meridian, the celestial equator (double line), and both tropics are all inscribed. The eight-shaped spherical curves for the full hours of mean local time are also visible. They are produced by a rotation of the hour line for the mean midday by a multiple of 15°.
By the way: Computer simulation allows us to solve the equation of time for any location of any planetary system, any numerical eccentricities, and any axial tilts. This produces many varying loop shapes, which may also be symmetric and do not necessarily have to possess a double point.

The *Bernhardt*-sundial

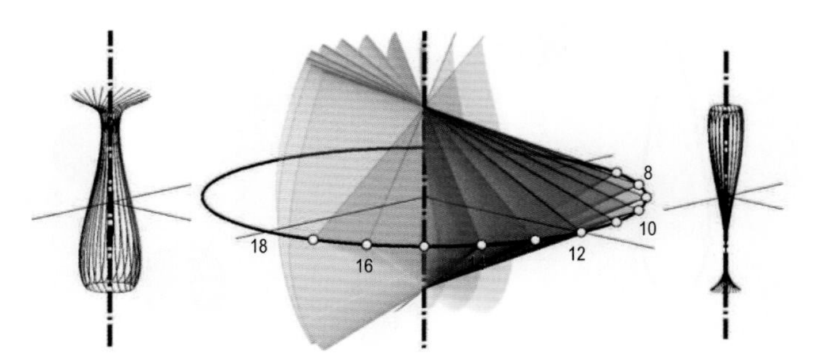

Fig. 11.45 The creation of the curved clock-hand

The 365 sunrays through a fixed point on the Earth throughout all precise middays of the year lie on a cone – the directrix of this cone is the eight-shaped loop on the celestial sphere derived from the equation of time ("Achterkegel").

Let us now consider a circle in an equator-parallel plane (Fig. 11.45). Its axis passes through the celestial pole. Let us now mark 24 points on the circle separated by 15° each and label them 0,1, . . . ,23,24 = 0.

Let us now inscribe rays towards the eight-shaped loop on the celestial sphere through point 12 for the mean midday. In order to prevent an intersection of this cone with the axis of the circle, point 12 is rotated from its deepest position slightly towards the east (the directions towards the eight-shaped loop are preserved).

During the day, a light ray which belongs to the eight-shaped cone through 12 rotates uniformly about the axis of the circle and in due course generally envelops a hyperboloid of revolution (Rolf *Wieland*: Equatorial sundials with automatic time compensation. Newsletter 26 and 27 of the workgroup concerning sundials of the Austrian Astronomical Society 2003/04). The surface of rotation enveloped by the rotating eight-shaped cone (or by all hyperboloids) can be employed as a curved clock-hand of a sundial for mean time. The surface of rotation is the *focal surface* of the *congruence of rays* formed by all light rays meeting the circle. The circle itself is a *focal curve* of this congruence.

According to construction, all light rays which touch the clock-hand at mean midday and meet the circle pass through point 12. The equation of time is thus observed. In order to read a zone time on the sundial (instead of local time), the circular hour index is rotated by the difference angle between the respective location's geographical longitude and that of the zone meridian.

The described idea dates back to *Oliver* in 1892 and was improved by *Bernhardt* in 1964. The problem that the occurring envelope surface of revolution is composed of two sheets due to its construction from an eight-shaped cone, is solved in this sundial by two different surfaces of revolution, which are simply swapped at the solstices. Each surface is employed for a half-year as a clock-hand. The boundary areas of the surfaces of revolution are especially critical at the solstices, because the enveloping hyperboloids overlap. At those moments of the year, the time can no longer be read to minute accuracy.

Riegler adapted the idea by *Oliver* (*Bernhardt*) by rigidly connecting both hulls of the surface of revolution to the plate of both clock-faces (Fig. 11.46 on the left). The outer sheet is pictured as a grid – the inner as a closed surface. Depending on the half-year in which one reads the time (each running from solstice to solstice), the shadow on the inner sheet or outer sheet is employed for the determination of time (Fig. 11.46 on the right). The shadow of the clock-face pane on the grid of the outer sheet further indicates the respective date.

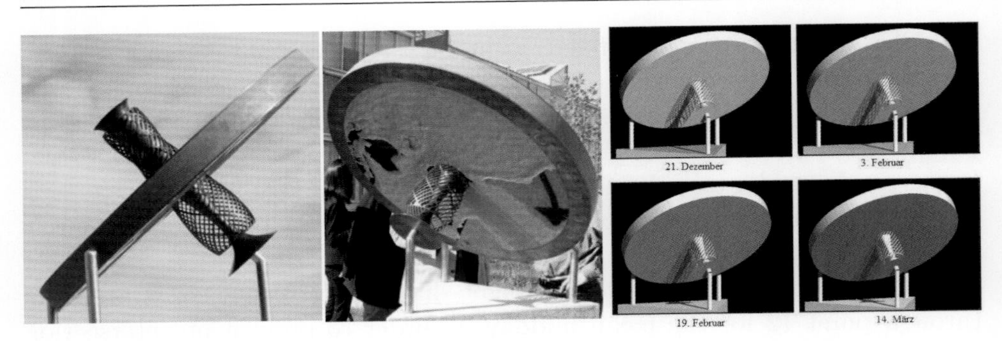

Fig. 11.46 *Riegler*-sundial ...

Cycloid sundial for mean time

From the U.S.A., a question was posed about the existence of a sundial, whose shadows on a clock-face parallel to the celestial axis display the mean time by straight hour lines. If date lines and eight-shaped hour lines are drawn on such a plane, the mean time might be read by means of a circular aperture – though this detail was not asked by the questioner.
A beautiful sundial that displays real time is described by Fred *Sawyer* (The Cycloid Polar Sundial. Compendium, North American Sundial Society, Volume 5, No. 4, pp. 21–24, 1998): If a circle rolls in a plane on a straight line, then a diameter of the circle envelops a cycloid (Fig. 11.47).
A cylinder – with a generatrix that is parallel to the celestial axis – that touches a hump of a common cycloid in an equator-parallel plane can, thus, serve as the clock-hand for real time if the hour index lies on the apex tangent of the cycloid hump.

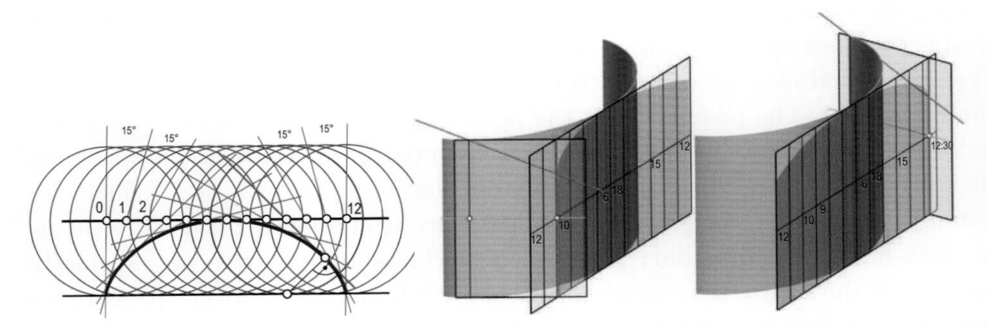

Fig. 11.47 Diameter ...   **Fig. 11.48** ... of cycloids and the consequences

A sundial can be positioned so that the apex tangent is directed from the east towards the west. The true local time between 6 o'clock to 12 o'clock in the morning can then be read on one half-tangent. The true local time between 12 o'clock to 6 o'clock in the evening can be read from the other half-tangent (Fig. 11.48).
The hour points are placed at equal distances. This allows the sundial to be adapted to the real local time of a zone meridian by simple translation of the hour index on its carrier straight line.

Now, one of both sun ray cones (as they have been considered for the *Bernhardt*-sundial) are directed as follows: Their tip is being moved along the hour scale of the cycloid sundial, its generatrices for the equation of time zero touch the envelope of the cycloid cylinder, and the ray of the equinox retains its declination of 0°.

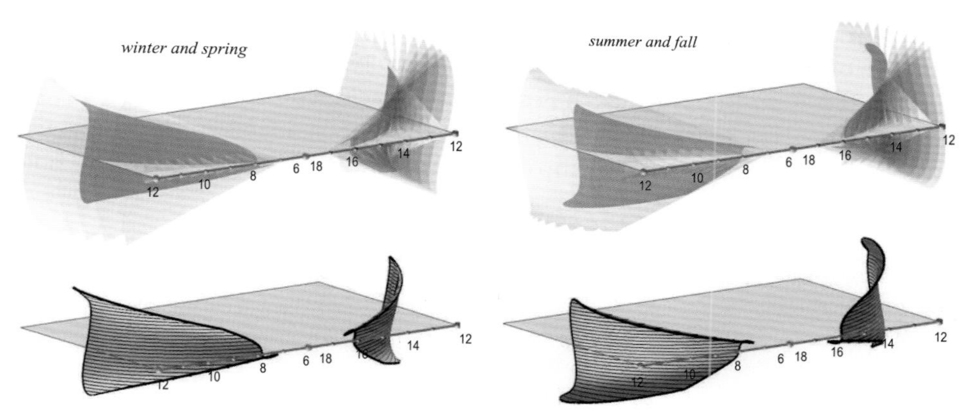

Fig. 11.49 Construction and appearance ... **Fig. 11.50** ... according to *Hofmann*

The result is a hull surface (Fig. 11.49, Fig. 11.50) which may be employed as the clock-hand for mean time. Self-intersections occur in the areas near the solstices, as well as near the cusps of the cycloid. A reading of the sundial is especially precise between 7 and 17 o'clock of mean time. The intermediate area of the clock-hand does not have to be especially adapted for this reading – in fact, the clock-hand and the clock-face may be connected.
Translations of the hour index allow a reading of zone times, especially during summer time. As with the *Bernhardt*-sundial, there are two clock-hand surfaces which are interchanged at the moments of the solstices.
The characteristics of the surface were calculated at uniform intervals by intersecting two adjacent eight-shaped cones – for instance, at one minute intervals. The surface was, thus, determined by a family of characteristics, and the connecting surface of two neighboring characteristics was interpolated.
While the sundial may be visually impressive, its plan for construction may only be provided as the answer to a given initial question. The construction would in any case be very expensive, and the mean times would be better displayed on other sundials.

The *Pilkington-Gibbs*-sundials

Another approach to a minute-precise heliochronometer was given in 1906 by *Pilkington* and *Gibbs* – *Gibbs* as the inventor and *Pilkington* as the manufacturer. They equipped a pivotable circular disc so that in areas outside of the horse latitudes bounded by the tropics it can be rotated in an equator-parallel manner immediately after it is being brought into line (Fig. 11.51). Near the equator, the carrier half-sphere must be positioned along a slope. The circular disk C (gray) that functions as the clock-face of a 24-hour-clock is twistable into itself (Fig. 11.52). An attachment A with a (red) reading mark is fixed to it. A second attachment B (yellow) is also fixed, but in a movable way: It can be restricted towards a circular arc (realized by a slot guide within the disk), whose center is the normal projection of the reading mark A on the circular disk (see also the exploded drawing Fig. 11.53 and Fig. 11.54). Two little drilled holes in the attachment B allow light rays to reach the reading mark. The attachments are dimensioned so that light may enter onto the reading mark through at least one of the two drilled holes regardless of the time of year.

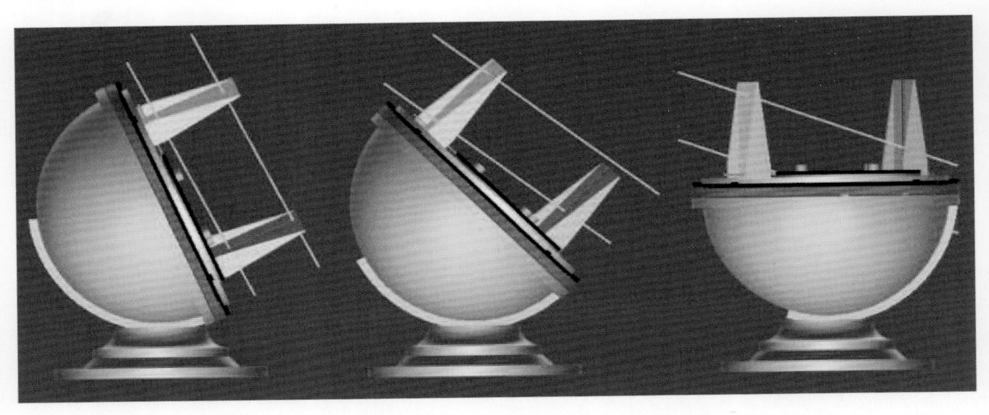

Fig. 11.51 Adjustment of the geographical latitude ...

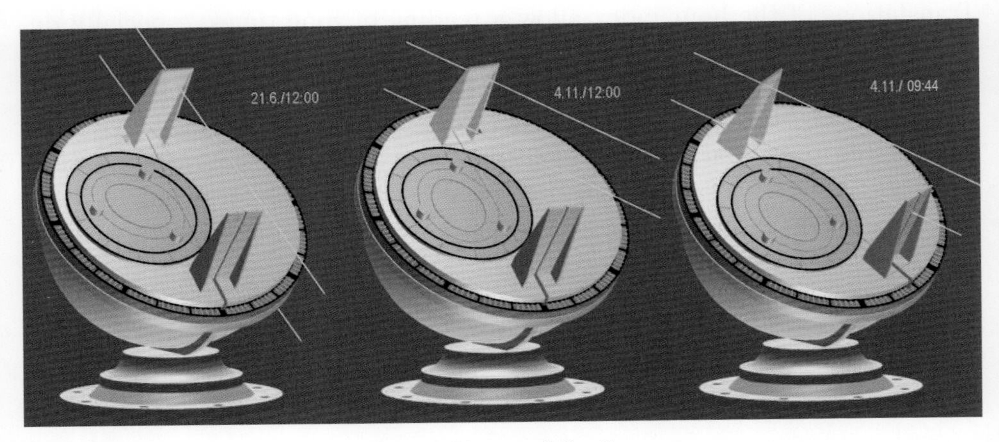

Fig. 11.52 ... and the time

The sundial is adjusted on location to its geographical latitude φ. A very simple instruction manual may now be given as follows: One may first adjust the current date on the small (green) wheel D (Fig. 11.52 center, for instance November 4th) and rotate the circular disk until one of the light rays hits the

reading mark (Fig. 11.52 on the right). Now, the exact time may be read on the hour index (in the particular example it is 09 : 44 – without the equation of time, it would merely indicate 10 : 00).
If both attachments were diametrical, then the equation of time would be disregarded, but if the circular disk is rotated far enough so that the incident light hits the reading mark, then the displayed time is pretty accurate: The deviation of each hour from the true midday corresponds to a rotation of the disk by 15°.
In order to involve the equation of time, the attachment B (with both drilling holes) must be slightly rotated – depending on the date. This occurs – as pictured in the above example – by means of an eccentric dial inside the half-sphere. Both attachments are connected by a rod (straight line). The rod is pressed towards the eccentric disk by a spring. A rotation of this disk causes the rod to be pressed in one or the other direction (Fig. 11.54).

Fig. 11.53 Exploded drawing

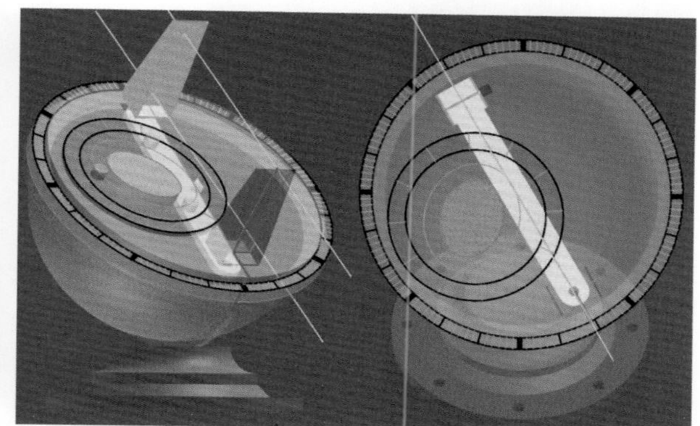

Fig. 11.54 The eccentric date dial

The dimensioning of the disk must be calculated so that the rotation angle corresponds to the required position with the corresponding date. This used to be done with the assistance of tables, and can today be calculated by using computers. Given an infinitely thin rod, the profile is the envelope of a family of straight lines – and the actual profile is a curve which is an offset (Fig. 11.54).
Internet addresses and literature about sundials:
About astronomy in general: Helmut Zimmermann, Alfred Weigert: *Lexicon of Astronomy.* Spektrum Akad. Verlag Heidelberg, 1999.

Relevant Internet addresses
About the equation of time loops: `http://www.uni-ak.ac.at/geom/sonnenstand`
About the Bernhardt-sundial: `http://www.praezisions-sonnenuhr.de/`
About the Riegler-sundial: `http://riegler.home.cern.ch/riegler/sundial/hauptseite.htm`
About the Pilkington-Gibbs-sundial: `http://homepage.ntlworld.com/jmikeshaw/page12.html`
With the consent of the publishers, the section about minute-precise sundials can also (essentially) be found in Georg Glaeser, Walter Hofmann: *About minute-precise sundials for mean time.* IBDG 1/2005.

Fig. 11.55 A strange view of the sun – due to different humidity conditions in the air; a possible explanation: Light rays can be curved when the atmosphere is inhomogeneous (see below).

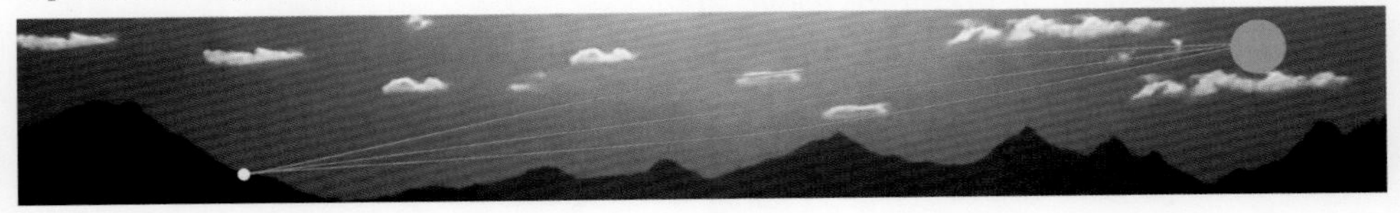

12 The multitude of filling patterns

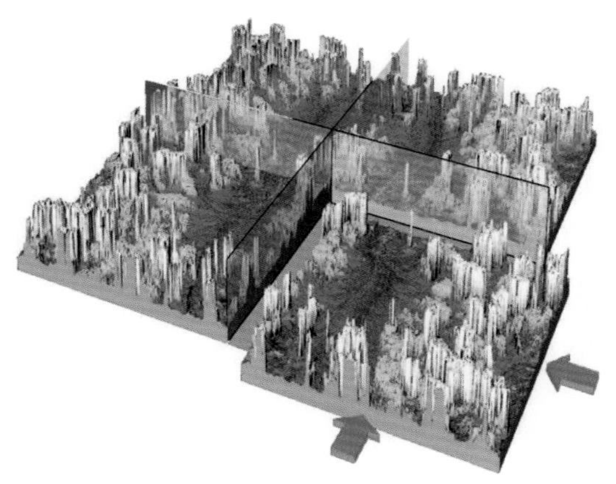

A tiling is a gap- and overlap-less coverage of a plane with specially formed panels or tiles. The simplest forms are triangles and quadrangles, which may be regular or irregular. Regular pentagons already cause problems, because a plane cannot be gaplessly covered by such shapes.
This is due to the violation of a basic condition:
In each apex point, k n-gons coincide, which means that the sum of interior angles in the n-gon must be $n \cdot 360°/k$. However, the sum of interior angles of an n-gon equals $(n-2) \cdot 180°$ (which can be proven inductively by decomposition of the n-gon into triangles). An equalization yields the condition $1/n + 1/k = 1/2$.
This is impossible for the regular pentagons, as there exists no natural number k for which this condition is met. Nevertheless, there are many possibilities of covering the plane with *irregular* pentagons.
The regular hexagon does not cause such problems – the cells of honeycombs are a classical example of a hexagon tiling. But this already exhausts all regular n-gons where tilings are possible.
There are tilings with only *one* basic form (monohedral) or with *many different* basic forms (n-hedral). The most important distinction is, however, whether or not the tiling is *periodical* or *non-periodical*. A coverage is periodical if there exists a translation in the plane that would move the pattern into itself.

Survey

12.1 Periodical tilings

Regular n-gons as the basic building blocks

Everybody knows the pattern of honeycombs (Fig. 12.1), which is based on regular hexagons. Since every such hexagon can be subdivided into equilateral triangles, this type of tiling also works with equilateral triangles.

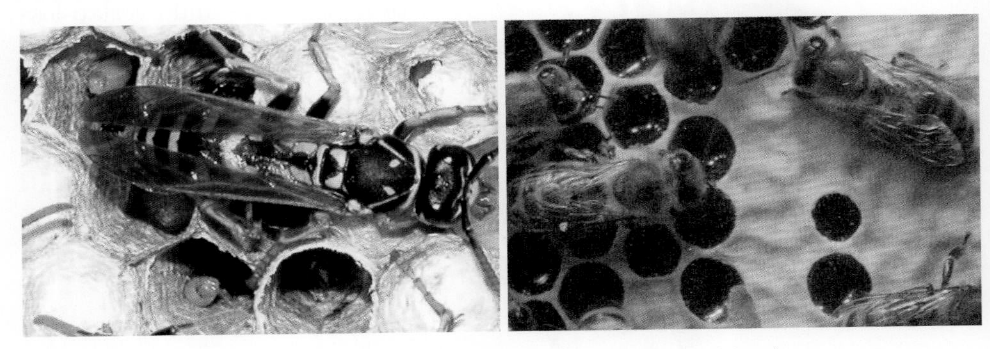

Fig. 12.1 Classic tiling with hexagons (wasps and bees)

It is also trivially possible to tile a floor using regular quadrangles (or squares). Other tilings with regular n-gons are impossible. In this case, nature shows itself to be generous and pragmatic as it often is (Fig. 12.2). In the left image, the elongated plates on the bottom side of the rattle snake's neck appear to be strongly shortened. In the planar tiling that is pictured, they barely get out of line.

Fig. 12.2 On the left: A hexagonal tiling occurs (at least on the top side). On the right: Certain "modifications" can be discerned on the bottom side.

Inspired by such pictures, one may immediately think of interpreting tilings as projections of spatial manifolds. A software developed for this purpose permits the interactive construction of shapes – based on magnetic repulsion – which, for instance, may very well be composed of regular pentagons, but which may, nevertheless, diverge "rhombusescally". Otherwise, the result would always be a pentagondodecahedron (Fig. 12.3).

.

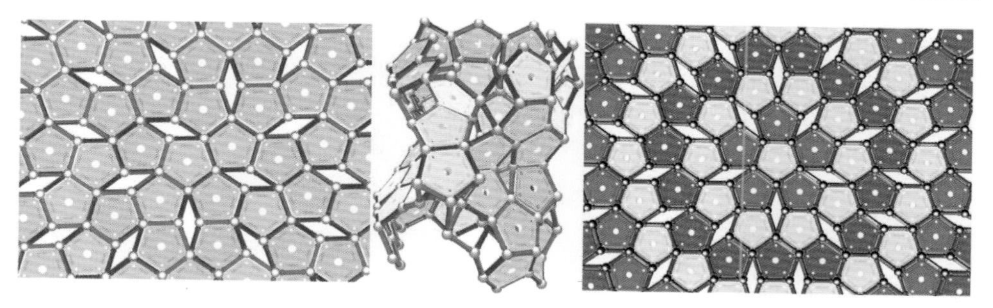

Fig. 12.3 Left and right: Planar "half-regular" shapes which are essentially composed of regular pentagons, but which nevertheless partially diverge. At the center: "Bending", thus, allows tilings of surfaces. The left – non-symmetrical – tiling by Franz Gruber seems to be unknown, and the right one was invented by no less a figure than Albrecht Dürer

• From shark skin to flow-optimized architecture

The arrangement of the facets on the skin of fish and reptiles is of practical importance to these animals. Investigations of shark skin have shown that water undergoes tiny turbulences as it moves accross it – an impressive achievement in streamlining.

Fig. 12.4 Analysis and architectural construction of shark skin tiling. The turbulences are produced deliberately for the purpose of ventilation.

In the Studio Hadid, Mario Gasser, Peter Schamberger, and Sebastian Gallnbrunner used these insights in architecture – an idea that won them the Austrian construction prize in 2005. ♠

• Tiling with paintings

An interesting "two-and-a-half-dimensional" tiling based on squares is due to Robert Lettner: He interprets square paintings as function graphs and thus creates artificial "artistic landscapes" (Fig. 12.5).
The color does give the tiling an extra dimension. A color value (red-green-blue) is assigned to each point in the plane, which is then somehow translated into a height value. A clever coloring of the corresponding spatial points produces interesting and often unintentional effects. ♠

• Alteration of the sides of a basic building block

Tiling with squares or rectangles seems to be trivial. By changing border lines, however, it can be varied such that the result is not trivial any more.

Fig. 12.5 Tiling with squares

Fig. 12.7 shows how the sides of the squares of *Florentine patterns* are replaced by line strips which are centrically similar with respect to the square side. ♠

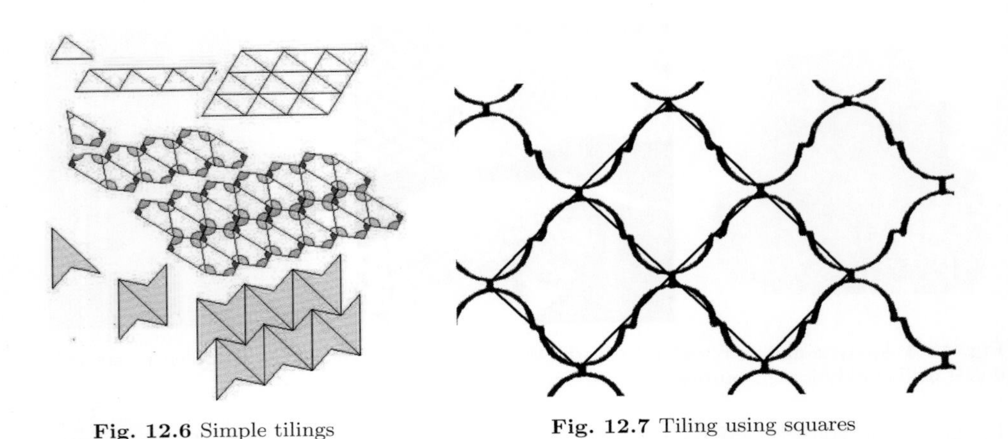

Fig. 12.6 Simple tilings

Fig. 12.7 Tiling using squares

Irregular polygons as basic building blocks

Every *triangle* can be used to tile a plane: For any arbitrary triangle, a reflection in a side's midpoint produces a parallelogram, from which whole strips can be built. Such strips, laid next to each other, can cover an entire plane.

Fig. 12.8 Non-trivial tilings using rhombuses, rectangles, and parallelograms

Tiling based on an arbitrary n-gon

Let an arbitrary quadrangle be reflected in a side's midpoint. Now position the quadrangle and the reflected quadrangle alternatingly next to each other (Fig. 12.6), producing a strip. Multiple such strips can be seamlessly arranged (this is best visualized by a color labelling of the quadrangle's interior angles). This method also works for non-convex quadrangles (Fig. 12.6 bottom left).

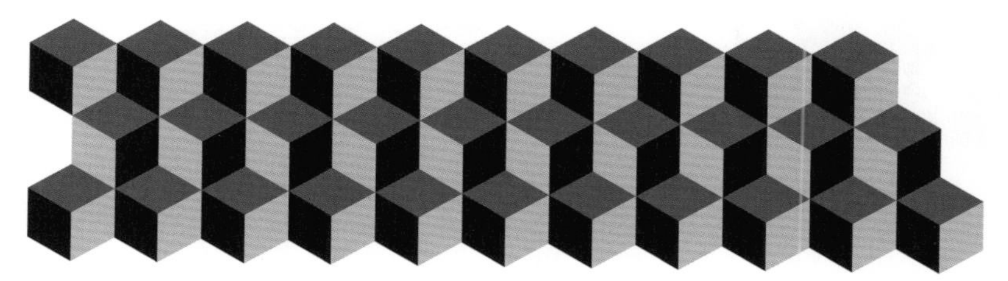

Fig. 12.9 Rhombus patterns with 3D effects (see also Fig. 2.16)

Special case: A rhombus from two equilateral triangles. Beautiful patterns can be produced from it, which may look like stacked cubes (for instance, in the Basilica di Santa Prassede in Rome).

- **Tiling with an arbitrary pentagon**

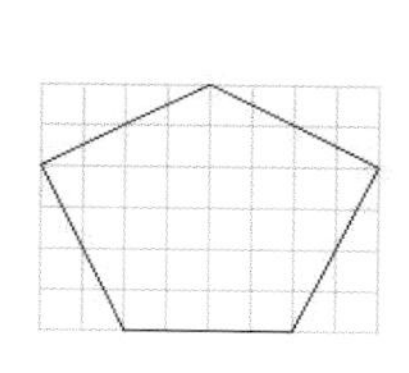

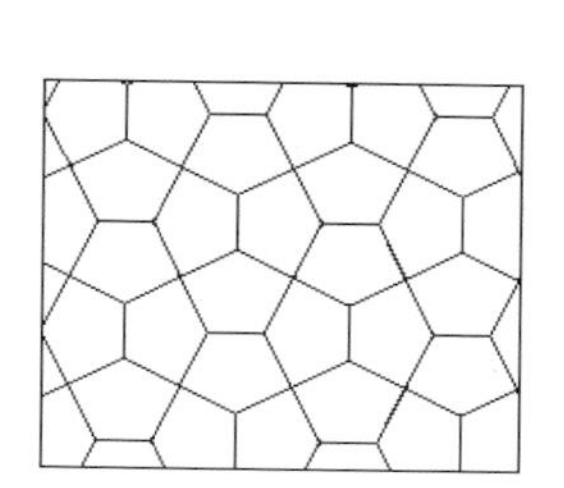

Fig. 12.10 Tiling with convex ...

Fig. 12.11 ... and non-convex pentagons

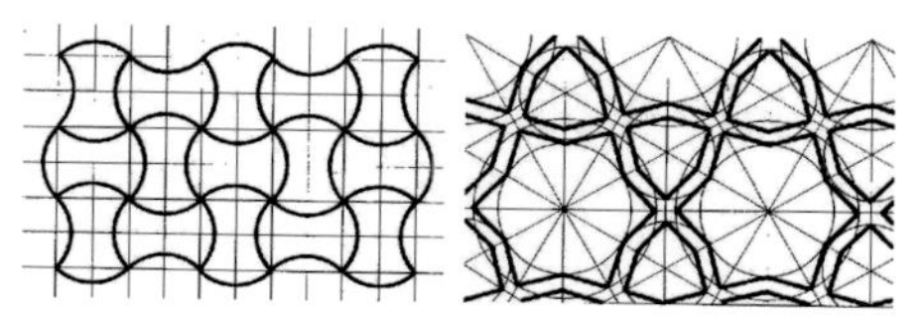

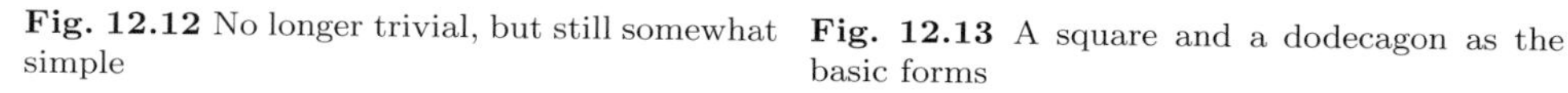

Fig. 12.12 No longer trivial, but still somewhat simple

Fig. 12.13 A square and a dodecagon as the basic forms

Fourteen types of convex pentagons that tile the plane are known today, and they were all discovered from 1918 to 1985. However, there may still be further undiscovered variants of tiling convex pentagons – it has not been proven that the search is complete.
Another type by which irregular tiling pentagon may be classified keeps the side lengths constant (convex and concave pentagons). Fig. 12.11 shows instructions by Walter Jank for the packing of skiing boots into the trunk of a car.

- Tiling with an arbitrary hexagon

Fig. 12.14 Sophie's tiling

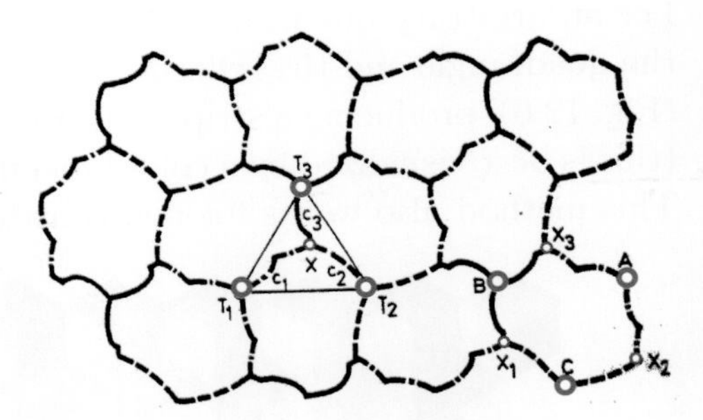

Fig. 12.15 Irregular hexagons

Three types of irregular convex tiling hexagons are known today. Fig. 12.15 shows a tessellation instruction which leads to irregular convex hexagons.

Fig. 12.16 Tiling of a spruce cone and a coral fish

Symmetry groups

Convex n-gons with $n \geq 7$ are impossible to tile based on a *single* basic form (a proof can be found in Reimann: Tilings, geometrically considered, in "Mathematical Mosaic", Cologne 1977).
Monohedral tilings may be classified topologically: All vertices of the adjacent edges are counted for the generating polygons. For the example of the chess board pattern, this would yield the number sequence (4,4,4,4). This produces 11 topological classes – the so-called *Laves-tilings* (F. Laves, 1931).
If a person chooses to classify monohedral tilings based on their symmetrical properties, one is left with 17 symmetry classes. Movements (transformations) of the plane constitute a so-called *Euclidean group*. They consist of the elements of identity, rotation about a point, translation, reflection at an axis, and glide reflection in an axis (= reflection + translation parallel to the reflection axis). A given monohedral tiling can be correlated with its symmetry group S, i.e., the kind of transformation in the plane that makes the pattern invariant (S should, then, include translations different from identity in at least two different directions). In this way, all monohedral patterns fall into 17 different symmetry groups!

Fig. 12.17 Tilings of curved surfaces in the animal kingdom

Periodical tilings with multiple basic forms

If all utilized basic forms are regular, one speaks of a semi-regular or *Archimedean tiling*. There are exactly two different types: Square and equilateral triangle (two variants), square and regular octagon, regular hexagon and equilateral triangle (two variants), a triplet of regular hexagon, equilateral triangle and square (Fig. 12.19), regular dodecagon and square – and finally – regular hexagon, regular dodecagon and equilateral triangle.
There are tilings with different convex heptagons (Steinhaus 1999, p. 77; Gardner 1984, pp. 248–249).

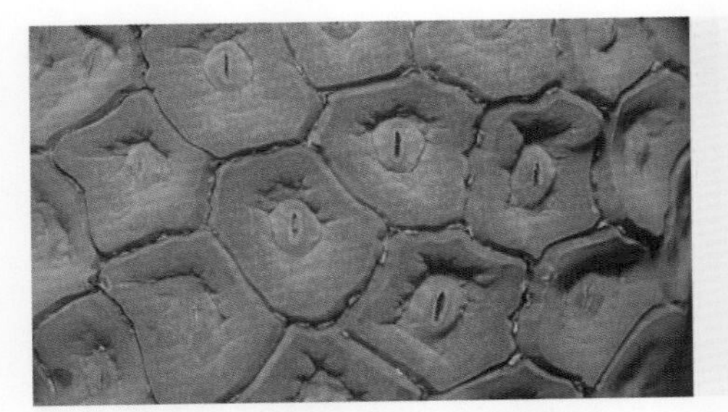

Fig. 12.18 Tiling on fruits (philodendron) and barks

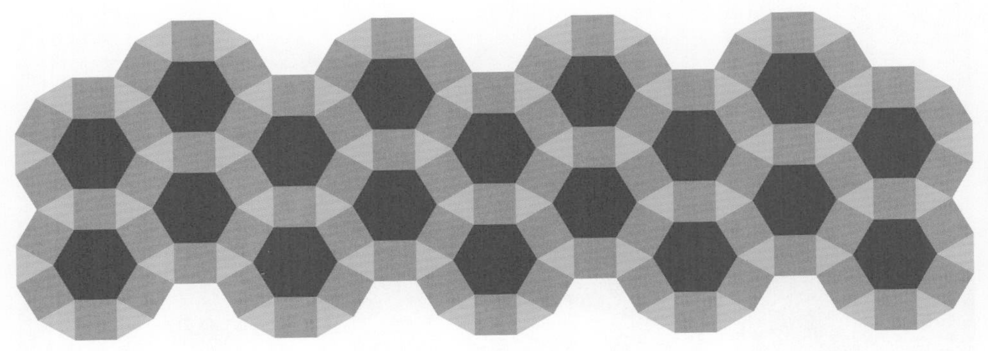

Fig. 12.19 Archimedean tilings

12.2 Non-periodical tilings

We can distinguish between two types of non-periodical tilings: *Aperiodical tilings*, where the basic forms are suited for periodical as well as non-periodical tilings, and *quasi-periodical tilings*, whose basic form is only suited for non-periodical tilings (and then only if it adheres to certain rules).

Self-similar aperiodical tilings

If many instances of a basic form can be fitted together so that an enlarged version of the basic form is produced, we speak of *inflation*. The inversion of this procedure – the decomposition of a form into scaled-down but otherwise equal versions – is called *deflation*. This pattern may then repeat itself at multiple scales, which is called self-similarity or scale invariance.
Some examples include the *chair tiling* (Fig. 12.20) or the Sphinx pattern (Fig. 12.21). From a single initial figure, the pattern is constructed by repeated deflation. The aperiodicity of the Sphinx pattern is caused by the greater form being composed of many mirrored and unmirrored smaller forms.

Central aperiodical tiling

Certain basic figures can be arranged radially, or in the manner of a spiral, about a center. The resulting tiling is non-periodical. A simple example is an equilateral triangle with an angle of 30° at the tip. It is then positioned 12 times about a center where the tips are located – and may, thus, be continued radially. If this pattern is cut into two halves and translated by one or many

Fig. 12.20 Chair tiling

Fig. 12.21 Sphinx pattern

steps, then a spirally arranged aperiodical tiling from equilateral triangles is created. The legs of the triangle can be replaced by congruent polygonal strips, which produces an infinite amount of central tilings.

Quasiperiodical tilings

Each periodical tiling possesses a basic form comprising a finite number of elements from which the whole may be patterned – but do quasiperiodical tilings also have a basic form?

In 1966, Robert *Berger* proved that there exists a finite number of basic elements which can tile the Euclidean plane only non-periodically. Berger, thus, went on a search and first found a set of generatrices of 20426 elements for a quasiperiodical tiling. This number was greatly diminished in the following years until finally (in 1973) Roger *Penrose* came upon a quasiperiodical tiling that is based on only two elements.

One of these pairs of generatices is renowned for its shapeliness and is known as "dragon and arrow". Another pair of generatrices is given by two rhombuses with equal side lengths. The thin rhombus has an interior angle of 36° and 144° – the thick rhombus is likewise defined by 72° and 108°. Tilings of this type are called Penrose patterns.

Fig. 12.22 Penrose patterns

The secret behind Penrose's pair of generatrices lies in the golden ratio. The dragon and the arrow are being created from a rhombus with the angles 72° and 108° by subdividing the main diagonal at a ratio of $\Phi : 1$, $\Phi := (1+\sqrt{5})/2$ and connecting the subdivision point to the blunt vertices. All sides of the dragon then have either a length of Φ or 1. If the respective edges of both shapes are connected by two circular arcs with the radii $1/\Phi$ (orange) and 1 (yellow) for the dragon and the radii $1/\Phi$ (yellow) and $1/\Phi^2$ (orange) for the arrow (Fig. 12.23), then the following rule has to be kept during the construction of a quasi-periodical tiling: Adjacent edges have to connect circular arcs of the same color.

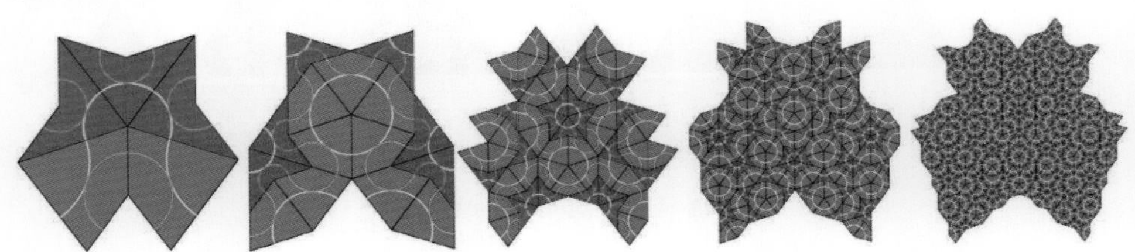

Fig. 12.23 Inflationary dragon

Penrose patterns may also be created by deflation or inflation (Fig. 12.22, Fig. 12.23, Fig. 12.24): A dragon deflates into two small dragons and a small arrow: $1D \to 2D + 1A$. An arrow deflates into a small dragon and a small arrow: $1A \to 1D + 1A$. (The inversion of this procedure is the inflation). In matrix notation we have

$$\begin{pmatrix} D \\ A \end{pmatrix} \to \begin{pmatrix} 2,1 \\ 1,1 \end{pmatrix} \cdot \begin{pmatrix} D \\ A \end{pmatrix}.$$

Thus, the iteration of this process yields

$$M^n = \begin{pmatrix} f_{2n+1} & f_{2n} \\ f_{2n} & f_{2n-1} \end{pmatrix},$$

where f_n is the n-th Fibonacci number. If one starts with a dragon shape and deflates it n times, or

$$\begin{pmatrix} 1 \\ 0 \end{pmatrix} \to M^n \cdot \begin{pmatrix} 1 \\ 0 \end{pmatrix} = (f_{2n+1}, f_{2n}),$$

then the ratio of the number of dragons to the number of arrows converges against Φ. If a whole plane is completely covered, this ratio is irrational. This is impossible for periodical patterns – and from that observation, the quasi-periodicity of the Penrose patterns directly follows.

Fig. 12.24 Inflationary rhombuses

It can be shown, that there exists an uncountable number of Penrose patterns from dragons and arrows! The same holds true for all other Penrose patterns, which, nevertheless, have the following bizarre characteristic: All patterns are locally isomorphic: Each neighborhood of a pattern is also present at an infinite number of other places!
How far away are the copies of a Penrose pattern's finite neighborhood? This answer is also surprising: If a circular neighborhood with a diameter of d is considered, then the distance between the boundary of the neighborhood and its next exact copy is never farther away than $d \cdot \Phi^3/2 \sim 2{,}12 \cdot d$.
A connection to physics: Up until 1984, only two forms of solid matter were known – amorphous solid bodies and periodically ordered crystals. In that year, however, Dan *Shechtman* and his colleagues discovered a quasi-periodical crystal formation in an aluminum-manganese alloy! Several hundred of such alloys are known today, and this third form of solid matter is now known as a quasicrystal. The above described Penrose pattern can be seen as a model for "two-dimensional" quasicrystals by imagining a crystal through periodically layered Penrose planes.
The great surprise of *Shechtman*'s discovery was that until then, a crystal was known as a solid body constructed from a periodically repeated initial cell – but this conception turned out to be too narrow. Now, a crystal did not merely have to exhibit 2-, 3-, 4-, or 6-fold rotational symmetry, but also

possess a discrete diffraction diagram (X-ray spectrum) with sharp reflections. It may also, for instance, exhibit 5-fold rotational symmetry, which is the case with Penrose patterns.
There exist many other tilings that permit the construction of quasi-periodical structures (e.g. the Ammann-grid, octagon-pattern, etc.). With these patterns, real solid bodies may be modelled, and questions concerning the stable arrangement of the atoms into these crystal symmetries may be researched.

12.3 Non-Euclidean tilings

Several models of the hyperbolic plane exist, such as the Poincaré disk model and the Poincaré half-space model. Let us here choose the disk model, which has the advantage that measuring hyperbolic angles is equivalent to measuring Euclidean angles. However, the metric is altogether different: The distances steadily increase towards the boundary of the circle. The circular edge itself is the set of all infinitely far-away points (horizon). The disk is produced by stereographic projection of the sphere model into the hyperbolic plane (Fig. 1.56).
Axiomatically, the hyperbolic plane differs from the Euclidean plane in that there exists an infinite number of parallels to a straight line through a point which does not lie on it. Another characterization may be given by the property that the sum of interrior angles of any triangle is less than 180°.
In the disk model, straight lines (shortest distance between two points) are circles which intersect the horizon orthogonally. Circles may look like circles – however, the circle midpoint is generally not at the (Euclidean) center. Equilateral polygons (bounded by orthogonal arcs) do not appear to be so, but are, instead, distorted due to a motion in the Poincaré disk.

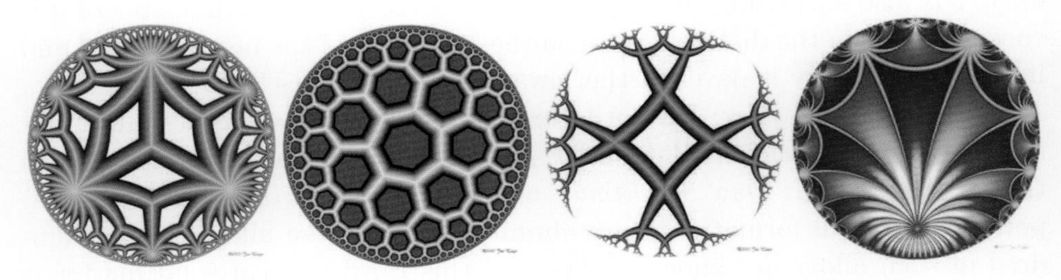

Fig. 12.25 Hyperbolical tilings by Jos *Leys* (`www.josleys.com`)

Which tilings with regular n-gons are possible?
Contrary to the situation in the Euclidean plane, the sum of angles of an n-gon is not equal to but smaller than $(n-2)\cdot 180°$. If k is the number of n-gons adjacent to an apex point, then the following must be true for a regular tiling with n-gons: $n\cdot 360/k < (n-2)\cdot 180$ or $1/n+1/k < 1/2$. This

is given for all $n \geq 3$ with an appropriate k, which means that there is an infinite number of regular tilings in the hyperbolic plane.
For each regular n-gon there is also a minimal value of k:
$(n, k_{min}) = (3{,}7), (4{,}5), (5{,}4), (6{,}4), (7{,}3)$, and $(n{,}3)$ for each $n > 7$.

Fig. 12.26 Hyperbolic tilings with a swordfish by Johannes Wallner

Fig. 12.25 and Fig. 12.26 show beautiful tilings of the hyperbolic plane. The figure in Fig. 12.26 is defined by Bézier curves.

What is the situation on the sphere?

Let us remember that a tiling of the sphere is, indeed, possible in certain special cases – it essentially concerned the Platonic solids, application p. 118.

• **From a dodecahedron to an original and easily manufacturable lamp.**
Fig. 12.27 shows a remarkable patent (probably from the Asian region – though the patent owner is unknown) inspired by such a tiling.

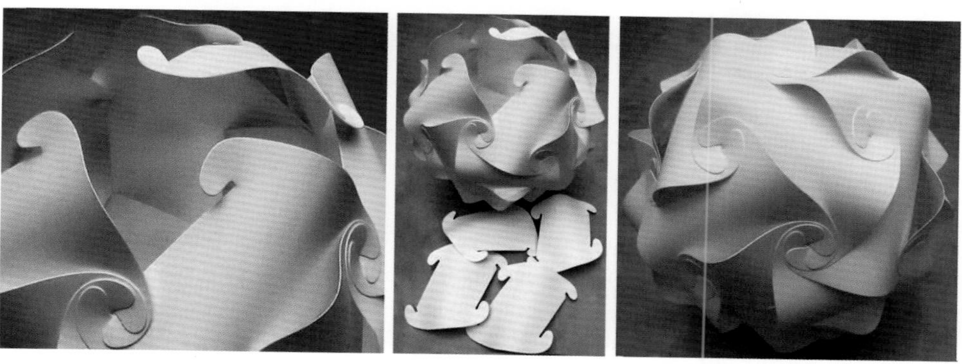

Fig. 12.27 A tiling of the sphere produces an aesthetical lamp.

In order to understand the issue, we must first abstract it. Let us consider a regular dodecahedron and subdivide the faces as in Fig. 12.28 into five equilateral triangles. Pairs of triangles connected by a dodecahedron edge form an "oblique rhombus", which is easy to manufacture. The $5 \times 12 = 60$ equilateral triangles are thus converted into $60/2 = 30$ congruent rhombuses.

All that is now required is a bit of imagination in order to make the rhombuses attach to each other (Fig. 12.27 at the center) – and we are done! ♠

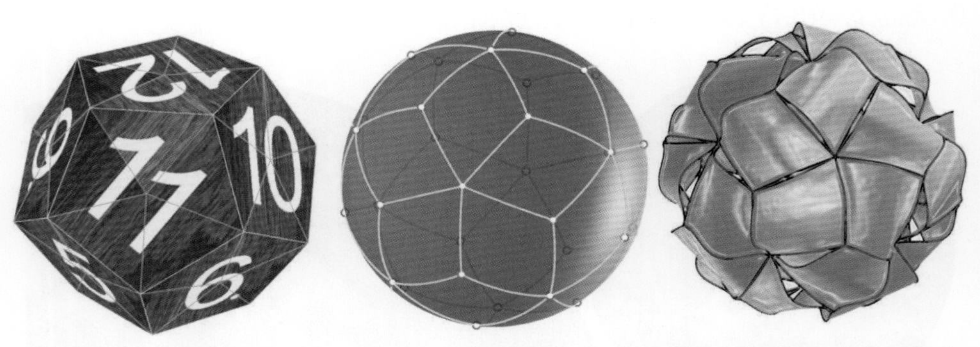

Fig. 12.28 From a dodecahedron to a tiling with triangles and "oblique rhombuses". It was also the basis of patents describing a novel football and an original lamp.

• **Puzzles in space**

The toy industry also undertook geometric deliberations and invented puzzle games which produce spheres. The fact that the building blocks are similar but not quite congruent lies at the heart of any good puzzle game. Take the triangulation of a sphere, as in application p. 118, combine two triangles into oblique quadrangles – and then construct a whole puzzle from this approach.

Fig. 12.29 Tiling a sphere with nearly-congruent puzzle pieces is quite a challenge.

To simplify matters for the frustrated player, the manufacturer provided a cylinder model of the result, which is in its developed form composed of a rectangular strip and two circular caps (in Fig. 12.29 on the right). ♠

13 The nature of geometry and the geometry of nature

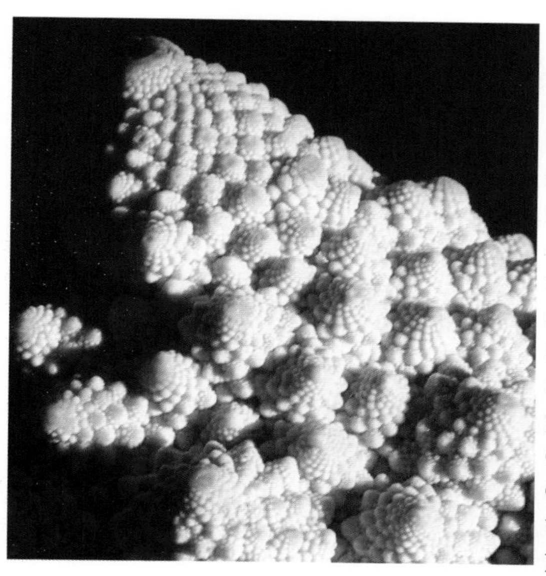

The title's wordplay is, perhaps, an indication of this chapter's intentions – to bridge the gap between nature and geometry.
It is commonly known that many of the surfaces or solids described by geometry can be found in nature – in both macrocosm and microcosm. A question arises: Is geometry a mere "plagiarization" of nature, or was it independently developed, with coincidental parallels to the results of evolution? Geometry helps us to understand, why, for example, the platonic solids are "obvious" solutions to many problems in nature and that if certain conditions are given (such as regularity or symmetry), no other possibilities actually exist. Nature usually produces regularity or symmetry if an equilibrium of forces or an optimal packing surface is sought. Evolution finds optimal forms on its own. The ancient Greeks *Plato* and *Archimedes* could not have known, more than two thousand years ago, that "their" solids are so frequently found at a molecular scale. *Pythagoras*, on the other hand, defined music by means of proportions. It is, therefore, no wonder that musical rules can be beautifully visualized geometrically. Another valid question may be: Is music, as defined by Pythagoras, rooted in "natural phenomena"?

Survey

Stanislaw Ulam [26] calls mathematics a self-contained microcosm which, nevertheless, possesses a strong capability of capturing and modeling arbitrary processes – not only of thought, but perhaps of the entire field of science. To some degree, this probably also applies to geometry.

13.1 The geometric basic forms in nature

We have already seen it throughout this book: Nature is full of certain basic forms. They are sometimes incredibly obvious, as in snail shells, and at other times rather "imprecise". Although we may not have an immediate answer for every case, we shall nevertheless attempt to get to the bottom of these issues.

Symmetry

Fig. 13.1 Antennae of a moth, a mosquito, and a beetle (all males)

In nature, symmetry is an essential design element. Biologists refer to *bilateral symmetry* when describing the regular arrangement of the parts of an organism – but only if it is symmetric with respect to a central plane (Fig. 13.1).

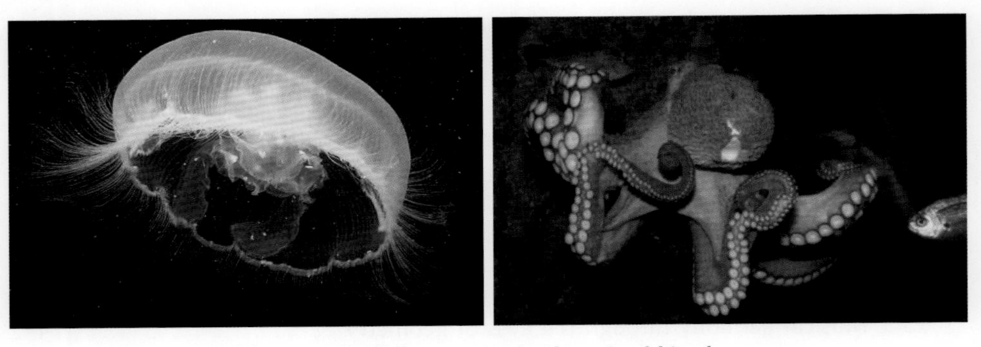

Fig. 13.2 Radial symmetry in the animal kingdom

If certain parts are arranged about an axis, then biologists speak of *radial symmetry* (Fig. 13.2). This way, the animal kingdom can be subdivided into

Radiata (jellyfish, starfish, radiolaria, etc.) and *Bilateria* (bilaterally symmetric classes of animals). In the kingdom of plants, blossoms exhibit radial symmetry (Fig. 13.23) much more often than bilateral symmetry (Fig. 13.3).

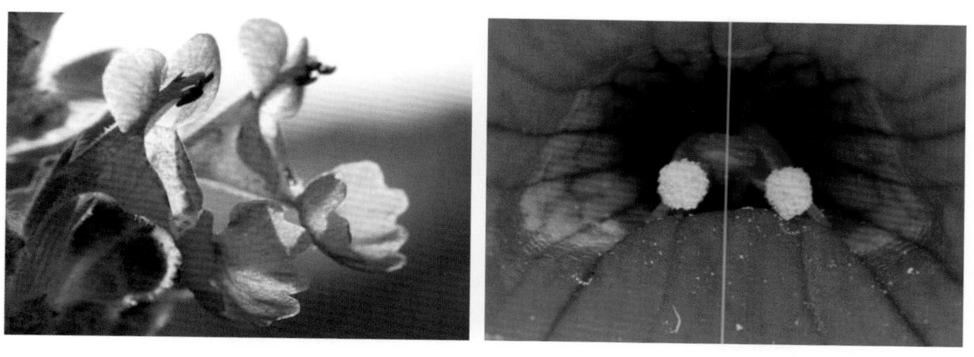

Fig. 13.3 Simple symmetry in the kingdom of plants

Fig. 13.4 shows the imitation of a bilaterally symmetrical plant by a tropical orb-weaver spider – a tactic that serves to attract insects (on the left: view from above, on the right: view from below).

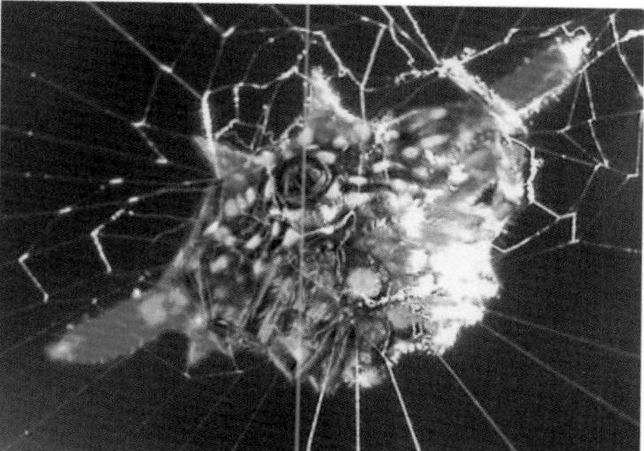

Fig. 13.4 Simple symmetry in the animal kingdom

Circles as distance curves and path curves

We have already explained that the most elemental and simplest curve in nature is not the straight line (which, indeed, almost never occurs in life forms), but the circle. These shapes are especially noticeable in propagating waves.

If we attempt to venture into the desert without being armed by either a compass or by ancient Bedouin tricks, then we will eventually run in rough circles. This is partly caused by the left stride length never being exactly equal to the right stride length, but also by our inability to account for the minute changes to the sun's position (ants and bees have this capability, which has been proven by many interesting experiments).

When a wasp cuts a circular hole into a leaf (Fig. 13.5), it applies its jaws at constant angles (analogous to a piece of ham being cut on the breakfast table). Precisely speaking, the jaws are always orthogonal to a pencil of rays through the center of the circle. The circle is naturally the *orthogonal trajectory* of this pencil of rays.

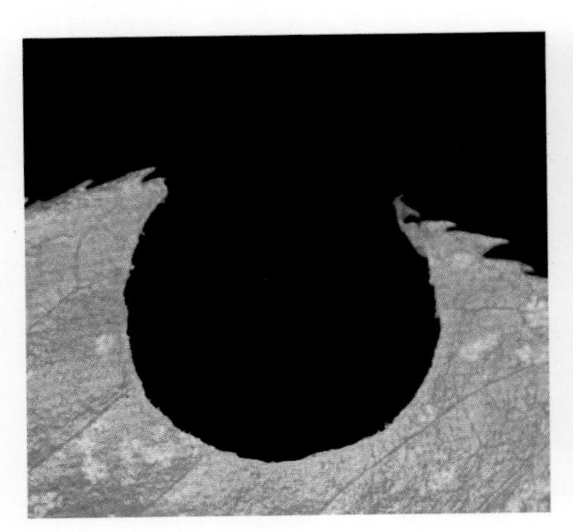

Fig. 13.5 The crime and the "perpetrator" ...

The wasp does not seem to follow this pencil of rays, which – after all – only exists in theory. It repeatedly cuts short segments of constant length into the leaf, which form an angle of slightly less than 90° with the moving direction. This procedure is repeated until the wasp reaches its starting point. Thus, it does not actually cut out a circle, but a regular polygon with so many vertices that it can hardly be distinguished from a circle.
Both the voyage in the desert and the cutting of the hole produce circles by a combination of movement and "work angle". Circles are also distance curves: All points are equidistant from a center.

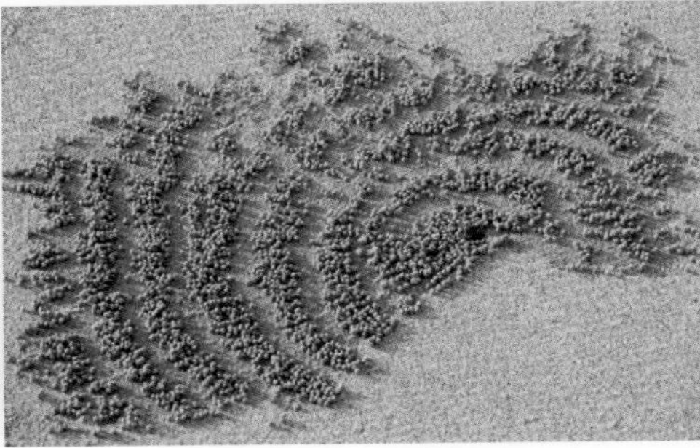

Fig. 13.6 The "building of a fortress" by a tiny crab in tidal times

This must be how the wonderful ring-like constructions of the tiny carcinus maenas (Fig. 13.6) are produced. The animal rolls up the excavated sand (which contains its food) into balls and moves it some steps away from the center. Once a single ring is finished, another concentric ring is made. This tedious work is all undone as the next tide sweeps it away ...
When plants, such as the circular lichen, grow on the surface of a stone, it makes sense that they should try to "augment mass" at their outer boundary.

If this happens regularly, then circular contours are automatically produced (Fig. 13.7).

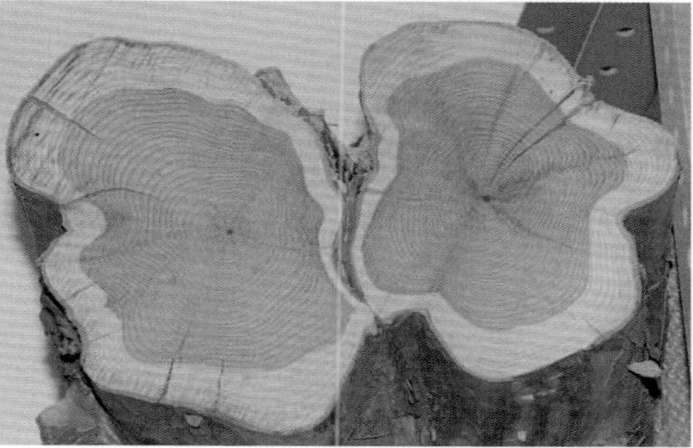

Fig. 13.7 Linear growth

The speed at which this lichen grows is not important for the geometrical principle. For a layperson, it would be difficult to detect whether the lichen stops expanding every few years. This is far easier to detect in trees: The tree produces new layers around the central trunk (Fig. 13.7). The coloring or the hardness of the cross-section suggests that, while this growth is periodical, it does not proceed at the same rate throughout the year. In tropical woods, these tree rings are not as pronounced, because the growth is more regular than in colder latitudes.
Although the tree's increase in thickness is more or less linear, it has no influence on the form of the cross-section.

Spheres and other convex basic forms

The circles in the plane correspond to spheres in three-dimensional space. If the carrier medium permits it, then the waves or impulses always propagate spherically. Natural objects (whether living or dead) at first also tend to expand spherically, even though after an initial growth phase, a change in strategy usually occurs. The reasons for this are manifold, though one particularly important reason may be as follows:
The surface of a sphere increases at a square of its radius, which is why (for instance) sonic strength or brightness decrease at a square of the distance. The volume that is included inside the spherical surface, however, increases *cubically* with the radius. If a sphere has reached twice its initial radius, then its volume (and mass) has increased eightfold, whereas its surface has only increased fourfold.

Minimal and maximal packets

The water droplets in Fig. 13.8 are forced into a spherical shape by the surface tension of the water (cohesion) – in opposition to the rules of gravity, it would seem. In order for such a droplet to remain on the leaf, it would have to touch the leaf by a large surface – ideally, a great-circle of the sphere. This makes it possible for adhesion to balance out the gravitational pull.

Fig. 13.8 Half-spherical water droplets

This aforementioned subtle equilibrium only works up to a certain scale. At some point, either the cohesion or the adhesion are too small to "counteract" the increasing mass.

A maximum of food for the younglings ...

Fig. 13.9 Oval insect eggs (egg-shaped ellipsoids of revolution)

The geometry of insect and bird eggs appears in an interesting geometrical light. Without going too much into detail, it may be said that bird eggs are neither spherical nor ellipsoidal – we know this not least because of our daily breakfast experience. It is most likely that the insect eggs in Fig. 13.9 are not entirely perfect geometrical surfaces either, even though they do seem to approximate spheres and ellipsoids of revolution much better than their chicken counterparts.

The "cute" ladybird in Fig. 13.9 on the left is, interestingly enough, almost shaped like half of an ellipsoid. The reason for this lies in its thereby optimal protection against agitated ants, which tend to protect "their" herds of aphides, from which they extract a valuable sugary syrup, with great ferocity. The plant lice are a favorite dish of the ladybird and its larvae ...

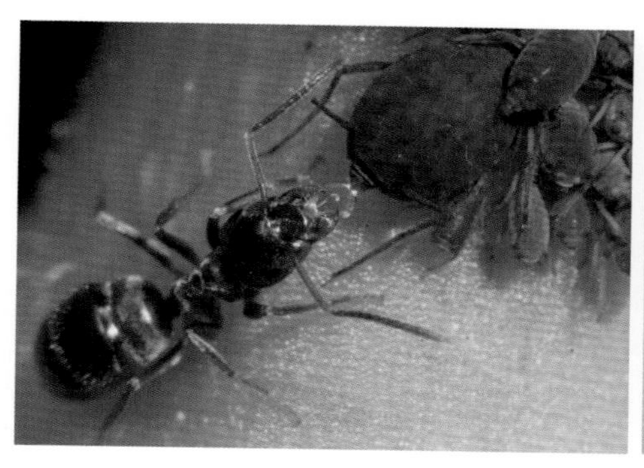

Fig. 13.10 Ants love their plant lice.

Don't roll away ...

In any case, the point of eggs having their particular shape is to contain a maximum of food for the offspring in the smallest amount of space. Bird eggs have the additional difficulty that they tend to roll out of their nests, whereas insect eggs mostly stick to one spot.

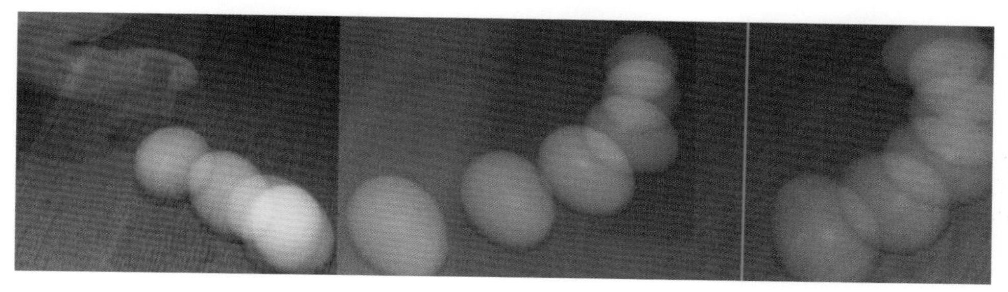

Fig. 13.11 "Laboratory experiment": An egg is being rolled on a wooden floor ("short distance")

In order to minimize this effect, nature simply deviates from the spherical form and thus, the eggs simply wobble around – if at all.
Fig. 13.11 shows an experiment in which the rolling behavior of a chicken egg with an angular momentum is tested. It is clearly visible that the egg moves in a very specific swerving motion, which may prevent it from rolling too far away (this, however, also has to do with the fluid contents of the egg).

Unrolling and development

As we have already seen, developable surfaces play a big role in technology, being so easy to manufacture. Whenever nature "goes to work" on leaves or blossoms, they are usually being unrolled – which is the geometrical equivalent of development.
The process of unrolling often happens over a longer period. In the meantime, the leaf may slightly alter its shape before partially hardening. Thus, the result is not necessarily a developable surface anymore. The many phases in the growth of a blossom in Fig. 13.12 and 13.13 show this very well.

Fig. 13.12 The many phases in the growth of a blossom I

Fig. 13.13 The many phases in the growth of a blossom II

13.2 Evolution and geometry

We have previously referred to the geometry of bird eggs. However, how do the birds "know" that non-spherical eggs are better for their offspring (assuming that spherical eggs roll away more often and thus break more frequently)? If a bird were to lay spherical eggs in inclined or otherwise unstable circumstances, then it might discover that an elliptical form would be more practical – if only the animal had the mental capacity for this analysis. And

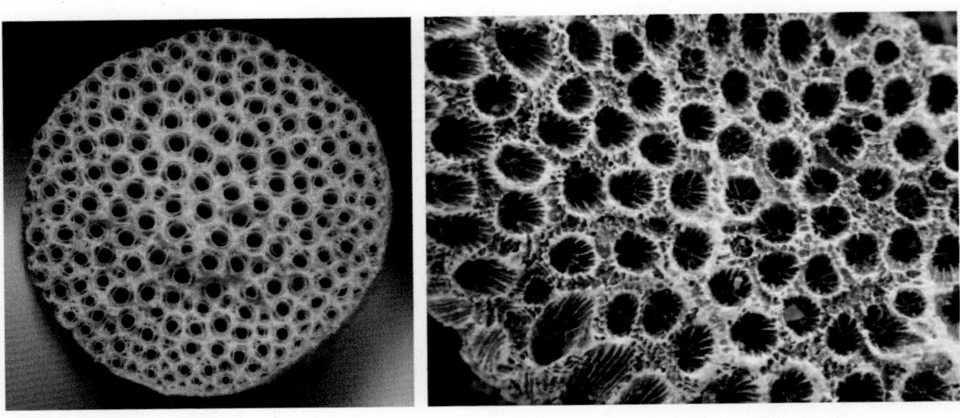

Fig. 13.14 On the left: The cross-section of a dried willow branch, on the right: coral colony

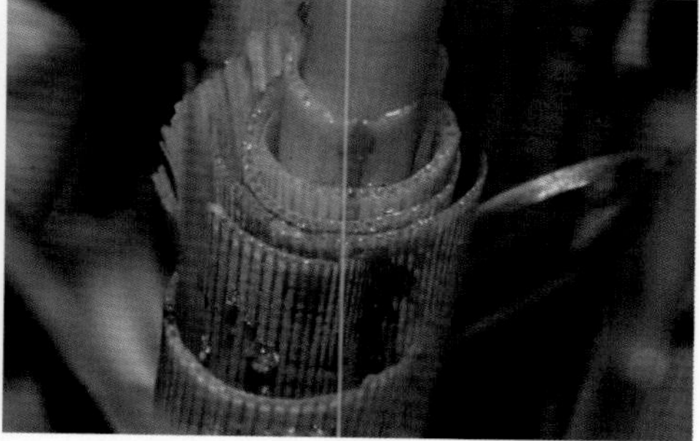

Fig. 13.15 Linear growth

Fig. 13.16 Feather duster worm and a moufflon

if only it could somehow affect the shapes of its own eggs in its next attempt, it certainly would.

While the *individual* may be incapable of such a feat, the *species* as a whole is able to surmount this challenge – when given evolutionary timescales. Evolution is as ingenious as it is simple. We must now deviate slightly from the main topic of this book – with good reasons, as a solid understanding of evolutionary principles is essential for the section that follows. Biologists are hereby asked for their indulgence with regard to my generalizations, but since evolutionary thinking does not yet enjoy widespread awareness – even 150 years after its landmark introduction – and is even the target of some hostility, it cannot hurt to attempt to clarify some of the basic issues.

Mutation

We shall first and foremost consider the principle of the "inexact replication" (*mutation*): No animal in nature produces offspring that are completely identical to itself, and no lifeform produces identical offspring all the time. This may in part be influenced by the physical difficulty of "repeating" a complicated genetic code without any error (one may, as a point of comparison, try to memorize the first 250 decimal places of π). The act of exact replication is further complicated by the varying states of the environment where such copying takes place. However, it stands to reason that even perfectly identical laboratory conditions would, sooner or later, be host to imprecise copying procedures.

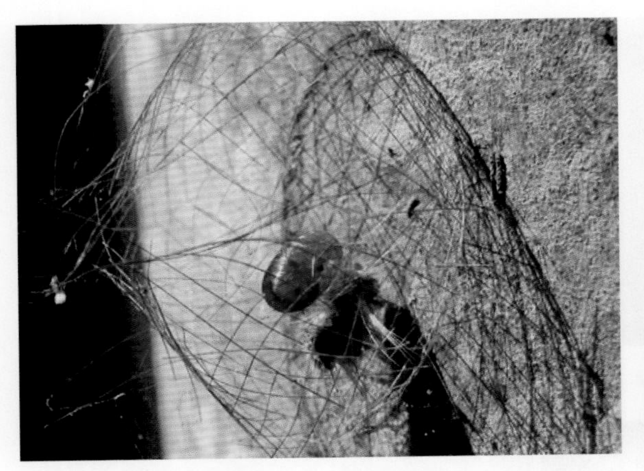

Fig. 13.17 On the left: An insect's geometrical construction, on the right: Drops of resin

The ancestors of reptiles and birds may, indeed, at some point, have "started" the colonization of land with spherical eggs like most fish eggs. If that was the case, then many of these eggs must have rolled away and were probably destroyed in the process. This waste may be compensated by a more prolific production of eggs, which of course costs more energy. However, due to the principle of inexact copying, some early reptiles may have produced more spherical eggs, while those of some other reptiles were less so. The non-spherical eggs, being more stable on the ground, were, thus, more likely to produce healthy offspring.

Inheritance

A second principle of nature now applies: Offspring usually inherit most of their phenotypic properties from their parents. If, therefore, more parents have been born from "less symmetric eggs", then – depending on the strength of this selective pressure – the next generation should exhibit a higher proportion of individuals laying non-symmetric eggs. If the selective pressure is strong, then it does not require all too many generations before the kinds of reptiles which lay "classically shaped eggs" – and not those who lay spherical eggs – become dominant.

The sum is what matters

It may now be argued – quite wrongly – that this principle should apply to every trait of the organism, and that it should take only a few generations before a perfect solution is found for every problem (which in the case of the eggs would be purely geometrical). However, this runs contrary to a third principle of nature: The *best sum of adaptations* is usually the successful one, and this also depends on a constantly changing environment. If some reptiles, for instance, continue to lay spherical eggs but, for reasons of another trait that is affected by the same genes, are "better at evading predators", then this might turn out to be a more successful adaptive strategy than "not laying spherical eggs".

Fig. 13.18 Tiling

Fig. 13.19 Exponential growth in snail and nautilus shells ...

It *may* occur, for example, that the individuals who are better at evading predators also adapt to lay non-symmetric eggs without any negative effects on their evasive strategies. For lack of a better term, this would kill two birds with one stone. It may simultaneously occur that other "better evaders" acquire a new trick by way of mutations that somehow lessens the disadvantage of spherical eggs. Whatever the situation might actually have been like, it seems that the advantage of more or less non-spherical eggs was so significant that this trait is now shared by virtually *all* oviparous land-based animals.

Phyllotaxis, or: How easy it is to be deceived?

What appears to be simple explanation for the non-spherical shape of eggs may require more profound scrutiny in other cases. A typical example is the arrangement of sunflower seeds, which is also known as *phyllotaxis*. Much has been written and theorized about this topic – and despite all of that, the solution turns out to be ingenious and easy in the manner of evolution itself.

Fig. 13.20 ... and bivalves

Fig. 13.21 Spiderweb

A premature conclusion

Let us briefly discuss the theories: Looking at a sunflower long enough, one will inevitably find spirals in the pattern. Two opposing spirals can in fact be discovered! It may further be noticed that the number of spirals in a pencil is always a so-called *Fibonacci* number (more on this topic can be found in [11]). The *golden mean* is closely linked to these famous numbers, as is exponential growth (by repeatedly multiplying a number with itself, we get a geometrical sequence which – if displayed in a coordinate system – yields an exponential curve). Thus, we might be easily impressed: It seems that sunflowers are "aware" of the golden mean and grow exponentially – while most other plants do not!

A completely different explanation

Now on to the most likely explanation (it may be misguided to claim to know the *definitive solution* for such matters):

A sunflower multiplies by means of its seeds, which are mostly eaten by birds and usually spread around because these birds tend to lose seeds from time to time. One thing is clear: For this strategy to work, a sunflower has to produce as many seeds as possible. This increases the probability that the species may survive in the long run.

Let us imagine a plant species (a predecessor of the modern sunflower) which may have adopted the following successful growth strategy during the course of its evolution: "Grow your seeds by creating a principal seed at the approximate center, and position each further seed such that the direction is changed and that the position is translated slightly away from the center". In order for this procedure to be quantifiable, the rotation angle would have to be constant, as would the translation distance. Up to a certain point, this strategy would, indeed, allow the flower to produce new seeds. The natural limit would have to be the situation where a new seed would no longer fit, the new space being occupied by an already existing seed.

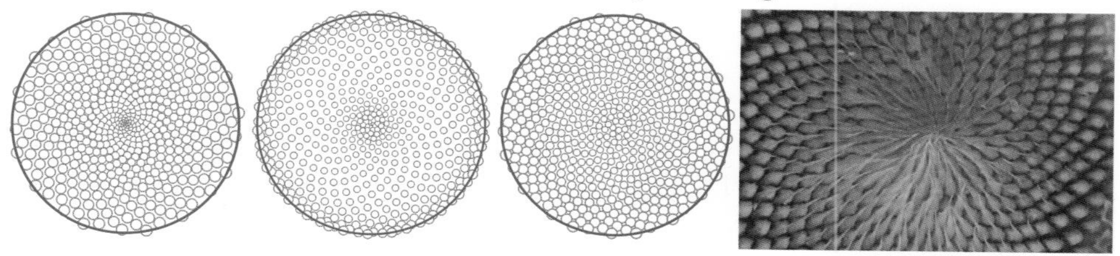

Fig. 13.22 Different "translation distances"

Among the infinite number of combinations of angles and distances, there is certainly an optimal one. As in the case of the egg shapes, evolution would eventually find it over the course of many generations: Mutation is constantly at work, changing angles and distances ever so slightly, and those flowers who are lucky to have hit upon a good combination would carry more seeds for their reproduction. "Their" advantageous combination would be passed on to their progeny, which have a chance of "improving" it further.

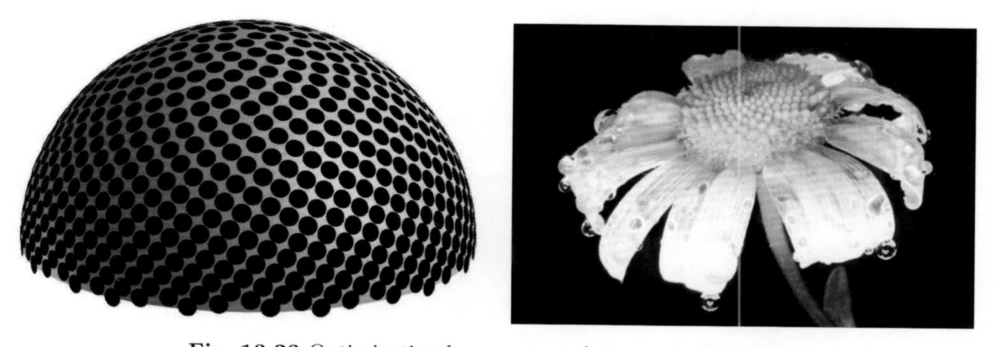

Fig. 13.23 Optimization by an extrapolation onto the sphere

The result is pretty impressive: A very "close" approximation of a modern sunflower (Fig. 13.22 on the left). The center is a little bit tight – there, the seeds would have to be smaller than at the far edges. The fact still remains, that this is a remarkably effective method of packing seeds into the blossom. We can also detect the aforementioned spirals. Indeed, this method creates them wholly without the need for exponential growth! The angle of rotation is very strange – in fact, it divides the whole revolution of 360° at the golden mean.

The images in Fig. 13.22 at the center and on the right show further possibilities if the translation distance is allowed to decrease by a certain percentage as the seed moves away from the center. This, however, is mere "cosmetics". If this procedure were to be "extrapolated" onto a sphere, as illustrated in Fig. 13.23, then the seed density would increase even further.

From a daisy to a meter-high sunflower, a common explanation for some of their specific proportions can often be found in the "golden mean" – a fact that has fascinated many numerologists. This is not meant to be derogatory: Even such deliberations may be mathematically valid. It can be shown that the golden mean is the "most irrational of all numbers":

$$\Phi = 1 + \cfrac{1}{1 + \cfrac{1}{1+\cfrac{1}{1+\cfrac{1}{1+\cdots}}}} \approx 1{,}6180339887.$$

The flower "needs" this irrationality, so that it may apply the rotation-translation-step as often as possible to the seeds.

- **The generation cannot always be seen so clear**

Fig. 13.24 Changing the "golden mean" by tenths of degrees ...

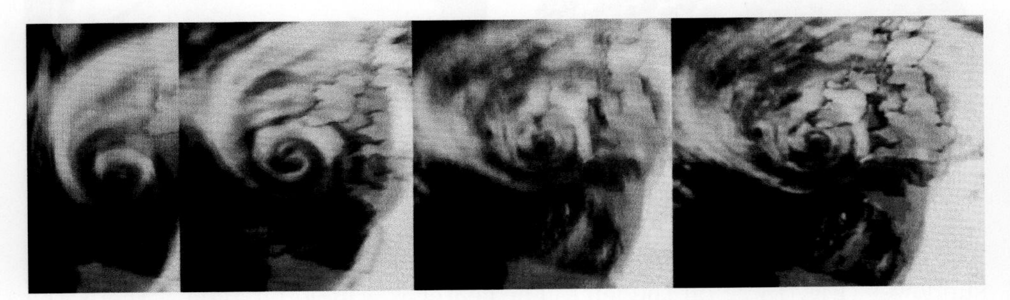

Fig. 13.25 Snapshot of a low pressure turbulence in the making

By changing the golden mean ever so slightly, we get patterns as in Fig. 13.24 on the right. Such figures are already familiar to us: For instance, the blossom in Fig. 13.24 on the left already resembles it very closely. Let us leave it at that, without inventing any further hypotheses: In geometry, spirals are relatively easy to produce – they may be very similar locally, but they may, in fact, be defined in a very different way.

The spiral pattern in Fig. 13.24 on the left may owe its appearance to a completely different principle than the filling pattern of the blossom.
Turbulences caused by low pressure areas are another class of phenomena where we should not too hastily infer the precise reasons for their appearance. Several parameters apply here. Conservation of angular momentum must surely play a central role, but so must lateral acceleration, which is connected to the north-south-expansion of the turbulence (different path velocities depending on the geographical latitude!) as well as widespread pressure drops. It is probable that there are several more important causes at play here. Given this complexity of cause and effect, it is an undeniably great accomplishment of modern meteorology that accurate predictions of such large-scale weather patterns can be made up to several days in advance.

13.3 Planetary paths and fish swarms

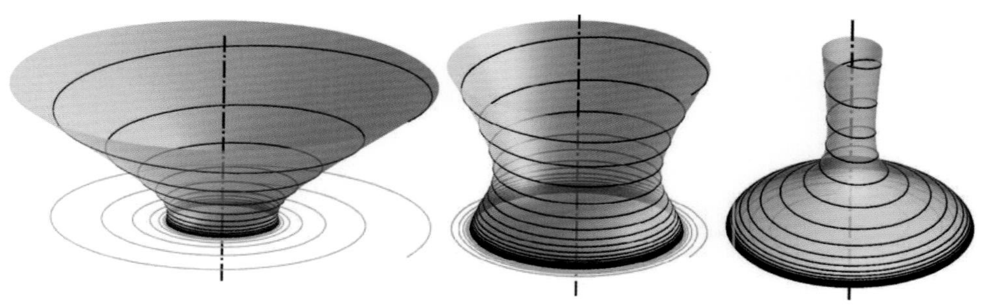

Fig. 13.26 Galactic spirals including their carrier surfaces of revolution

The forms of galaxies may very well be inevitable: At a large scale, outer space is essentially dominated by only a few laws – especially by gravitational attraction and the law of inertia.

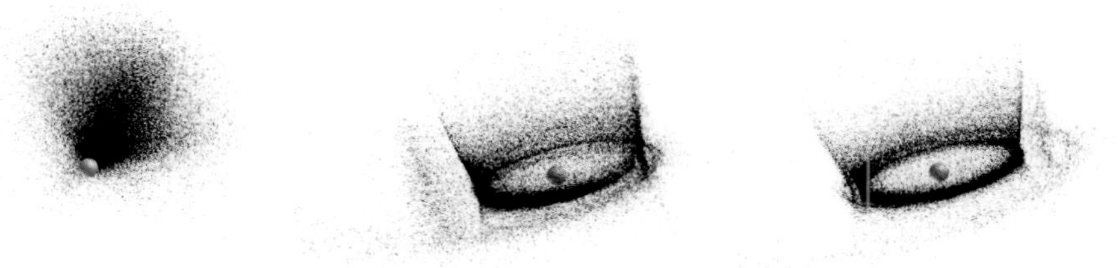

Fig. 13.27 A cosmic catastrophe (devastating impact) ...

Let us hypothetically consider a galaxy like our own Milky Way, which forms as a result of the compaction of interstellar mass and, according to the law of conservation of angular momentum, is finally left with a constant angular velocity or momentum.
Due to gravity, the center of this system will gradually increase in density (and may eventually form a black hole which will continue to "accrue" mass until it evaporates). The clusters of stars in the system will continue to move towards the center and increase their angular velocity in the process.

No matter what parameters one applies: One can always find a galaxy that corresponds to a snapshot of the simulation – and this simulation will always contain spirals made of star clusters.

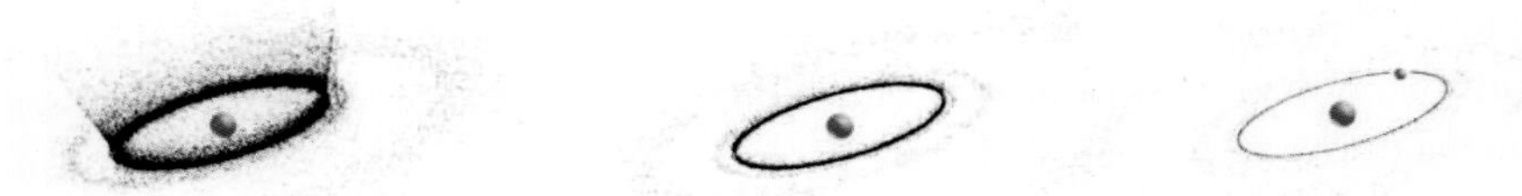

Fig. 13.28 . . . and the subsequent consolidation (moon)

Our moon seems to have been formed 4 to 5 billion years ago by a collision of the newly formed volcanic Earth with another massive celestial body. Following the impact, glowing chunks abounded in our planet's neighborhood, each of which possessed Earth's angular momentum and drifted along in spiral motion before being "caught" on a circular orbit around the planet. A gradual accretion of mass followed – probably due to lack of space and gravitational forces – and the moon was thus gradually formed.

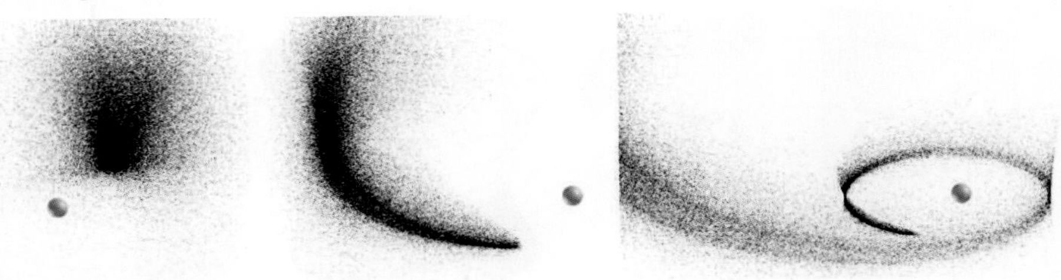

Fig. 13.29 The explosion cloud of a supernova passes our sun . . .

It may "only" have been hundreds of millions of years prior to those tumultuous events that the remains of a supernova passed our still very young sun, which managed to catch part of this debris by its gravitational pull.

Fig. 13.30 . . . and "produces" a planet

The laws of physics then acted their part, and it did not take long until the supernova remains (which are mostly metal) were circling our star on elliptical paths (with the sun being at one focal point). Collisions of the orbiting matter caused its gradual accretion until the planets were born. Our asteroid belt is probably nothing more than a not-yet-consolidated gift of that long gone supernova . . .

One may now argue that these topics do not have that much to do with geometry, and that such speculation is best left to physicists. On the other hand, one may also consider it a beautiful example of separate academic disciplines supporting one another: Without computer simulations, which depend on computer geometry, the results would certainly not look as spectacular as they do, and the layman who is not able to solve differential equations would have to take the physicists at their word.

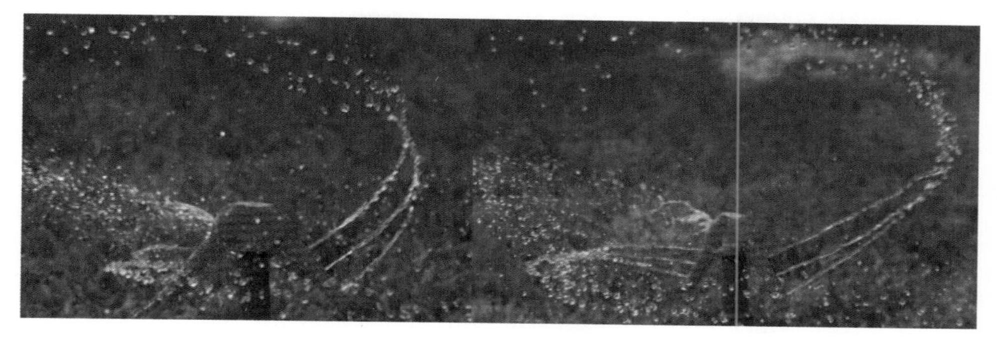

Fig. 13.31 "Spiral lawn sprinkler"

The following question is closer to home: When a lawn sprinkler rotates using the *repulsion* of the exiting water streams, spirals are formed which resemble their galactic counterparts: As the water moves away from the axis of rotation, its angular velocity decreases by a square of the distance according to the law of conservation of momentum. Other components apply here, such as air resistance and gravity, but it need not concern us here, as so much of this phenomenon is simply due to chance.

Fig. 13.32 Impending danger and escape

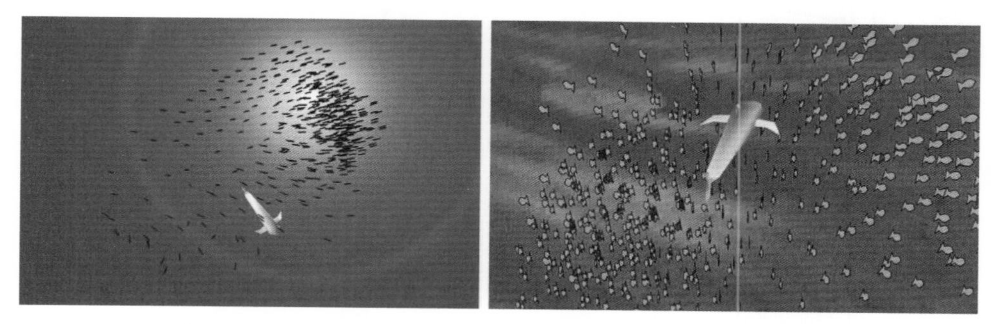

Fig. 13.33 Back to the swarm ... and immediately away from it!

Let us examine another water-based scenario, but from a different perspective: The seemingly chaotic developments in outer space are not that far

removed from, for instance, the simulation of a swarm of fish. This time, let us focus on the discoveries of biology. Small fish often exhibit a very sophisticated swarming behavior – not unlike many species of birds.
Interestingly enough, there is no leader animal which determines the direction of the swarm. This whole choreography is rather the product of many singular reactions. The goal of a single fish is only to remain close to its neighbors (about a single body length). When foraging, individuals tend to move more leisurely in the direction of more food.
When a predatory fish takes interest in the swarm, the individuals in its vicinity will naturally want to flee from the place of danger. Their neighbors may not even have realized the cause of disturbance, but they still reflexively follow their neighbors on their paths of escape. This leads to a chain reaction (shock wave), which may even involve individuals that are far away from the predator.
Once the danger has passed, the fish which are farthest away will once again seek the close company of their relatives. After a certain amount of time – and possibly until the next predatory attack – "order" will have returned to the swarm.
This supposed chaos is actually governed by only a few parameters, which, nevertheless, barely restrict the variety of shapes that a swarm can have at any given moment.

- **Artistic inspiration by geometry in nature**

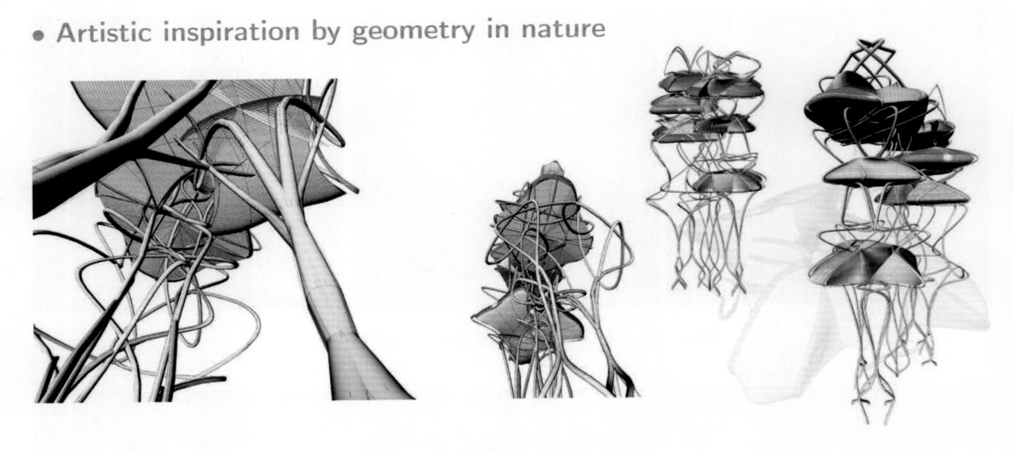

Fig. 13.34 Art and nature

A very interesting avant-garde architectural sketch by Andreas Krainer ("the Emperor and the Queen", Fig. 13.34) does not resemble jellyfish by mere accident, since it consequently follows a geometrical plan of construction that is often found in nature:
This entire sketch is based on *a single spatial curve*, which defines both the position of the beams and girders as well as the form of the hull. Beams and girders in the traditional sense no longer exist, and the transition between them is completely smooth. By a multiplication and a deliberate positioning of the generating curves into a specific pattern (such as a hexagon whose

center and corners contain one copy each), a radial system is created which defines the tower.
Each tower has the same relation to the next as one supporting-beam-element to the next. This produces a repetition of the same principle of arrangement, only at a higher scale. A change of the generating space curve changes the whole sketch, which allows for an unimaginable variety of possibilities. ♠

- Organic architecture

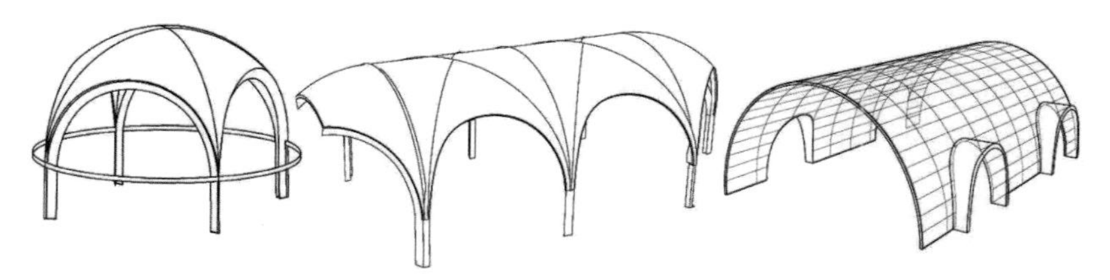

Fig. 13.35 Dome, rib vault, barrel vault

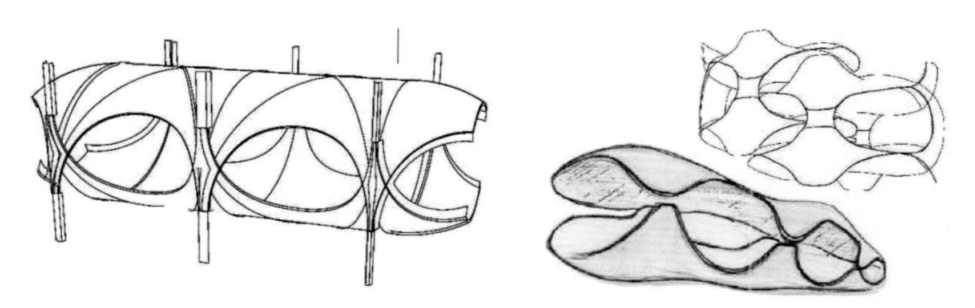

Fig. 13.36 Inspired by the shapes of plants

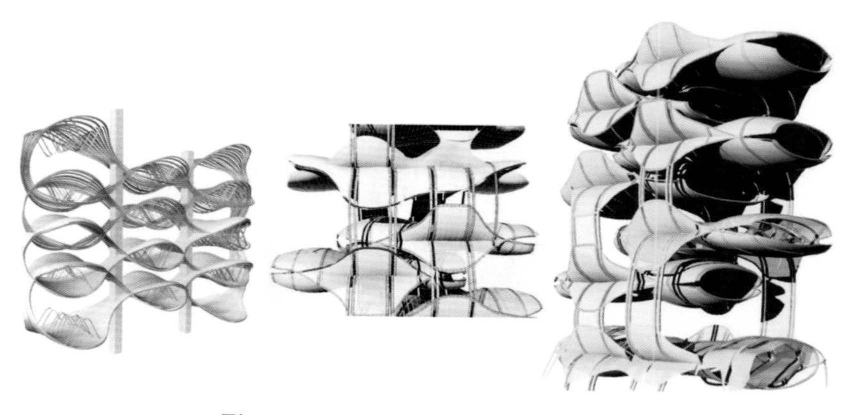

Fig. 13.37 New architectural formations

A geometrically interesting sketch was created by Matthäus Wasshuber. Proceeding from classical vault shapes (Fig. 13.35), which themselves were inspired by nature, he connected his vault-like modules following examples found in the kingdom of plants (Fig. 13.36).

After a longer process of experimentation, he ended up with the sketches shown in Fig. 13.37, which combine shapeliness with the capacity to carry large weights. ♠

• "Amphibious architecture"

The Jules Verne Foundation for Submarines and Deep Sea Robotics is a novel combination of showroom, research facility, and robotic submarine-manufacturing center in Le Havre, France. This unconventional waterfront project (Fig. 13.38, Fig. 13.39) brings spacial elements from the underwater world to our terrestrial surface. Visitors are able to experience firsthand the connection between manufacturing, research, and marine conservation in a highly functional building with an exotic, aquatic atmosphere.

Fig. 13.38 Organic structures inside a building (Sebastian Kaus)

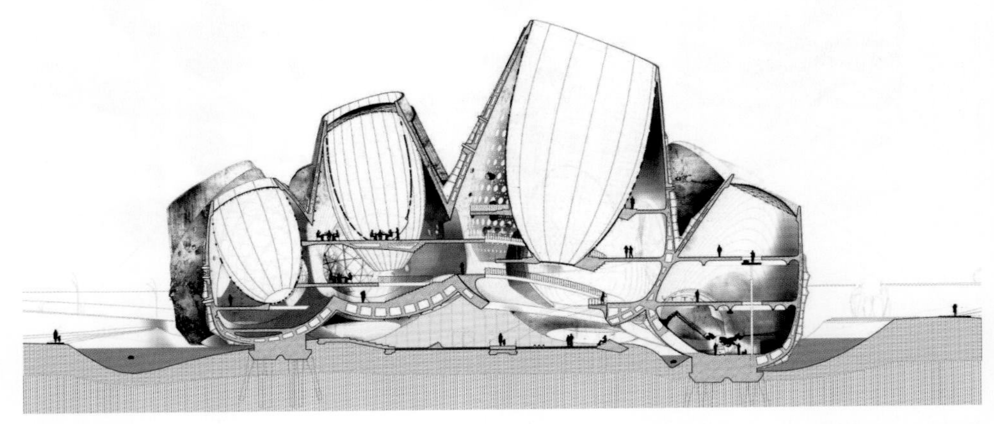

Fig. 13.39 In the virtual cross section of the building, one can see how much geometry is behind the architectural design.

By aggregating, intersecting, and blending highly articulated volumes with a monolithic pier-plinth that interacts with natural tides, the building absorbs a variety of complex programmatic functions while maintaining a high degree of openness and connectivity between key spaces.

Another example of "amphibious architecture" (Fig. 13.40, Science Center Bremen) shows the extent to which architects adapt forms from nature – both on the level of small detail as well as in their large-scale conceptions – in order to create new, *aesthetically pleasing and simultaneously purposeful* forms.

Fig. 13.40 Science Center Bremen (Thomas Klumpp): It not only resembles the shape of a seashell on the outside, but even its inside is organically intertwined (on the left). ♠

• Interplay of massive surfaces and fragile structures

Fig. 13.41 Modern architecture "in context" with landscapes – a challenge (Stephan Sobl)

Fig. 13.41 shows that it can be a challenging task to combine nature and architecture (Hoover Dam Bypass between Arizona and Nevada) and at the same time have an interplay of opaque, massive surfaces that embrace light-weight fragile structures. ♠

13.4 Scaling behavior in nature

To some degree, it is possible to conclude from "looking" at a neutral photograph (without the background or other objects in the vicinity being visible) how big an animal must be: An elephant's legs are so massive because they need to be able to carry its own weight. The tiger, being the largest land predator, also needs to have characteristically large muscle diameters (the tiger's paw is infamous for its sturdiness).

Fig. 13.42 Largest herbivore and carnivore on land: Proportionally speaking, both animals need higher muscle diameters.

Reducing the size by two orders of magnitude (Fig. 13.43) brings us into the world of insects and other articulate animals. Their muscle diameters need not be so large anymore: The weight of the animal has, in fact, diminished cubically in relation to the scaling factor, while the muscular force has only been reduced in square proportion to the diameter.
Interestingly enough, both the tiger and the wasp employ striped patterns – as do many other animals. In the case of the wasp, this may be explained as a warning signal, but on the tiger it is a form of camouflage. Comparable geometrical patterns are here employed in different ways.

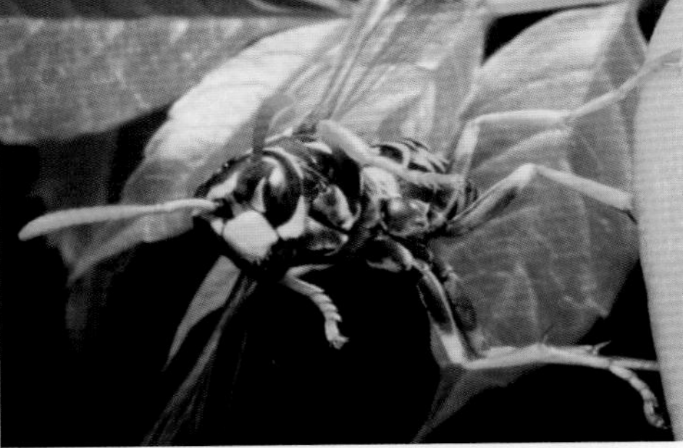

Fig. 13.43 Large herbivores and carnivores at the macro scale: Proportionally speaking, both animals are content with far smaller muscle diameters.

It may be arguable to what degree a giraffe actually resembles the seahorse in Fig. 13.44, but consider the following: Geometrically speaking, there is a limited amount of possibilities of "fitting" volumes and functionality into a confined space. Given the immense variety of lifeforms, one should, therefore, expect to stumble upon certain similarities eventually. Apart from the curious shape of their heads, however, the giraffe and the seahorse share very little in common. They both live in completely different conditions, to say nothing of their different scales.

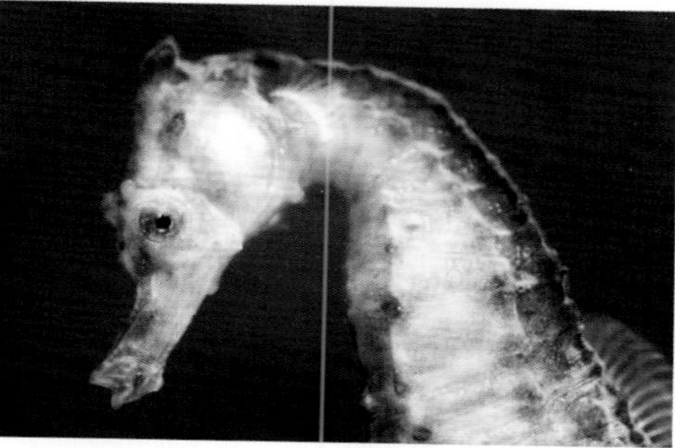

Fig. 13.44 A geometrical resemblance is certainly there – including the mane or the dorsal fin – but that is about it ...

Geometrical similarities often give the misleading impression that the underlying laws that produced these shapes must have been comparable. This may not have been so. We do not even have to mention the strange laws that govern physics at the quantum scale to conclude that, from a mathematical and physical perspective, events in the world of the small do not have to proceed as they do in the world of the large.

Fig. 13.45 Scaling behavior: It is difficult to determine the size of such a stone formation. This causes the left image to look like the feeding of a big lizard.

This scaling behavior is less obvious in inorganic objects: As in Fig. 13.45, it does not seem to matter for a weathered rock whether it is several meters or just a few centimeters tall.
Fractal appearance – the "frayed silhouette", which seems to repeat itself over and over again – occurs both at large and at small scales. It has been proven,

Fig. 13.46 Ice crystals – only slightly enlarged

that it is virtually impossible to ascertain the approximate size of a cloud from its mere photograph.

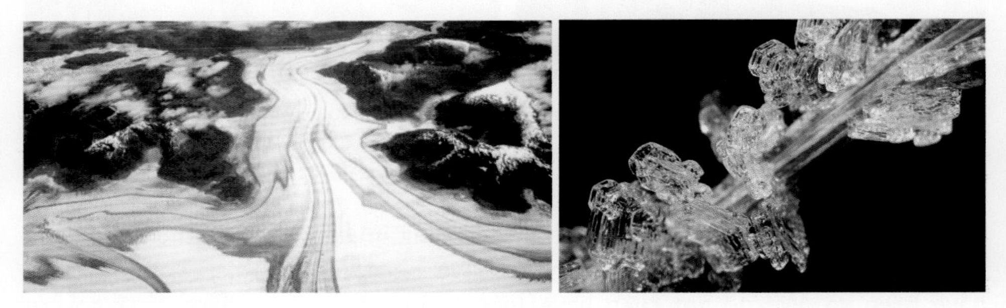

Fig. 13.47 Glaciers and "screwed" ice crystals

When comparing Fig. 13.46 to Fig. 13.47, we may notice that frozen water tends to produce fractal contours of some aesthetic value at every scale. It is no wonder that such forms have always inspired artists to produce novel creations (Fig. 13.48).

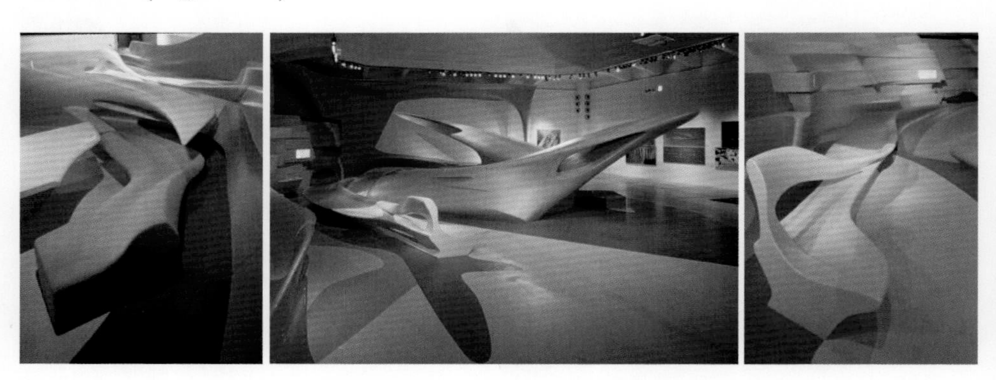

Fig. 13.48 Art and nature: Zaha Hadid: Ice-Storm

While modern architecture is *certainly* concerned with the creation of attractive forms, whose variety is often taken from nature, the ideal case combines mere aesthetics with a realization of certain useful insights. Fig. 13.49 shows the implementation of the "shark skin principle" (application p. 405) in the form of a whole architectural object – in this case, the forms of separate scales of shark skin have been taken as templates. Their advantages in aerodynamics have been proven by realistic simulations.

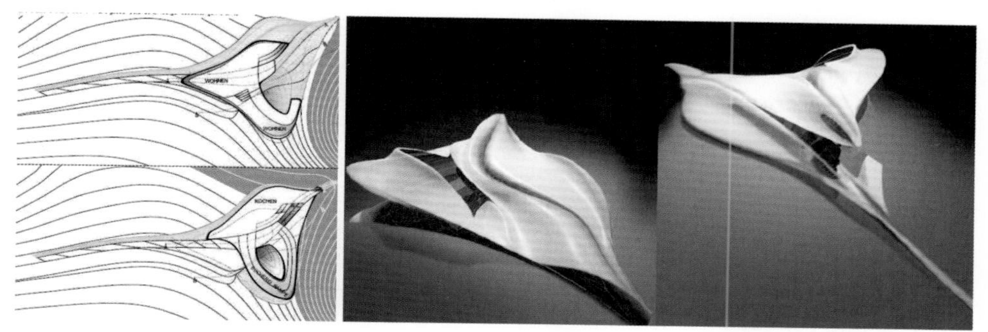

Fig. 13.49 Something else from the Studio Hadid: The scales of shark skin as the underlying principle which manifests itself both in detail (application p. 405) as well as in the whole building, and which is supposed to yield aerodynamic advantages both at great and small scales.

In the architectural sketch by Martin Zangerl for the Centre for Digital and Media Technology in Los Angeles, a prototype of a spatial knot (a minimal surface) is used to transform a voluminous state into "bony", open state (Fig. 13.50 bottom left) by means of variation. This procedure occurs both at a small and a large scale (top left, right) so that parts of buildings with either completely closed or highly open hulls may emerge.

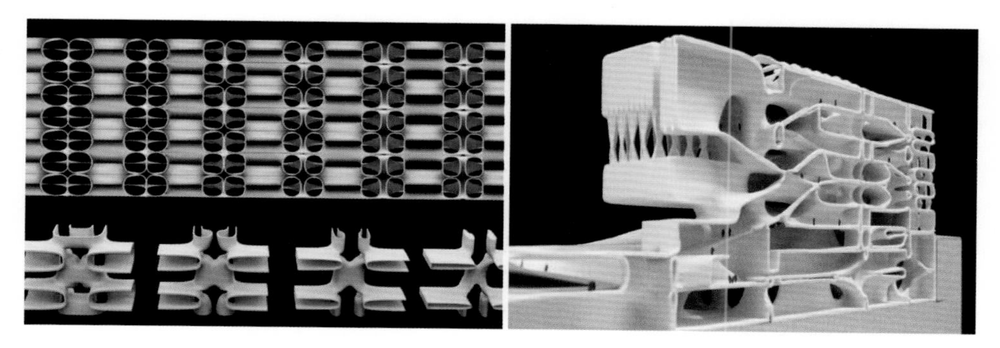

Fig. 13.50 From voluminous museum halls to open offices.

13.5 Musical harmony through the eyes of geometry

Many mathematical and geometrical relationships can be found in music, and we shall here focus on three particular examples. Along with parts of the text itself, these examples are taken from the book *Lexicon of harmonic theory* by Reinhard Amon, Doblinger/Metzler - March 2005, (which in 2006 received the German Music Edition Prize, `http://personal.mdw.ac.at/amon`). A whole appendix of [11] is dedicated to the topic of music. The images were created by the programming library *Open Geometry*.
Readers who do not possess any prior knowledge of music theory are simply invited to enjoy the mathematical balance inherent in music, and the geometric visualizations that it enables[1].

• **Displaying a tonal system on a torus**

A torus is cut into 12 slices (Fig. 13.51). On each slice there are 12 spaces (for reasons of clarity, these are only present once in the image), which may either be interpreted as chromatic-enharmonic locations of the tonal system (12 major and minor keys) or as enharmonically changed tones.

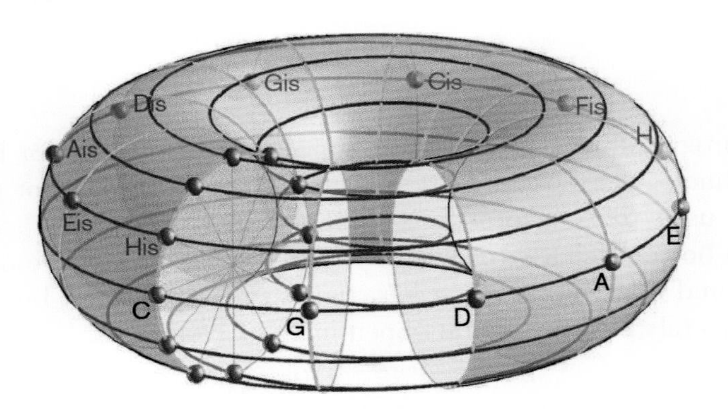

Fig. 13.51 Displaying a tonal system on a torus

In 12 convolutions, the torus is being revolved by a rotoid, which intersects the slides in every point and which produces 12 enharmonically closing circles of fifths. ♠

• **The "grid of keys"**

Fig. 13.52, in the form of a grid of congruent circles, visualizes a circle of fifths with all notes and chords for the respective key, including the circle of the parallel key as well as the subdominant, tonic and dominant scale degrees.

[1]There is a significant difference between English and German notation (`http://en.wikipedia.org/wiki/Key_signature_names_and_translations`):
C, *D*, *E*, *F*, *G*, *A* have the same signature. The letter *H*, however, denotes *B* in English; Instead of *Cis*, *Dis*, *Fis*, *Gis*, *Ais* (German), one writes $C\sharp$, $D\sharp$, $F\sharp$, $G\sharp$, $A\sharp$,
instead of *Ces*, *Des*, *Es*, *Ges*, *As*, *B* (German), one writes $C\flat$, $D\flat$, $E\flat$, $G\flat$, $A\flat$, and $B\flat$.

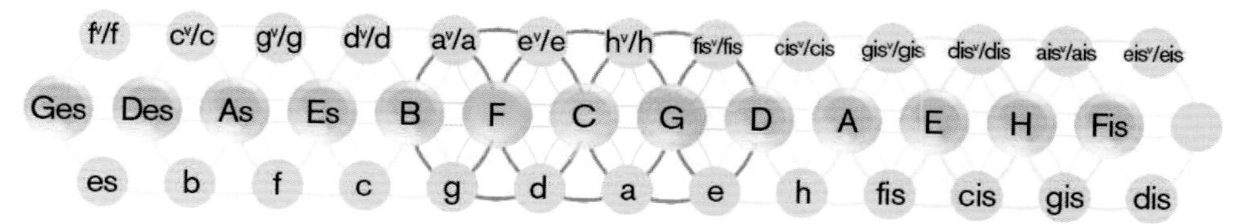

Fig. 13.52 Grid of keys in the plane

The enharmonic closure is attempted in Fig. 13.53. Geometrically speaking, this involves the development of Fig. 13.52 on a cylinder of revolution. A transformation onto a torus or a sphere is also possible. ♠

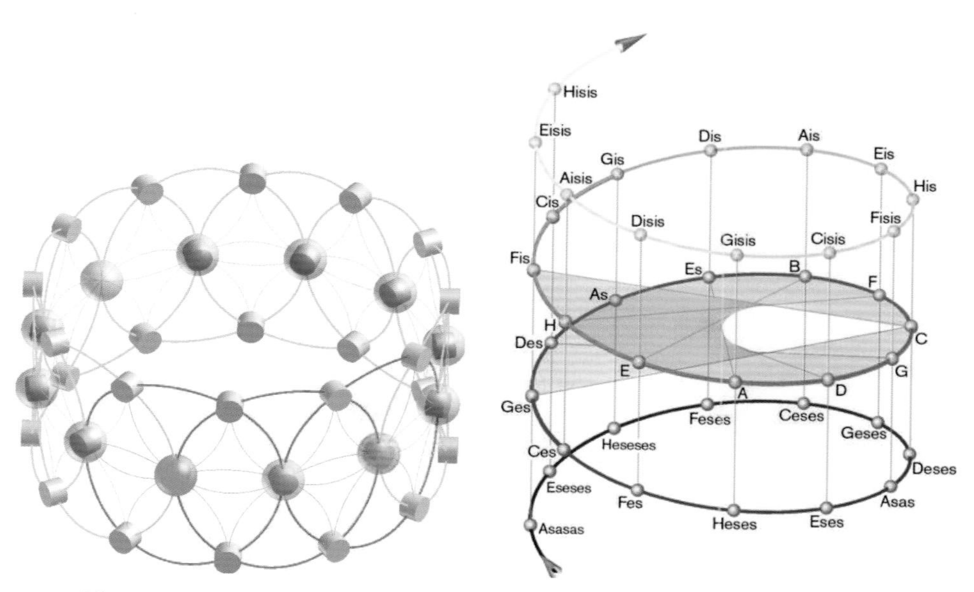

Fig. 13.53 Closed grid of keys

Fig. 13.54 tonal system

- **Three-dimensional circle of fifths**

Fig. 13.54 shows the attempt to connect the closed circle of fifths of a well temperament (red surface) with the "spiral of fifths", which is open upwards and downwards (it is actually a helix, to be geometrically precise). The base tones lie on the central plane (red), as well as above in the ascending spirals of the ♯ and 𝄪 areas (light beige, cream) and in the descending spirals of the ♭ and 𝄫 areas (dark red).
The enharmonically changing tones lie on the vertical green lines. The connecting straight lines of corresponding tones form two oblique, closed helical ruled surfaces (see Fig. 7.27). ♠

- **Tonal centers**

Fig. 13.55 shows the C-major tonality as an example. The columns in the inner circle display the triads which correspond to this scale. In the outer

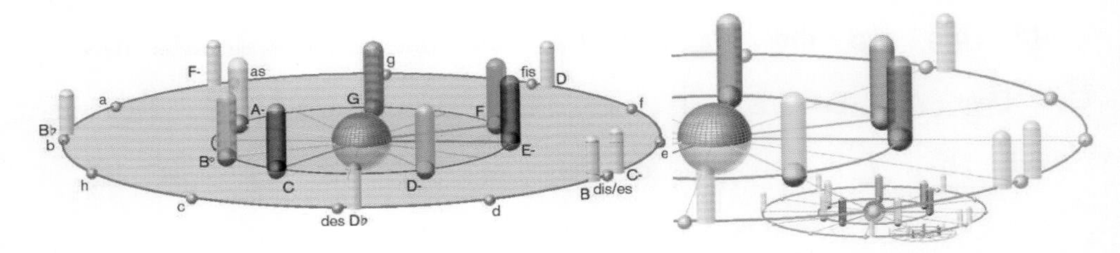

Fig. 13.55 A tonal system with columns. On the right: New tonal center

circle, the typical sounds of the extended C-major tonality can be found on top of the chromatic tones (D-flat-major as the Neapolitan chord, B-flat-major as the double subdominant, F-minor as the minor form of the subdominant, D-major as the double dominant, C-minor as a variation of the tonic and B-major as the dominant of the dominant parallel E-minor). Every sound or tone represented by a column or a sphere may become the new tonal center (Fig. 13.55 on the right), which happens by a procedure of modulation. ♠

• **Inverted retrograde canons**

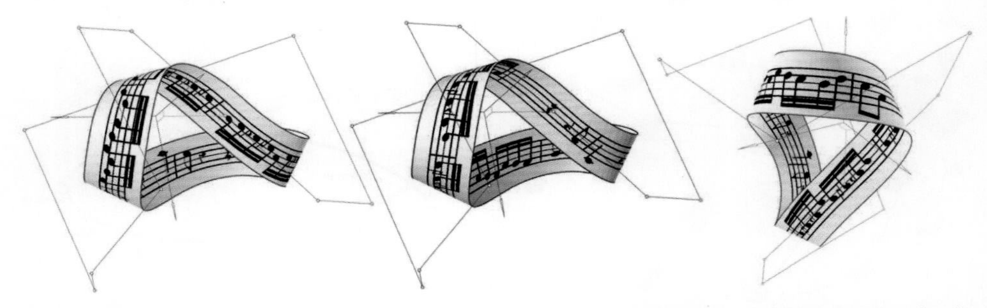

Fig. 13.56 A canon by J.S. *Bach* which may be played forwards and backwards but also by inverting the notes on the central line. After a single playing through the score, the notes have been inverted. The second melody nevertheless harmonizes with the first and should, in fact, be played simultaneously.

Fig. 13.56 shows a Möbius strip designed by Franz Gruber using our own in-house software, on which a "marquee" displaying a canon by J.S. *Bach* has been visualized by means of a texture. An animation of this intriguing object can be found on the website to the book. ♠

A Geometrical free-hand drawing

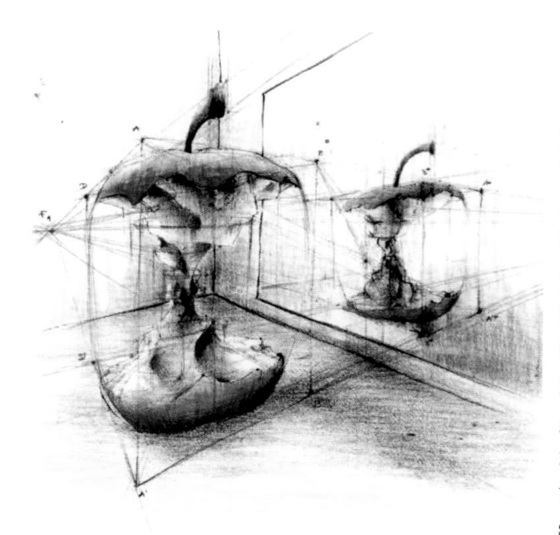

This appendix is intended for people who believe – as I myself do – that despite the quality of today's computer "renderings", correct free-hand drawing is still of principal importance for geometric communication. I would even go a step further than that: Due to perfect computer renderings being the "status quo", great free-hand drawings are becoming an increasingly rare skill – and therefore increasingly more impressive!

This appendix is written so that it may be read *before* acquiring the knowledge contained in the rest of the book. We will, from time to time, make reference to previous contents, but even without these references, the course is largely self-contained. This should permit a "geometric layperson" to produce qualitative correct drawings within a short period of time.

A simple but useful saying goes as follows: "When a problem takes till dawn, an auxiliary figure must be drawn!" As a first step, the solution should be inscribed into this auxiliary figure. The more useful this figure thus becomes (due to geometrical correctness), the easier it is to figure out which steps must be undertaken in order to reach the existing solution.

Many drawings in this appendix were done by the participants of my course about "artistic perspective" – a lot of them received a final artistic finish by Christian Perrelli. An experienced eye may even detect slight errors in the drawings: This, too, contributes to the charm of free-hand drawings, and owes to their origin in our imperfect imagination – and besides, we have already seen enough perfect renderings anyway ...

Survey

A.1 Normal view vs. oblique view

In this appendix, we shall illuminate three types of visualization more closely: The oblique view, the normal view, and the perspective (central view) – the latter being far more difficult than the oblique and the normal views.

Why the oblique view is not our ultimate goal

The section about the oblique view will be the shortest by far.

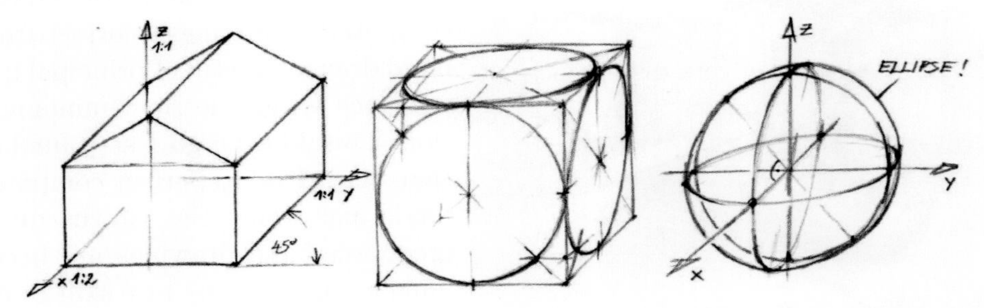

Fig. A.1 A house, a cube with inscribed circles, and a sphere in the oblique projection

Most people have drawn "oblique views" at some points in their lives – most probably oblique projections, also called front views: Assuming a non-distorted frontal view, "polyhedral forms" are thus easy to construct. Straight lines proceeding towards the back are drawn at a constant angle α (such as $\alpha = 45°$), and depths are always shortened by the same factor (such as by half). Thus, the image of a house is quickly constructed by means of a cuboid (Fig. A.1 on the left). What's more, it is easily possible to inscribe the images of circles into the images of squares as in Fig. A.1 at the center. The foremost and the backmost circles remain circular. The ellipses appearing as the images of the other circles are each defined by the four edge midpoints – the edges' images are the tangents. It should, however, be noted that the apexes of the ellipse are not known. It is further not easily possible to specify the contour of a sphere, and if we simply inscribe it without construction (Fig. A.1 on the right), then it is surely wrong!

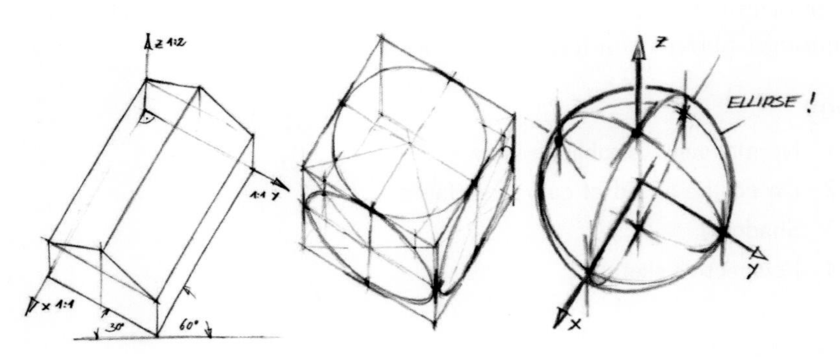

Fig. A.2 A house, a cube with inscribed circles, and a sphere in bird's eye view

A second approach to producing visually transparent oblique views is based on building objects above their undistorted, oblique top views (angle α, such as $\alpha = 30°$). The results of this method can be seen in Fig. A.2. The heights are either shortened (by half, for instance) or undistorted. This yields a bird's eye view.
In the case of cubes with circles inscribed to their faces, the lowest and the highest circles are undistorted, and the remaining are displayed as ellipses bounded by circumscribed parallelograms. As before, the axes of the ellipses are unknown. Spheres leave us with the same problem as before.
As practical and as easily explicable as the above mentioned approaches are – they all share a severe disadvantage:

The usual oblique views (oblique projection and bird's eye view) yield undistorted images. They are, thus, ill-suited to visualize spheres and other surfaces of revolution.

Isometry

In theory, given an oblique view, one may freely choose the images of the coordinate axes and the distortion ratios along them. These assumptions relatively successfully compensate the aforementioned disadvantage.
Let us start with a special assumption which produces a normal view, the so-called *isometry*: Here, any pair of axes' images encloses an angle of 120°, and all segments in the direction of the coordinate axes remain undistorted (actually equally distorted, since isometry literally means "equal metric").

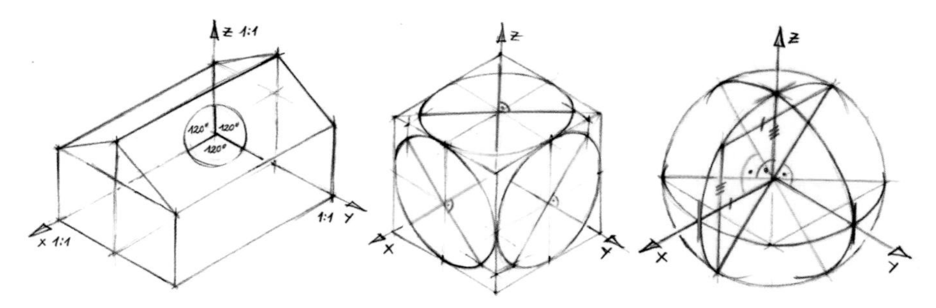

Fig. A.3 A house, a cube with inscribed circles, and a sphere in isometry

In Fig. A.3, our test objects are displayed isometrically. The house now appears to be "more realistic", and the sphere on the right now possesses a circular contour, which obviously is the most significant improvement. The depiction of the cube now has the disadvantage that the images of two opposite vertices of the cube coincide at the center of the image. The ellipses appearing as the images of circles are even congruent. The principal axes of the ellipses are orthogonal to the images of the coordinate axes which gives the artist a great advantage during construction, because the symmetry axes of the ellipses are known.

Due to the strict definition of the axis angles, isometry leaves one no choice when depicting an object. The horizontal base plane is always seen at the same angle of elevation. Let us summarize:

Isometric depictions are well suited for quick and transparent sketches of polyhedral as well as round bodies, as long as they are not exactly like cubes. Furthermore, here is no space to vary the viewing angle.

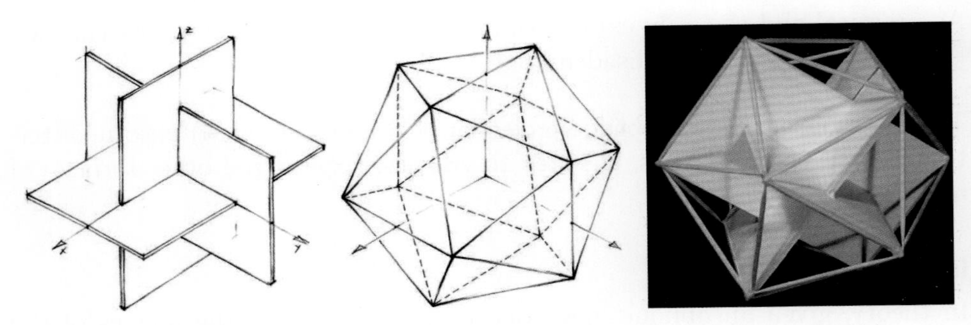

Fig. A.4 Isometry of an icosahedron (on the right: Spaghetti-model by Sophie)

As an exercise, let us draw the isometric view of an icosahedron. Without prior knowledge about its geometric construction, such a body is practically impossible to draw. If one is aware, however, that an icosahedron may be defined by three mutually orthogonal "golden rectangles" (side ratios of $\approx 8:5$), then an optimal isometric view may finally be attempted with confidence.

The advantage of a normal projection

From the theory, we know that the images of the axes may be chosen (almost) freely. In this case, there always exist precisely defined distortion ratios for each axis direction, by which a normal view may be accomplished, and not just a general parallel view. The advantage:

Each normal projection depicts circles as ellipses which are symmetrical to the image of the circle axis – and sphere contours are circles. This causes a natural impression of the objects in view.

In general, one deals with three different distortion ratios for the three axis directions. Apart from the increased difficulty of constructing objects in such a way, the determination of these ratios requires a construction, which a free-hand artist cannot do with an adequate enough amount of accuracy.
We are already aware of a very special type of normal projection: the isometry. There, all distortion ratios are identical due to equal angles enclosed by the images of the axes. This makes a shortening unnecessary, since the result would just be scaled.

However, there exist other choices of angles between the images of the axes which correspond to very simple distortion ratios.

The engineer's view – dimetrical and descriptive

If one tries to find the axes' angles where the distortion ratios correspond to certain values, such as $1 : 1/2 : 1$, then it turns out that (e.g.) the image of the y-axis needs to deviate slightly less than 45° from the horizontal axis – about 42°, to be precise. The image of the x-axis and that of the z-axis are symmetric to that of the y-axis due to identical distortion ratios along x- and z-direction. The image of the x-axis thus deviates by about 7° from the horizontal (Fig. A.5).

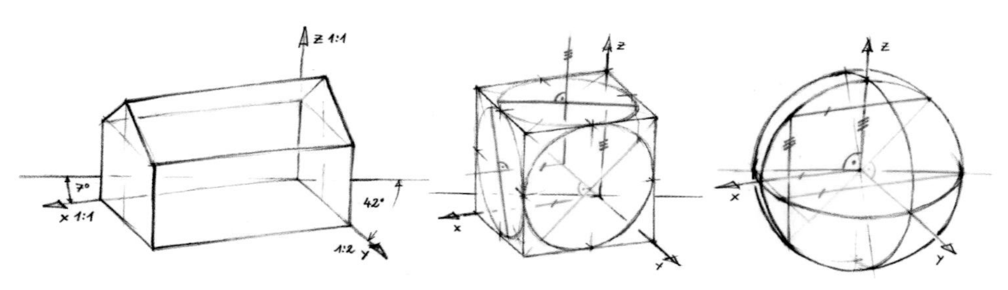

Fig. A.5 A house, a cube with inscribed circles, and a sphere in the engineer's view

This configuration is known as the engineer's view and offers similar advantages to isometry, but it is also well suited for "cube"-shaped objects (Fig. A.5 at the center). Of course, there also exists a symmetric solution where the x-axis forms the aforementioned 42°. The distortion ratio along the x-direction is thus about one half, while distortion in the y-direction does not take place at all.

The general solution

The engineer's view is also a well-defined normal view. Should neither the isometry nor the engineer's view provide a "useful impression" of a specific object, then there is also a general method that, even when drawn free-hand, yields relatively accurate results. The expected result may be well estimated prior to drawing, which gives the artist an additional advantage: It is not our desire to simply iterate through several techniques by "trial and error", but to proceed towards our goal in an economic way.

One starts with the image of a "unit circle" in the horizontal base plane (midpoint U). The principal axis of the ellipse being the circle's image is orthogonal to the image of the z-axis according to the rules of normal projection (Fig. A.6). According to the extent to which the circle's image is "bulgy", one is dealing with a more or less oblique base plane.

In the next step, the image of the x-axis is chosen arbitrarily.

The image of the y-axis is given automatically by the third step: There exists a circle tangent t where the x-axis intersects the unit circle (point X) – in the image, it is a tangent to the ellipse. Due to the fact that the horizontal

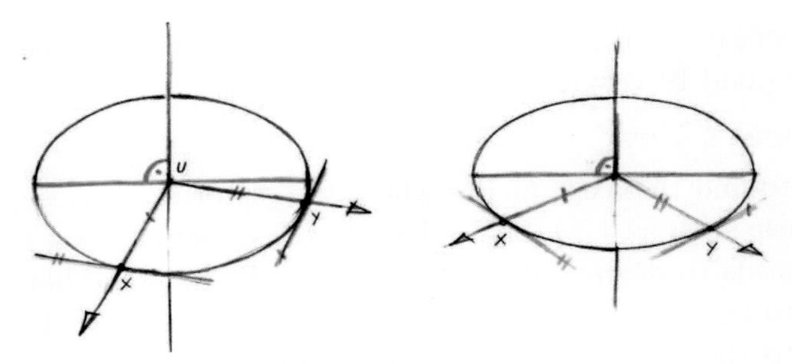

Fig. A.6 Different images of the unit circle

tangent t is orthogonal to the x-axis (both considered as lines in space), it is parallel to the y-axis. The y-axis intersects the unit circle at the point Y. The segments $\overline{UX}$ and $\overline{UY}$ are the "unit segments" on the axes. This already makes it possible to inscribe arbitrary figures to the base plane.

If the images of a horizontal circle and one horizontal axis direction are given, then the image of the missing horizontal axis direction and the distortion of the top view are also defined.

What still remains is the degree by which the heights are distorted. From a qualitative point of view, we can state the following: The more bulgy the ellipse (image of the base circle) is (axis lengths $2a = 2r$ and $2b$), the stronger the z-values are shortened. This statement is instantly quantifiable, since we already know the silhouette of the *unit sphere* about U – it is the circle about U's image that touches the image of the unit circle in the principal vertices.

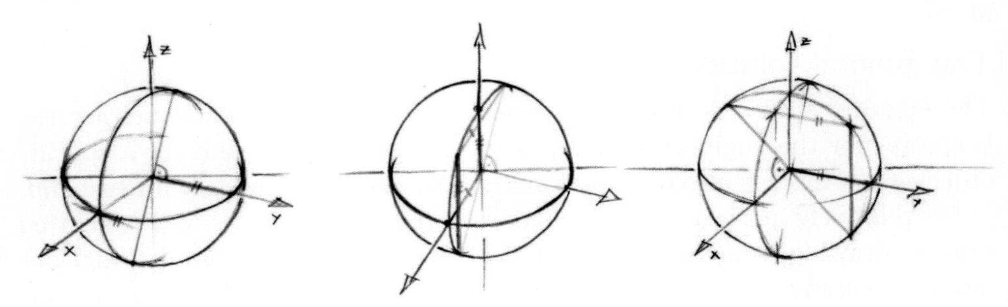

Fig. A.7 General views of the globe

The unit circle is the equator of the sphere. By analogy, we can inscribe the sphere's *null meridian* through the point X. Its axis is the y-axis, and the principal vertices of its image are orthogonal to the image of its axis – just like in the case of the equator. The position of the point X determines the "bulginess" of the ellipse being the meridians image. The image of the north pole Z can be found at the intersection of this image ellipse with the image of the z-axis. $\overline{UZ}$ is the unit segment for the y-coordinates. (For those interested: The image of the north pole can also be found by pivoting a focal point of the image of the unit circle by 90° or by intersecting the sphere's silhouette with the parallels to the image of the z-axis at a distance of $\pm b$ (semiminor axis of the equator's image) – the north pole has an equal height.

A general normal projection is best defined by the image of the unit sphere's equator. By inscribing the sphere's null meridian, its north pole can be determined – and thus, one has full control over all distortion ratios.

In order to consolidate the aforementioned in our minds and to improve our skill of drawing ellipses, we shall now sketch general views of the globe (Fig. A.7).
This type of "construction" that considers the right distortion ratios has the advantage that we can specify the direction from which the drawn object is seen. The images of the equator and the null meridian of the unit sphere give a good impression about the appearance of the rest of the image. Such a test sketch can also be undertaken for the image of the "unit cube".

- Circles with axes parallel to the coordinate axes

A scene is to be sketched that consists of several cylinders of revolution (Fig. A.8).

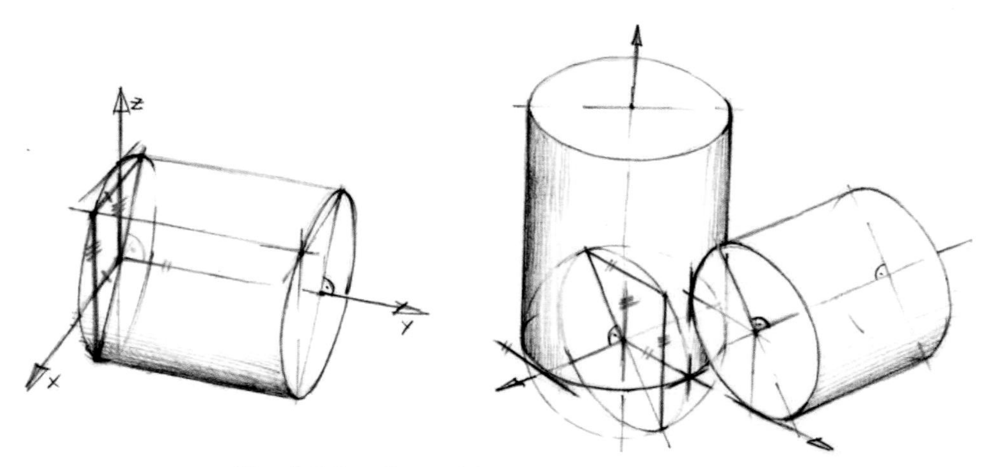

Fig. A.8 Standing and lying cylinders of revolution

The images of the top circles of an upright cylinder of revolution are always ellipses, whose principal axes are orthogonal to the image of the z-axis. Lying cylinders are to be depicted by the same way:

- The images of the boundary circles are ellipses, whose *principal axes are orthogonal to the image of the axis of the cylinder.*
- The diameter of the cylinder is equal to the length of the principal axis.
- The search for one further image point will show how bulgy the circles' images actually are. If, for instance, the bottom cap or the top cap lie in a

plane parallel to the yz-plane, then such a point can be found by parallel translation of the images of the y- and z-axes through the principal vertices of the ellipse showing up as the circle's image.

This construction is plausible: Let A and B be those endpoints of a diameter of the circle k whose images are the principal vertices of the ellipse k^n. Both axes y and z are orthogonal lines in space so that the intersection of the parallels through the endpoints A and B lies on k according to *Thales'* theorem. We thus get a point on the ellipse appearing as the image of k. In cases of symmetric axis positions, we have also found the auxiliary vertices of the circle's image. ♠

A.2 Do not be afraid of curved surfaces

At first unfamiliar ...

Free-hand drawing is strange at first: Instead of employing a ruler at every turn, we force ourselves to draw straight lines by hand. No doubt, we would also feel more confident if our circles were made by compass. It does, in fact, require a considerable amount of experience to draw a circle with a given radius by free-hand.

... but soon highly advantageous!

Except for sphere contours, perfect circles rarely occur during free-hand drawing. One most often deals with ellipses and is thus, as a free-hand artist, at a definitive advantage compared to those who have to rely on rulers and compasses. The true advantage lies in the fact that due to some spatial imagination and geometric aptitude, we can quickly surmount many obstacles that would needlessly occupy a "constructing" artist for a much longer time. Further, it is often a matter of fact that an exact construction is much easier *in hindsight*, having previously "estimated" the result in a sketch.

Free-hand drawing that is based on geometric facts is an important aspect of the later derivation of exact solutions.

Viewed in this light, every person that deals with geometry should be trained in this craft – even in the absence of any artistic aspirations.

It does not get any more difficult

The next advantage is that the transition from polyhedral building blocks to elegant, curved surfaces can proceed smoothly – no pun intended – without a very apparent increase in the degree of difficulty.

During construction with ruler and compass, we quickly hit a brick wall and usually fail due to our "tools"! During free-hand drawing, however, we may concentrate on our geometrical thinking.

• One-sheeted hyperboloid of revolution

The surface can be generated by the rotation of a straight line about a skew axis. There exists a second family of straight lines on the surface. The contour and the two families of straight lines are to be depicted.

Let us start with a horizontal base circle, which we want to subdivide into eight equal parts. This may be done by first selecting an arbitrary x-axis. The orthogonal y-axis is parallel to the tangents of the x-axis-points on the base circle. Both axes define a square which is circumscribed to the circle. Its diagonals, along with the coordinate axes, yield the evenly distributed points $1, \ldots, 8$ on the base circle.

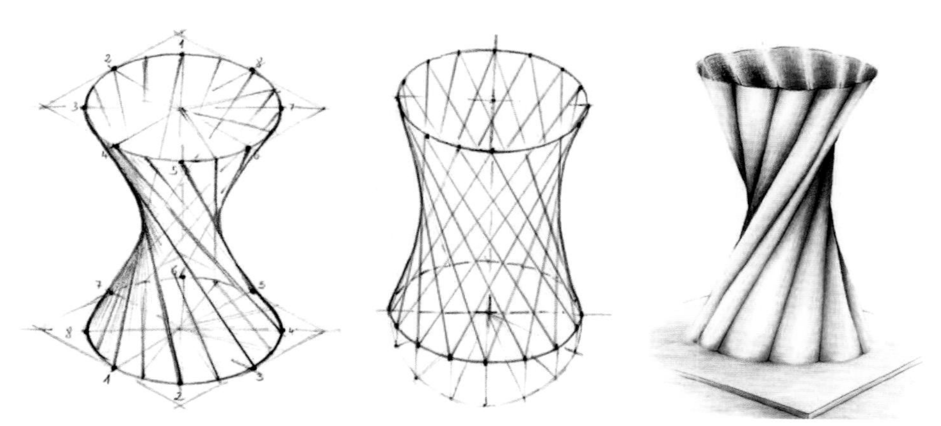

Fig. A.9 Variations of a one-sheeted hyperboloid of revolution

Let us depict a congruent cap circle at an arbitrary height. Then let's lift the points $1, \ldots, 8$ to (the image of) the cap circle and shift the labels of the "upper" points by – for instance – three points. This yields eight straight lines on the surface and eight tangents onto the contour hyperbola. By lifting the eight points into the other direction, we may derive eight further tangents. Thus, the contour seems to be adequately defined. During the final composition of the image, one should consider that a change of visibility occurs at the contour.

A variation may be given by using the affinity of the base circle image with the parallel-rotated circle position. A circle may easily be decomposed into equal parts. ♠

• Beam connection (Fig. A.10)

Two recumbent cylinders of revolution with equal radii and orthogonally intersecting axes are to be depicted.

Let the x- and the y-axis be the axes of two cylinders. It turns out to be helpful if the images of the axes of the cylinder axes are chosen as unsymmetrical as possible relative to the image of the z-axis – just as in the engineer's view. There is a sphere inscribed to both cylinders simultaneously, which is centered at U. We shall try to depict this sphere at first.

The contour generatrices of both cylinders touch the circular contour of the sphere. Let us cap them off with circles by picking their midpoints on the cylinder axes at arbitrary positions. The principal axes of the corresponding ellipses appearing as the images of the circles are orthogonal to the images of the cylinder axes – further points may be derived by a parallel translation of the axes.

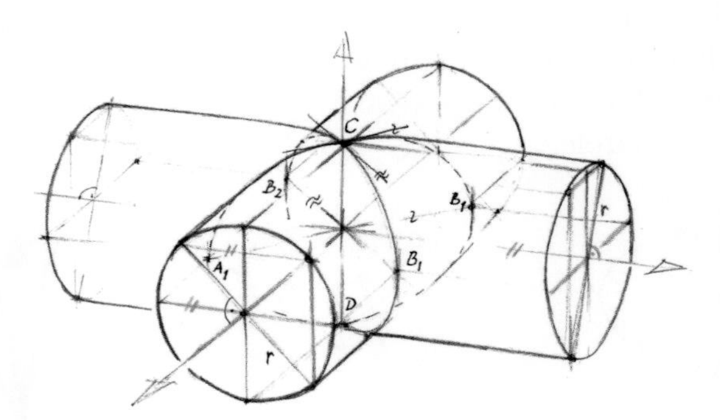

Fig. A.10 Beam connection

The intersection of the cylinders consists of two congruent ellipses. One must consider that an arbitrary (non-circular) ellipse in space may also be depicted as an ellipse, but that – in general – no axis of the ellipse will be orthogonal to the image of any coordinate axis. The objective will, thus, consist of determining as many points and tangents of the ellipse as possible.
Let us consider the highest and the lowest generatrices of both cylinders. They intersect each other in two points C and D. The generatrices which lie in the xy-plane intersect in the points A_1, A_2 and B_1, B_2, which are symmetric with respect to the origin. At those locations, the ellipses which form the intersection of the cylinders possess z-parallel tangents. C and D belong to both ellipses. The corresponding tangents are parallel to A_1B_1 and A_2B_2. As an additional information, it should be considered that the ellipses' images touch the contour generatrices of the cylinder. The respective contact points are of great importance, because they announce the transition from the visible to the non-visible part of the curve. ♠

• **Pencil tip** (Fig. A.11)
A regular hexagonal prism is to be conically "tapered" (which happens when the pencil is sharpened).
The vertical faces of the prism will intersect the cone of revolution along to hyperbolas. From these, we shall – at the very least – locate the highest points.
Let us choose the horizontal base circle of the cone of revolution, into which we shall inscribe a regular hexagon. The tip may be chosen freely on the z-axis. The tangents from the image of the tip to the image of the base circle are the contour generatrices of the cone. The choice of the x-axis (point X on the

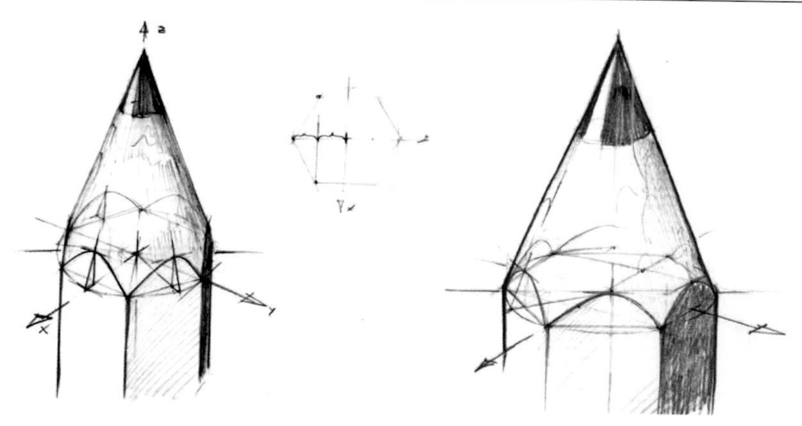

Fig. A.11 Pencil tip

circle) entails the direction of the y-axis (parallel to the tangent in X). This time, the images of the axes may very well be symmetric. The neighboring points of the hexagon can be derived – relying on a well known property of the regular hexagon – by intersecting the y-parallel with the circle through the midpoint M of UX. The remaining points are given by mirror-symmetry with regard to U.
In the yz-plane, the highest point H of the section hyperbola can be constructed directly: It lies precisely above the side midpoint and on the generatrix of the cone. All other apexes are located equally high above the remaining midpoints of the sides. The corresponding tangents are parallel to the hexagon sides. During sketching, it should be considered that two of the six arcs touch the contour of the cone and alter from the visible to the non-visible part of the surface.
The pencil lead shares circle with the cone, which is located at the appropriate height and is coaxial with the base circle.
During free-hand drawing – as well as during precise construction – one may always stumble upon inconvenient perspectives (such as projecting intersections). For instance, the attempt in Fig. A.11 on the left may not be optimal. One may be consoled by the fact that free-hand drawing tends to be quick. There is never a lot of time lost, and one may, therefore, always start anew. ♠

• Helical stairs (Fig. A.12)
In a front view, a helix with an upright axis appears as a sine curve. It is easy to draw such a curve: It is created by means of a harmonic oscillation and a simultaneous uniform translation in an orthogonal direction.
Let us draw such a curve. In order to simplify our task, let us also draw the contour generatrices of the carrier cylinder, whose perpendicular bisector is the image of the axis of the helix (Fig. A.12 on the left). The curve's period length equals the *pitch* of the helical motion. As an exercise, let us also depict a second helix, which is created by reducing the radius of the original helix to about a quarter. It appears as a sine curve with the same frequency but

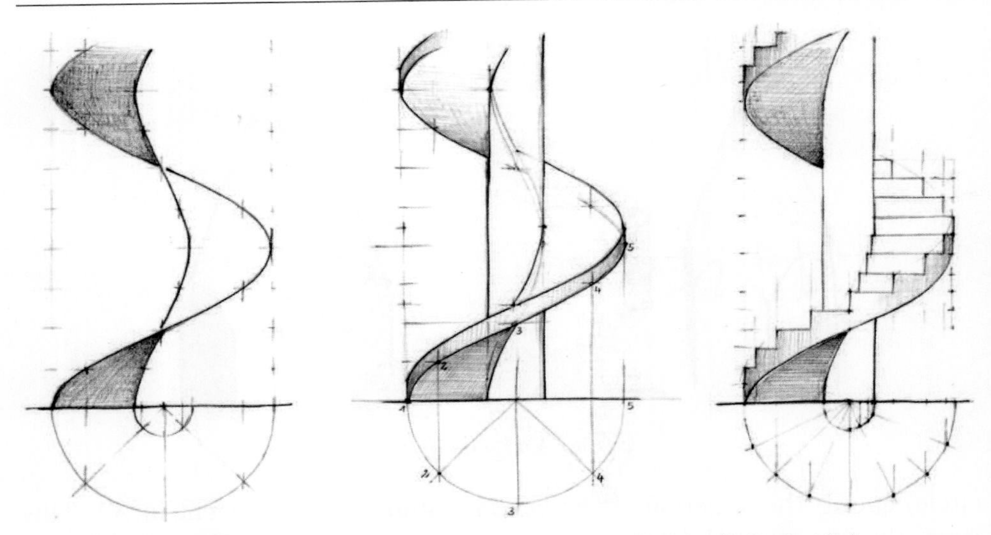

Fig. A.12 Normal view and front view of varying helices (helical stairs)

a smaller amplitude. In Fig. A.12 at the center, a variation of the original is visible – a helical staircase appears on the right. In order to accomplish roughly 16 equally distributed points on the helix, we assume the planes that correspond to a height difference of 1/16 of the pitch. The 16-part-subdivision can be found during construction by repeatedly bisecting the pitch. ♠

• **Surfaces of revolution**

Surfaces of revolution may be produced by a rotation of the meridian, during which the points on the meridian move on the circles of latitude. Tangents of the meridian circle generally generate cones of revolution with their apexes on the axis of rotation. (In some cases, the cone degenerates to a cylinder of revolution which is coaxial with the surface of revolution. The cone may also collapse to a plane which is orthogonal to the axis of rotation.)

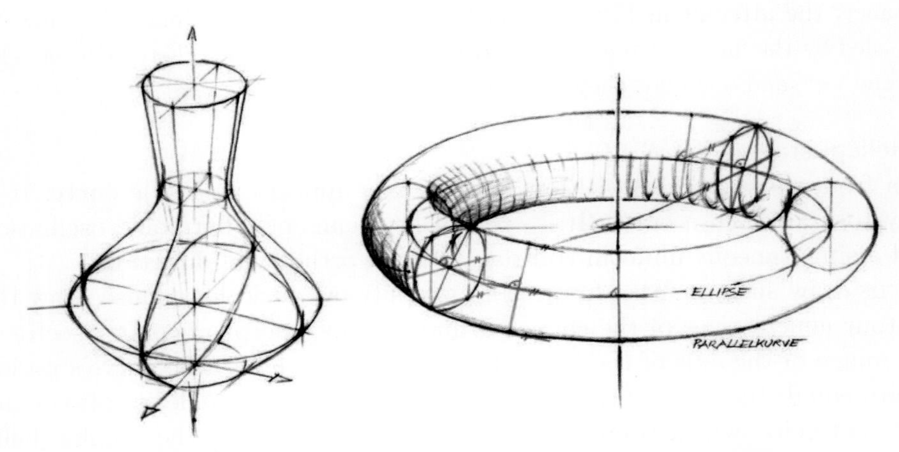

Fig. A.13 Surfaces of revolution: a torus can be seen on the right.

In order to sketch the given surface of revolution in a visually accurate way, it is paramount to inscribe the meridian first including the axis of rotation (Fig. A.13 on the left) – which may, for the purpose of control, include the mirror-symmetric meridian –, and to denote special points on it: These are, above all, the points with axis-parallel tangents but also any inflection points which may arise. The tangents in the boundary points are also important. The normal views of the respective circles of latitude are now drawn in the normal projection (Fig. A.13 on the right). The respective heights are shortened in relation to the obliqueness of the axis. If the obliqueness is only slight, one may even use the same heights during free-hand drawing. We use the images of the apexes of the cones to draw the tangents at the ellipses that are the images of the corresponding circles of latitude. The contact points correspond to the contour of the surface of rotation – the tangents of the ellipses are also tangents of the outline of the surface. At the points with minimal or maximal distance from the axis of rotation (where touching cylinders of revolution occur), the principal apexes of the image ellipse are the contour points of the surface. ♠

- Blossom of a lily

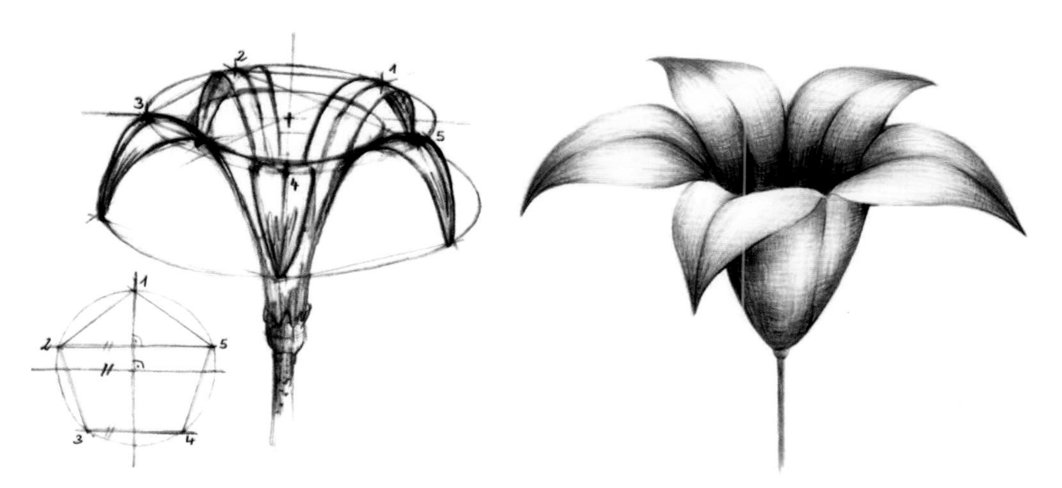

Fig. A.14 Blossom of a Lily

The advanced or the artistically ambitious may try their hand at the following example: A bell-shaped surface of revolution (Fig. A.14) should be "serrated" so that something similar to a bellflower or Lily emerges – and since nature itself rarely produces absolutely perfect surfaces, we are allowed to be a little bit imprecise in our free-hand drawing. It is, nevertheless, advantageous to first sketch the surface of revolution without the indentations, and to fix the silhouette thus.

In another step, we subdivide the boundary curve into five approximately equal sections (nature seems to have a penchant for five-fold symmetry)

and estimate the corresponding meridian through the points $1, \cdots, 5$ of the subdivision.
In a last step, we inscribe the circle of latitude, which shall limit our indentations. The midpoints of the points defined by the intersections with the special meridian are the endpoints of the section lines. With some care, these points are now to be connected to the starting points. The images of the section curves may in some cases touch the silhouette. If a silhouette of the flower remains, then it is formed by parts of the original silhouette of the surface of revolution. ♠

A.3 Shadows

In order to add a realistic appearance to our free-hand sketches, we may now add shadows, but only carefully: Shadow constructions are complicated and therefore "dangerous". A realistically drawn *false* shadow can do more damage than its desired contribution towards the clarity of the image. We shall thus limit ourselves to shadows of simple scenes. Furthermore, it is better to start one's shadow exercises with sunlight (parallel shadows), instead of using light sources directly within the scene.
Convex bodies possess neither ridges nor indentations. Thus, they do not cast shadows on themselves. Cuboids, cylinders of revolution, and cones of revolution are simple examples. For these three prototypes, we shall now attempt to sketch parallel shadows in simple scenes.

- Parallel shadows of a cuboid

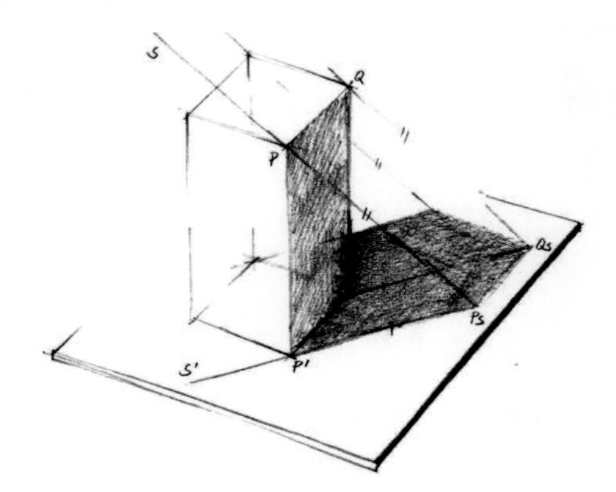

Fig. A.15 Cuboid with shadow

Let us choose the light position – ideally, by choosing the shadow P_s of a point P of the top surface. All light rays are then parallel to the direction PP_s.
Furthermore, all top views of the light rays are parallel to the connection of the top view P' of P with the shadow P_s in the base plane. Thus, we can

construct the shadow of a further point Q of the cap surface by drawing the parallel to the light ray through Q and the parallel to the light ray's top view through Q'. The triangles $PP'P_s$ and $QQ'Q_s$ are similar in any case and even congruent in the special case of equal heights. PQ and P_sQ_s should be parallel and of equal length, because P is parallel to the base plane. (The "light plane" through the edge is intersected by two parallel planes.)
The shadow provides important information about the cuboid's *penumbra*. We will try to rely on our spatial imagination during its construction.
A slight hint for the shading of a cuboid: Surfaces in penumbra are "slightly darker", but they, neverless, exhibit the surface structure (such as a building wall). Shadows, on the other hand, are "no longer aware" of their origin and exhibit the structure of the surface on which they're being cast (such as a lawn or asphalt). ♠

• Parallel shadows of cylinders and cones

Shadows of cylinders of revolution and cones of revolution are almost as easy to draw as those of cuboids.

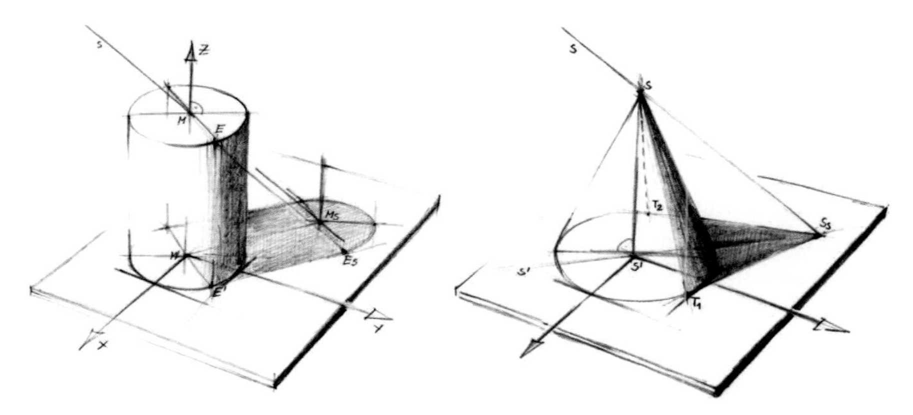

Fig. A.16 Cylinder and cone with shadow

The direction of the light can be given by defining the shadow of the center M of the cap circle or the cone's apex S. We thus obtain the characteristic triangle once more, which can be employed for the construction of further shadow points. In the case of the cylinder of revolution, the shadow of the cap circle should be centered at the shadow M_s of M: It is congruent to the image of the cap circle. The shared tangents of the elliptic image of the base circle and of the shadow of the top surface provide the straight-line component of the shadow boundary. The penumbral shadows that are parallel to the axis start in the contact points with the base circle. In the case of the cone, one only needs to draw the tangents to the image of the base circle from the shadow S_s of the apex S, producing the straight shadow boundaries. From the contact points, the straight penumbral boundaries then run towards the cone's apex. ♠

A.4 Perspective sketching

There is an entrance exam for the University of Applied Arts Vienna, where, among other parameters of artistry, the drawing skill of the applicant is to be determined. Years ago, there used to be an exercise that such applicants had to accomplish: A staircase of the University was to be sketched. Fig. A.17 shows a fisheye photograph of this staircase (from above). Is it even possible to sketch it correctly?

Fig. A.17 Overview with the fisheye lens – general and special perspective

Reducing the level of difficulty

A wide-angle photograph can be seen in the middle. The task does not seem to have been made easier. Slight errors during construction may not even catch the eye. It surely does no harm to simplify the view slightly (on the right). No examiner would require that every point be geometrically transparent on a candidate's first attempt. first attempt. On the other hand, systematical considerations may very well improve the result.

The method of renaissance artists

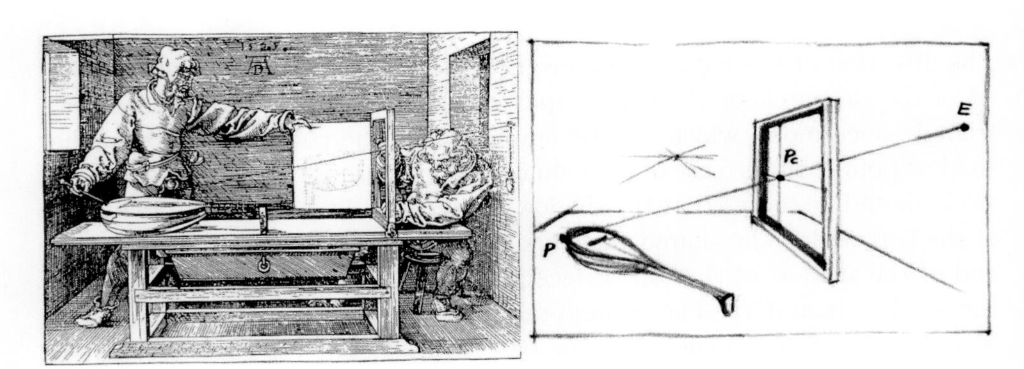

Fig. A.18 Wood engraving for the *intersection method*

Perspectives (central views) are the result of a central projection from an *eye point* to the image plane. This serves to simulate the impression of natural sight.

In a simple wood engraving, Albrecht Dürer showed to his contemporaries, and to the generations that would follow how lines of sight may be intersected with the image plane:
The viewing ray appears with help of a taut thread through a loop E – the eye point. One person places the thread at the point that is to be displayed. Another person marks the intersection point with the image frame by means of a crosshair. By closing the lid for a short period of time, the image point may be fixed.

The most important terms

In order to proceed with our analysis, we need to employ some special "terminology": The eye point E is located at a distance d in front of the *principal point* H of the perspective. The straight line EH is the *viewing axis* or *principal ray*. A straight line parallel to the image plane π is called the *frontal line*, and a straight line orthogonal to π (and thus parallel to the viewing axis) is called the *depth line*.
The following theorem is important for drawing and may be applied here:

Perspective images of straight lines intersect, with the exception of frontal lines. Depth lines vanish in the principal point.

Frontal lines are very practical during perspective:

Affine ratios are preserved on frontal lines. The angle between two frontal lines remains undistorted in perspective.

The horizontal plane, on which we stand, is called *base plane* – its intersection with the image plane is called *base line*. The location of our feet is called *standpoint*, and the height of the projection center is called *eye height*. The *horizon* is produced when all points at infinity of the base plane are projected to the image plane. Thus:

The perspective images of horizontal parallel straight lines intersect at the horizon. In cases of horizontal perspective, the horizon contains the principal point.

Frontal perspective

Let us now connect the scene to a Cartesian coordinate system with three axis directions x, y ,and z, whith the xy-plane serving as the base plane. The z-axis is then upright.
If the principal ray (the viewing direction) has the direction of a coordinate axis, the image plane is parallel to the respective perpendicular coordinate plane. This is called a *frontal perspective*. Two axes are frontal lines and span a *frontal plane*, in which figures are undistorted (Fig. A.19). The third axis is

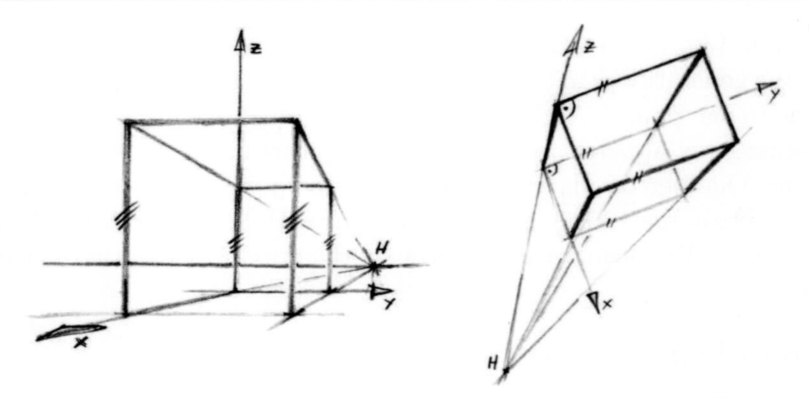

Fig. A.19 Two special positions of the coordinate system. On the left, the yz-plane is frontal, at the right the xy-plane is frontal.

the depth line and vanishes in the principal point. Aside from the "contents" of the frame, Fig. A.18 is a classical example of frontal perspective.

- **The "square-grid method"**

A method that was popular in the Renaissance approximated a complex top view by means of a square grid.

Let us tile the base plane with square tiles, starting at the base line. This makes the sides of the squares either parallel or orthogonal to the image plane. In perspective, the image of this grid is to be drawn as follows: Firstly, the points on the base line equal their perspective images – and secondly, the depth lines all vanish in H. We may still freely choose the vanishing point D_1 of a diagonal direction – the other vanishing point D_2 is symmetric with respect to the principal point H.

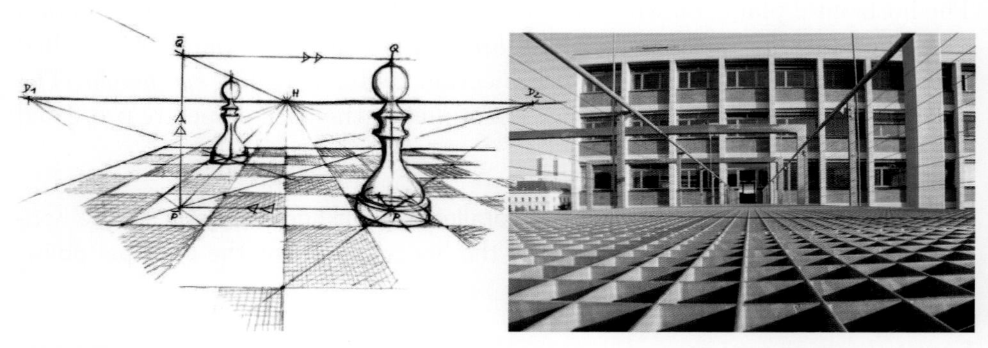

Fig. A.20 Square-grid method. On the left: The classical method from the Renaissance. On the right: A wide-angle photograph which clearly depicts the vanishing points of the square diagonals (the photo exhibits a slight "barrel distortion", which often occurs in lenses of a very short focal length).

The distance of the points D_1 and D_2 from H is equal to the distance $\overline{EH}$ of the image plane's eye point (we have drawn parallels to all 45°-straight-lines through the eye point E). For this reason, these points are called *distance points*.

We may now "interpolate" the top views of arbitrary points. The appropriate height may be drawn as follows: The top view is to be translated orthogonally to the base line (which – in perspective – equals a projection from H to g). Above the top view that was projected to g in this way, we may now measure the undistorted height and finally project it back. ♠

Perspectives with a horizontal viewing axis

We no longer wish to look in the direction of a coordinate axis, but nevertheless let the viewing axis remain in a horizontal position. The upright straight lines still remain as frontal lines, and their images are mutually parallel. The x-axis and the y-axis vanish at two points F_1 and F_2 on the horizon.

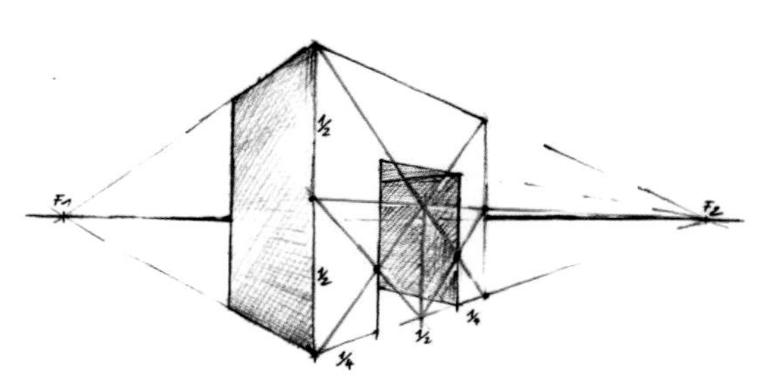

Fig. A.21 Subdivision of segments (special multiples)

Fig. A.22 Cuboid objects

Cuboid objects, which are positioned with many edges parallel to one coordinate axis (Fig. A.22), are very easy to sketch. Vertical segments should be drawn, as previously described, by using orthogonal projection to the image plane. They can also easily be subdivided. One particular wish may be difficult to grant during free-hand drawing: The desire for *exact* drawing of segments along the coordinate directions. However, since there are simple methods for drawing affine ratios in a perspective view, this may not be a show stopper.

Fig. A.21 shows how fast midpoints or the doublings of horizontal segments may be derived by drawing diagonals into overlying vertical rectangles.

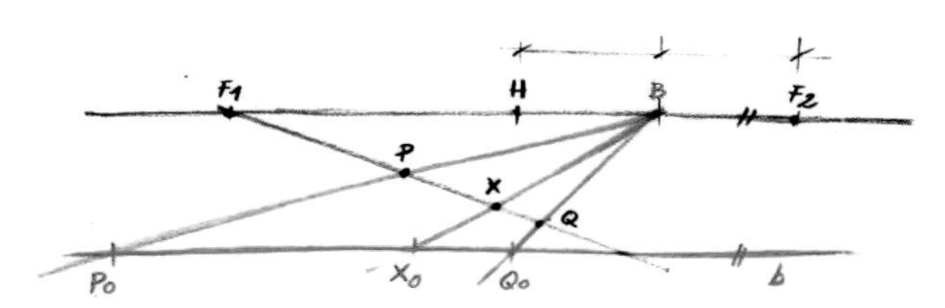

Fig. A.23 General subdivision of segments

The general method is described in Fig. A.23: An arbitrary auxiliary vanishing point B is to be selected on the horizon, and the segment PQ is to be translated on a parallel to the base line. The result is a segment P_0Q_0 which may not be seen undistorted (in general), but which may be employed to inscribe the desired affine ratio. One may afterwards move back again. (This method works because it can be interpreted as a parallel projection in space.) A hint: P_0Q_0 is the more likely to appear undistorted, the nearer B is to the so-called *measurement point* of the segment (it is the vanishing point of the rotational chords during parallel rotations with respect to the image plane). During sketching, one may approach the measurement point of the x-direction (with its vanishing point at F_1) by approximately bisecting the segment between the principal point H and the vanishing point of the y-direction (F_2).

Fig. A.24 Roman aqueduct and a variation by Zorica Nicolic

We may now already draw the images as in Fig. A.24 on the right – though we should not be afraid of curved lines. The elliptic images of the circular arcs are well determined by their vertical starting tangents and the horizontal tangent at the highest point, which in the image passes through a main vanishing point!

When sketching curves, one should take care, if possible, to draw the tangents at endpoints first, as well as the tangents of special points (such as the highest and lowest point).

In case the perspective needs to be shaded: It is always an advantage to follow the directions that lead to the main vanishing points. This strengthens the "vanishing point appearance", which is so crucial for our spatial understanding.
The "hanging bridge" in Fig. A.25 was produced in the same way. For illustrative purposes, the tangents at the points P and Q of the catenary and its intersections T are inscribed exactly below the lowest point.

Scaffolding

The capturing of curved lines by means of points and tangents (line elements) is an important and legitimate part of sketching. Whenever we wish to depict

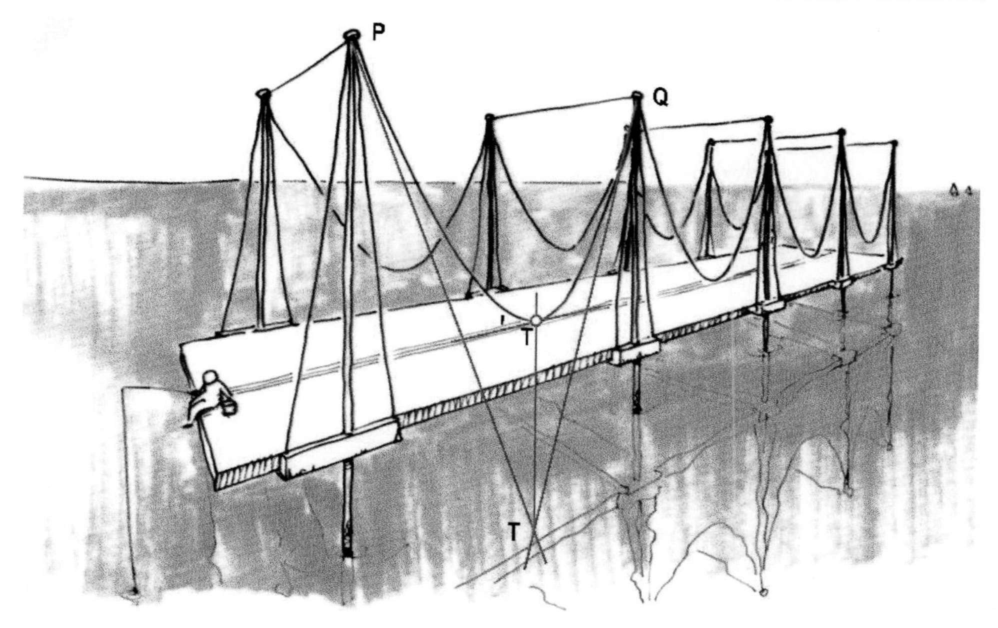

Fig. A.25 Catenaries, wherever one looks

complicated objects, it is always advisable to imagine a sort of scaffolding that captures the locations of important points of the object.

Fig. A.26 On the left: An almost undistorted bicycle (frontal perspective). Center: A usual view. On the right: Strongly distorted (being photographed by a lens of very short focal length).

Let us consider a bicycle (Fig. A.26), which – to be honest – is not even that complicated, since it essentially "lies in a single plane". If the bicycle pole is imagined to be horizontal, then its end points, along with the midpoints of the wheels, form almost an equilateral trapezoid. The pedals are located roughly halfway between the hubs. This is already a good starting point, and once it is sketched, the "rest of it" appears almost by itself. Fig. A.27 shows two bicycle variants by Stefan Wirnsperger and Otmar Öhlinger. The following must be considered when drawing the wheels:

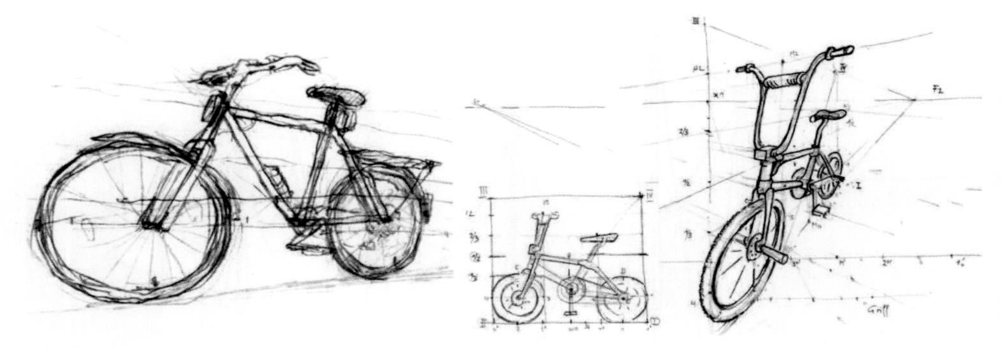

Fig. A.27 Free-hand drawing of bicycles

Rules for the depiction of circles in perspective:

- The image of a circle is an ellipse (in general, and sometimes also part of a hyperbola), and thus, it possesses two symmetry axes. The center of the circle is *not* seen at the center of the image curve (ellipse or hyperbola).
- One should, if possible, sketch the images of the foremost and backmost points of the circle, including their perpendicular tangents, and also the highest and lowest points. The images of the corresponding horizontal tangents pass through the vanishing point of the horizontal diameter.

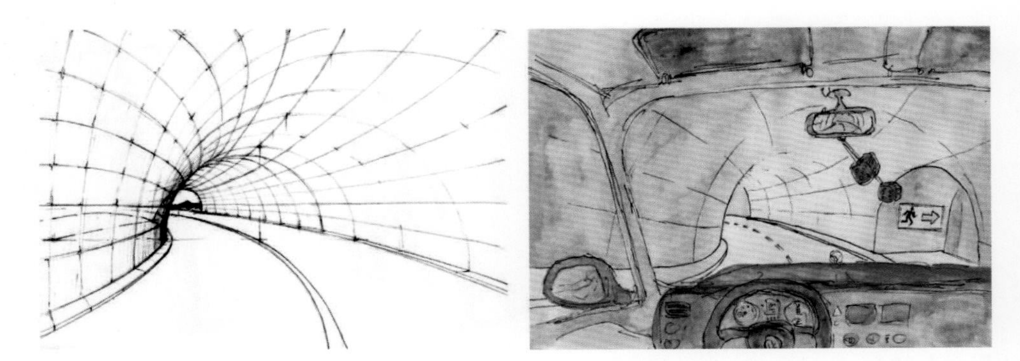

Fig. A.28 Interior views of a tunnel

Fig. A.28 is meant to illustrate that images of circles may very well be parts of hyperbolas. When viewing a torus-shaped tunnel (see also Fig. 6.62), one is, in fact, located within the right circular road edge and is viewing the scene such that nearly all visible circles become parts of hyperbolas. If the tunnel exhibits a high curvature, then the inner edge of the road may be an ellipse. This happens whenever the plane which is perpendicular to the viewing direction through the eye point does not intersect the circle.

Fig. A.29 shows further perspectives of surfaces of rotation but, this time, without construction lines. In the case of the wine glass on the right, it can

Fig. A.29 Rotational surfaces in perspective

clearly be seen that the parallel circles of the surface become steadily bulgier towards the bottom.

• A helical staircase in perspective

In 1531 (five years after Albrecht Dürer) Hieronymus *Rodler* published a book about perspective drawing. His explanations sometimes had little geometrical background, which lead – besides remarkable drawings – to a strange extreme perspective shown in Fig. A.30 on the right. On the left, an alternative by Christian Perrelli is displayed.

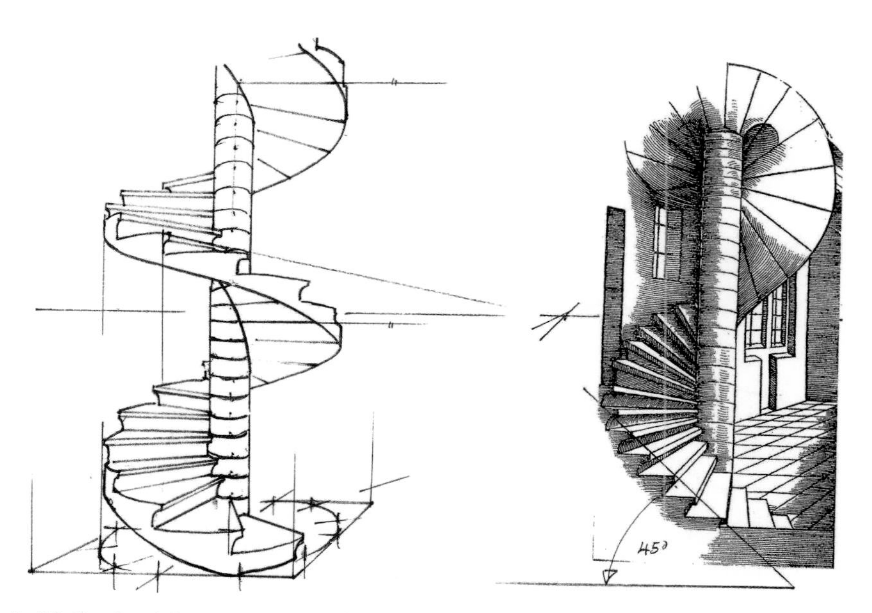

Fig. A.30 On the right a perspective drawing by Hieronymus *Rodler*, on the left an alternative.

It is a good exercise for the reader to find the essential differences between the two drawings. One might also consider the differences to a fisheye photo (Fig. A.31). ♠

Fig. A.31 This fisheye photography comes close to Rodler's drawing. There are, however, essential differences.

Shadows in perspective

Shadow constructions of central views are no more complicated than those of normal views. It may be repeated, however, that the novice should not attempt to draw shadows of complex scenes. This recommendation is, again, due to the noticeable and distracting nature of incorrectly drawn shadows – especially, as it goes against the primary goal of perspective sketching: Visual descriptiveness.

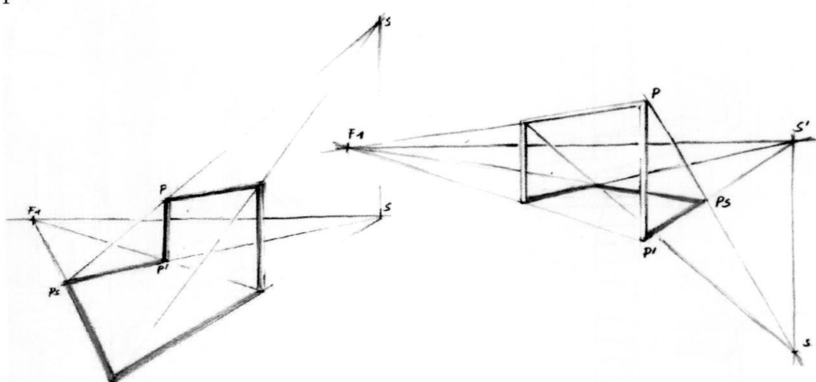

Fig. A.32 Frontal lighting on the left, back lighting on the right

The following theorem is of fundamental importance:

A light ray through any given point passes through the sun point, and its top view passes through the sun foot point.

Let us differentiate between frontal lighting, back lighting, and side lighting (the limiting case).

The situation of frontal lighting is the easiest to understand (Fig. A.32 on the left): The sun S is depicted as a point above the horizon. In cases of a horizontal viewing axis, the image of its foot point S' lies exactly below S on the horizon h. We may derive the shadow P_s of a point P by intersecting PS with $P'S'$. In the image, it also follows that $P_s^c = P^cS^c \cap P'^cS'^c$. In the drawing, we may omit the symbol for the central projection (the ^{c}) as long as we are able to differentiate – at least in the mind – between the spatial situation and the central view (perspective).

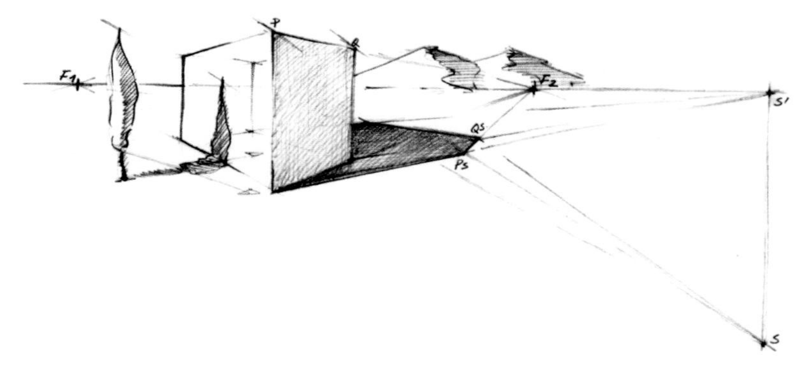

Fig. A.33 Simple scene with shadows

Now let us take a look at the back lighting situation (Fig. A.32 on the right): Nothing seems to change from the perspective of construction – if we accept that the sun point, as a theoretical point, may also be *below the horizon*, without the scene being set at night. We can, in fact, get the image of the sun by connecting the sun to the eye point (this would be a ray of sunlight that would hit the back of our head) and intersect the connecting straight line with the image plane (which is always vertical and *in front of* us).

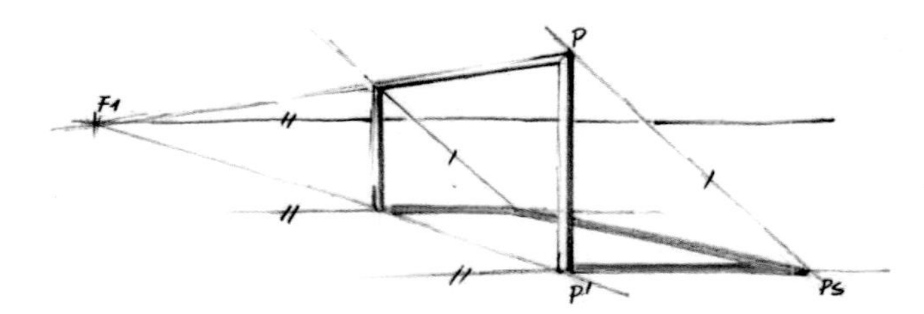

Fig. A.34 Side lighting: Easier to construct and visually appealing

In sketching practice, the sun's mid- and foot points are usually far outside the range of the drawing sheet. This is no problem for free-hand artists: As long as we know the approximate position of these points, we may still inscribe the important respective sections of the connecting straight lines.

Reflections

Only very small sleights of hand are needed in order to enrich our perspectives by simple reflections. Let us restrict our attention to horizontal and vertical

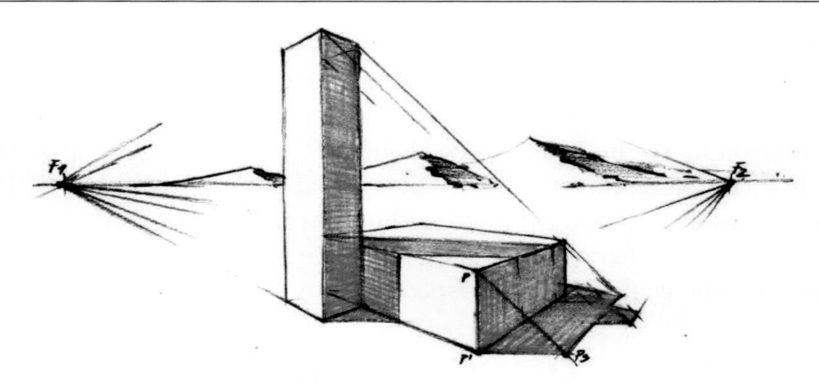

Fig. A.35 Simple scene with side lighting

reflection planes (water and the "classical" mirror on the wall). We would do well to remember the following rule:

Each mirror produces a *virtual counterworld*, which we may observe through the mirror window.

There, thus, exists an "opposite point" P^* for each point P, which by the rules of reflection is located on the normal of the mirror plane. The distances of both points from the mirror plane are equal.
The *reflection in water* is especially easy to draw (Fig. A.36): We simply have to find the top view P' *in the water plane* of a point P. The segment $\overline{P'P}$ may now be inscribed downwards. If the viewing axis is horizontal, then we may immediately recognize this "doubling" in the image.

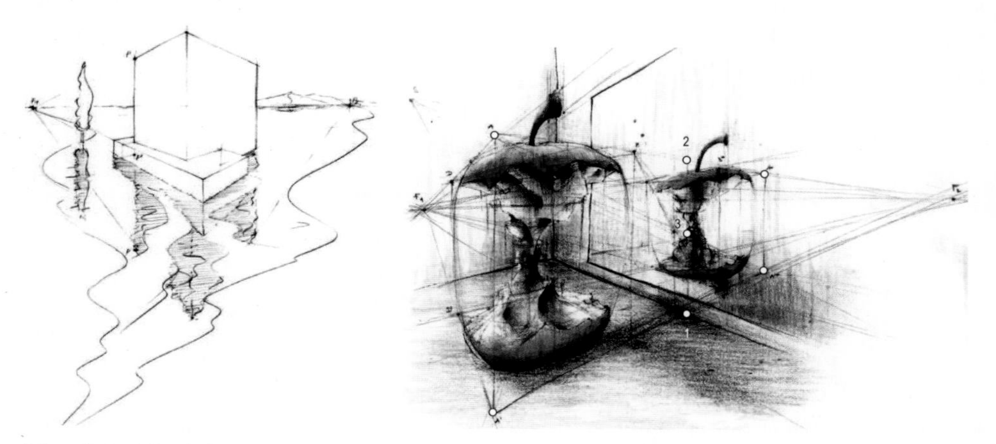

Fig. A.36 The reflection in a water surface is easy.

Fig. A.37 Vertical mirror. The circumscribed cuboid is the secret to our success ...

The *reflection in a perpendicular wall* (Fig. A.37) is no more difficult. The path s of the reflection plane and the direction of the reflection normal (vanishing point S) must be known. Let P be, once again, a point with a top view of P'. Let us project the segment PP' in the direction of the reflection normal. This gives the points 1 and 2 in the mirror plane (1 at the intersection with the path s and 2 above it). Above the midpoint 3 of the segment $\overline{12}$ (whose half is also visible in the image) we can find the top view of the reflected point. The reflection point P^* is located directly above it. Let us keep $1 \to 2 \to 3$ in mind.

Three principal vanishing points

So far, we have only considered perspectives with a horizontal viewing axis. Let us now deal with the general case.

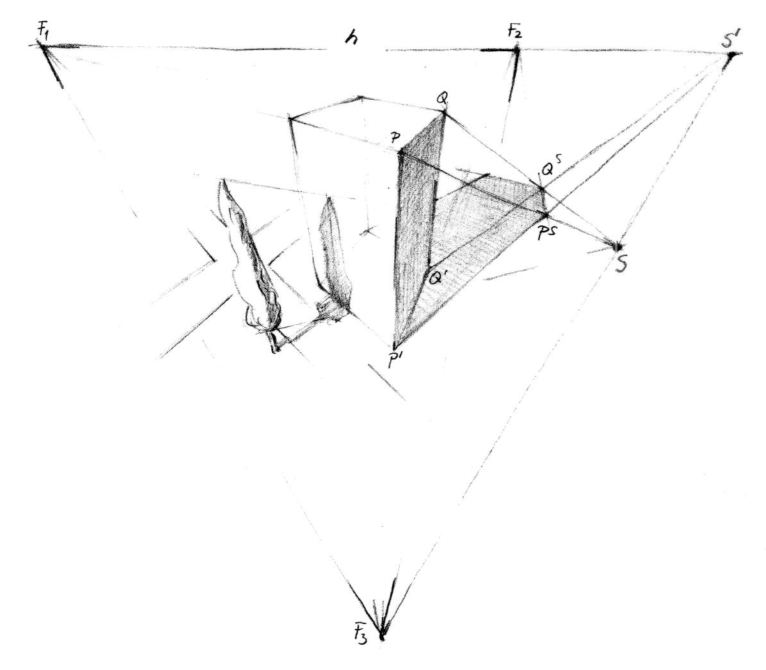

Fig. A.38 Three principal vanishing points

The three coordinate directions vanish in the image through the *principal vanishing points* F_1, F_2, and F_3. The horizon $h = F_1F_2$, if viewed from above or below, tilts in the direction of the upper or lower edge of the page. Perpendicular straight lines no longer appear to be parallel. A cuboid can quickly be sketched (Fig. A.38), but what about its shadow on the base plane? Now we have to now adapt the previously defined rules, and never forget that the sun and its top view are also connected by a vertical, so that:

In a general perspective, sun point S and sun foot S' lie on a straight line through the vanishing point F_3 of all vertical lines.

We may again distinguish between back lighting (Fig. A.38) and frontal lighting (Fig. A.32), depending on whether or not the sun point lies above or below the horizon. The special case of side lighting does not seem to bring as many advantages as with a horizontal viewing axis, since now the light rays appear parallel only in the top view (according to our rule, the sun point lies on a horizontal through F_3).

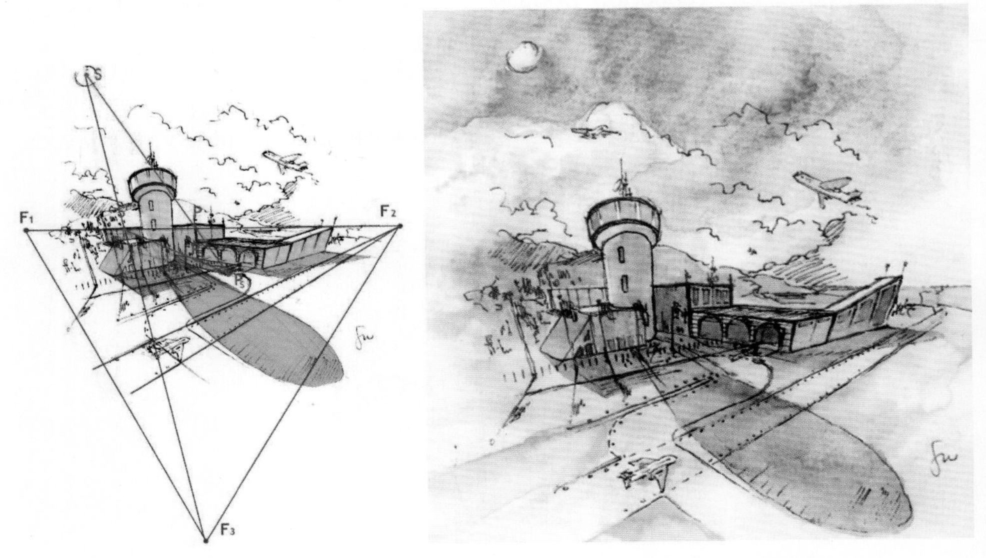

Fig. A.39 Frontal lighting: Different styles, with the same theory behind them ...

Fig. A.39 shows a sketch and the "finished" general perspective (Stefan Wirnsperger), both of which show correct cast shadows. The perspective image on the right shows no construction lines. The sketch clearly shows that it is necessary to think things through very carefully.

Fig. A.40 Always pack sophisticated objects into cuboids, and give the mere suggestion of shadows!

Fig. A.40 shows other perspectives with three principal points. In such sophisticated drawings it would be too risky to to draw cast shadows accurately; so just give a mere suggestion of them. Nevertheless, it is always useful to pack more complicated objects into minimal bounding boxes, since we know best about dealing with cuboids.

B A geometry-based photography course

As we have already seen, geometry is not just a theoretical mental game for a few chosen ones. The applications are manifold, and one of the geometrically closest is photography. Photos are usually "classical" perspectives. Our knowledge about perspective helps us to accomplish certain effects deliberately. For those readers who may only have flipped over the theoretical parts of the book, some of the geometrical statements of the previous chapters will be repeated.

Conversely, many readers may already be aware of some of the topics and techniques that will be discussed. It might still serve a purpose to illuminate these matters from a geometrical standpoint, if only for the sake of clarity: How may we employ simple tricks to accomplish images which otherwise might only have been taken by using expensive shift lenses? Why is the choice of the focal length not only a question of "view", but should also concern the further use of the photograph? When are ultra-perspectives not only permitted but actually necessary? Why does the depth of field depend on the focal ratio? What is different in underwater photography?

The many illustrations for these questions are distributed throughout the book and are not merely limited to this appendix. This way, the interrelations between the not-so-separate subjects are strengthened. Anyone who wishes to approach geometry with the help of photography (and vice versa) is most welcome to do so!

Survey

B.1 Focal lengths and viewing angles

Fig. B.1 The famous Piazza di San Marco in Venice: 17mm focal length, full frame sensor. The image seems to be unnaturally distorted. However, we will see later that this may under some circumstances be necessary.

Fig. B.2 Left: Somehow the same motive. With 400mm focal length, however, you need to go far away, and the impression is completely different. Right: The same motive as above with a 15mm fisheye and a full frame sensor.

Today's lenses are enormously complicated and are in most cases zoom-lenses. Before the advent of digital photography, by far the most popular types of negatives were 24mm ×36mm in size. The lenses of non-digital reflex cameras usually had focal lengths in the approximate range of 20 – –400mm (with some expensive exceptions). Modern digital cameras very often have a sensor that is far smaller than 24mm ×36mm, and the range of the focal length is accordingly smaller (the only exception are some professional "full frame cameras").

Focal lengths and viewing angle are connected

The physical term of *focal length* used to be (and still is, given certain conditions) a measure for the geometrical term of the *angle of view* – the angle between the lines of sight from the lens center through the endpoints of the image diagonal. The smaller the focal length, the greater this angle is (Fig. B.3). If the angle is greater than 50° (which in full frame cameras corresponds to a focal length of about 35mm), then one is usually faced with a *wide-angle lens*. Lenses with angles above 85° (focal length of about 24mm) are already counted among the *super-wide-angle* or *ultra-wide-angle* lenses, and those with angles up to above 180° (yes, this is possible!) are usually called *fisheye lense*s (below 14–16mm). If, on the other hand, the image angle is smaller than 20 degrees (about 85mm), then one is dealing with a so-called *telescopic lens*.

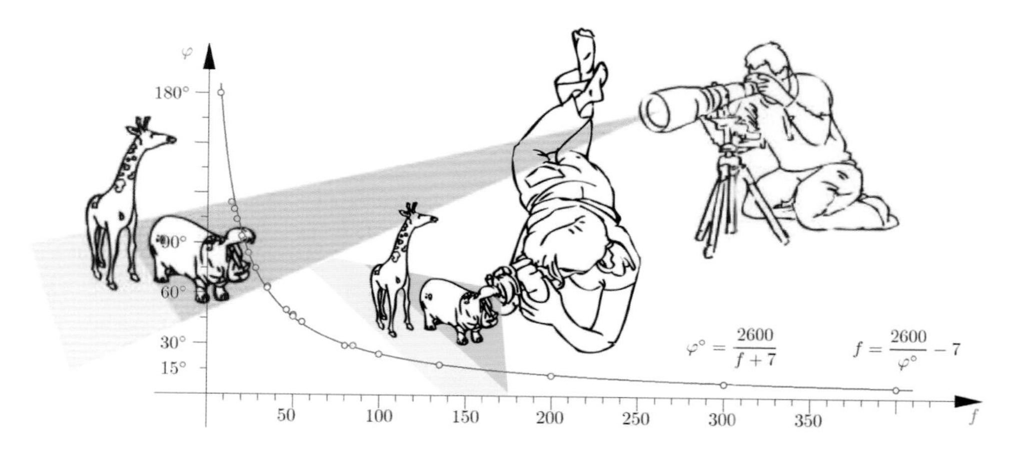

Fig. B.3 Different viewing angles φ (measured in degrees) correspond to different focal lengths f (measured in millimeters). The formulas to be seen are interpolating data given by various camera producers. The general rule is: the smaller the focus the larger the viewing angle.

By now, the focal length has become a confusing measure for the image angle. Non-full-frame digital cameras based on the reflex camera principle often have a "crop factor" between 1,5 and 1,6 (sometimes even 2). If a zoom lens is labeled with 18−−200mm, then on a full frame camera, this corresponds to 18×1,5−200×1,5mm = 27 − 300 mm. This is due to the smaller size of the CCD (the light-sensitive layer or chip).

In digital cameras which are not reflex cameras (and are, thus, incompatible with the "good old" lenses), one may usually find focal lengths that correspond to about one fourth of the values of full frame cameras. One might, nevertheless, produce comparable image angles. Here, the size of the CCD appears to be *significantly* smaller.

One may now conjecture that lenses with different focal lengths or different image angles produce different images. From the perspective of geometry, however, the following important theorem holds:

The chosen perspective depends only on the position of the lens center and on the direction of the optical axis (inclination of the camera), and not on the chosen focal length. A telescopic lens photo is merely a "magnification" of the image center.

• **The entire objective-lens system works like one single converging lens**
An objective lens consists of several lenses and lens groups. If we attempt to analyze the direction of light in a typical modern lens with a complicated optical arrangement (as in Fig. B.4), we have to admit that the matter does not seem to be as trivial as it first appeared.

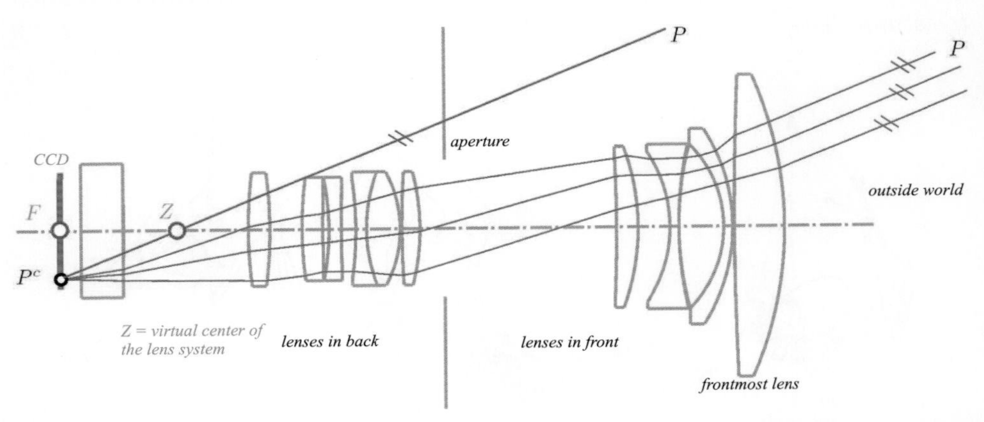

Fig. B.4 The complicated light path through a modern lens. Still, the whole lens system works like one single lens (center C).

It may further be stated that these rays, which hit the lens system in parallel, are finally focused on the light-sensitive layer (assuming a distance of ∞ has been set). If we translate the direction to the far distant point P through the image point P^c, then it has to pass through a fixed point C on the optical axis – if this were not the case, then we would not be dealing with a central projection. We may, thus, have confidence in the lens manufacturer's ability to arrange the dozen or so lenses in precisely such a way.

The question of *why* lens manufacturers go to such elaborate ends is easily answered: It serves to correct the many deviations from the perfect image. It starts with the simple physical fact that different spectral colors are refracted in a different way. A simple lens, thus, produces "rainbow edges". ♠

B.2 3D-images in photography?

The Gaussian collineation

Very much simplified, we usually say: Photography is a projection of space from the lens center onto a chosen image plane (the light-sensitive layer).

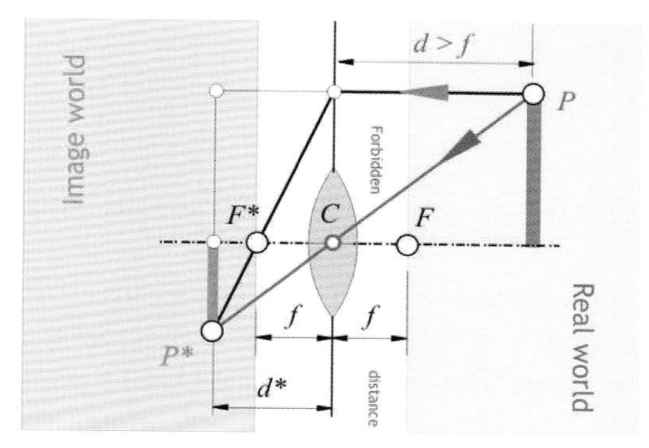

Fig. B.5 The Gaussian collineation. The closer to the focus F, the larger the image gets. For $d < f$, the image point P^* is on the "wrong side".

There is one major thing, however, that has to be considered: Physically speaking, we cannot exactly speak of a projection of three-dimensional space onto a plane. Fig. B.5 shows how a space point P is actually transformed into a virtual space point P^* behind the lens, which by no means has to lie in our projection plane (compare also the application p. 309 and especially Fig. 9.31). This relation between real space and virtual image space is called the *Gaussian collineation.* The transformation is linear since one can easily prove that straight lines are transformed into virtual straight lines[1], and it is invertible – contrary to a mere projection on a plane.

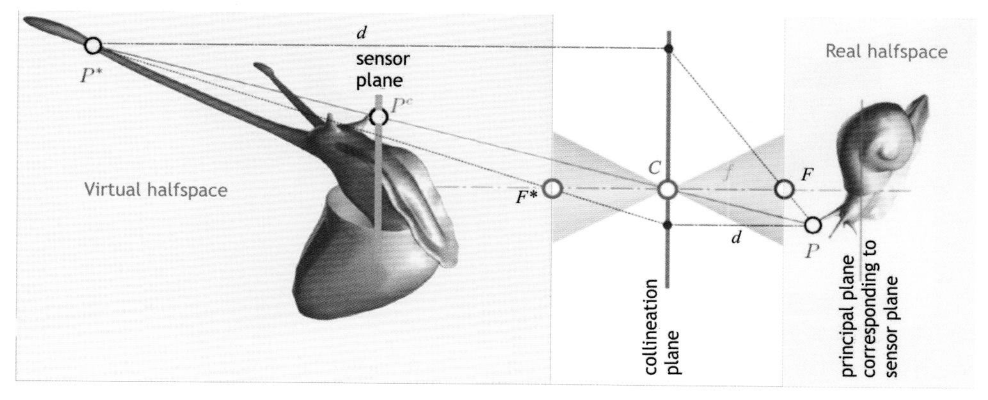

Fig. B.6 The Gaussian collineation, interpreted in three-dimensional space. To each object on space, we get a collinear virtual object. The sensor plane corresponds to a plane perpendicular to the optical axis. Perfectly sharp points in a photo lie in that plane.

[1] For more details, see `http://www1.uni-ak.ac.at/geom/files/3d-images-in-photography.pdf`

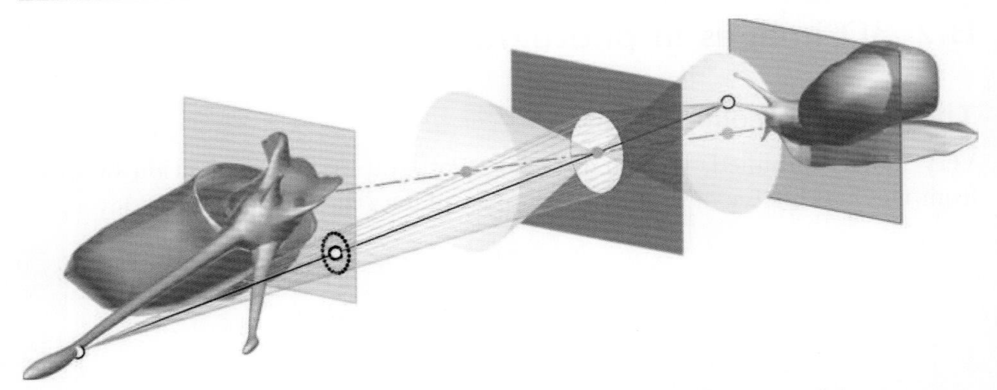

Fig. B.7 The diameters of the aperture and the circle of confusion (COC) correspond linearly

Assume we are given a distance setting. If the point P is not too far away, then the light rays that emanate from it hit the system of lenses but not exactly on the photo layer. The point is thus pictured as a "spot" called *circle of confusion (COC)* (Fig. B.7) which causes its image to become blurry. If P is an important point, then the distance between the frontmost and the backmost lens block should be altered so that a single image point P^c on the photo layer is created (Fig. B.8).

The locus of all sharply imaged points

Fig. B.8 The three garter snakes move around quickly. If lucky, the photographer takes a photo when all the three eye pairs are approximately in the focal plane.

The following theorem may come as a bit of a surprise to some but nevertheless, follows directly from the laws of geometrical optics, where the object distance, which is measured on the optical axis is important:

In a photo – given a distance setting of x meters/centimeters – all points which lie in a plane parallel to the camera's back wall at distance x are depicted sharply. This *focal plane* corresponds to the sensor plane in the Gaussian collineation.

Let's say, we want to depict a point P off the optical axis sharply. The "distance" x of P is then not the "true distance" PC to the lens center, but rather the orthogonal distance of P to the reference plane through the camera's back wall that roughly contains the sensor. A perfect depth of field then exists in the entire plane through P parallel to the sensor plane.

The reason why the distance is measured to the back wall of the camera and not to the principal plane γ that contains the lens center C is a practical one: It is very hard to tell where the lens center C really is – the position changes with the distance x, and with zoom lenses, it also depends on the focal length that is chosen.

It all depends on the relative size of our objects ...

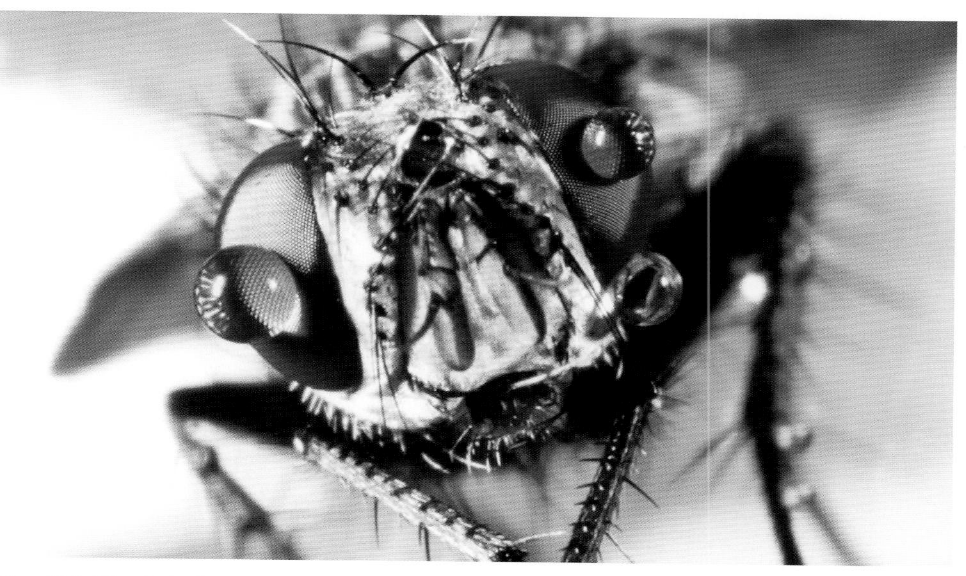

Fig. B.9 The fly's distance to the lens focus is approaching the single focal length. Due to the Gaussian collineation, the distortion of the virtual 3D-image is extreme and the depth of field is thus tiny.

When we take a look at Fig. B.8, we have the impression that the depth of field is already very small. This is mainly due to the large diameter of the aperture (almost *maximum aperture*). Fig. B.9 also looks very blurred to some extent – only the really important parts of the object appear sharp. However, in this case, the diameter of the aperture was set to the minimum. Why is the image not overall sharper?

Fig. B.10 shall explain what happens when we fix an object with small distance from the lens center. A snail is slowly creeping towards the lens, and – due to the Gaussian collineation – the virtual image behind the lens cen-

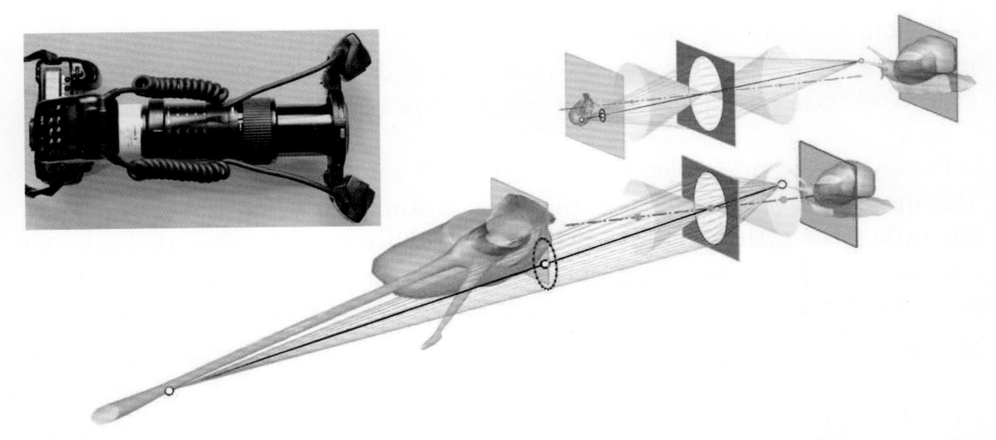

Fig. B.10 The size of the COC (blur disk) heavily depends on the distance of the space point. Upper left: A macro lens that allows the large virtual image to fit in ...

ter "explodes". The very limit is reached when the protruding eyes reach the "forbidden plane" in distance f from the lens center. Each millimeter change of distance can now make a huge difference, and therefore, the depth of field "breaks down".

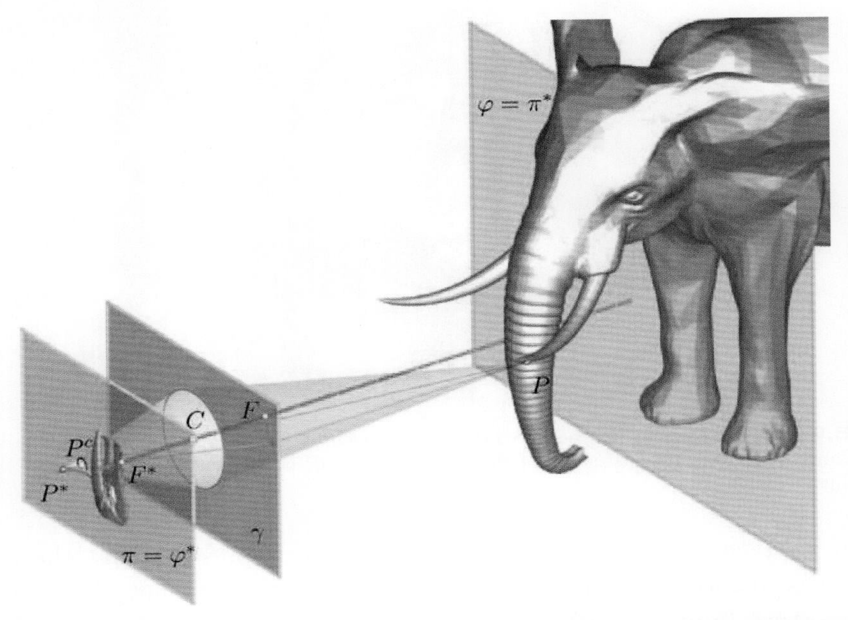

Fig. B.11 If the depicted object is far enough away from the camera, the virtual spatial image gets flat and thus "almost two-dimensional", and the problems with the depth of field play a subordinate roll.

What a difference when we take pictures of large objects (Fig. B.11): The distance is then *much* larger than f, and according to our collineation, this part of space is "harmless": The virtual image of the large object is almost

flat and comes close to what we maybe thought so far photography does: Depicting 3D-space on a plane.
We can now take advantage of our knowledge and state that if we were able to use a much smaller lens with accordingly much smaller focus, we could take a picture of a fly or wasp just like we normally take a picture of an elephant. All seems to be relative ...
To sum it up, we should by now understand the following:

The range of an acceptable sharp focus depends on three parameters: 1. The aperture diameter, 2. the focal length f (the smaller f, the larger the distance d of the space point is in comparison to f), and 3. the distance from the lens: Due to the properties of the collineation, the distance of the virtual space point P^* and the image point P^c increases dramatically when our object comes close to the "forbidden space" (where the distance to the center gets close to f).

B.3 When to use which focal length?

The important difference

In full awareness of the insight that the focal length or the image angle have no effect on the perspective, it does not seem to be as important anymore whether we take our pictures with a wide-angle or a telescopic lens. This approach is correct: A telescopic lens only helps us to exclude objects from the image which are not important to us and instead to fill the frame with our primary photographic objective. If, on the other hand, the goal is to get "as much into the picture as possible", it may be advisable to use a wide-angle lens.

Fig. B.12 Change of distance (constant principal ray)

If we approach an object closely (small distance), then we need a big viewing angle (a wide-angle lens) in order to picture it completely.
However, the object will then appear to be highly distorted, which may in some cases be the intention of the photographer. If, for instance, the viewing

angle is 50°, then this entails an "appropriate" distance (about twice the diameter of the object) and the use of a "normal lens" (with a focal length of about 50 to 60mm).
The farther away we step from the object, the less noticeable the distortion becomes: The perspective moves ever closer to the principal point. We would now have to enlarge the image considerably (by the use of a telescopic lens) in order to picture the object at the same size.

Fig. B.13 An ultra-wide-angle-photo (focal length of 17mm) is on the left, where the trunk is already extremely close to the lens. A telescopic photo is pictured on the right for comparison (see also Fig. B.12), and a flattening of distances becomes noticeable.

In ultra-telescopic photographs (which are very common in nature photography – the gnu in the African savanna being just one example), the image becomes close to a normal view. This has both advantages and disadvantages: The drama of a wide-angle photograph may be gone, but the distances are also shortened. One may get the false impression that the gnus and the gazelles – which, indeed, tend to graze together – are practically on the same spot (Fig. B.13 on the right).

Fig. B.14 Depth amplification with a wide-angle lens (28mm) and depth shortening with a telescopic lens (670mm). The boat is about 400 meters away (this may be estimated by the diameter of the sun, which spans an image angle of about 0,5° [11]), but is certainly many kilometers in front of the background islands.

Subjective change of proportions

The advantage is that the proportions of the animals in the picture are much closer to their "true proportions". A gnu may have a back which slants downwards, but the true angle of inclination may best be seen in Fig. B.12 on the far right. Sometimes, it may also be an advantage to be a bit further away from the animals. The elephant in Fig. B.13 on the left must have mistaken the camera for food. The next photo was already done at a more respectable distance.

From one extreme to the other ...

The normal projection is an extreme case of a telescopic photo. In such images, the depth values – the distances from the observer – are no longer important (Fig. B.14).

... or: How a broom closet may be turned into a dining hall

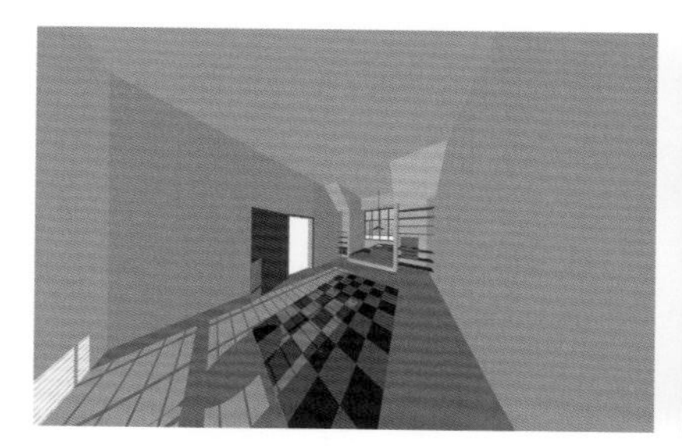

Fig. B.15 Wide-angle lens vs. regular lens for interior photography

When taking photos of small rooms, one may no longer have the choice of an arbitrary focal length. The dilemma persists that while a wide-angle lens enables the photographer to encompass as large a portion of the room as possible, it gives the room a strange appearance (Fig. B.15 on the left): It looks as though one were standing in a long and drawn-out hall. This may be of interest for real estate agents who wish to sell a 10m^2-room for a steep price ...

Unintended caricatures and passport photographs

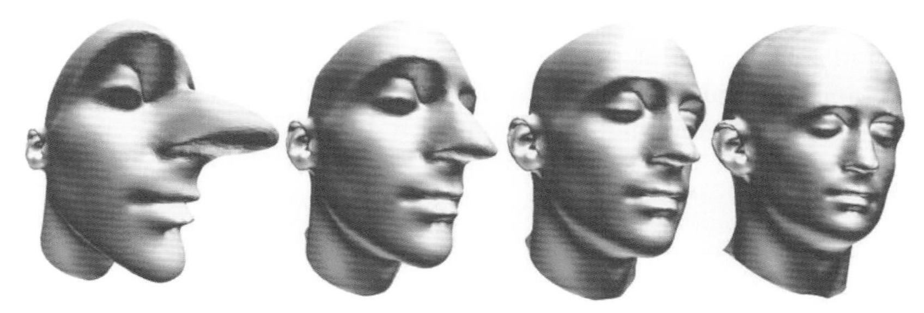

Fig. B.16 Wide-angle vs. telescopic lens: Portraits should not be done with wide-angle lenses!

When taking pictures of human faces, the "magnification effect" of proportions inherent in wide-angle lenses becomes very noticeable (Fig. B.16). Thus, when we are asked to take a passport photograph, we should definitely employ a telescopic lens!
Even though the heads of most people are not precisely spherical, the picture in Fig. B.16 may give an indication of the distortive devastation which may be wrought when using a wide-angle lens.

Which focal length to use under water?

Fig. B.17 The extreme small focus of 14mm was used for this picture, and the adult dolphin merely touched the lens.

When taking pictures of large underwater creatures, wide-angle lenses have to be employed in any case (Fig. B.17). Each meter of distance affects the visibility. At a distance of 10 or 20 meters from the camera, just about every color dissolves into gray and drab, and not even the best flash gun can alter this fact!
Telescopic lenses are virtually never used under water (Fig. B.17) – and for many good reasons. First of all, the refraction between water and light (in the lens and the camera) already "lenghtens" the focal length by about a third so that even wide-angle lenses quickly become "normal lenses".

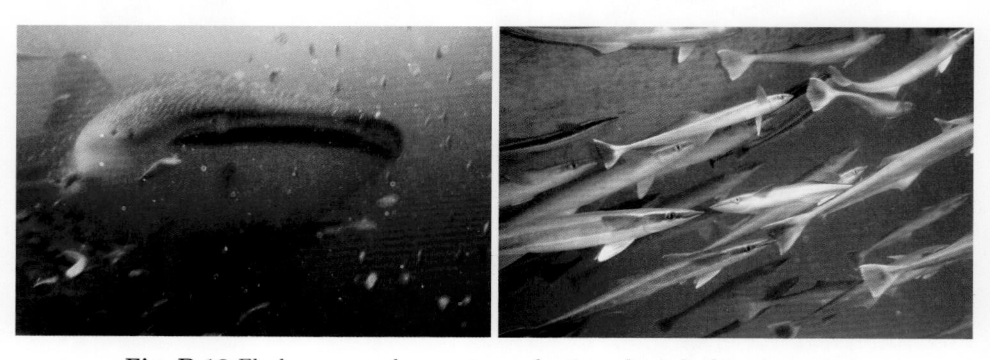

Fig. B.18 Flash guns can be counterproductive when plankton is present.

Secondly, one is only capable of seeing a few meters ahead, even in crystal clear water – and finally, color fastness decreases rapidly with distance.

Fig. B.19 When the water is full with plankton, the only way to get good photos is to use ultra-wide-angle lenses (14mm focal length on a full frame sensor)

Colors appear as they actually are only when being very close to the observer, and only when employing a flashlight or being close to the surface: A distance of only one meter in one meter depth is enough for the light to have to travel for two meters (one meter to the pictured object and another in return) through water – which already causes parts of the red color components to be "swallowed". However, due to the fact that artificial light sources are not present in nature, red color tones may appear under water as camouflage colors.
In Fig. B.18, a whale shark plows through plankton, which also causes the view to be highly limited. Flash guns are completely useless (and may in fact illuminate floating particles which would block the view). Fig. B.18 on the right: Remoras which accompany the whale shark can be captured from up close and appear in their true colors as they are illuminated by a flashlight.

Fig. B.20 The same turtle – on the surface and in 10 meters depth

The solution for the "whale shark-problem": Go as close as possible and use an ultra-wideangle lens (Fig. B.19).

Depth is crucial

Under water and from a distance of more than two meters, flash guns are useless or even counterproductive (Fig. B.18). Thus, one has to rely on sunlight. Fig. B.20 on the left illustrates that more or less true colors are to be expected on the surface if the distance of the photographer is not too large (1,5 meter). Fig. B.20 on the right shows the same turtle in 10 meters depth. Now, the sunlight has to travel 11,5 meters.
Fig. B.21 on the left illustrates that in 20 meters depth all red and orange colors work like camouflage and that a weak flash gun does not help either.

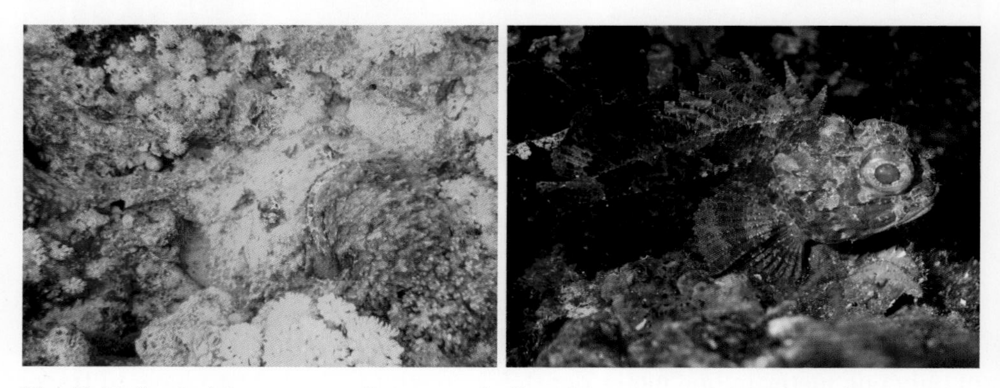

Fig. B.21 On the left: one meter distance to the Stonefish, using a weak flash gun (picture taken during the day in 20 m depth). On the right: A flashlight from close distance helps to detect otherwise nearly invisible creatures.

Macro photography under water

Macro photography under water may be difficult for a diver because one has to be able to stay very stable without the help of gravity. Since one gets very close to the object, however, and almost always uses flash lights or flash guns due to the lack of direct sun light, the colors are almost "true" on the photograph (Fig. B.21 on the right, Fig. B.22).

Fig. B.22 The closer one gets to the predator (on the left a Scorpionfish, on the right the same Stonefish as in Fig. B.21), the more one begins to understand the secret of the camouflage.

Fig. B.23 illustrates that, when taking pictures at night (or in greater depths), the flash gun is the only light source. "True colors" are then easier to achieve from less than half one meter distance.

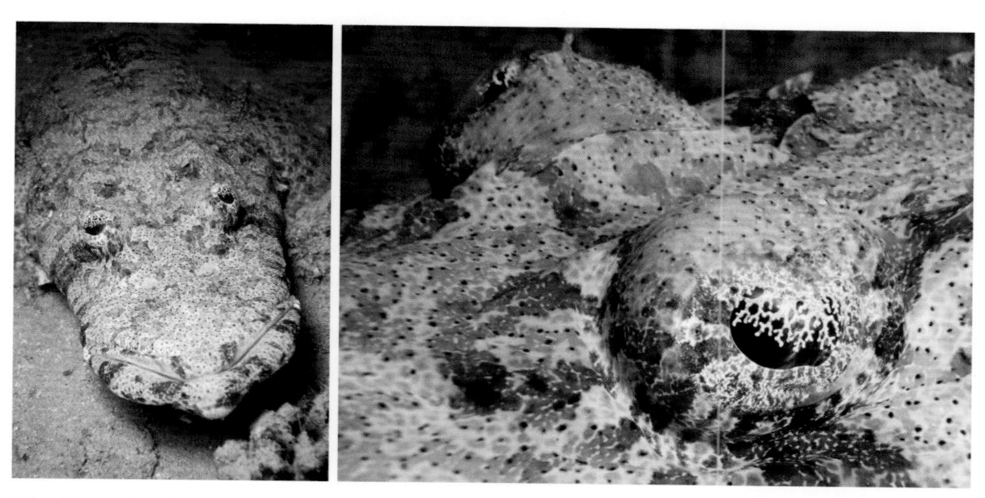

Fig. B.23 On the left: At night, the only light source is the flash gun. On the right: Remarkably camouflaged eyes of the Crocodilefish (note the side view of the eye in the back).

Fig. B.24 and Fig. B.25, finally, shall illustrate that when we take pictures of really small objects, there is almost no difference to macro photography on land: We have a very shallow depth of field and should try to get as many points of interest into focus and/or use a high focal ratio.

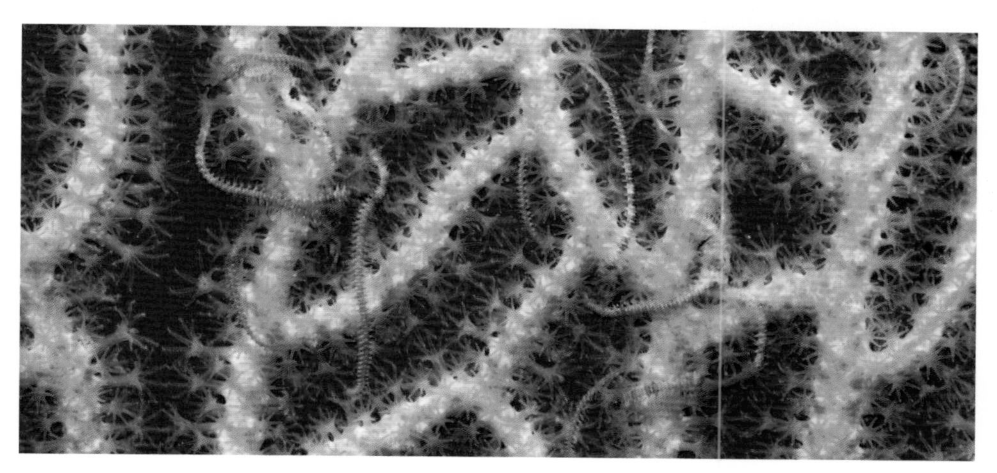

Fig. B.24 A "real" closeup. What makes it easier: The principal plane contains many polyps on the hard coral.

In Fig. B.25, this was easier said than done. The intelligent and curious cuttlefish, however, allowed a very calm approach to the necessary close distance. Eyes of cephalopods are, by the way, completely different from fish eyes and have – due to their sine-shaped opening, two yellow spots.

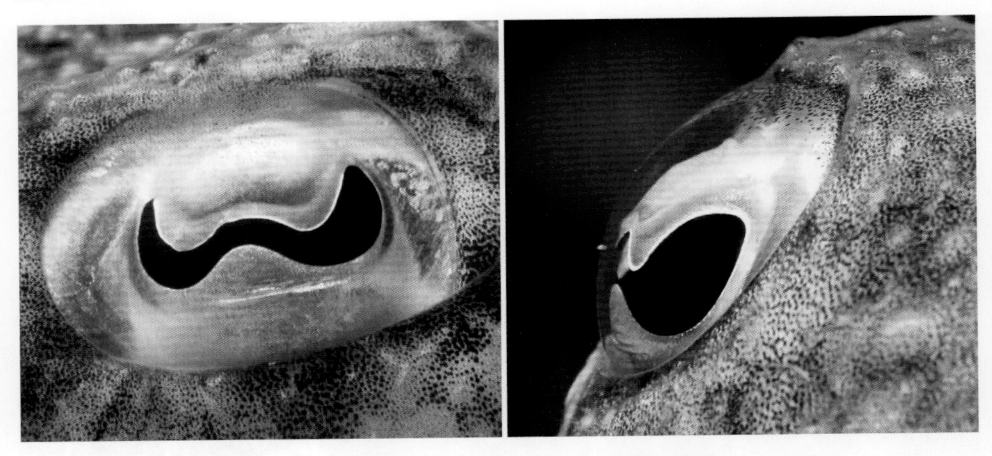

Fig. B.25 The eye of a sepia (cuttlefish) in different viewing angles

B.4 Primary and secondary projection

Ultra-wide-angle pictures are sometimes necessary!

Fig. B.26 Primary perspective with an ultra-wideangle lens

At first, the extreme wide-angle picture in Fig. B.26 may appear to be exaggerated – and it is, because we usually look at the image from a distance of 50cm or more.
You are encouraged to try the following experiment: Shut one eye and position the other eye as close as possible above the center of the image (at a distance of about 10 to 15cm – because your eye could not focus if the

picture was closer). You may now agree that the image no longer appears so distorted.

This may be explained by the fact that it is, indeed, possible to approach an object this closely in nature. From this proximity, it would no longer be possible to view the whole scene with a single image – you would have to move your head or at least your eyes in order to capture other details of your surroundings. In relation to the very large object, however, the eyes remain in practically the same spot.

Let us now go back to the significantly downscaled image: If you try to move your uncovered eye in its extreme position, close above the image center, its motion would be equivalent to the described head motion – and because the other eye is covered, you would not be able to assess distances anyway. We have thus managed to scale down the same situation that actually happened in three-dimensional space. Let us call the spatial situation *primary perspective* and the image situation the *secondary perspective.*

By accepting these assumptions, the following rather obscure theorem holds true:

In order to get a realistic impression of a photograph, primary and secondary perspectives should coincide. In other words: The ambitious photographer would already have to ponder, at the moment that the picture is taken, whether the person looking at the photo would be able to assume the camera's original position.

Fig. B.28 on the right may – in some circumstances – be a necessary method of depiction: Imagine a tree exhibit in a museum through which visitors would be allowed to walk. If the massive tree should be pictured on a high wall, so that a visitor passing by closely may be given the impression that it would be possible to look up at the canopy, the trunk of the tree in the lower part of the image would *have* to be thick enough to produce a realistic spatial effect!

Cinema screens and portable TVs

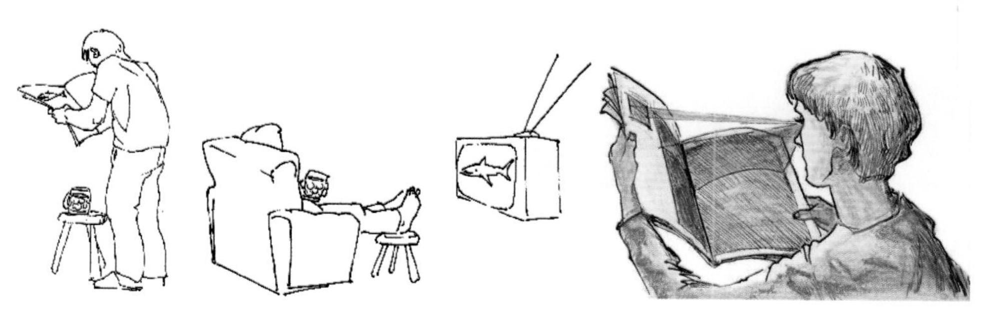

Fig. B.27 Unrealistic secondary perspectives: The photo of the shark looks strangely distorted on the left. At the center: A distortion also occurs due to the small image angle. On the right: The size of the printed image in a magazine is important. The shark requires a whole spread!

Let us come back to the underwater-example. An underwater photographer has taken pictures of sharks (two primary perspectives can be seen in Fig. B.26). The animals appear to be highly distorted as they move closer to inspect the ultra-wide-angle lens. The pictures are then presented on a huge screen. Spectators in the front rows are forced into a secondary perspective that closely resembles the primary.
Some time later, the film is also broadcast on television. The small picture in the TV guide (Fig. B.27 on the left) may first raise interest: It shows an unnaturally distorted shark with giant teeth. The viewers that look at the picture not only on a small television screen but also from a large distance easily miss the dramatic effect of the original picture (Fig. B.26). They may even sneer at the shark's snout which moves too quickly from side to side ...

If, therefore, you intend to print pictures on a small scale for reasons of space (as in this book, for instance), then you should choose a greater distance (and a telescopic lens) when taking the photos. Smaller geometrical illustrations are best done in normal projection. If, however, the pictures are to be printed on a large-scale (a photography magazine is a good example), then you may *and ought to* employ a wide-angle lens (Fig. B.27 on the right).

This insight may not be so useful for paparazzi: They may want to publish full-spread photos of their victim's latest and greatest secrets. So it seems that they would have to use an ultra-wide-angle-lens ...

In Fig. B.28, the extreme wide-angle photo makes for an impressive state room. Our knowledge about columns and chandeliers helps us to estimate its true dimensions. In the right image, the regular tree resembles an African baobab with an extremely thick trunk. This might be interpreted as a "distortion of facts". On the other hand, trees do look this way if we approach them closely and look up. If a poster-sized print of this photograph were to hang at a height of two meters, one may indeed get the impression, when walking under it, that one was walking through the forest.

Fig. B.28 Twice an extreme wide-angle lens. On the left, we may be capable of "rectifying" the scene by using our existing knowledge. On the right, we may make an estimation about the thickness of the trunk.

Fig. B.29 shows two further extreme perspectives, which both express a certain "message" and may, for instance, be used for stylistic reasons. If one wishes to "drive a point home" and to demonstrate the enormous performance of one's creation, it may be valid to exaggerate the visuals slightly ...

Fig. B.29 Extreme wide-angle photos may also be employed in order to send a message or to express a stylistic statement. They elicit responses in either case, since objects which are experienced as being up close usually generate some kind of reaction. Things which may actually be minute or of "regular size" appear to be gargantuan.

Leonardo's dilemma

Fig. B.30 Leonardo's dilemma: Should he distort the heads in extreme perspectives, or picture them "normally"? Leonardo wisely chose the latter, because he knew that people would walk along the 10m painting and only rarely remain at the "ideal position".

Leonardo da Vinci was faced with the following problem as he laid out his "Last Supper": The painting is a whole ten meters long. The observers would walk next to it at a relatively short distance and look at the various details of the picture. If they stood still, then they would need an "ultra-wide-angle-lens". It would seem that all objects would have to be highly distorted – not only the table and the room, but also the heads of the apostles!
Leonardo knew that human beings are most perceptive of unnatural distortions when it comes to heads and facial features. Thus, he decided on a compromise: He distorted the still life environment to a very significant degree, but left the heads completely unaffected!

B.5 From below or from above?

Fig. B.31 Frog's perspective, but from a bird's eye view, or a bird's perspective, but from a frog's eye view?

A horizontal lens axis approximates our familiar way of seeing pretty well. Our brain evolved in the "savanna" and is used to process stereoscopic images of objects which lie at a comfortable distance and are approximately of "our own size" – not too big, not too small.

Fig. B.32 Frog's perspective and bird's eye view in on of the elliptic towers of the *Sagrada Família* in Barcelona

If the size of an object exceeds our field of vision, then we have to tilt our visual apparatus – either upwards (when looking up at skyscrapers) or down-

wards (when looking at small objects before our feet). Over time, the terms *frog's eye view* and *bird's eye view* seem to have established themselves. This is fine as long as we remember that many animals must perceive images very differently from ourselves and that this is not merely a difference of scale.
According to the rules of central projection, our scene is augmented by one further principal vanishing point: The vanishing point of all vertical straight lines.
It seems to be the case that images which exhibit a very apparent vanishing point appear as being "more artistic" to critical eyes. The following "message" is transmitted by the photographer: The object itself, or my position above it, is very high.

Fig. B.33 Architecture photos: "Slightly vanishing" (tilted) on the left. The bottom half, which was also photographed on the right, is later cropped from the image.

The so-called *slightly tilted edges* are almost "painful" to the trained eye. They occur when the vanishing point of the vertical straight line lies far beyond the image (Fig. B.33 on the left). In such cases, the vertical edges of the object appear to be "nearly" parallel. For example, the left edge images turn out to be not quite parallel to the left paper edge, and the right ones, which are rotated in contrary motion, are not quite parallel to the right paper edge.

Fig. B.34 Two extreme perspectives: On the right the optical axis is horizontal.

Either a shift-lens ...

Professional photographers are well aware of this problem. In the era of analogue photography, professionals or amateurs with thick wallets helped themselves by employing so-called *shift lenses* and a tripod: This special lens enables the photographer not only to take pictures in the direction of the optical axis, but also to "shift" this direction slightly. From a geometrical point of view, a regular central projection is still being produced. However, the camera axis may still be horizontally fixed while simultaneously "looking up" towards the edges of the building. According to the laws of geometry, the edges remain parallel in the image.

... or some geometrical know-how

Fig. B.35 Strongly vanishing vertical edges

The dawn of digital photography has changed various things, and arguably for the better: A primary perspective is captured digitally and may later be changed in an image editing software – a process which may be similar, but is still a lot easier than the darkroom of yore, where a comparable amount of photographic "trickery" took place. Even rectangular images can be deformed in such a way that the edges appear as a general quadrangle, whereas the images of vertical corners appear to be parallel. If we now crop the edges into a rectangle, our wizardry should go unnoticed.
Unnoticed? A "geometrician" might easily prove that the new image must have been altered. A small number of tests would be enough – for instance, spatial angles become oblique angles if the distortion is corrected properly.
The change which has taken place is, in fact, two-dimensional and thus had no regard for the depth distances of the singular points. Thus, it *cannot*

Fig. B.36 An attempt was made three times to keep the optical axis horizontal by pointing the camera at a far distant point which lies on eye level.

possibly be exactly correct. For this reason, we should only resort to this trick "in emergencies" – if, for instance, we have returned from our vacation with a very special photo of architecture which we cannot possibly replicate.

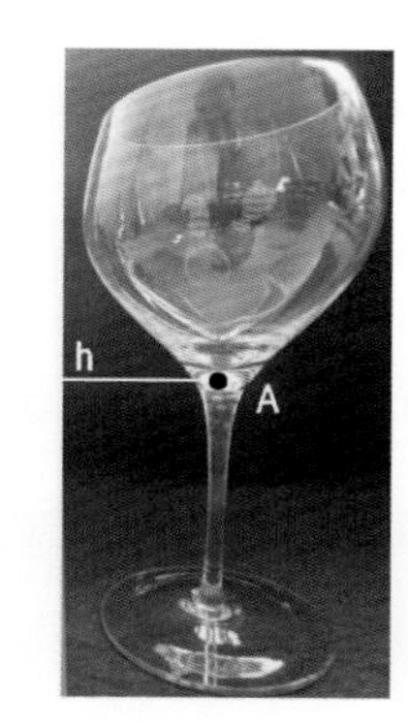

Fig. B.37 The "evacuation" of the principal point

If the following principle is considered *before* taking the photo, then we may be spared a lot of work later on: We should not only take the picture of the relevant object (such as the tall building in Fig. B.33), but also of many other important areas next to it – as long as the optical axis is horizontal. This may even be accomplished without fancy equipment like a bubble level by simply pointing the camera at the heads of people who reside at a constant height in front of the building. In post-processing, we only have to crop away the "digital junk" ...

- An ad showing long female legs

We know by now that objects appear to be more distorted, the further away they are from the optical axis and the closer they are to the lens center. In an ultra-wide-angle photo, heads should, therefore, always be near the principal point. However, do the same rules apply to legs?

Let us first consider Fig. B.37. The original picture of the wine glass is seen on the left, and only a section of it is pictured on the right. If the picture were to be put on a billboard, then the relevant section would certainly be chosen – the rest would not be interesting, and besides, it might be a good idea to catch people's attention through the extreme distortion.

Fig. B.38 Let us now use an ultra-wide-angle lens. The legs should be turned towards it. The length at which the legs appear now only depends on the image section being displayed. The flippers strengthen the impression of elongation.

Which brings us back on topic. Commercials often show women with unusually long legs – sometimes almost "grotesquely" so. Fig. B.38 shows which tricks may be employed to accomplish just that. ♠

Collinear distortion of a photo

Fig. B.39 A swimmer's perspective on the "Badeschiff" at the Viennese Danube Channel

Let us mention a second method by which effects like Fig. B.38 may be accomplished: We can let photos undergo collinear transformations, which "turn" arbitrary quadrangles into different but equally arbitrary quadrangles (such as rectangles). This collineation is linear, which means that straight lines are always transformed into straight lines.
Fig. B.39 on the left shows an intriguing photo which was taken half below and half above water. It can instantly be seen that both halves are central projections, although with different parameters. The verticals in the upper half vanish in a principal vanishing point, while the vertical edges of the bottom half swimming pool are nearly parallel. On a computer, a collinear transformation was applied to the right image so that the upper vertical lines of the skyscraper contours and the tree are "bent parallel".

Now, however, a section of the image which fits entirely into a trapezoid must be chosen (image on the right with a red border). It would be difficult to conclude, based on the final result, that the photograph has been manipulated to a considerable degree, and actually depicts an illusory world which no longer bears an exact resemblance to reality. The fact that the edges under water are not exactly parallel is merely one indication.

Fig. B.40 On the left: A snail crawls across the perspective image of a cube. On the right: A collineation was applied to the entire photo so that the cube once again appears "normally"; but look at the snail ...

In Fig. B.40 on the left, a snail crawls across the perspective image of a cube. The projection of the cube undergoes a second collineation (secondary perspective) by the mere act of photography. On the right, this collineation is "canceled out" by applying another collineation to the entire photo. This leads to the cube looking "normal" again, but completely distorts the image of the snail.

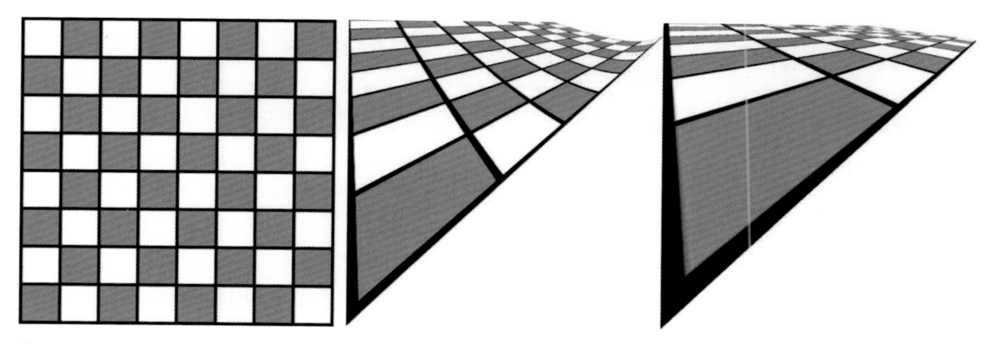

Fig. B.41 The extreme distortion of a square grid. At the center: With the help of an ordinary image editing software – not line-preserving (collinear). On the right: Collinear by the exact application of mathematical equations.

Such inconsistencies are pretty common these days and may hint at the software which must have been employed for post-processing. Fig. B.41 shows the admittedly extreme transformation of a square grid by using a renowned software package. The result is clearly not linear, and the grid subdivisions do not appear where they should be if their location is based on reality. The right image could be the photo of a grid from an extreme position.

Finally, it may be said that a collinear distortion or rectification of a *planar figure* may be done very precisely. Fig. B.42 illustrates this claim by a practical example (photo of a medical lecture by Zvi Ram). A geometrically interesting fact: The virus has the form of an icosahedron with a diameter of 1/10000mm!

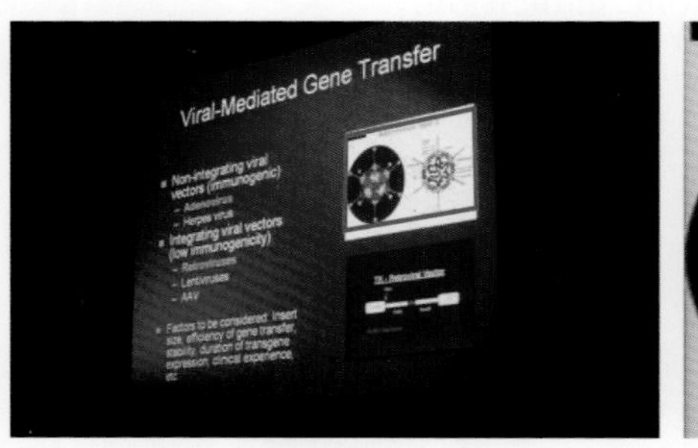

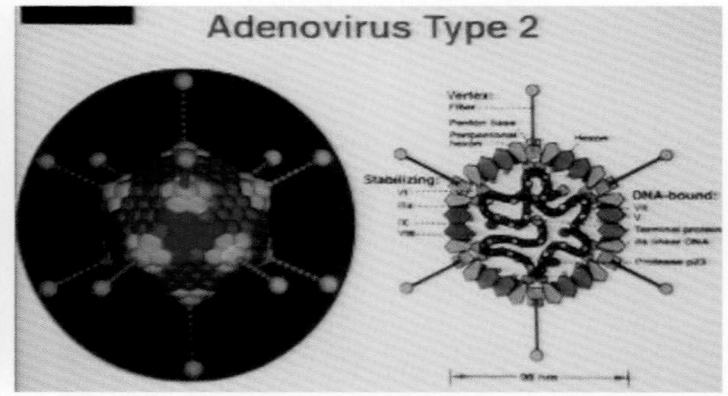

Fig. B.42 Taking many pictures of a whole projection screen: If one does not stand immediately next to the projector, then the projected image will be distorted in a collinear way. Nevertheless, the rectification of a general *planar* quadrangle can be done precisely. This allows the detail marked by the red border to be reconstructed with remarkable accuracy, despite the limited resolution of the projection!

Fig. B.43 On the left: An interesting snapshot of the historical site at night taken with a compact camera (format 4 : 3). On the right: A geometrically acceptable manipulation of the photo (format 3 : 2). The slight kink in the edge of the tower on the right is also visible in the original on the left.

Let us consider Fig. B.43 as one last example: If you have captured an original photo of the baptistery in Florence – with a moderately priced compact camera – then surely nobody will reproach you for your slight "alterations" (the full moon is visible behind the dome, see also p. 293). After all, another night stop in the center of Florence may cost you as much as your camera …

Literature

The book is built up in such a way that one ideally does not need other sources immediately (if this is necessary, the source is mentioned directly in the text). Here we list additional literature.

[1] G.Bär: *Geometrie – Eine Einführung für Ingenieure und Naturwissenschaftler*. B.G.Teubner, 2001.

[2] H.Brauner: *Lehrbuch der konstruktiven Geometrie*. Springer Wien, 1986.

[3] G.Farin: *NURBS – From Projective Geometry to Practical Use*. A.K.Peters Ltd., 1999.

[4] G.Farin: *Curves and Surfaces for CAGD*. The Morgan Kaufmann Series in Computer Graphics, 2001.

[5] T.Fruchterman,E. Reingold: *Graph Drawing by Force-directed Placement*. Software - Practice and Experience **21**, 1991, pp. 1129-1164.

[6] O.Giering, J.Hoschek : *Geometrie und ihre Anwendungen*. Hanser Verlag, München 1994.

[7] G.Glaeser: *Fast Algorithms for 3D-Graphics*. Springer Verlag New York, 1994.

[8] G.Glaeser, H.Stachel: *Open Geometry – Open GL + Advanced Geometry*. Springer N.Y., 1997.

[9] G.Glaeser, H.P.Schröcker: *Geometric Programming Using Open Geometry GL*. Springer N.Y., 2002.

[10] G.Glaeser: *Geometrie und ihre Anwendungen in Kunst, Natur und Technik, 2. edition*. Spektrum akademischer Verlag, Heidelberg 2007.

[11] G.Glaeser: *Der mathematische Werkzeugkasten - Anwendungen in Natur und Technik, 3. edition*. Spektrum akademischer Verlag, Heidelberg 2008.

[12] G.Glaeser: *Wie aus der Zahl ein Zebra wird*. Spektrum akademischer Verlag, Heidelberg 2010.

[13] G.Glaeser, K. Polthier: *A Mathematical Picturebook*. Springer Berlin Heidelberg New York, Heidelberg 2013.

[14] G. Glaeser, F. Gruber: *Developable surfaces and contemporary architecture*. Mathematics and the Arts, Vol. **1**/2007, 1–15.

[15] B.Green: *The elegant Universe. Superstrings, Hidden Dimensions, and the Quest for the Ultimate Theory*. WW.W. Norton & Company Ltd., 1999.

[16] Douglas R. Hofstadter: *Gödel, Escher, Bach. An Eternal Golden Braid*. Basic Books, 1979.

[17] D.Marsh: *Applied Geometry for Computer Graphics and CAD*. Springer, 2000.

[18] R.Müllner, H.Löffler, A.Asperl: *Darstellende Geometrie I, II*. oebv-hpt, 1999.

[19] G.Pillwein, A.Asperl, R.Müllner, M.Wischounig: *Raumgeometrie – Konstruieren und Visualisieren*. oebv-hpt, 2006.

[20] H.Pottmann, J.Wallner: *Computional Line Geometry*. Springer 2001.

[21] H.Pottmann, A. Asperl, M. Hofer and A. Kilian: *Architectural Geometry*. Bentley Institute Press 2007.

[22] H.Prautzsch et.al.: *Bézier and B-Spline Techniques*. Springer, 2002.

[23] H.Scheid: *Elemente der Geometrie*. Elsevier, Heidelberg, 1994.

[24] J.Stewart: *Pentagonien, Andromeda und die gekämmte Kugel*. Elsevier, Heidelberg, 1994.

[25] D´À. Thompson: *On Growth and Form*. Cambridge University Press, 1961. Canto Edition Reprint 2000.

[26] S.Ulam: *Mathematics and Logic*. Dover Publ., 1992.

[27] J.Warren, H.Weimer: *Subdivision Methods for Geometric Design*. Academic Press, 2002.

[28] W.Wunderlich: *Darstellende Geometrie I und II*. Bibliographisches Institut Mannheim, 1966/67.

[29] W.Wunderlich: *Ebene Kinematik*. Bibliographisches Institut Mannheim, 1971.

Picture credits

The book contains more than 2000 computer renderings, photos and hand drawings, distributed on almost 900 figures. The following figures have been used (with courtesy of the authors), without displaying the names of the photographer / artist:
2.12 (on the left) Reiner Zettl
3.74 photo left: Eberhard von Nellenburg, photo on the right: Cédric Thévenet
5.17 Gerald Zugmann
5.18 Gerald Zugmann, Paula Goldman
6.40, 10.25 Statikbüro Bollinger/Grohmann/Schneider, Metallbau Pagitz
6.79 Matthias Ecker
6.96 on the right Karin Pfaffstetter-Odehnal
6.98 ISOCHROM.com, Vienna
7.62 Photo on the right: Christian Perrelli
p. 273 Architectural design: Rüpert Zallmann, Photo Reiner Zettl
7.69 on the right, 9.4, 9.69 Harald Andreas Korvas (`sodwana.uni-ak.ac.at/korvas`)
8.32 Photo: Kasimir Reimann
9.12 Photo on the left: Othmar Glaeser
9.28 Photo left: Elisabeth Halmer
9.28 On the right: Torre Annunziata / Villa di Poppea, credit: B. Andreae, Neue Forschungen in Pompeji und den anderen vom Vesuvausbruch 79 n.Chr. verschütteten Städten (1975) Abb. 26, Database: HeidICON - Klassische Archäologie, Ruprecht-Karls-Universität Heidelberg
A.32 on the left Othmar Glaeser
9.74, 9.75 artpool.cc (`www.artpool.cc/marianne`, `www.artpool.cc/grisei`)
13.48 Helen Binet Studio, 24a Bartholomew Villas, London NW5 2LL, U.K.
B.43 Romana König

Index

C

D

E

H

I

J

K

L

P

Q

R

S

T

U

V

W

X

Y

Z